Lecture Notes in Computer Science 1442

Edited by G. Goos, J. Hartmanis and J. van Leeuwen

Springer
Berlin
Heidelberg
New York
Barcelona
Budapest
Hong Kong
London
Milan
Paris
Singapore
Tokyo

Amos Fiat Gerhard J. Woeginger (Eds.)

Online Algorithms

The State of the Art

Springer

Series Editors

Gerhard Goos, Karlsruhe University, Germany
Juris Hartmanis, Cornell University, NY, USA
Jan van Leeuwen, Utrecht University, The Netherlands

Volume Editors

Amos Fiat
Tel-Aviv University, Department of Computer Science
Tel-Aviv 69978, Israel
E-mail: fiat@math.tau.ac.il

Gerhard J. Woeginger
Technical University of Graz, Institute of Mathematics
A-8010 Graz, Austria
E-mail: gwoegi@opt.math.tu-graz.ac.at

Cataloging-in-Publication data applied for

Die Deutsche Bibliothek - CIP-Einheitsaufnahme

Online algorithms : the state of the art / Amos Fiat ; Gerhard J. Woeginger (ed.). - Berlin ; Heidelberg ; New York ; Barcelona ; Budapest ; Hong Kong ; London ; Milan ; Paris ; Singapore ; Tokyo : Springer, 1998
(Lecture notes in computer science ; 1442)
ISBN 3-540-64917-4

CR Subject Classification (1991): C.2, D.1.3, F.2, E.1, G.2

ISSN 0302-9743
ISBN 3-540-64917-4 Springer-Verlag Berlin Heidelberg New York

Typesetting: Camera-ready by author
SPIN 10638041 06/3142 – 5 4 3 2 1 0 Printed on acid-free paper

Preface

A workshop on the competitive analysis of on-line algorithms was held at Schloss Dagstuhl (Germany) during the last week of June 1996. Many survey talks were presented at this workshop. We then asked the speakers to prepare survey chapters that covered and extended the contents of their talks. This volume now is the compilation of the survey chapters. We thank all contributors who so generously gave of their time to make the book a success.

Amos Fiat and Gerhard J. Woeginger
May 1998

Sooner or later in life everyone discovers that perfect happiness is unrealizable, but there are few who pause to consider the antithesis that perfect unhappiness is equally unattainable. The obstacles preventing the realization of both these extreme states are of the same nature; they derive from our human condition which is opposed to everything infinite. Our ever-insufficient knowledge of the future opposes it; and this is called in the one instance, hope, and in the other, uncertainty of the following day.

(Primo Levi comments on the on-line nature of human life.
From: Primo Levi, *Se questo è un uomo*, 1958)

Contributors

SUSANNE ALBERS
Max-Planck-Institut für Informatik
Im Stadtwald
66123 Saarbrücken, Germany
E-mail: albers@mpi-sb.mpg.de

JAMES ASPNES
Department of Computer Science
Yale University
51 Prospect Street
P.O. Box 208285
New Haven, CT 06520-8285, USA
E-mail: aspnes-james@cs.yale.edu

YOSSI AZAR
Department of Computer Science
Tel-Aviv University
Tel-Aviv 69978, Israel
E-mail: azar@math.tau.ac.il

YAIR BARTAL
International Computer Science Institute
U.C. Berkeley
Berkeley, CA 94704-1198, USA
E-mail: yairb@icsi.berkeley.edu

PIOTR BERMAN
Department of Computer Science
and Engineering
The Penn State University
University Park, PA 16802, USA
E-mail: berman@cse.psu.edu

AVRIM BLUM
Carnegie Mellon University
Pittsburgh, PA 15213, USA
E-mail: avrim@cs.cmu.edu

MAREK CHROBAK
Department of Computer Science
University of California
Riverside, CA 92521, USA
E-mail: marek@cs.ucr.edu

JÁNOS CSIRIK
Department of Computer Science
University of Szeged
Aradi vértanúk tere 1
6720 Szeged, Hungary
E-mail: csirik@inf.u-szeged.hu

RAN EL-YANIV
Institute of Computer Science
Hebrew University
Jerusalem, Israel
E-mail: raniy@math.huji.ac.il

AMOS FIAT
Department of Computer Science
Tel-Aviv University
Tel-Aviv 69978, Israel
E-mail: fiat@math.tau.ac.il

SANDY IRANI
Information and
Computer Science Department
University of California
Irvine, CA 92697, USA
E-mail: irani@ics.uci.edu

BALA KALYANASUNDARAM
Computer Science Department
University of Pittsburgh
Pittsburgh, PA 15260, USA
E-mail: kalyan@cs.pitt.edu

Anna R. Karlin
Department of Computer Science
University of Washington
Box 352350
Seattle, WA 98103, USA
E-mail: karlin@cs.washington.edu

Hal A. Kierstead
Department of Mathematics
Arizona State University
Tempe, AZ 85287, USA
E-mail: kierstead@math.la.asu.edu

Lawrence L. Larmore
Department of Computer Science
University of Nevada
Las Vegas, NV 89154-4019, USA
E-mail: larmore@cs.unlv.edu

Stefano Leonardi
Dipartimento di Informatica Sistemistica
Università di Roma "La Sapienza"
Via Salaria 113
00198-Roma, Italia
E-mail: leon@dis.uniroma1.it

John Noga
Department of Computer Science
University of California
Riverside, CA 92521, USA
E-mail: jnoga@cs.ucr.edu

Kirk Pruhs
Computer Science Department
University of Pittsburgh
Pittsburgh, PA 15260, USA
E-mail: kirk@cs.pitt.edu

Jiří Sgall
Mathematical Institute
Žitná 25
115 67 Praha 1, Czech Republic
E-mail: sgall@math.cas.cz

Jeffery Westbrook
AT&T Labs - Research
Murray Hill, NJ 07974, USA
E-mail: jeffw@research.att.com

Gerhard J. Woeginger
Institut für Mathematik
Technische Universität Graz
Steyrergasse 30
8010 Graz, Austria
E-mail:
gwoegi@opt.math.tu-graz.ac.at

Contents

CHAPTER 3:
Competitive Analysis of Paging 52
Sandy Irani

CHAPTER 4:
Metrical Task Systems, the Server Problem and the Work Function Algorithm 74
Marek Chrobak & Lawrence L. Larmore

CHAPTER 8: On-line Load Balancing 178

Yossi Azar

CHAPTER 9: On-Line Scheduling 196

Jiří Sgall

CHAPTER 11: On-line Network Routing 242

Stefano Leonardi

CHAPTER 12: On-line Network Optimization Problems 268

Bala Kalyanasundaram & Kirk Pruhs

CHAPTER 13: Coloring Graphs On-Line 281

Hal A. Kierstead

CHAPTER 14: On-Line Algorithms in Machine Learning 306

Avrim Blum

CHAPTER 15:
Competitive Solutions for On-line Financial Problems 326

Ran El-Yaniv

CHAPTER 16: On the Performance of Competitive Algorithms in Practice 373

Anna R. Karlin

CHAPTER 17: Competitive Odds and Ends 385

Amos Fiat & Gerhard J. Woeginger

1

Competitive Analysis of Algorithms

AMOS FIAT
GERHARD J. WOEGINGER

1 Competitive analysis

When Hannibal crossed the Alps into Italy, it was no longer as simple to respond to a Roman naval threat in Spain as it was before crossing the Alps. Had Hannibal known the entire future Roman strategy before crossing the Alps then (assuming appropriate computational ability) he could have computed an optimal strategy to deal with the Rome/Cartago crisis. Classical computational complexity of algorithms deals with the questions as to what computational resources would Hannibal have needed to perform this computation. **Competitive analysis** of algorithms deals with the question as to whether Hannibal's decisions were reasonable given his partial understanding of Roman strategy.

Decision making can be considered in two different contexts: making decisions with complete information, and making decisions based on partial information. A major reason for the study of algorithms is to try to answer the question "which is the better algorithm". The study of the computational complexity of algorithms is useful for distinguishing the quality of algorithms based on the computational resources used and the quality of the solution they compute. However, the computational complexity of algorithms may be irrelevant or a secondary issue when dealing with algorithms that operate in a state of uncertainty. **Competitive analysis** of algorithms has been found useful in the study of such algorithms.

If we were to compare two arbitrary algorithms, A and B, on an instance by instance basis, we could expect that sometimes A beats B and sometimes B

beats A. Thus, such a comparison does not impose a total ordering on the quality of the two algorithms and does not answer the basic question "which is the better algorithm?". There are many ways to choose a quality function that imposes a total ordering on the quality of the algorithms. One obvious possibility is simply to consider the worst case behavior of the algorithm, another is to make some assumption on the input distribution. As discussed hereinafter, both measures are problematic, especially in the context of algorithms that make decisions in a state of uncertainty.

Polynomial time approximation algorithms for hard problems guarantee the quality of the poly-time solution, when compared to the optimal solution. *Competitive analysis* can be viewed as the information theoretic analogue to the computational concept of a poly-time approximation algorithm. Like approximation algorithms, competitive analysis bounds the ratio between the worst case behavior of the algorithm on a problem instance and the behavior of the optimal algorithm on the *same* problem instance. In this book several generalizations and variants of this measure are considered. The "optimal algorithm" itself is not necessarily well defined and variants are considered when the "optimal algorithm" is optimal within some limited class of algorithms.

Competitive analysis is useful in the analysis of systems that have some notion of a time progression, that have an environment, that respond in some way to changes in the environment, and that have a memory state. *I.e.*, there is some notion of a configuration that varies over time and dealing with future environmental change depends in some way on the current configuration. There are a great many problems that can be phrased in this manner, whether or not timeliness is an inherent part of the problem. *E.g.*, for stock transactions timeliness of events is inherent: the bid/ask price of a share is valid for a very short period of time and needs to be addressed in a timely fashion, for Lunar exploration by a mobile Lunar probe timeliness of events depends on the probe behavior itself: an impassable crevice in the Lunar terrain may be revealed only by going around the boulder, if the probe does not go around the boulder then it will not learn of this problem until (possibly) later on.

Competitive analysis is certainly useful for so-called "on-line algorithms" that have to respond to events over time (fluctuating stock prices), but can be used in many other contexts as well. Competitive analysis is used whenever the nature of the problem is such that decisions have to be made with incomplete information. This could be because some event has not yet occurred (stock price tomorrow at noon is unknown now), because obtaining the missing information requires action by the algorithm (move probe around boulder), or because the algorithm is distributed and no single node has global information. Nonetheless, the greater body of work involving competitive analysis deals with on-line problems. So much so that any use of competitive analysis is mistakenly refered to as on-line.

The following chapters of this book give numerous examples of competitive analysis of algorithms and of problems amiable to competitive analysis.

2 Applications of competitive analysis

The problems considered herein are all optimization variants of NP decision problems. Usually, a somewhat artificial distinction is made between problem parameters and the problem instance. This distinction is used to distinguish between what is known to the algorithm in advance, *i.e.*, the problem parameters, and what is not known ahead of time, *i.e.*, the problem instance. Various portions of the problem instance may become known over time, either as a consequence of the algorithm's own actions or because the nature of the problem is such that parts of the input are time dependent.

Problem parameters are typically values like the number of pages in memory, the number of machines available for processing scheduling requests, the network on which routing demands are served, etc. The piecemeal revelation of the problem instance could be because certain events occur (the Romans invade Spain) and they should be addressed, or because the actions of the algorithm directly reveal more about the problem instance (Hannibal learns the geography of Italy simply by being there and seeing it with his own eyes).

A general model for the type of algorithms and problems we consider is that an algorithm A is always associated with some configuration, some new portion of the input becomes available, and the algorithm moves to a new configuration. We consider two basic types of problems: maximization or benefit problems and minimization or cost problems. For benefit problems there is some benefit function that depends on the input and on the sequence of configurations occupied by A over time that represents the benefit that A obtains from the problem instance. For cost problems there is some cost function that depends on the input and on the sequence of configurations occupied by A over time and represents the cost associated with A to deal with the problem instance. We remark that this informal model is a generalization of the task system formalization by Borodin, Linial & Saks [2] to include benefit problems and other partial information problems that are not necessarily on-line problems.

Typical maximization goals are to route the largest number of non-interfering routing requests in a network, to deal with the largest subgraph with some property, to gain the greatest profit in a currency trading scenario, to maximize the throughput in a computing system, etc.

Typical minimization goals are to minimize the number of page faults, to minimize the highest relative load on a network communication link, to minimize the number of colors used to color a graph, to minimize the number of bins used in a bin packing problem, to minimize the distance traversed by a robot in some terrain with obstacles so as to reach point B while starting from point A, etc.

The major issue with all the mentioned problems is that incomplete knowledge of the problem instance may lead the algorithm to perform very poorly. In fact, for many of these problems it is possible to ensure that every decision made by the algorithm is the worst possible decision. For example, irrespective of the paging algorithm used, the page just ejected from memory could be the page required on the very next memory access. Thus, a standard worst case analysis seems to imply that all paging algorithms are equally bad. This definitely contradicts both intuition and practical experience.

One alternative to worst case analysis is to make probabilistic assumptions on the input distribution. This can certainly be done but leads to the very difficult questions as to how the distribution was selected and what evidence suggests that this distribution is either typical or representative. Certainly, there may be different distributions that represent different classes of "typical" scenarios and it is entirely uncertain as to how one can hope to give any significant argument as to why such distributions are relevant or exhaustive.

To avoid the problems of probabilistic models we seek a worst case model that will hold for any distribution. To avoid the problem that the standard worst case measure may be entirely insensitive to the algorithm used and gives us no useful information, we must refine the worst case measure. The problem with worst case behavior as exhibited by the paging problem above is that it can be terribly bad. However, what if we could show that whenever the behavior of the algorithm is bad on a specific input then the behavior of *any* algorithm on that input is bad? This means that the algorithm might indeed behave poorly on some specific input but only if the algorithm has a "good excuse" for doing so because no other algorithm could do much better. Just how much better is the subject of this book. This is the underlying idea of competitive analysis: we do not consider the absolute behavior of the algorithm but rather the ratio between the algorithm's behavior and the optimal behavior on the same problem instance.

Let $\text{cost}_A(I)$ denote the cost of algorithm A on problem instance I. We define the competitive ratio of an algorithm A for a cost problem P to be the value

$$\inf\{c \mid \text{cost}_A(I) \leq c \cdot \text{cost}_B(I),\ \forall I \in P, \forall B\}.$$

Similarly, let $\text{benefit}_A(I)$ denote the benefit of algorithm A on problem instance I, then we define the competitive ratio of algorithm A on benefit problem P to be

$$\sup\{c \mid c \cdot \text{benefit}_A(I) \geq \text{benefit}_B(I),\ \forall I \in P, \forall B\}.$$

In fact, the definition of the competitive ratio usually allows an additive constant in the definitions above whereas the definitions above (without the constant) imply that A is *strictly competitive* with a competitive ratio of c. In the literature, the competitive ratio is sometimes also called the worst case ratio or the worst case performance guarantee.

One class of problems that seems inherently suitable for competitive analysis is the class of on-line problems where a sequence of events $\sigma = \sigma_1, \sigma_2, \ldots$ appears over time and have to be dealt with immediately. An on-line algorithm has to make a decision in response to an event without knowing what events the future holds.

A typical example of such a problem is the paging problem: when a page fault occurs the paging algorithm has to decide what page is to be evicted from memory without knowing the future page access pattern. Many different flavors of on-line algorithms are considered in the chapters of this book; these include data structure problems, paging problems, distributed problems, packing problems, routing problems, graph problems, scheduling problems, load balancing problems, financial problems, and others. Of course, in general there may be several reasons for uncertainty about the problem instance, both temporal and geographical.

Partial information about the input instance is not only limited to not knowing about future events. In various exploration and navigation problems, the issue is how to obtain the goal (search for a missing child, reach the beach, leave Buckingham palace blindfolded, etc.) while traversing the terrain efficiently without a map. Depending on the sensors allowed (vision, touch, etc.) some of the missing information may possibly be obtained by performing some movement.

Another class of problems where only partial information is available is in a distributed setting where the global configuration may be unavailable to individual processors that are only aware of their own local view of the world. Again, as with exploration/navigation in an unknown terrain, it may be possible to improve the local understanding of the global environment at the extra expense of performing appropriate communications.

Because decisions have to be made under uncertainty, and because we consider a worst case measure, good algorithms with respect to the competitive ratio seem indecisive. Certainly, such algorithms seem to procrastinate before making any decision which is difficult to repair if it proves to be wrong. In the famous k-server problem, algorithms with good competitive ratios will move multiple servers to an active area only after a great length of time, even if no other activity takes place. However, we note that a misunderstanding of the worst case measure can lead to difficulties. Back to the 3rd century B.C., Hannibal did leave behind in Spain a great many forces in expectation of a possible Roman landing. This is an example of a (wrong) decision motivated by a misunderstanding of competitive analysis. Because of the winner-take-all nature of the conflict, a better understanding of competitive analysis might have led Hannibal to move all his troops across the Alps or to decide not to attack Rome at all.

It may be useful to consider the relationship between good approximation algorithms for hard problems and good competitive algorithms. Both seek to obtain a good approximation to some optimal solution. The difference lies in that approximation algorithms are limited in the computational resources available whereas competitive algorithms are limited in their knowledge of the problem instance. It follows that competitive analysis of algorithms is an information theoretic measure, not a computational complexity measure. Nonetheless, most of the interesting on-line problems have polynomial time algorithms and non poly-time on-line algorithms are almost never considered.

It is natural to extend the competitive ratio in many ways, and several such extensions are dealt with in the chapters of this book. One natural extension is to define bicriteria problems with both cost and benefit functions. Here, the goal could be to obtain a simultaneous minimization of the cost and maximization of the benefit, or find some interesting points along the tradeoff curve. A trivial but interesting example is to compare the number of page faults for a paging algorithm with k pages in memory to the number of page faults for an optimal algorithm with fewer pages in memory. In fact, decreasing the memory size of the optimal algorithm makes the resulting competitive ratio drop very quickly. For some problems (like the most general case of distributed paging) one can obtain interesting non-trivial results only if one allows the on-line algorithm to use more resources than the algorithm against which it is compared.

Another extension is to limit the class of possible problem instances representing some interesting subset. Much work in this direction has been done in the context of paging where the attempt has been to restrict attention to input sequences representing locality of reference.

The analysis of the various algorithms is often performed through the use of adversary arguments. One assumes that an adversary has control over the yet unknown parts of the input problem and that the adversary determines the yet unknown parts of the problem instance so as to increase the competitive ratio as much as possible.

Another important extension of the competitive ratio is done to deal with randomized on-line algorithms, where the goal is to use randomization so as to improve the competitive ratio. In such a case, one can only talk of the expected cost or of the expected benefit. Various adversary models have been defined in the context of randomized algorithms (Ben-David et al. [1]) but the most useful model is that of the *oblivious adversary* that is unaware of the coin tosses performed by the algorithm. The assumption of an oblivious adversary essentially states that the problem instance has been predetermined before the algorithm began execution and that the algorithm's own actions do not influence the choice of problem instance. Of course, in some cases this is false model: the Roman decision to attack Spain was definitely influenced by Hannibal's decision to cross the Alps. Other standard adversary models for randomized on-line algorithms are the *adaptive on-line adversary* (who is allowed to observe the coin tosses, but must produce its own suboptimal solution on-line) and the *adaptive off-line adversary* (who is allowed to observe the coin tosses, and produces the optimal solution off-line).

3 Historical background

Perhaps the first competitive analysis of an algorithm was performed in 1966 by Graham [4]. Graham introduces a simple deterministic greedy algorithm (the so-called "List Scheduling" algorithm) for a scheduling problem on parallel machines, and he performs a complete worst-case analysis of this on-line algorithm. Graham's paper does neither use the term "on-line" nor "competitive". (Remarkably, Graham also analyzes one of the first polynomial time approximation algorithms for an NP-hard optimization problem, years before the term "NP-hard" was introduced).

In Volume 2 of *"The Art of Computer Programming"* (Section 4.7 in [10]), Knuth discusses the computation of the Cauchy product of two power series. The nth coefficient of the Cauchy product can be computed solely based on the first n coefficients of the two multiplicands. In the first edition of [10] in 1968, Knuth calls an algorithm "sequential" (and then adds in parentheses "also called on-line") if it reads the coefficients from two operands in sequence and outputs the coefficients of the result as soon as they are determined. Moreover, Knuth defines "off-line" to be the complement of "on-line". Knuth's concept of an on-line algorithm is similar to today's concept of an on-line algorithm in that an on-line algorithm must be able to produce its output without knowing the "future". However, Knuth's definition seems to require that the output is an exact solution.

The first use of the term "on-line" in the context of approximation algorithms was in the early 1970s and used for bin-packing algorithms. The first on-line approximation algorithms that are also called on-line approximation algorithms can be found in the Ph.D. thesis [5] and in the journal article [6] of Johnson. Johnson [7] suggests that the origin of the words "on-line" and "off-line" lies in cryptographic systems, in which decryption was either done as part of the communications system (*i.e.*, *on* the communication *line*) or after the fact by using other facilities (*i.e.*, *off* the communication *line*).

The adversary method for deriving lower bounds on the competitive ratio has been implicitly used by Woodall [15] in the analysis of the so-called Bay Restaurant problem (which is essentially a linear storage allocation problem with the objective of annoying the owner of the restaurant as much as possible). Kierstead and Trotter [9] use the adversary method in their investigation of on-line interval graph coloring. Yao [16] formulates a theorem that starts with the words *"For any on-line algorithm ..."* and that proves the impossibility of an on-line bin-packing algorithm with a competitive ratio strictly better than 3/2. This seems to be the first result stated on the class of all on-line algorithms for a certain optimization problem, thus exploiting the distinction between on-line and off-line algorithms.

In 1985, Sleator and Tarjan published two papers that triggered the on-line boom in Theoretical Computer Science: In [13] they give competitive algorithms for the list update problem and for the paging problem. In [14] they introduce self adjusting binary trees and the (in)famous dynamic optimality conjecture for splay trees.

The phrase *competitive analysis* itself was coined by Karlin, Manasse, Rudolph, and Sleator in their paper [8] on competitive snoopy caching. The paper [2] by Borodin, Linial and Saks on task systems gives a general formalism for a great many on-line cost problems. In 1988, Manasse, McGeoch, and Sleator published their k-server paper [11] along with their (in)famous k-server conjecture. While the terminology is quite different, the first competitive analysis of on-line financial problems can definitely be associated with Cover in [3]. The first set of problems to be studied in terms of competitive analysis that are not inherently on-line problems are the navigation and exploration problems of Papadimitriou and Yannakakis [12].

Much of the work in the field of on-line algorithms has been motivated because of two open problems — the dynamic optimality conjecture for splay trees of Sleator and Tarjan [14] and the k-server conjecture of Manasse, McGeoch, and Sleator [11]. Both of these problems are still open although it seems that very considerable progress has been made. Even if these problems are very far away from being decided, it is undoubtedly the case that the study of these problems has inadvertently resulted in a great body of research, parts of which are described in this book. What is so special about both these conjectures is that both can be explained in a few minutes yet are seemingly very hard to prove (or disprove). These two conjectures and their misleading simplicity goes a long way towards explaining why so much work has been done in this field over the past few years.

4 Topics surveyed

Certainly not all studies in competitive analysis of algorithms have been surveyed in this book. However, the list of topics that have been surveyed is fairly large and representative of much of the work that has been going on. Any report on a research topic that has so many researchers engaged in ongoing research will suffer in that at least some of the material is yesterdays news. Nonetheless, we hope that at very least the material covered will be found useful as an advanced introduction to tomorrows research.

Chapter 2 by Susanne Albers and Jeffery Westbrook deals with competitive analysis of dynamic data structures. This chapter deals at length with the list update problem. It defines the standard algorithms like Transpose, Move-To-Front, and Frequency-Count, and discusses worst-case and average-case questions for deterministic and randomized versions. The chapter also deals with several issues relating to self adjusting binary search trees (including splay trees, dynamic monotone trees, WPL-trees, and D-trees), and with the application of self-organizing data structures to data compression.

Chapter 3 by Sandy Irani deals explicitly with paging. The chapter includes a variety of deterministic, randomized and memoryless randomized results in the "standard paging model". It also includes results on various extensions of competitive analysis to deal with locality of reference (e.g. paging with access graphs, Markov paging), and it includes results on variations of paging (e.g. weighted caching, paging with multi-size pages).

Chapter 4 by Marek Chrobak and Lawrence Larmore deals with the famous k-server conjecture by Manasse, McGeoch, and Sleator. Specifically, it deals extensively with the so-called *work function algorithm* (WFA) whose applications seem to extend far beyond the k-server problem. The chapter gives unified proofs for several results in this area. E.g. WFA if shown to be $(2n-1)$-competitive for n-state metrical task systems (metrical task systems form a pretty general class of optimization problems and cover e.g. the notorious problem of running an ice cream machine). While Chapter 4 deals entirely with deterministic algorithms, the editors have added a short discussion of randomized algorithms and variants on the models to Chapter 17.

Chapter 5 by Yair Bartal deals with distributed data management problems. Such problems arise e.g. in multiprocessor systems as memory management problems for a globally addressed shared memory, or in distributed networks of processors where data files are kept in different sites and are accessed for information retrieval by dispersed users and applications. The main problems discussed in this chapter are the file migration problem (where multiple copies of the same file are forbidden), the file allocation problem (where files may be copied and deleted arbitrarily), and the distributed paging problem (where memory capacity limitations are taken into account). The algorithms described in this chapter are competitive against an adversary that not only knows the future, but also has global knowledge of the network.

Chapter 6 by James Aspnes deals with a different type of competitive analysis which is specifically geared towards a distributed setting. Like on-line algorithms, distributed algorithms must deal with limited information and with unpredictable user and system behavior. Unlike on-line algorithms, in many distributed algorithms the primary source of difficulty is the possibility that components of the underlying system may fail or behave badly. In this chapter, the distributed nature of the computation is used extensively so as to obtain the upper bounds. The discussion also includes tools for building competitive algorithms by composition, and for obtaining more meaningful competitive ratios by restricting the set of possible adversaries.

Chapter 7 by János Csirik and Gerhard Woeginger deals with competitive algorithms for packing and covering problems. All results discussed are centered about bin-packing, a fundamental problem in optimization that models problems like packing a suitcase, assigning newspaper articles to newspaper pages, loading trucks, packet routing in communication networks, assigning commercials to station breaks on television, etc. The items to be packed in bin-packing problems are real numbers, vectors, rectangles, or squares. The chapter discusses worst case and average case results for bin-packing problems and for several closely related covering problems.

Chapter 8 by Yossi Azar deals with competitive algorithms for load balancing. The goal in load balancing is to divide the work as equally as possible between numerous processors whereas jobs that come in have to be assigned immediately to some processor. The literature contains many different variants of this problem: Jobs may stay permanently in the system, or they may disappear again after some time. Every job may entail the same work on every machine to which it is assigned, or the work may depend on the assignment. A job may be assigned to a machine once and for all, or reassignment may be possible. Chapter 8 deals with these variants at length and summarizes some major open problems.

Chapter 9 by Jiří Sgall deals with on-line scheduling problems. The main difference between load-balancing problems (discussed in Chapter 8) and scheduling problems is the following: In load balancing the jobs have two dimensions, since they stay in the system for a certain duration and they cause a certain amount of work to the machines. In scheduling the jobs have only one dimension, which is the time needed to process the job (= amount of work caused by the job). Moreover, in scheduling problems jobs may have release dates, deadlines, precedence constraints, they may be preemptable, etc. This chapter is a colorful compilation of various on-line results in scheduling.

Chapter 10 by Piotr Berman deals with on-line searching and navigation problems. Here, one or more agents (e.g. robot, human, or hybrid) have to move in an unknown terrain and attain some goal such as reaching a destination, mapping the terrain, meeting each other, etc. The difficulties arise because the agents do not have a map of the terrain and only acquire knowledge while moving through the terrain. This definitely is one of the most basic and one of the oldest on-line problems, first encountered by Theseus at Crete. The chapter surveys navigation, searching, and exploration problems in this area.

Chapter 11 by Stefano Leonardi deals with network routing problems. Two main models in this area are packet routing and circuit routing. In packet routing one is allowed to store transmissions in transit and to forward them later, whereas in circuit routing this is not allowed. The main constraints arise from limited link bandwidth and from slow switching speeds. The chapter considers many flavors of routing problems including load minimization and throughput maximization. In the second part of the chapter optical routing is considered, a problem that is motivated by wavelength division multiplexing in all-optical networks. Optical routing has its own set of rules and both cost and benefit versions of this problem are considered.

Chapter 12 by Bala Kalyanasundaram and Kirk Pruhs deals with a variety of on-line network problems. The chapter discusses on-line versions of classical problems in the area of graph algorithms that arise from learning the structure of the underlying graph step by step. Among the considered problems are the minimum spanning tree problem, the transportation problem, matching problems (weighted and unweighted), Steiner tree problems, and the traveling salesman problem.

Chapter 13 by Hal Kierstead deals with on-line graph coloring: The graph is revealed vertex by vertex, and every vertex must be colored immediately. A trivial on-line algorithm would be to color every vertex with another color. Interestingly, on general graphs the best possible on-line algorithm only beats the trivial algorithm by poly-log factors (and even getting this poly-log improvement turns out to be quite hard). The chapter also discusses the performance of on-line algorithms on various classes of special graphs like interval graphs, trees, bipartite graphs, cocomparability graphs, etc. Special emphasis is put on the performance of the First-Fit coloring algorithm, a very simple greedy coloring heuristic. Moreover, the connection between a dynamic storage allocation problem and coloring of interval graphs is explained.

Chapter 14 by Avrim Blum deals with on-line learning. The chapter surveys a collection of results in computational learning theory that fit nicely into the competitive framework. The problems considered in this chapter include the well known predicting from expert advice problem (e.g. the weighted majority algorithm), several standard models of on-line learning from examples (e.g. agnostic learning and learning in the presence of random noise), several algorithms for on-line learning (e.g. the Winnow algorithm and an algorithm for learning decision lists), and other issues such as attribute-efficient learning and the infinite attribute model, and learning target functions that change over time.

Chapter 15 by Ran El-Yaniv surveys work related to financial problems. The family of these problems and their related applications is very broad; they originate in a number of disciplines including economics, finance, operations research, and decision theory. The four main problems discussed in the chapter are search problems (where one searches for the minimum price in a sequence of prices that are revealed step by step), replacement problems (where one has to switch repeatedly from one activity to another so that the total cost is minimized), portfolio selection (where one has to repeatedly reallocate the current capital among the available investment opportunities), and leasing problems (where one needs to lease some equipment during a number of time periods).

Chapter 16 by Anna Karlin deals with real world experimentation with algorithms designed via competitive analysis. The applications considered include cache coherence, virtual circuit holding times, paging, routing, and admission control. It turns out that in general algorithms that are optimized to work against a worst-case adversary do not necessarily work well in practice. However, competitive analysis gives valuable insights and often leads to the "correct" algorithms.

Chapter 17 is an afterword by the editors that collects some competitive odds and ends. Finally, Appendix A by Marek Chrobak and John Noga contains a bibliography on on-line algorithms with over 800 references.

References

1. S. Ben-David, A. Borodin, R.M. Karp, G. Tardos, and A. Wigderson. On the power of randomization in on-line algorithms. *Algorithmica*, 11:2–14, 1994.
2. A. Borodin, N. Linial, and M. Saks. An optimal online algorithm for metrical task systems. In *Proc. 19th Annual ACM Symposium on Theory of Computing*, pages 373–382, 1987.
3. T.M. Cover. Universal portfolios. *Mathematical Finance*, 1(1):1–29, January 1991.
4. R. L. Graham. Bounds for certain multiprocessing anomalies. *Bell System Technical Journal*, 45:1563–1581, 1966.
5. D. S. Johnson. *Near-optimal bin packing algorithms.* PhD thesis, MIT, Cambridge, MA, 1973.
6. D. S. Johnson. Fast algorithms for bin packing. *J. Comput. System Sci.*, 8:272–314, 1974.
7. D. S. Johnson. Private communication, 1998.
8. A. Karlin, M. Manasse, L. Rudolph, and D. Sleator. Competitive snoopy caching. *Algorithmica*, 3(1):79–119, 1988.
9. H. A. Kierstead and W. T. Trotter. An extremal problem in recursive combinatorics. *Congressus Numerantium*, 33:143–153, 1981.
10. D.E. Knuth. *The Art of Computer Programming; Volume 2: Seminumerical Algorithms.* Addison-Wesley, 1st edition, 1968.

11. M. Manasse, L. A. McGeoch, and D. Sleator. Competitive algorithms for online problems. In *Proc. 20th Annual ACM Symposium on Theory of Computing*, pages 322–333, 1988.
12. C. H. Papadimitriou and M. Yannakakis. Shortest paths without a map. *Theoretical Computer Science*, 84:127–150, 1991.
13. D. Sleator and R. E. Tarjan. Amortized efficiency of list update and paging rules. *Communications of the ACM*, 28:202–208, 1985.
14. Daniel D. Sleator and Robert Endre Tarjan. Self-adjusting binary search trees. *Journal of the ACM*, 32:652–686, 1985.
15. D. R. Woodall. The bay restaurant - a linear storage problem. *American Mathematical Monthly*, 81:240–246, 1974.
16. A. C. C. Yao. New algorithms for bin packing. *J. Assoc. Comput. Mach.*, 27:207–227, 1980.

2

Self-Organizing Data Structures

SUSANNE ALBERS
JEFFERY WESTBROOK

1 Introduction

This chapter surveys results in the design and analysis of self-organizing data structures for the search problem. The general search problem in pointer data structures can be phrased as follows. The elements of a set are stored in a collection of nodes. Each node also contains $O(1)$ pointers to other nodes and additional state data which can be used for navigation and self-organization. The elements have associated key values, which may or may not be totally ordered (almost always they are). Various operations may be performed on the set, including the standard dictionary operations of searching for an element, inserting a new element, and deleting an element. Additional operations such as set splitting or joining may be allowed. This chapter considers two simple but very popular data structures: the unsorted linear list, and the binary search tree.

A self-organizing data structure has a rule or algorithm for changing pointers and state data after each operation. The self-organizing rule is designed to respond to initially unknown properties of the input request sequence, and to get the data structure into a state that will take advantage of these properties and reduce the time per operation. As operations occur, a self-organizing data structure may change its state quite dramatically.

Self-organizing data structures can be compared to static or constrained data structures. The state of a static data structure is predetermined by some strong knowledge about the properties of the input. For example, if searches are generated according to some known probability distribution, then a linear list may be sorted by decreasing probability of access. A constrained data structure must satisfy some structural invariant, such as a balance constraint in a binary search tree. As long as the structural invariant is satisfied, the data structure does not change.

Self-organizing data structures have several advantages over static and constrained data structures [64]. (a) The amortized asymptotic time of search and update operations is usually as good as the corresponding time of constrained structures. But when the sequence of operations has favorable properties, the performance can be much better. (b) Self-organizing rules need no knowledge of the properties of input sequence, but will adapt the data structure to best suit the input. (c) The self-organizing rule typically results in search and update algorithms that are simple and easy to implement. (d) Often the self-organizing rule can be implemented without using any extra space in the nodes. (Such a rule is called "memoryless" since it saves no information to help make its decisions.)

On the other hand, self-organizing data structures have several disadvantages. (a) Although the total time of a sequence of operations is low, an individual operation can be quite expensive. (b) Reorganization of the structure has to be done even during search operations. Hence self-organizing data structures may have higher overheads than their static or constraint-based cousins.

Nevertheless, self-organizing data structures represent an attractive alternative to constraint structures, and reorganization rules have been studied extensively for both linear lists and binary trees. Both data structures have also received considerable attention within the study of on-line algorithms. In Section 2 we review results for linear lists. Almost all previous work in this area has concentrated on designing *on-line algorithms* for this data structure. In Section 3 we discuss binary search trees and present results on *on-line* and *off-line algorithms*. Self-organizing data structures can be used to construct effective data compression schemes. We address this application in Section 4.

2 Unsorted linear lists

The problem of representing a dictionary as an unsorted linear list is also known as the *list update problem*. Consider a set S of items that has to be maintained under a sequence of *requests*, where each request is one of the following operations.

Access(x). Locate item x in S.
Insert(x). Insert item x into S.
Delete(x). Delete item x from S.

Given that S shall be represented as an unsorted list, these operations can be implemented as follows. To access an item, a list update algorithm starts at the front of the list and searches linearly through the items until the desired item is found. To insert a new item, the algorithm first scans the entire list to verify that the item is not already present and then inserts the item at the end of the list. To delete an item, the algorithm scans the list to search for the item and then deletes it.

In serving requests a list update algorithm incurs cost. If a request is an access or a delete operation, then the incurred cost is i, where i is the position of the requested item in the list. If the request is an insertion, then the cost is $n+1$, where n is the number of items in the list before the insertion. While processing a request sequence, a list update algorithm may rearrange the list. Immediately after an access or insertion, the requested item may be moved at no extra cost to any position closer to the front of the list. These exchanges are called *free exchanges.* Using free exchanges, the algorithm can lower the cost on subsequent requests. At any time two adjacent items in the list may be exchanged at a cost of 1. These exchanges are called *paid exchanges.*

The cost model defined above is called the *standard model.* Manasse *et al.* [53] and Reingold *et al.* [61] introduced the P^d cost model. In the P^d model there are no free exchanges and each paid exchange costs d. In this chapter, we will present results both for the standard and the P^d model. However, unless otherwise stated, we will always assume the standard cost model.

We are interested in list update algorithms that serve a request sequence so that the total cost incurred on the entire sequence is as small as possible. Of particular interest are *on-line* algorithms, i.e., algorithms that serve each request without knowledge of any future requests. In [64], Sleator and Tarjan suggested comparing the quality of an on-line algorithm to that of an *optimal off-line* algorithm. An optimal off-line algorithm knows the entire request sequence in advance and can serve it with minimum cost. Given a request sequence σ, let $C_A(\sigma)$ denote the cost incurred by an on-line algorithm A in serving σ, and let $C_{OPT}(\sigma)$ denote the cost incurred by an optimal off-line algorithm OPT. Then the on-line algorithm A is called c-competitive if there is a constant a such that for all size lists and all request sequences σ,

$$C_A(\sigma) \leq c \cdot C_{OPT}(\sigma) + a.$$

The factor c is also called the *competitive ratio.* Here we assume that A is a deterministic algorithm. The competitive ratio of a randomized on-line algorithm has to be defined in a more careful way, see Section 2.2. In Sections 2.1 and 2.2 we will present results on the competitiveness that can be achieved by deterministic and randomized on-line algorithms.

At present it is unknown whether the problem of computing an optimal way to process a given request sequence is NP-hard. The fastest optimal off-line algorithm currently known is due to Reingold and Westbrook [60] and runs in time $O(2^n n! m)$, where n is the size of the list and m is the length of the request sequence.

Linear lists are one possibility to represent a dictionary. Certainly, there are other data structures such as balanced search trees or hash tables that, depending on the given application, can maintain a dictionary in a more efficient way. In general, linear lists are useful when the dictionary is small and consists of only a few dozen items [15]. Furthermore, list update algorithms have been used as subroutines in algorithms for computing point maxima and convex hulls [14, 31]. Recently, list update techniques have been very successfully applied in the development of data compression algorithms [18]. We discuss this application in detail in Section 4.

2.1 Deterministic on-line algorithms

There are three well-known deterministic on-line algorithms for the list update problem.

- **Move-To-Front**: Move the requested item to the front of the list.
- **Transpose**: Exchange the requested item with the immediately preceding item in the list.
- **Frequency-Count**: Maintain a frequency count for each item in the list. Whenever an item is requested, increase its count by 1. Maintain the list so that the items always occur in nonincreasing order of frequency count.

Other deterministic on-line algorithms that have been proposed in the literature are variants of the above algorithms, see [17, 32, 35, 42, 47, 62, 40, 64, 74]. Rivest [62], for instance, introduced a move-ahead-k heuristic that moves a requested item k positions ahead. Gonnet *et al.* [32] and Kan and Ross [41] considered a k-in-a-row rule, where an item is only moved after it is requested k times in a row. This strategy can be combined both with the Move-To-Front and Transpose algorithms.

The formulations of list update algorithms generally assume that a request sequence consists of accesses only. It is obvious how to extend the algorithms so that they can also handle insertions and deletions. On an insertion, the algorithm first appends the new item at the end of the list and then executes the same steps as if the item was requested for the first time. On a deletion, the algorithm first searches for the item and then just removes it.

In the following, we concentrate on the three algorithms Move-To-Front, Transpose and Frequency-Count. We note that Move-To-Front and Transpose are *memoryless* strategies, i.e., they do not need any extra memory to decide where a requested item should be moved. Thus, from a practical point of view, they are more attractive than Frequency-Count. Sleator and Tarjan [64] analyzed the competitive ratios of the three algorithms.

Theorem 1. *The Move-To-Front algorithm is 2-competitive.*

Proof. Consider a request sequence $\sigma = \sigma(1), \sigma(2), \ldots, \sigma(m)$ of length m. First suppose that σ consists of accesses only. We will compare simultaneous runs of Move-To-Front and OPT on σ and evaluate on-line and off-line cost using a potential function Φ. For an introduction to amortized analysis using potential functions, see Tarjan [70].

The potential function we use is the number of inversions in Move-To-Front's list with respect to OPT's list. An *inversion* is a pair x, y of items such that x occurs before y Move-To-Front's list and after y in OPT's list. We assume without loss of generality that Move-To-Front and OPT start with the same list so that the initial potential is 0.

For any t, $1 \leq t \leq m$, let $C_{MTF}(t)$ and $C_{OPT}(t)$ denote the actual cost incurred by Move-To-Front and OPT in serving $\sigma(t)$. Furthermore, let $\Phi(t)$ denote the potential after $\sigma(t)$ is served. The *amortized cost* incurred by Move-To-Front on $\sigma(t)$ is defined as $C_{MTF}(t) + \Phi(t) - \Phi(t-1)$. We will show that for any t,

$$C_{MTF}(t) + \Phi(t) - \Phi(t-1) \leq 2C_{OPT}(t) - 1. \tag{1}$$

Summing this expression for all t we obtain $\sum_{t=1}^{m} C_{MTF}(t) + \Phi(m) - \Phi(0) \leq \sum_{t=1}^{m} 2C_{OPT}(t) - m$, i.e., $C_{MTF}(\sigma) \leq 2C_{OPT}(\sigma) - m + \Phi(0) - \Phi(m)$. Since the initial potential is 0 and the final potential is non-negative, the theorem follows.

In the following we will show inequality (1) for an arbitrary t. Let x be the item requested by $\sigma(t)$. Let k denote the number of items that precede x in Move-To-Front's and OPT's list. Furthermore, let l denote the number of items that precede x in Move-To-Front's list but follow x in OPT's list. We have $C_{MTF}(t) = k + l + 1$ and $C_{OPT}(t) \geq k + 1$.

When Move-To-Front serves $\sigma(t)$ and moves x to the front of the list, l inversions are destroyed and at most k new inversions are created. Thus

$$\begin{aligned} C_{MTF}(t) + \Phi(t) - \Phi(t-1) &\leq C_{MTF}(t) + k - l \;=\; 2k + 1 \\ &\leq 2C_{OPT}(t) - 1. \end{aligned}$$

Any paid exchange made by OPT when serving $\sigma(t)$ can increase the potential by 1, but OPT also pays 1. We conclude that inequality (1) holds.

The arguments above can be extended easily to analyze an insertion or deletion. On an insertion, $C_{MTF}(t) = C_{OPT}(t) = n + 1$, where n is the number of items in the list before the insertion, and at most n new inversions are created. On a deletion, l inversions are removed and no new inversion is created. □

Bentley and McGeoch [15] proved a weaker version of Theorem 1. They showed that on any sequence of accesses, the cost incurred by Move-To-Front is at most twice the cost of the *optimum static off-line algorithm*. The optimum static off-line algorithm first arranges the items in order of decreasing request frequencies and does no further exchanges while serving the request sequence.

The proof of Theorem 1 shows that Move-To-Front is $(2 - \frac{1}{n})$-competitive, where n is the maximum number of items ever contained in the dictionary. Irani [37] gave a refined analysis of the Move-To-Front rule and proved that it is $(2 - \frac{2}{n+1})$-competitive.

Sleator and Tarjan [64] showed that, in terms of competitiveness, Move-To-Front is superior to Transpose and Frequency-Count.

Proposition 2. *The algorithms Transpose and Frequency-Count are not c-competitive for any constant c.*

Recently, Albers [4] presented another deterministic on-line algorithm for the list update problem. The algorithm belongs to the Timestamp(p) family of algorithms that were introduced in the context of randomized on-line algorithms and that are defined for any real number $p \in [0,1]$. For $p = 0$, the algorithm is deterministic and can be formulated as follows.

Algorithm Timestamp(0): Insert the requested item, say x, in front of the first item in the list that precedes x in the list and that has been requested at most once since the last request to x. If there is no such item or if x has not been requested so far, then leave the position of x unchanged.

Theorem 3. *The Timestamp(0) algorithm is 2-competitive.*

Note that Timestamp(0) is not memoryless. We need information on past requests in order to determine where a requested item should be moved. In fact, in the most straightforward implementation of the algorithm we need a second pass through the list to find the position where the accessed item must be inserted. Often, such a second pass through the list does not harm the benefit of a list update algorithm. When list update algorithms are applied in the area of data compression, the *positions* of the accessed items are of primary importance, see Section 4.

The Timestamp(0) algorithm is interesting because it has a better overall performance than Move-To-Front. The algorithm achieves a competitive ratio of 2, as does Move-To-Front. However, as we shall see in Section 2.3, Timestamp(0) is considerably better than Move-To-Front on request sequences that are generated by probability distributions.

El-Yaniv [29] recently presented a new family of deterministic on-line algorithms for the list update problem. This family also contains the algorithms Move-To-Front and Timestamp(0). The following algorithm is defined for every integer $k \geq 1$.

Algorithm MRI(k): Insert the requested item, say x, just after the last item in the list that precedes x in the list and was requested at least $k+1$ times since the last request to x. If there is no such item or if x has not been requested so far, then move x to the front of the list.

El-Yaniv [29] showed that MRI(1) and Timestamp(0) are equivalent and also proved the following theorem.

Theorem 4. *For every integer $k \geq 1$, the MRI(k) algorithm is 2-competitive.*

Bachrach and El-Yaniv [10] recently presented an extensive experimental study of list update algorithms. The request sequences used were derived from the Calgary Compression Corpus [77]. In many cases, members of the MRI family were among the best algorithms.

Karp and Raghavan [42] developed a lower bound on the competitiveness that can be achieved by deterministic on-line algorithms. This lower bound implies that Move-To-Front, Timestamp(0) and MRI(k) have an optimal competitive ratio.

Theorem 5. *Let A be a deterministic on-line algorithm for the list update problem. If A is c-competitive, then $c \geq 2$.*

Proof. Consider a list of n items. We construct a request sequence that consists of accesses only. Each request is made to the item that is stored at the last position in A's list. On a request sequence σ of length m generated in this way, A incurs a cost of $C_A(\sigma) = mn$. Let OPT$'$ be the optimum static off-line algorithm. OPT$'$ first sorts the items in the list in order of nonincreasing request frequencies and then serves σ without making any further exchanges. When rearranging the list, OPT$'$ incurs a cost of at most $n(n-1)/2$. Then the requests in σ can be served at a cost of at most $m(n+1)/2$. Thus $C_{OPT}(\sigma) \leq m(n+1)/2 + n(n-1)/2$. For long request sequences, the additive term of $n(n-1)/2$ can be neglected and we obtain

$$C_A(\sigma) \geq \tfrac{2n}{n+1} \cdot C_{OPT}(\sigma).$$

The theorem follows because the competitive ratio must hold for all list lengths. □

The proof shows that the lower bound is actually $2 - \frac{2}{n+1}$, where n is the number of items in the list. Thus, the upper bound given by Irani on the competitive ratio of the Move-To-Front rule is tight.

Next we consider list update algorithms for other cost models. Reingold *et al.* [61] gave a lower bound on the competitiveness achieved by deterministic on-line algorithms.

Theorem 6. *Let A be a deterministic on-line algorithm for the list update problem in the P^d model. If A is c-competitive, then $c \geq 3$.*

Below we will give a family of deterministic algorithms for the P^d model. The best algorithm in this family achieves a competitive ratio that is approximately 4.56-competitive. We defer presenting this result until the discussion of randomized algorithms for the P^d model, see Section 2.2.

Sleator and Tarjan considered another generalized cost model. Let f be a nondecreasing function from the positive integers to the nonnegative reals. Suppose that an access to the i-th item in the list costs $f(i)$ and that an insertion costs $f(n+1)$, where n is the number of items in the list before the insertion. Let the cost of a paid exchange of items i and $i+1$ be $\Delta f(i) = f(i+1) - f(i)$. The function f is *convex* if $\Delta f(i) \geq \Delta f(i+1)$ for all i. Sleator and Tarjan [64] analyzed the Move-To-Front algorithm for convex cost functions. As usual, n denotes the maximum number of items contained in the dictionary.

Theorem 7. *If f is convex, then*

$$C_{MTF}(\sigma) \leq 2 \cdot C_{OPT}(\sigma) + \sum_{i=1}^{n-1}(f(n) - f(i))$$

for all request sequences σ that consist only of accesses and insertions.

The term $\sum_{i=1}^{n-1}(f(n) - f(i))$ accounts for the fact that the initial lists given to Move-To-Front and OPT may be different. If the lists are the same, the term can be omitted in the inequality. Theorem 7 can be extended to request sequences that include deletions if the total cost for deletions does not exceed the total cost incurred for insertions. Here we assume that a deletion of the i-th item in the list costs $f(i)$.

2.2 Randomized on-line algorithms

The competitiveness of a randomized on-line algorithm is defined with respect to an adversary. Ben-David *et al.* [13] introduced three kinds of adversaries. They differ in the way a request sequence is generated and how the adversary is charged for serving the sequence.

- **Oblivious Adversary**: The oblivious adversary has to generate a complete request sequence in advance, before any requests are served by the on-line algorithm. The adversary is charged the cost of the optimum off-line algorithm for that sequence.
- **Adaptive On-line Adversary**: This adversary may observe the on-line algorithm and generate the next request based on the algorithm's (randomized) answers to all previous requests. The adversary must serve each request on-line, i.e., without knowing the random choices made by the on-line algorithm on the present or any future request.
- **Adaptive Off-line Adversary**: This adversary also generates a request sequence adaptively. However, it is charged the optimum off-line cost for that sequence.

A randomized on-line algorithm A is called c-competitive against any oblivious adversary if there is a constant a such that for all size lists and all request sequences σ generated by an oblivious adversary, $E[C_A(\sigma)] \leq c \cdot C_{OPT}(\sigma) + a$. The expectation is taken over the random choices made by A.

Given a randomized on-line algorithm A and an adaptive on-line (adaptive off-line) adversary ADV, let $E[C_A]$ and $E[C_{ADV}]$ denote the expected costs incurred by A and ADV in serving a request sequence generated by ADV. A randomized on-line algorithm A is called c-competitive against any adaptive on-line (adaptive off-line) adversary if there is a constant a such that for all size lists and all adaptive on-line (adaptive off-line) adversaries ADV, $E[C_A] \leq c \cdot E[C_{ADV}] + a$, where the expectation is taken over the random choices made by A.

Ben-David *et al.* [13] investigated the relative strength of the adversaries with respect to on-line problems that can be formulated as a request-answer game, see [13] for details. They showed that if there is a randomized on-line algorithm that is c-competitive against any adaptive off-line adversary, then there is also a c-competitive deterministic on-line algorithm. This immediately implies that no randomized on-line algorithm for the list update problem can be better than 2-competitive against any adaptive off-line adversary. Reingold *et al.* [61] proved a similar result for adaptive on-line adversaries.

Theorem 8. *If a randomized on-line algorithm for the list update problem is c-competitive against any adaptive on-line adversary, then $c \geq 2$.*

The optimal competitive ratio that can be achieved by randomized on-line algorithms against oblivious adversaries has not been determined yet. In the following we present upper and lower bounds known on this ratio.

Randomized on-line algorithms against oblivious adversaries The intuition behind all randomized on-line algorithms for the list update problem is to move requested items in a more conservative way than Move-To-Front, which is still the classical deterministic algorithm.

The first randomized on-line algorithm for the list update problem was presented by Irani [37, 38] and is called Split algorithm. Each item x in the list maintains a pointer $x.split$ that points to some other item in the list. The pointer of each item either points to the item itself or to an item that precedes it in the list.

Algorithm Split: The algorithm works as follows.
Initialization:

For all items x in the list, set $x.split \leftarrow x$.

If item x is requested:

For all items y with $y.split = x$, set $y.split \leftarrow$ item behind x in the list.
With probability 1/2:
Move x to the front of the list.
With probability 1/2:
Insert x before item $x.split$.
If y preceded x and $x.split = y.split$, then set $y.split \leftarrow x$.
Set $x.split$ to the first item in the list.

Theorem 9. *The Split algorithm is* (15/8)*-competitive against any oblivious adversary.*

We note that $(15/8) = 1.875$. Irani [37, 38] showed that the Split algorithm is not better than 1.75-competitive in the $i-1$ cost model. In the $i-1$ cost model, an access to the i-th item in the list costs $i-1$ rather than i. The $i-1$ cost model is often useful to analyze list update algorithms. Compared to the standard i cost model, where an access to the i-th items costs i, the $i-1$ cost model always incurs a smaller cost; for any request sequence σ and any algorithm A, the cost difference is m, where m is the length of σ. Thus, a lower bound on the competitive ratio developed for a list update algorithm in the $i-1$ cost model does not necessarily hold in the i cost model. On the other hand, any upper bound achieved in the $i-1$ cost model also holds in the i cost model.

A simple and easily implementable list update rule was proposed by Reingold *et al.* [61].

Algorithm Bit: Each item in the list maintains a bit that is complemented whenever the item is accessed. If an access causes a bit to change to 1, then the requested item is moved to the front of the list. Otherwise the list remains unchanged. The bits of the items are initialized independently and uniformly at random.

Theorem 10. *The Bit algorithm is 1.75-competitive against any oblivious adversary.*

Reingold *et al.* analyzed Bit using an elegant modification of the potential function given in the proof of Theorem 1. Again, an inversion is a pair of items x, y such that x occurs before y in Bit's list and after y in OPT's list. An inversion has *type 1* if y's bit is 0 and *type 2* if y's bit is 1. Now, the potential is defined as the number of type 1 inversions plus twice the number of type 2 inversions.

The upper bound for Bit is tight in the $i-1$ cost model [61]. It was also shown that in the i cost model, i.e. in the standard model, Bit is not better than 1.625-competitive [3].

Reingold *et al.* [61] gave a generalization of the Bit algorithm. Let l be a positive integer and L be a non-empty subset of $\{0.1. \ldots, l-1\}$. The algorithm Counter(l, L) works as follows. Each item in the list maintains a mod l counter. Whenever an item x is accessed, the counter of x is decremented by 1 and, if the new value is in L, the item x is moved to the front of the list. The counters of the items are initialized independently and uniformly at random to some value in $\{0.1. \ldots, l-1\}$. Note that Bit is Counter(2,$\{1\}$). Reingold *et al.* chose parameters l and L so that the resulting Counter(l, L) algorithm is better than 1.75-competitive. It is worthwhile to note that the algorithms Bit and Counter(l, L) make random choices only during an initialization phase and run completely deterministically thereafter.

The Counter algorithms can be modified [61]. Consider a Counter($l, \{0\}$) algorithm that is changed as follows. Whenever the counter of an item reaches 0, the counter is reset to j with probability p_j, $1 \leq j \leq l-1$. Reingold *et al.* [61] gave a value for l and a resetting distribution on the p_j's so that the algorithm achieves a competitive ratio of $\sqrt{3} \approx 1.73$.

Another family of randomized on-line algorithms was given by Albers [4]. The following algorithm works for any real number $p \in [0.1]$.

Algorithm Timestamp(p): Each request to an item, say x, is served as follows. With probability p execute Step (a).

(a) Move x to the front of the list.

With probability $1-p$ execute Step (b).

(b) Insert x in front of the first item in the list that precedes x and

(i) that was not requested since the last request to x

or

(ii) that was requested exactly once since the last request to x and the corresponding request was served using Step (b) of the algorithm.

If there is no such item or if x is requested for the first time, then leave the position of x unchanged.

Theorem 11. *For any real number $p \in [0, 1]$, the algorithm Timestamp(p) is c-competitive against any oblivious adversary, where $c = \max\{2-p, 1+p(2-p)\}$.*

Setting $p = (3-\sqrt{5})/2$, we obtain a Φ-competitive algorithm, where $\Phi = (1+\sqrt{5})/2 \approx 1.62$ is the Golden Ratio. The family of Timestamp algorithms also includes two deterministic algorithms. For $p = 1$, we obtain the Move-To-Front rule. On the other hand, setting $p = 0$, we obtain the Timestamp(0) algorithm that was already described in Section 2.1.

In order to implement Timestamp(p) we have to maintain, for each item in the list, the times of the two last requests to that item. If these two times are stored with the item, then after each access the algorithm needs a second pass through the list to find the position where the requested item should be inserted. Note that such a second pass is also needed by the Split algorithm. In the case of the Split algorithm, this second pass is necessary because pointers have to be updated.

Interestingly, it is possible to combine the algorithms Bit and Timestamp(0), see Albers *et al.* [6]. This combined algorithm achieves the best competitive ratio that is currently known for the list update problem.

Algorithm Combination: With probability 4/5 the algorithm serves a request sequence using Bit, and with probability 1/5 it serves a request sequence using Timestamp(0).

Theorem 12. *The algorithm Combination is 1.6-competitive against any oblivious adversary.*

Proof. The analysis consists of two parts. In the first part we show that given any request sequence σ, the cost incurred by Combination and OPT can be divided into costs that are caused by each unordered pair $\{x, y\}$ of items x and y. Then, in the second part, we compare on-line and off-line cost for each pair $\{x, y\}$. This method of analyzing cost by considering pairs of items was first introduced by Bentley and McGeoch [15] and later used in [4, 37]. In the following we always assume that serving a request to the i-th item in the list incurs a cost of $i - 1$ rather than i. Clearly, if Combination is 1.6-competitive in this $i-1$ cost model, it is also 1.6-competitive in the i-cost model.

Let $\sigma = \sigma(1), \sigma(2), \ldots, \sigma(m)$ be an arbitrary request sequence of length m. For the reduction to pairs we need some notation. Let S be the set of items in the list. Consider any list update algorithm A that processes σ. For any $t \in [1, m]$ and any item $x \in S$, let $C_A(t, x)$ be the cost incurred by item x when A serves $\sigma(t)$. More precisely, $C_A(t, x) = 1$ if item x precedes the item requested by $\sigma(t)$ in A's list at time t; otherwise $C_A(t, x) = 0$. If A does not use paid exchanges, then the total cost $C_A(\sigma)$ incurred by A on σ can be written as

$$\begin{aligned} C_A(\sigma) &= \sum_{t \in [1,m]} \sum_{x \in S} C_A(t, x) = \sum_{x \in S} \sum_{t \in [1,m]} C_A(t, x) \\ &= \sum_{x \in S} \sum_{y \in S} \sum_{\substack{t \in [1,m] \\ \sigma(t) = y}} C_A(t, x) \\ &= \sum_{\substack{\{x,y\} \\ x \neq y}} \Big(\sum_{\substack{t \in [1,m] \\ \sigma(t) = x}} C_A(t, y) + \sum_{\substack{t \in [1,m] \\ \sigma(t) = y}} C_A(t, x) \Big). \end{aligned}$$

For any unordered pair $\{x, y\}$ of items $x \neq y$, let σ_{xy} be the request sequence that is obtained from σ if we delete all requests that are neither to x nor to y. Let $C_{BIT}(\sigma_{xy})$ and $C_{TS}(\sigma_{xy})$ denote the costs that Bit and Timestamp(0) incur in serving σ_{xy} on a two item list that consist of only x and y. Obviously, if Bit serves σ on the long list, then the relative position of x and y changes in the same way as if Bit serves σ_{xy} on the two item list. The same property holds for Timestamp(0). This follows from Lemma 13, which can easily be shown by induction on the number of requests processed so far.

Lemma 13. *At any time during the processing of σ, x precedes y in Timestamp(0)'s list if and only if one of the following statements holds: (a) the last requests made to x and y are of the form xx, xyx or xxy; (b) x preceded y initially and y was requested at most once so far.*

Thus, for algorithm $A \in \{\text{Bit, Timestamp(0)}\}$ we have

$$C_A(\sigma_{xy}) = \sum_{\substack{t\in[1,m]\\ \sigma(t)=x}} C_A(t,y) + \sum_{\substack{t\in[1,m]\\ \sigma(t)=y}} C_A(t,x)$$

$$C_A(\sigma) = \sum_{\substack{\{x,y\}\\ x\neq y}} C_A(\sigma_{xy}). \tag{2}$$

Note that Bit and Timestamp(0) do not incur paid exchanges. For the optimal off-line cost we have

$$C_{OPT}(\sigma_{xy}) \leq \sum_{\substack{t\in[1,m]\\ \sigma(t)=x}} C_{OPT}(t,y) + \sum_{\substack{t\in[1,m]\\ \sigma(t)=y}} C_{OPT}(t,x) \; + \; p(x,y)$$

and

$$C_{OPT}(\sigma) \geq \sum_{\substack{\{x,y\}\\ x\neq y}} C_{OPT}(\sigma_{xy}), \tag{3}$$

where $p(x,y)$ denotes the number of paid exchanges incurred by OPT in moving x in front of y or y in front of x. Here, only inequality signs hold because if OPT serves σ_{xy} on the two item list, then it can always arrange x and y optimally in the list, which might not be possible if OPT serves σ on the entire list. Note that the expected cost $E[C_{CB}(\sigma_{xy})]$ incurred by Combination on σ_{xy} is

$$E[C_{CB}(\sigma_{xy})] = \frac{4}{5}E[C_{BIT}(\sigma_{xy})] + \frac{1}{5}E[C_{TS}(\sigma_{xy})]. \tag{4}$$

In the following we will show that for any pair $\{x,y\}$ of items $E[C_{CB}(\sigma_{xy})] \leq 1.6C_{OPT}(\sigma_{xy})$. Summing this inequality for all pairs $\{x,y\}$, we obtain, by equations (2),(3) and (4), that Combination is 1.6-competitive.

Consider a fixed pair $\{x,y\}$ with $x \neq y$. We partition the request sequence σ_{xy} into phases. The first phase starts with the first request in σ_{xy} and ends when, for the first time, there are two requests to the same item and the next request is different. The second phase starts with that next request and ends in the same way as the first phase. The third and all remaining phases are constructed in the same way as the second phase. The phases we obtain are of the following types: x^k for some $k \geq 2$; $(xy)^k x^l$ for some $k \geq 1, l \geq 2$; $(xy)^k y^l$ for some $k \geq 1, l \geq 1$. Symmetrically, we have y^k, $(yx)^k y^l$ and $(yx)^k x^l$.

Since a phase ends with (at least) two requests to the same item, the item requested last in the phase precedes the other item in the two item list maintained by Bit and Timestamp(0). Thus the item requested first in a phase is always second in the list. Without loss of generality we can assume the same holds for OPT, because when OPT serves two consecutive requests to the same item, it cannot cost more to move that item to the front of the two item list after the first request. The expected cost incurred by Bit, Timestamp(0) (denoted by TS(0)) and OPT are given in the table below. The symmetric phases with x and y interchanged are omitted. We assume without generality that at the beginning of σ_{xy}, y precedes x in the list.

Phase	*Bit*	*TS*(0)	*OPT*
x^k	$\frac{3}{2}$	2	1
$(xy)^k x^l$	$\frac{3}{2}k+1$	$2k$	$k+1$
$(xy)^k y^l$	$\frac{3}{2}k+\frac{1}{4}$	$2k-1$	k

The entries for OPT are obvious. When Timestamp(0) serves a phase $(xy)^k x^l$, then the first two request xy incur a cost of 1 and 0, respectively, because x is left behind y on the first request to x. On all subsequent requests in the phase, the requested item is always moved to the front of the list. Therefore, the total cost on the phase is $1+0+2(k-1)+1=2k$. Similarly, Timestamp(0) serves $(xy)^k y^l$ with cost $2k-1$.

For the analysis of Bit's cost we need two lemmata.

Lemma 14. *For any item x and any $t \in [1,m]$, after the t-th request in σ, the value of x's bit is equally likely to be 0 or 1, and the value is independent of the bits of the other items.*

Lemma 15. *Suppose that Bit has served three consecutive requests yxy in σ_{xy}, or two consecutive requests xy where initially y preceded x. Then y is in front of x with probability $\frac{3}{4}$. The analogous statement holds when the roles of x and y are interchanged.*

Clearly, the expected cost spent by Bit on a phase x^k is $1+\frac{1}{2}+0(k-2)=\frac{3}{2}$. Consider a phase $(xy)^k x^l$. The first two requests xy incur a expected cost of 1 and $\frac{1}{2}$, respectively. By Lemma 15, each remaining request in the string $(xy)^k$ and the first request in x^l have an expected cost of $\frac{3}{4}$. Also by Lemma 15, the second request in x^l costs $1-\frac{3}{4}=\frac{1}{4}$. All other requests in x^l are free. Therefore, Bit pays an expected cost of $1+\frac{1}{2}+\frac{3}{2}(k-1)+\frac{3}{4}+\frac{1}{4}=\frac{3}{2}k+1$ on the phase. Similarly, we can evaluate a phase $(xy)^k y^l$.

The Combination algorithm serves a request sequence with probability $\frac{4}{5}$ using Bit and with probability $\frac{1}{5}$ using Timestamp(0). Thus, by the above table, Combination has an expected cost of 1.6 on a phase x^k, a cost of $1.6k+0.8$ on a phase $(xy)^k x^l$, and a cost $1.6k$ on a phase $(xy)^k y^l$. In each case this is at most 1.6 times the cost of OPT.

In the proof above we assume that a request sequence consists of accesses only. However, the analysis is easily extended to the case that insertions and deletions occur, too. For any item x, consider the time intervals during which x is contained in the list. For each of these intervals, we analyze the cost caused by any pair $\{x, y\}$, where y is an item that is (temporarily) present during the interval. □

Teia [72] presented a lower bound for randomized list update algorithms.

Theorem 16. *Let A be a randomized on-line algorithm for the list update problem. If A is c-competitive against any oblivious adversary, then* $c \geq 1.5$.

An interesting open problem is to give tight bounds on the competitive ratio that can be achieved by randomized on-line algorithms against oblivious adversaries.

Results in the P^d cost model As mentioned in Theorem 6, no deterministic on-line algorithm for the list update problem in the P^d model can be better than 3-competitive. By a result of Ben-David *et al.* [13], this implies that no randomized on-line algorithm for the list update problem in the P^d model can be better than 3-competitive against any adaptive off-line adversary. Reingold *et al.* [61] showed that the same bound holds against adaptive on-line adversaries.

Theorem 17. *Let A be a randomized on-line algorithm for the list update problem in the P^d model. If A is c-competitive against any adaptive on-line adversary, then* $c \geq 3$.

Reingold *et al.* [61] analyzed the Counter$(l, \{l-1\})$ algorithms, l being a positive integer, for list update in the P^d model. As described before, these algorithms work as follows. Each item maintains a mod l counter that is decremented whenever the item is requested. When the value of the counter changes to $l-1$, then the accessed item is moved to the front of the list. In the P^d cost model, this movement is done using paid exchanges. The counters are initialized independently and uniformly at random to some value in $\{0, 1, \ldots, l-1\}$.

Theorem 18. *In the P^d model, the algorithm Counter$(l, \{l-1\})$ is c-competitive against any oblivious adversary, where* $c = \max\{1 + \frac{l+1}{2d}, 1 + \frac{1}{l}(2d + \frac{l+1}{2})\}$.

The best value for l depends on d. As d goes to infinity, the best competitive ratio achieved by a Counter$(l, \{l-1\})$ algorithm decreases and goes to $(5+\sqrt{17})/4 \approx 2.28$.

We now present an analysis of the deterministic version of the Counter$(l, \{l-1\})$ algorithm. The deterministic version is the same as the randomized version, except that all counters are initialized to zero, rather than being randomly initialized.

Theorem 19. *In the P^d model, the deterministic algorithm Counter$(l, \{l-1\})$ is c-competitive, where* $c = \max\{3 + \frac{2l}{d}, 2 + \frac{2d}{l}\}$.

Proof. The analysis is similar in form to that of Combination. Consider a pair of items $\{x, y\}$. Let $c(x)$ and $c(y)$ denote the values of the counters at items x and

y, respectively. We define a potential function Φ. Assume w.l.o.g. that OPT's list is ordered (x, y). Then

$$\Phi = \begin{cases} (1 + 2d/l)c(y) & \text{if Counter's list is ordered } (x, y) \\ k + d - c(x) + (1 + 2d/l)c(y) & \text{if Counter's list is ordered } (y, x) \end{cases}$$

The remainder of the proof follows by case analysis. For each event in each configuration, we compare the amortized cost incurred by Counter to the actual cost incurred by OPT. (See the proof of competitiveness of MTF.) □

As in the randomized case, the optimum value of l for the deterministic Counter algorithm depends on d. As d goes to infinity, the best competitive ratio decreases and goes to $(5 + \sqrt{17})/2 \approx 4.56$, exactly twice the best randomized value.

2.3 Average case analyses of list update algorithms

In this section we study a restricted class of request sequences: request sequences that are generated by a probability distribution. Consider a list of n items $x_1, x_2, \ldots, x_n$, and let $\mathbf{p} = (p_1, p_2, \ldots, p_n)$ be a vector of positive probabilities p_i with $\sum_{i=1}^{n} p_i = 1$. We study request sequences that consist of accesses only, where each request it made to item x_i with probability p_i, $1 \leq i \leq n$. It is convenient to assume that $p_1 \geq p_2 \geq \cdots \geq p_n$.

There are many results known on the performance of list update algorithms when a request sequence is generated by a probability distribution, i.e. by a discrete memoryless source. In fact, the algorithms Move-To-Front, Transpose and Frequency-Count given in Section 2.1 as well as their variants were proposed as heuristics for these particular request sequences.

We are now interested in the asymptotic expected cost incurred by a list update algorithm. For any algorithm A, let $E_A(\mathbf{p})$ denote the asymptotic expected cost incurred by A in serving a single request in a request sequence generated by the distribution $\mathbf{p} = (p_1, \ldots, p_n)$. In this situation, the performance of an on-line algorithm has generally been compared to that of the *optimal static ordering*, which we call STAT. The optimal static ordering first arranges the items x_i in nonincreasing order by probabilities and then serves a request sequence without changing the relative position of items. Clearly, $E_{STAT}(\mathbf{p}) = \sum_{i=1}^{n} i p_i$ for any distribution $\mathbf{p} = (p_1, \ldots, p_n)$.

As in Section 2.1, we first study the algorithms Move-To-Front(MTF), Transpose(T) and Frequency-Count(FC). By the strong law of large numbers we have $E_{FC}(\mathbf{p}) = E_{STAT}(\mathbf{p})$ for any probability distribution $\mathbf{p}$ [62]. However, as mentioned in Section 2.1, Frequency-Count may need a large amount of extra memory to serve a request sequence. It was shown by several authors [17, 19, 35, 45, 55, 62] that

$$E_{MTF}(\mathbf{p}) = 1 + 2 \sum_{1 \leq i < j \leq n} \frac{p_i p_j}{(p_i + p_j)}$$

for any $\mathbf{p} = (p_1, \ldots, p_n)$. A simple, closed-form expression for the asymptotic expected cost of the Transpose rule has not been found. The expression for

$E_{MTF}(\mathbf{p})$ was used to show that $E_{MTF}(\mathbf{p}) \leq 2E_{STAT}(\mathbf{p})$ for any distribution $\mathbf{p}$. However, Chung *et al.* [22] showed that Move-To-Front performs better.

Theorem 20. *For any probability distribution* $\mathbf{p}$, $E_{MTF}(\mathbf{p}) \leq \frac{\pi}{2}E_{STAT}(\mathbf{p})$.

This bound is tight as was shown by Gonnet *et al.* [32].

Theorem 21. *For any* $\epsilon > 0$, *there exists a probability distribution* $\mathbf{p}_\epsilon$ *with* $E_{MTF}(\mathbf{p}_\epsilon) \geq (\frac{\pi}{2} - \epsilon)E_{STAT}(\mathbf{p}_\epsilon)$.

The distributions used in the proof of Theorem 21 are of the form

$$p_i = 1/(i^2 H_n^2) \qquad i = 1, \ldots, n$$

where $H_n^2 = \sum_{i=1}^n 1/i^2$. These distributions are called *Lotka's Law*. There are probability distributions $\mathbf{p}_0$ for which the ratio of $E_{MTF}(\mathbf{p}_0)/E_{STAT}(\mathbf{p}_0)$ can be smaller than $\pi/2 \approx 1.58$. Let $p_i = 1/(iH_n), 1 \leq i \leq n$, with $H_n = \sum_{i=1}^n 1/i$. This distribution is called *Zipf's Law*. Knuth [45] showed that for this distribution $\mathbf{p}_0$, $E_{MTF}(\mathbf{p}_0) \leq (2\ln 2)E_{STAT}(\mathbf{p}_0)$. We note that $2\ln 2 \approx 1.386$.

Rivest [62] proved that Transpose performs better than Move-To-Front on distributions.

Theorem 22. *For any distribution* $\mathbf{p} = (p_1, \ldots, p_n)$, $E_T(\mathbf{p}) \leq E_{MTF}(\mathbf{p})$. *The inequality is strict unless* $n = 2$ *or* $p_i = 1/n$ *for* $i = 1, \ldots, n$.

Rivest conjectured that Transpose is optimal among all *permutation rules.* A permutation rule, when accessing an item at position j, applies a permutation π_j to the first j positions in the list. However, Anderson *et al.* [8] found a counterexample to this conjecture. Bitner [17] showed that while $E_T(\mathbf{p}) \leq E_{MTF}(\mathbf{p})$, the Move-To-Front rule converges faster to its asymptotic expected cost than Transpose.

The algorithms Move-To-Front, Transpose and Frequency-Count were also analyzed experimentally [10, 15, 62, 73]. Rivest [62] generated request sequences that obeyed Zipf's law. On these sequences, Transpose indeed performed better than Move-To-Front. In contrast, Bentley and McGeoch [15] considered request sequences that came from word counting problems in text and Pascal files. In their tests, Transpose always performed worse than Move-To-Front and Frequency Count, with Move-To-Front usually being better then Frequency-Count. In general, STAT achieved a smaller average search time than the three on-line algorithms.

Finally, we consider the Timestamp(0) algorithm that was also presented in Section 2.1. It was shown in [5] that Timestamp(0) has a better performance than Move-To-Front if request sequences are generated by probability distributions. Let $E_{TS}(\mathbf{p})$ denote the asymptotic expected cost incurred by Timestamp(0).

Theorem 23. *For any probability distribution* $\mathbf{p}$, $E_{TS}(\mathbf{p}) \leq 1.34E_{STAT}(\mathbf{p})$.

Theorem 24. *For any probability distribution* $\mathbf{p}$, $E_{TS}(\mathbf{p}) \leq 1.5E_{OPT}(\mathbf{p})$.

Note that $E_{OPT}(\mathbf{p})$ is the asymptotic expected cost incurred by the optimal off-line algorithm OPT, which may dynamically rearrange the list while serving a request sequence. Thus, this algorithm is much stronger than STAT. The algorithm Timestamp(0) is the only algorithm whose asymptotic expected cost has been compared to $E_{OPT}(\mathbf{p})$.

The bound given in Theorem 24 holds with high probability. More precisely, for every distribution $\mathbf{p} = (p_1, \ldots, p_n)$, and $\epsilon > 0$, there exist constants c_1, c_2 and m_0 dependent on $\mathbf{p}, n$ and ϵ such that for any request sequence σ of length $m \geq m_0$ generated by $\mathbf{p}$,

$$Prob\{C_{TS}(\sigma) > (1.5 + \epsilon)C_{OPT}(\sigma)\} \leq c_1 e^{-c_2 m}.$$

2.4 Remarks

List update techniques were first studied in 1965 by McCabe [55] who considered the problem of maintaining a sequential file. McCabe also formulated the algorithms Move-To-Front and Transpose. From 1965 to 1985 the list update problem was studied under the assumption that a request sequence is generated by a probability distribution. Thus, most of the results presented is Section 2.3 were developed earlier than the results in Sections 2.1 and 2.2. A first survey on list update algorithms when request sequences are generated by a distribution was written by Hester and Hirschberg [36]. The paper [64] by Sleator and Tarjan is a fundamental paper in the entire on-line algorithms literature. It made the competitive analysis of on-line algorithms very popular. Randomized on-line algorithms for the list update problem have been studied since the early nineties. The list update problem is a classical on-line problem that continues to be interesting both from a theoretical and practical point of view.

3 Binary search trees

Binary search trees are used to maintain a set S of elements where each element has an associated key drawn from a totally ordered universe. For convenience we assume each element is given a unique key, and that the n elements have keys $1, \ldots, n$. We will generally not distinguish between elements and their keys.

A binary search tree is a rooted tree in which each node has zero, one, or two children. If a node has no left or right child, we say it has a *null* left or right child, respectively. Each node stores an element, and the elements are assigned to the nodes in *symmetric order*: the element stored in a node is greater than all elements in descendents of its left child, and less than all elements in descendents of its right child. An inorder traversal of the tree yields the elements in sorted order. Besides the elements, the nodes may contain additional information used to maintain states, such as a color bit or a counter.

The following operations are commonly performed on binary search trees.

Successful-Access(x). Locate an element $x \in S$.
Unsuccessful-Access(x). Determine that an element x is **not** in S.

Insert(x). Add a new element x to S. The tree is modified by adding a new node containing the element as a leaf, so that symmetric order is maintained.

Delete(x). Remove element x from S. The resultant tree has one fewer nodes. There are several different deletion algorithms for binary search trees.

Split(x). Split S into two sets: $S_1 = \{y \mid y \in S, y \leq x\}$ and $S_2 = \{y \mid y \in S, y > x\}$. S_1 and S_2 must be contained in two search trees.

Meld(S_1, S_2). The inverse of a split (for all $x \in S_1, y \in S_2$, $x < y$).

Adel'son-Vel'skii and Landis [2] introduced a primitive operation for restructuring a binary search tree, the **edge rotation**. Figure 1 shows examples of left and right rotation of edge $\langle x, y\rangle$. Rotation preserves symmetric order.

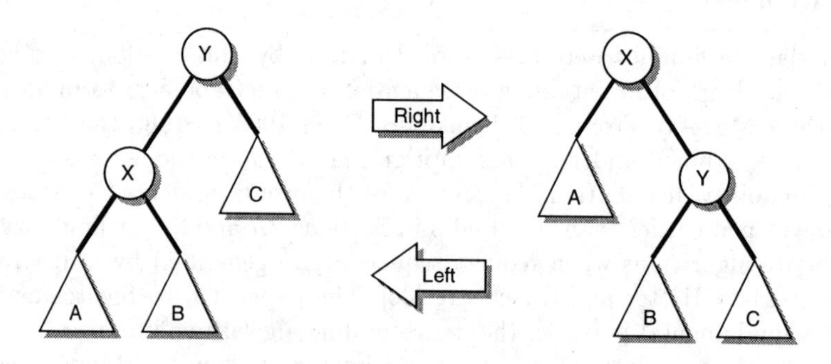

Fig. 1. Right and left rotations of $\langle x, y\rangle$.

In the *standard search tree model* [65, 76] a search tree algorithm has the following behavior:

> The algorithm carries out each (successful) access by traversing the path from the root to the node containing the accessed item, at a cost of one plus the depth of the node containing the item. Between accesses the algorithm performs an arbitrary number of rotations anywhere in the tree, at a cost of one per rotation.[1]

The path from the root to a node is called the *access path*. If the path consists of only left edges, it is called a *left path*. A *right path* is defined analogously.

We expand the model to include unsuccessful searches, insertions and deletions as follows. Let T_0 be a search tree on n elements. T_0 is extended to search tree T by replacing any null child in the original tree by a leaf. Each leaf has an associated key range. If leaf l replaces the null left child of node x, then the range of l is the set of key values that are less than x but greater than the predecessor of x in the tree. If x is the least key in the original tree, then the range of l is all keys less than x. Similarly, if l replaces a null right child of x, then its

[1] This definition is a slightly modified form of the one given in [65].

range is all keys greater than x but less than the successor of x, if one exists. This is a well-known extension, and it is easy to see that the leaf ranges are disjoint. Successful searches are carried out as before. Between any operation, any number of rotations can be performed at cost 1 per rotation.

> The algorithm carries out an unsuccessful access to key i by traversing the path from the root to the leaf whose range contains i. The cost is 1 plus the length of the path.

> The algorithm carries out an insertion of a new element x (not already in the tree by assumption) by performing an unsuccessful search for x. Let l be the leaf reached by the search. Leaf l is replaced by a new node containing x. The new node has two new leaves containing the two halves of the range originally in l. The cost is 1 plus the length of the path.

The model for deletion is rather more complicated, as deletion is itself a more complicated operation and can be done in a number of ways.

> The algorithm deletes an element x (already in the tree by assumption) in several phases. In the first phase, a successful search for x is performed, at the usual cost. In the second phase, any number of rotations are performed at cost 1 per rotation. In the third phase, let S_x be the subtree rooted at x after phase two. Let p be the predecessor of x in S_x, or x itself if x is the least element in S_x. Similarly, let s the successor of x in S_x, or x itself if x has no successor. The algorithm chooses one of p or s, say p w.l.o.g., and traverses the path from x to p, at cost 1 plus the path length. In phase four, the algorithm changes pointers in x, p, and their parents, to construct two search trees, one consisting only of x, and the other containing all the remaining elements. The singleton tree is discarded. The cost of phase four is 1.

The successful and unsuccessful access can be implemented in a comparison model with a well-known recursive procedure (see [27, Chapter 13]): the search key is compared to the element in the root of the tree. If they are equal, the root element is returned. Otherwise the left subtree or right subtree of the root is recursively searched, according to whether the key is less than or greater than the root element, respectively. If the desired subtree is null, the procedure returns an unsuccessful search indicator. Examples of various deletion routines algorithms can be found in [27] or [65].

For what follows, we will restrict our attention to request sequences that consist only of successful accesses, unless otherwise stated. For an algorithm, A, in the standard search model, let $A(\sigma, T_0)$ denote the sum of the costs of accesses and rotations performed by A in response to access sequence σ starting from tree T_0, where T_0 is a search tree on n elements and σ accesses only the elements in T_0. Let $\mathrm{OPT}(\sigma, T_0)$ denote the minimum cost of servicing σ, starting from search tree T_0 containing elements $1, \ldots, n$, including both access costs and rotation costs.

Definition 25. An algorithm A is $f(n)$-competitive if for all σ, T_0

$$A(\sigma, T_0) \leq f(n) \cdot \mathrm{OPT}(\sigma, T_0) + O(n). \tag{5}$$

Let S denote the *static* search algorithm, i.e., the algorithm which performs no rotations on the initial tree T_0.

Definition 26. An algorithm A is $f(n)$-static-competitive if for all σ, T_0

$$A(\sigma, T_0) \leq f(n) \cdot \min_T S(\sigma, T) + O(n) \tag{6}$$

where T is a search tree on the same elements as T_0.

Note that $S(\sigma, T)$ is given by

$$\sum_{i=1}^{n} f_\sigma(i) d_T(i)$$

where $f_\sigma(i)$ is the number of times element i is accessed in σ, and $d_T(i)$ denotes the depth of element i in search tree T.

A final definition deals with probabilistic request sequences, in which each request is chosen at random from among the possible requests according to some distribution $\mathcal{D}$. That is, on each request element i is requested with probability p_i, $i = 1, \ldots, n$. For a fixed tree T, the expected cost of a request is

$$1 + \sum_{i=1}^{n} p_i d_T(i).$$

We denote this $S(\mathcal{D}, T)$, indicating the cost of distribution $\mathcal{D}$ on static tree T. Finally, let $\bar{S}(\mathcal{D}) = \min_T S(\mathcal{D}, T)$ denote the expected cost per request of the optimal search tree for distribution $\mathcal{D}$. For a search tree algorithm A, let $\bar{A}(\mathcal{D}, T_0)$ denote the *asymptotic expected cost* of servicing a request, given that the request is generated by $\mathcal{D}$ and the algorithm starts with T_0.

Definition 27. An algorithm A is called $f(n)$-distribution-competitive if for all n, all distributions $\mathcal{D}$ on n elements and all initial trees T_0 on n elements,

$$\bar{A}(\mathcal{D}, T_0) \leq f(n) \cdot \bar{S}(\mathcal{D}). \tag{7}$$

Definitions 26 and 27 are closely related, since both compare the cost of an algorithm on sequence σ to the cost of a fixed search tree T on σ. The fixed tree T that achieves the minimum cost is the tree that minimizes the *weighted path length* $\sum_{i=1}^{n} w(i) d_T(i)$, where the weight of node i is the total number of accesses to i, in the case of static competitiveness, or the probability of an access to i, in the case of distribution optimality. Knuth [44] gives an $O(n^2)$ algorithm for computing a tree that minimizes the weighted path length. By information theory [1], for a request sequence σ, $m = |\sigma|$,

$$S(\sigma, T) = \Omega(m + \sum_{i=1}^{n} f_\sigma(i) \log(m / f_\sigma(i))).$$

In the remainder of this section, we will first discuss the off-line problem, in which the input is the entire request sequence σ and the output is a sequence of rotations to be performed after each request so that the the total cost of servicing σ is minimized. Little is known about this problem, but several characterizations of optimal sequences are known and these suggest some good properties for on-line algorithms.

Next we turn to on-line algorithms. An $O(\log n)$ competitive ratio can be achieved by any one of several balanced tree schemes. So far, no on-line algorithm is known that is $o(\log n)$-competitive. Various adaptive and self-organizing rules have been suggested over the past twenty years. Some of them have good properties against probability distributions, but most perform poorly against arbitrary sequences. Only one, the *splay* algorithm, has any chance of being $O(1)$-competitive. We review the known properties of splay trees, the various conjectures made about their performance, and progress on resolving those conjectures.

3.1 The off-line problem and properties of optimal algorithms

An off-line search tree algorithm takes as input a request sequence σ and an initial tree T_0 and outputs a sequence of rotations, to be intermingled with servicing successive requests in σ, that achieves the minimum total cost $\mathrm{OPT}(\sigma, T_0)$.

It is open whether $\mathrm{OPT}(\sigma, T_0)$ can be computed in polynomial time. There are $\frac{1}{n+1}\binom{2n}{n}$ binary search trees on n nodes, so the dynamic programming algorithm for solving metrical service systems or metrical task systems requires exponential space just to represent all possible states.

A basic subproblem in the dynamic programming solution is to compute the rotation distance, $d(T_1, T_2)$, between two binary search trees T_1 and T_2 on n nodes. The rotation distance is the minimum number of rotations needed to transform T_1 into T_2. It is also open whether the rotation distance can be computed in polynomial time. Upper and lower bounds on the worst-case rotation distance are known, however.

Theorem 28. *The rotation distance between any two binary trees is at most 2n-6, and there is an infinite family of trees for which this bound is tight.*

An upper bound of $2n - 2$ was shown by Crane [28] and Culik and Wood [46]. This bound is easily seen: a tree T can be converted into a right path by repeatedly rotating an edge hanging off the right spine onto the right spine with a right rotation. This process must eventually terminate with all edges on the right spine. Since there are $n - 1$ edges, the total number of rotations is $n - 1$. To convert T_1 to T_2, convert T_1 to a right path then compute the rotations needed to convert T_2 to a right path, and apply them in reverse to the right spine into which T_1 has been converted. At most $2n - 2$ rotations are necessary. Sleator, Tarjan and Thurston [66] improved the upper bound to $2n-6$ using a relation between binary trees and triangulations of polyhedra. They also demonstrated the existence of an infinite family in which $2n - 6$ rotations were required. Makinen [52] subsequently showed that a weaker upper bound of $2n-5$ can be simply proved with elementary tree concepts. His proof is based on using either the left path or right path as an intermediate tree, depending on which is closer. Luccio and Pagli [51] showed that the tight bound $2n - 6$ could be achieved by adding one more possible intermediate tree form, a root whose left subtree is a right path and whose right subtree is a left path.

Wilber [76] studied the problem of placing a lower bound on the cost of the optimal solution to specific families of request sequences. He described two techniques for calculating lower bounds, and used them to show the existence of request sequences on which the optimal cost is $\Omega(n \log n)$. We give one of his examples below. Let i and k be non-negative integers, $i \in [0, 2^k - 1]$. The *k-bit reversal* of i, denoted $br_k(i)$, is the integer j given by writing the binary representation of i backwards. Thus $br_k(6) = 3$ ($110 \Rightarrow 011$). The bit reversal permutation on $n = 2^k$ elements is the sequence $B^k = br_k(0), br_k(1), \ldots, br_k(n-1)$.

Theorem 29. [76] *Let k be a nonnegative integer and let $n = 2^k$. Let T_0 be any search tree with nodes $0, 1, \ldots, n-1$. Then* $\mathrm{OPT}(B^k, T_0) \geq n \log n + 1$.

Lower bounds for various on-line algorithms can be found using two much simpler access sequences: the *sequential access sequence* $\sigma^S = 1, 2, \ldots, n$, and the reverse order sequence $\sigma^R = n, n-1, \ldots, 1$. It is easy to see that $\mathrm{OPT}(\sigma^S, T_0) = O(n)$. For example, T_0 can be rotated to a right spine in $n - 1$ time, after which each successive accessed element can be rotated to the root with one rotation. This gives an amortized cost of $2 - 1/n$ per access. The optimal static tree is the completely balanced tree, which achieves a cost of $\Theta(\log n)$ per access. The sequential access sequence can be repeated k times, for some integer k, with the same amortized costs per access.

Although no polynomial time algorithm is known for computing $\mathrm{OPT}(\sigma, T_0)$, there are several characterizations of the properties of optimal and near-optimal solutions. Wilber [76] and Lucas [50] note that there is a solution within a factor of two of optimum in which each element is at the root when it is accessed. The near-optimal algorithm imitates the optimal algorithm except that it rotates the accessed item to the root just prior to the access, and then undoes all the rotations in reverse to restore the tree to the old state. Hence one may assume that the accessed item is always at the root, and all the cost incurred by the optimum strategy is due to rotations.

Lucas [50] proves that there is an optimum algorithm in which rotations occur prior to the access and the rotated edges form a connected subtree containing the access path. In addition, Lucas shows that if in the initial tree T_0 there is a node x none of whose descendents (including x) are ever accessed, then x need never be involved in a rotation. She also studies the "rotation graph" of a binary tree and proves several properties about the graph [48]. The rotation graph can be used to enumerate all binary search trees on n nodes in $O(1)$ time per tree.

Lucas makes several conjectures about the off-line and on-line search tree algorithms.

Conjecture 1 (Greedy Rotations) *There is a c-competitive off-line algorithm (and possibly on-line) such that each search is to the element at the root and all rotations decrease the depth of the next element to be searched for.*

Observe that the equivalent conjecture for the list update problem (that all exchanges decrease the depth of the next accessed item) is true for list update, since an off-line algorithm can be 2-competitive by moving the next accessed item to the front of the list just prior to the access. The cost is the same as is incurred by the move-to-front heuristic, since the exchanges save one unit of access cost. But for list update, there is no condition on relative position of elements (i.e. no requirement to maintain symmetric order), so the truth of the conjecture for search trees is non-obvious.

Lucas proposes a candidate polynomial-time off-line algorithm, and conjectures that it provides a solution that is within a constant factor of optimal, but does not prove the conjecture. The proposed "greedy" algorithm modifies the tree prior to each access. The accessed element is rotated to the root. The edges that are rotated off the path during this process form a collection of connected subtrees, each of which is a left path or a right path. These paths are then converted into connected subtrees that satisfy the heap property where the heap key of a node is the time it or a descendent off the path will be next accessed. This tends to move the element that will be accessed soonest up the tree.

3.2 On-line algorithms

An $O(\log n)$-competitive solution is achievable with a balanced tree; many balanced binary tree algorithms are known that handle unsuccessful searches, insertions, deletions, melds, and splits in $O(\log n)$ time per operation, either amortized or worst-case. Examples include AVL-trees, red/black trees, and weight-balanced trees; see the text [27] for more information. These data structures make no attempt to self-organize, however, and are concerned solely with keeping the maximum depth of any item at $O(\log n)$. Any heuristic is, of course, $O(n)$-competitive.

A number of candidate self-organizing algorithms have been proposed in the literature. These can generally be divided into memoryless and state-based algorithms. A memoryless algorithm maintains no state information besides the current tree. The proposed memoryless heuristics are:

1. Move to root by rotation [7].
2. Single-rotation [7].
3. Splaying [65].

The state-based algorithms are

1. Dynamic monotone trees [17].
2. WPL trees [21].
3. D-trees [57, 58].

3.3 State-based algorithms

Bitner [17] proposed and analyzed *dynamic monotone trees.* Dynamic monotone trees are a dynamic version of a data structure suggested by Knuth [45] for approximating optimal binary search trees given a distribution $\mathcal{D}$. The element with maximum probability of access is placed at the root of the tree. The left and right subtrees of the root are then constructed recursively. In dynamic monotone trees, each node contains an element and a counter. The counters are initialized to zero. When an element is accessed, its counter is incremented by one, and the element is rotated upwards while its counter is greater than the counter in its parent. Thus the tree stores the elements in symmetric order by key, and in max-heap order by counter. A similar idea is used in the "treap," a randomized search tree developed by Aragon and Seidel [9].

Unfortunately, monotone and dynamic monotone trees do poorly in the worst case. Mehlhorn [56] showed that the distribution-competitive ratio is $\Omega(n/\log n)$. This is easily seen by assigning probabilities to the elements $1, \ldots, n$ so that $p_1 > p_2 > \ldots > p_n$ but so that all are very close to $1/n$. The monotone tree corresponding to these probabilities is a right path, and by the law of large numbers a dynamic monotone tree will asymptotically tend to this form. The monotone tree is also no better than $\Omega(n)$-competitive; repetitions of sequential access sequence give this lower bound. Bitner showed, however, that monotone trees do perform better on probability distributions with low entropy (the bad sequence above has nearly maximum entropy). Thus as the distributions become more skewed, monotone trees do better.

Bitner also suggested several conditional modification rules. When an edge is rotated, one subtree decreases in depth while another increases in depth. With conditional rotation, an accessed node is rotated upwards if the total number of accesses to nodes in the subtree that moves up is greater than the total number of accesses to the subtree that moves down. No analysis of the conditional rotation rules is given, but experimental evidence is presented which suggests they perform reasonably well. No such rule is better than $\Omega(n)$-competitive, however, for the same reason single exchange is not competitive.

Oommen *et al.* [21] generalized the idea of conditional rotations by adding two additional counters. One stores the number of accesses to descendents of a node, and one stores the total weighted path length (defined above) of the subtree rooted at the node. After an access, the access path is processed bottom-up and successive edges are rotated as long as the weighted path length (WPL) of the whole tree diminishes. The weight of a node at time t for this purpose is taken to be the number of accesses to that node up to time t. Oommen *et al.* claim that their WPL algorithm asymptotically approaches the tree with minimum weighted path length, and hence is $O(1)$-distribution-competitive. By the law of large numbers, the tree that minimizes the weighted path length will asymptotically approach the optimal search tree for the distribution. The main contribution of Oommen *et al.* is to show that changes in the weighted path length of the tree can be computed efficiently. The WPL tree can be no better than $\Omega(\log n)$-competitive (once again using repeated sequential access) but it is unknown if this bound is achieved. It is also unknown whether WPL-trees are $O(1)$-static-competitive.

Mehlhorn [57, 58] introduced the D-tree. The basic idea behind a D-tree is that each time an element is accessed the binary search tree is extended by adding a dummy node as a leaf in the subtree rooted at accessed element. The extended tree is then maintained using a weight-balanced or height-balanced binary tree (see [58] for more information on such trees). Since nodes that are frequently accessed will have more dummy descendents, they will tend to be higher in the weight-balanced tree. Various technical details are required to implement D-trees in a space-efficient fashion. Mehlhorn shows that the D-tree is $O(1)$-distribution-competitive.

At the end of an access sequence the D-tree is near the optimal static binary tree for that sequence, but it is not known if it is $O(1)$-static-competitive, since a high cost may be incurred in getting the tree into the right shape.

3.4 Simple memoryless heuristics

The two memoryless heuristics *move-to-root* (MTR) and *simple exchange* (SE) were proposed by Allen and Munro as the logical counterparts in search trees of the move-to-front and transpose rules of sequential search.

In simple exchange, each time an element is accessed, the edge from the node to its parent is rotated, so the element moves up. With the MTR rule, the element is rotated to the root by repeatedly applying simple exchanges. Allen and Munro show that MTR fares well when the requests are generated by a probability distribution, but does poorly against an adversary.

Theorem 30. [7] *The move-to-root heuristic is $O(1)$-distribution-competitive.*

Remark: the constant inside the $O(1)$ is $2\ln 2 + o(1)$.

Theorem 31. [7] *On the sequential access sequence $\sigma = (1, 2, \ldots, n)^k$ the MTR heuristic incurs $\Omega(n)$ cost per request when $k \geq 2$.*

Remark: Starting from any initial tree T_0, the sequence $1, 2, \ldots, n$ will cause MTR to generate the tree consisting of a single left path. Thereafter the cost of a request to i will have cost $n - i + 1$.

Corollary 32. *The competitive ratio of the MTR heuristic is $\Theta(n)$, and the static competitive ratio of MTR is $\Omega(n/\log n)$.*

The corollary follows from the observation of Section 3.1 that the sequential access sequence can be satisfied with $O(1)$ amortized cost per request, and with $O(\log n)$ cost per request if a fixed tree is used.

Although MTR is not competitive, it does well against a probability distribution. The simple exchange heuristic, however, is not even good against a probability distribution.

Theorem 33. [7] *If $p_i = 1/n$, $1 \leq i \leq n$, then the asymptotic expected time per request of the simple exchange algorithm is $\sqrt{\pi n} + o(\sqrt{n})$.*

For this distribution the asymptotic expected cost of a perfectly balanced tree is $O(\log n)$. This implies:

Corollary 34. *The distribution competitive ratio of SE is $\Omega(\sqrt{n}/\log n)$.*

Corollary 35. *The competitive ratio of SE is $\Theta(n)$, and the static competitive ratio of SE is $\Omega(n/\log n)$.*

Corollary 35 can be proved with a sequential access sequence, except that each single request to i is replaced by enough consecutive requests to force i to the root. After each block of consecutive requests to a given element i, the resulting tree has the same form as the tree generated by MTR does after a single request to i.

3.5 Splay trees

To date, the only plausible candidate for a search tree algorithm that might be $O(1)$-competitive is the splay tree, invented by Sleator and Tarjan [65]. A splay tree is a binary search tree in which all operations are performed by means of a primitive called *splaying*. A splay at node x is a sequence of rotations on the path that moves x the root of the tree. The crucial difference between splaying and simple move-to-root is that while move-to-root rotates each edge on the path from x to the root in order from top to bottom, splaying rotates some higher edges before some lower edges. The order is chosen so that the nodes on the path decrease in depth by about one half. This halving of the depth does not happen with the move-to-root heuristic.

The splaying of node x to the root proceeds by repeatedly determining which of the three cases given in Fig. 2 applies, and performing the diagramed rotations. Sleator and Tarjan call these cases respectively the zig, zig-zig, and zig-zag cases. Note that in the zig and zig-zag cases, the rotations that occur are precisely those that would occur with the move-to-root heuristic. But the zig-zig is different. Simple move-to-root applied to a long left or right path leads to another long left or right path, while repeatedly executing zig-zig makes a much more balanced tree.

The fundamental lemma regarding splay trees is the *splay access lemma.* Let $w_1, w_2, \ldots, w_n$ be a set of arbitrary real weights assigned to elements 1 through n respectively, and let $W = \sum_{i=1}^{n} w_i$.

Lemma 36. [65] *The cost of splaying item i to the root is $O(\log(W/w_i))$.*

This result is based on the following potential function (called a *centroid* potential function by Cole [24, 25]). Let s_i be the sum of the weights of the elements that are descendents of i in the search tree, including i itself. Let $r_i = \log s_i$. Then $\Phi = \sum_{i=1}^{n} r_i$. The amortized cost of a zig-zig or zig-zag operation is $3(r_z - r_x)$ while the cost of a zig operation is $3r_y$ (with reference to Fig. 2).

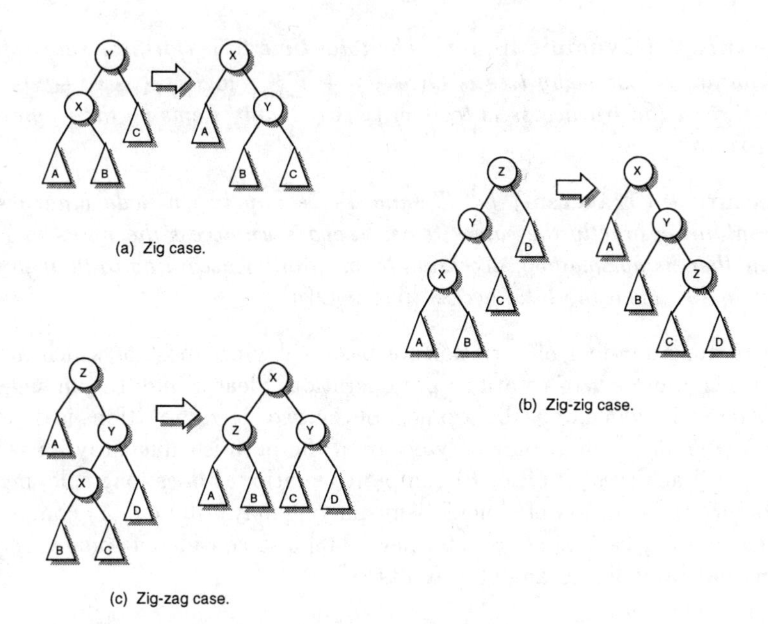

Fig. 2. Three cases of splaying. Case (a) applies only when y is the root. Symmetric cases are omitted.

Theorem 37. [65] *The splay tree algorithm is $O(\log n)$-competitive.*

This follows from Lemma 36 with $w_i = 1$ for all i, in which case the amortized cost of an access is $O(\log n)$. Therefore splay trees are as good in an asymptotic sense as any balanced search tree. They also have other nice properties. Operations such as delete, meld, and split can be implemented with a splay operation plus a small number of each pointer changes at the root. Splay trees also adapt well to unknown distributions, as the following theorem shows.

Theorem 38. [65] *The splay tree is $O(1)$-static-competitive.*

This theorem follows by letting $w_i = f_\sigma(i)$, the frequency with which i is accessed in σ, and comparing with the information theoretic lower bound given at the beginning of this section.

Sleator and Tarjan also made several conjectures about the competitiveness of splay trees. The most general is the "dynamic optimality" conjecture.

Conjecture 2 (Dynamic optimality) *The splay tree is $O(1)$-competitive.*

Sleator and Tarjan made two other conjectures, both of which are true if the dynamic optimality conjecture is true. (The proofs of these implications are non-trivial, but have not been published. They have been reported by Sleator and Tarjan [65], Cole *et al.* [26] and Cole [24, 25].)

Conjecture 3 (Dynamic finger) *The total time to perform m successful accesses on an n-node splay tree is $O(m + n + \sum_{j=1}^{m-1} \log(|i_{j+1} - i_j| + 1)$, where for $1 \leq i \leq m$ the jth access is to item i_j (we denote items by their symmetric order position).*

Conjecture 4 (Traversal) *Let T_1 and T_2 be any two n-node binary search trees containing exactly the same items. Suppose we access the items in T_1 one after another using splaying, accessing them in order according to their preorder number in T_2. Then the total access time is $O(n)$.*

There are a number of variations of basic splaying, most of which attempt to reduce the number of rotations per operation. Sleator and Tarjan suggested *semisplaying*, in which only the topmost of the two zig-zig rotations is done, and *long splaying*, in which a splay only occurs if the path is sufficiently long. Semi-splaying still achieves an $O(\log n)$ competitive ratio, as does long splaying with an appropriate definition of "long." Semisplaying may still be $O(1)$-competitive, but long splaying cannot be. Klostermeyer [43] also considered some variants of splaying but provides no analytic results.

3.6 Progress on splay tree conjectures

In this section we describe subsequent progress in resolving the original splay tree conjectures, and several related conjectures that have since appeared in the literature.

Tarjan [71] studied the performance of splay trees on two restricted classes of inputs. The first class consists of sequential access sequences, $\sigma = 1, 2, \ldots, n$. The dynamic optimality conjecture, if true, implies that the time for a splay tree to perform a sequential access sequence must be $O(n)$, since the optimal time for such a sequence is at most $2n$.

Theorem 39. [71] *Given an arbitrary n-node splay tree, the total time to splay once at each of the nodes, in symmetric order, is $O(n)$.*

Tarjan called this the *scanning theorem*. The proof of the theorem is based on an inductive argument about properties of the tree produced by successive accesses. Subsequently Sundar [67] gave a simplified proof based on a potential function argument.

In [71] Tarjan also studied request sequences consisting of double-ended queue operations: PUSH, POP, INJECT, EJECT. Regarding such sequences he made the following conjecture.

Conjecture 5 (Deque) *Consider the representation of a deque by a binary tree in which the ith node of the binary tree in symmetric order corresponds to the ith element of the deque. The splay algorithm is used to perform deque operations on the binary tree as follows:* POP *splays at the smallest node of the tree and removes it from the tree;* PUSH *makes the inserted item the new root, with null left child and the old root as right child;* EJECT *and* INJECT *are symmetric. The* *cost of performing any sequence of m deque operations on an arbitrary n-node binary tree using splaying is $O(m+n)$.*

Tarjan observed that the dynamic optimality conjecture, if true, implies the deque conjecture. He proved that the deque conjecture is true when the request sequence does not contain any EJECT operations. That is, new elements can be inserted at both ends of the queue, but only removed from one end. Such a deque is called *output-restricted.*

Theorem 40. [71] *Consider a sequence of m* PUSH, POP, *and* INJECT *operations performed as described in the deque conjecture on an arbitrary initial tree T_0 containing n nodes. The total time required is $O(m+n)$.*

The proof uses an inductive argument.

Lucas [49] showed the following with respect to Tarjan's deque conjecture.

Theorem 41. [49] *The total cost of a series of ejects and pops is $O(n\alpha(n,n))$ if the initial tree is a simple path of n nodes from minimum node to maximum node.*[2]

Sundar [67, 68] came within a factor of $\alpha(n)$ of proving the deque conjecture. He began by considering various classes of restructurings of paths by rotations. A right 2-turn on a binary tree is a sequence of two right rotations performed on the tree in which the bottom node of the first rotation is identical to the top node of the second rotation. A 2-turn is equivalent to a zig-zig step in the splay algorithm. (The number of single right rotations can be $\Omega(n^2)$. See, for example, the remark above following Thm. 31.) As reported in [71], Sleator conjectured that the total number of right 2-turns in any sequence of right 2-turns and right rotations performed on an arbitrary n-node binary tree is $O(n)$. Sundar observed that this conjecture, if true, would imply that the deque conjecture was true. Unfortunately, Sundar disproved the turn conjecture, showing examples in which $\Omega(n \log n)$ right 2-turns occur.[3]

Sundar then considered the following generalizations of 2-turns.

1. *Right twists.* For $k > 1$, a right k-twist arbitrarily selects k different edges from a left subpath of the binary tree and rotates the edges one after another in top-to-bottom order. From an arbitrary initial tree, $O(n^{1+1/k})$ right twists can occur and $\Omega(n^{1+1/k}) - O(n)$ are possible.
2. *Right turns:* For any $k > 1$ a right k-turn is a right k-twist that converts a left subpath of k edges in the binary tree into a right subpath by rotating the edges of the subpath in top-to-bottom order. $O(n\alpha(k/2,n))$ right twists can occur if $k \neq 3$ and $O(n \log\log n)$ can occur if $k=3$. On the other hand, there are trees in which $\Omega(n\alpha(k/2,n)) - O(n)$ k-twists are possible if $k \neq 3$ and $\Omega(n \log\log n)$ are possible when $k=3$.

---

[2] $\alpha(i,j)$ is the functional inverse of the Ackermann function, and is a very slowly growing function. See, for example, [69] for more details about the inverse Ackermann function.

[3] Sundar reports that S.R. Kosaraju independently disproved the turn conjecture, with a different technique.

3. *Right cascade:* For $k > 1$, a right k-cascade is a right k-twist that rotates every other edge lying on a left subpath of $2k - 1$ edges in the binary tree. The same bounds hold for right k-cascades as for right turns.

(Symmetric definitions and results hold for left twists, turns, and cascades.)

Using these results, Sundar proved the following theorem.

Theorem 42. [67, 68] *The cost of performing an intermixed sequence of m deque operations on an arbitrary n-node binary tree using splaying is $O((m + n)\alpha(m + n, m + n))$.*

Sundar added one new conjecture to the splay tree literature. A *right ascent* of a node x is a maximal series of consecutive right rotations of the edge connecting a node and its parent.

Conjecture 6 (Turn-ascent) *The maximum number of right 2-turns in any intermixed series of right 2-turns and r right ascents performed on an n-node binary search tree is $O(n + r)$.*

Sundar observes that if this conjecture is true, it implies the truth of the deque conjecture.

The greatest success in resolving the various splay conjectures is due to Cole [24, 25], who was able to prove the truth of the dynamic finger conjecture. His paper is quite complex, and builds both on the work of Sundar and of Cole *et al.* [26] on splay sorting.

Theorem 43. Dynamic finger theorem. [65] *The total time to perform m successful accesses on an n-node splay tree is $O(m+n+\sum_{j=1}^{m-1} \log(|i_{j+1}-i_j|+1))$, where for $1 \leq i \leq m$ the jth access is to item i_j (we denote items by their symmetric order position).*

The proof of this theorem is very intricate, and we will not attempt to summarize it here.

In recent work, Chaudhuri and Hoft [20] prove if the nodes of an arbitrary n-node binary search tree T are splayed in the preorder sequence of T then the total time spent is $O(n)$. This is a special case of the traversal conjecture. Cohen and Fredman [23] give some further evidence in favor of the truth of the splay tree conjecture. They analyze several classes of request sequences generated from a random distribution, and show the splay tree algorithm is $O(1)$-competitive on these sequences.

3.7 Remarks

While exciting progress has been made in resolving special cases of the dynamic optimality conjecture for splay trees, it is unclear how this work will impact the full conjecture. In competitive analysis one usually compares the performance of an on-line algorithm to the performance of an (unknown) optimal off-line algorithm by means of some form of potential function. None of the results on the splay tree conjectures use such a potential function. Rather than comparing the splay tree algorithms to an optimal off-line algorithm, the proofs directly analyze properties of the splay tree on the special classes of requests. Finding some potential function that compares on-line to off-line algorithms is perhaps the greatest open problem in the analysis of the competitive binary search trees.

Splay trees have been generalized to multiway and k-ary search trees by Martel [54] and Sherk [63]. Some empirical results on self-adjusting trees and splay trees in particular have appeared. Moffat *et al.* [59] give evidence that sorting using splay trees is quite efficient. On the other hand, Bell and Gupta [11] give evidence that on random data that is not particularly skewed, self-adjusting trees are generally slower than standard balanced binary trees. There still remains a great deal of work to be done on empirical evaluation of self-adjusting trees on data drawn from typical real-life applications.

4 Data compression: An application of self-organizing data structures

Linear lists and splay trees, as presented in Section 3.5, can be used to build locally adaptive data compression schemes. In the following we present both theoretical and experimental results.

4.1 Compression based on linear lists

The use of linear lists in data compression recently became of considerable importance. In [18], Burrows and Wheeler developed a data compression scheme using unsorted lists that achieves a better compression than Ziv-Lempel based algorithms. Before describing their algorithm, we first present a data compression scheme given by Bentley *et al.* [16] that is very simple and easy to implement.

In data compression we are given a string S that shall be *compressed*, i.e., that shall be represented using fewer bits. The string S consists of *symbols*, where each symbol is an element of the alphabet $\Sigma = \{x_1, \ldots, x_n\}$. The idea of data compression schemes using linear lists it to convert the string S of symbols into a string I of integers. An *encoder* maintains a linear list of symbols contained in Σ and reads the symbols in the string S. Whenever the symbol x_i has to be compressed, the encoder looks up the current position of x_i in the linear list, outputs this position and updates the list using a list update rule. If symbols to be compressed are moved closer to the front of the list, then frequently occurring symbols can be encoded with small integers.

A *decoder* that receives I and has to recover the original string S also maintains a linear list of symbols. For each integer j it reads from I, it looks up the symbol that is currently stored at position j. Then the decoder updates the list using the same list update rule as the encoder. Clearly, when the string I is actually stored or transmitted, each integer in the string should be coded again using a variable length prefix code.

In order to analyze the above data compression scheme one has to specify how an integer j in I shall be encoded. Elias [30] presented several coding schemes that encode an integer j with essentially $\log j$ bits. The simplest version of his schemes encodes j with $1 + 2\lfloor \log j \rfloor$ bits. The code for j consists of a prefix of $\lfloor \log j \rfloor$ 0's followed by the binary representation of j, which requires $1 + \lfloor \log j \rfloor$ bits. A second encoding scheme is obtained if the prefix of $\lfloor \log j \rfloor$ 0's followed by the first 1 in the binary representation of j is coded again using this simple scheme. Thus, the second code uses $1 + \lfloor \log j \rfloor + 2\lfloor \log(1 + \log j) \rfloor$ bits to encode j.

Bentley *et al.* [16] analyzed the above data compression algorithm if encoder and decoder use Move-To-Front as list update rule. They assume that an integer j is encoded with $f(j) = 1 + \lfloor \log j \rfloor + 2\lfloor \log(1 + \log j) \rfloor$ bits. For a string S, let $A_{MTF}(S)$ denote the average number of bits needed by the compression algorithm to encode one symbol in S. Let m denote the length of S and let m_i, $1 \leq i \leq n$, denote the number of occurrences of the symbol x_i in S.

Theorem 44. *For any input sequence S,*

$$A_{MTF}(S) \leq 1 + H(S) + 2\log(1 + H(S)),$$

where $H(S) = \sum_{i=1}^{n} \frac{m_i}{m} \log(\frac{m}{m_i})$.

The expression $H(S) = \sum_{i=1}^{n} \frac{m_i}{m} \log(\frac{m}{m_i})$ is the "empirical entropy" of S. The empirical entropy is interesting because it corresponds to the average number of bits per symbol used by the optimal static Huffman encoding for a sequence. Thus, Theorem 44 implies that Move-To-Front based encoding is almost as good as static Huffman encoding.

Proof of Theorem 44. We assume without loss of generality that the encoder starts with an empty linear list and inserts new symbols as they occur in the string S. Let $f(j) = 1 + \lfloor \log j \rfloor + 2\lfloor \log(1 + \log j) \rfloor$. Consider a fixed symbol x_i, $1 \leq i \leq n$, and let $q_1, q_2, \ldots, q_{m_i}$ be the positions at which the symbol x_i occurs in the string S. The first occurrence of x_i in S can the encoded with $f(q_1)$ bits and the k-th occurrence of x_i can be encoded with $f(q_k - q_{k-1})$ bits. The m_i occurrences of x_i can be encoded with a total of

$$f(q_1) + \sum_{k=1}^{m_i} f(q_k - q_{k-1})$$

bits. Note that f is a concave function. We now apply Jensen's inequality, which states that for any concave function f and any set $\{w_1, \ldots, w_n\}$ of positive reals whose sum is 1, $\sum_{i=1}^{n} w_i f(y_i) \leq f(\sum_{i=1}^{n} w_i y_i)$ [34]. Thus, the m_i occurrences of x_i can be encoded with at most

$$m_i f(\frac{1}{m_i}(q_1 + \sum_{k=2}^{m_i}(q_k - q_{k-1}))) = m_i f(\frac{q_{m_i}}{m_i}) \leq m_i f(\frac{m}{m_i})$$

bits. Summing the last expression for all symbols x_i and dividing by m, we obtain

$$A_{MTF}(S) = \sum_{i=1}^{n} \frac{m_i}{m} f(\frac{m}{m_i}).$$

The definition of f gives

$$\begin{aligned} A_{MTF}(S) &\leq \sum_{i=1}^{n} \frac{m_i}{m} + \sum_{i=1}^{n} \frac{m_i}{m} \log(\frac{m}{m_i}) + \sum_{i=1}^{n} \frac{m_i}{m} 2\log(1 + \log(\frac{m}{m_i})) \\ &\leq \sum_{i=1}^{n} \frac{m_i}{m} + \sum_{i=1}^{n} \frac{m_i}{m} \log(\frac{m}{m_i}) + 2\log(\sum_{i=1}^{n} \frac{m_i}{m} + \sum_{i=1}^{n} \frac{m_i}{m} \log(\frac{m}{m_i})) \\ &= 1 + H(S) + 2\log(1 + H(S)). \end{aligned}$$

The second inequality follows again from Jensen's inequality. □

Bentley *et al.* [16] also considered strings that are generated by probability distributions, i.e., by discrete memoryless sources $\mathbf{p} = (p_1, \ldots, p_n)$. The p_i's are positive probabilities that sum to 1. In a string S generated by $\mathbf{p} = (p_1, \ldots, p_n)$, each symbol is equal to x_i with probability p_i, $1 \leq i \leq n$. Let $B_{MTF}(\mathbf{p})$ denote the expected number of bits needed by Move-To-Front to encode one symbol in a string generated by $\mathbf{p} = (p_1 \ldots, p_n)$.

Theorem 45. *For any* $\mathbf{p} = (p_1, \ldots, p_n)$,

$$B_{MTF}(\mathbf{p}) \leq 1 + H(\mathbf{p}) + 2\log(1 + H(\mathbf{p})),$$

where $H(\mathbf{p}) = \sum_{i=1}^{n} p_i \log(1/p_i)$ *is the entropy of the source.*

Shannon's source coding theorem (see e.g. Gallager [31]) implies that the number $B_{MTF}(\mathbf{p})$ of bits needed by Move-To-Front encoding is optimal, up to a constant factor.

Albers and Mitzenmacher [5] analyzed the data compression algorithm if encoder and decoder use Timestamp(0) as list update algorithm. They showed that a statement analogous to Theorem 44 holds. More precisely, for any string S, let $A_{MTF}(S)$ denote the average number of bits needed by Timestamp(0) to encode one symbol in S. Then, $A_{TS}(S) \leq 1 + H(S) + 2\log(1 + H(S))$, where $H(S)$ is the empirical entropy of S. For strings generated by discrete memoryless sources, Timestamp(0) achieves a better compression than Move-To-Front.

Theorem 46. *For any* $\mathbf{p} = (p_1, p_2, \ldots, p_n)$,

$$B_{TS}(\mathbf{p}) \leq 1 + \overline{H}(\mathbf{p}) + 2\log(1 + \overline{H}(\mathbf{p})),$$

where $\overline{H}(\mathbf{p}) = \sum_{i=1}^{n} p_i \log(1/p_i) + \log(1 - \sum_{i \leq j} p_i p_j (p_i - p_j)^2/(p_i + p_j)^2)$.

Note that $0 \leq \sum_{i \leq j} p_i p_j (p_i - p_j)^2/(p_i + p_j)^2 < 1$.

The above data compression algorithm, based on Move-To-Front or Timestamp(0), was analyzed experimentally [5, 16]. In general, the algorithm can be implemented in two ways. In a *byte-level* scheme, each ASCII character in the input string is regarded as a symbol that is encoded individually. In contrast, in a *word-level* scheme each word, i.e. each longest sequence of alphanumeric and nonalphanumeric characters, represents a symbol. Albers and Mitzenmacher [5] compared Move-To-Front and Timestamp(0) based encoding on the Calgary Compression Corpus [77], which consists of files commonly used to evaluate data compression algorithms. In the byte-level implementations, Timestamp(0) achieves a better compression than Move-To-Front. The improvement is typically 6–8%. However, the byte-level schemes perform far worse than standard UNIX utilities such as pack or compress. In the word-level implementations, the compression achieved by Move-To-Front and Timestamp(0) is comparable to that of the UNIX utilities. However, in this situation, the improvement achieved by Timestamp(0) over Move-To-Front is only about 1%.

Bentley *et al.* [16] implemented a word-level scheme based on Move-To-Front that uses a linear list of limited size. Whenever the encoder reads a word from the input string that is not contained in the list, the word is written in non-coded form onto the output string. The word is inserted as new item at the front of the list and, if the current list length exceeds the allowed length, the last item of the list is deleted. Such a list acts like a cache. Bentley *et al.* tested the compression scheme with various list lengths on several text and Pascal files. If the list may contain up to 256 items, the compression achieved is comparable to that of word-based Huffman encoding and sometimes better.

Grinberg *et al.* [33] proposed a modification of Move-To-Front encoding, which they call *Move-To-Front encoding with secondary lists.* They implemented this new compression scheme but their simulations do not show an explicit comparison between Move-To-Front and Move-To-Front with secondary lists.

As mentioned in the beginning of this section, Burrows and Wheeler [18] developed a very effective data compression algorithm using self-organizing lists that achieves a better compression than Ziv-Lempel based schemes. The algorithm by Burrows and Wheeler first applies a reversible transformation to the string S. The purpose of this transformation is to group together instances of a symbol x_i occurring in S. The resulting string S' is then encoded using the Move-To-Front algorithm.

More precisely, the transformed string S' is computed as follows. Let m be the length of S. The algorithm first computes the m rotations (cyclic shifts) of S and sorts them lexicographically. Then it extracts the last character of these rotations. The k-th symbol of S' is the last symbol of the k-th sorted rotation. The algorithm also computes the index J of the original string S in the sorted list of rotations. Burrows and Wheeler gave an efficient algorithm to compute the original string S given only S' and J.

In the sorting step, rotations that start with the same symbol are grouped together. Note that in each rotation, the initial symbol is adjacent to the final symbol in the original string S. If in the string S, a symbol x_i is very often followed by x_j, then the occurrences of x_j are grouped together in S'. For this reason, S' generally has a very high locality of reference and can be encoded very effectively with Move-To-Front. The paper by Burrows and Wheeler gives a very detailed description of the algorithm and reports of experimental results. On the Calgary Compression Corpus, the algorithm outperforms the UNIX utilities compress and gzip and the improvement is 13% and 6%, respectively.

4.2 Compression based on splay trees

Splay trees have proven useful in the construction of dynamic Huffman codes, arithmetic codes and alphabetic codes [33, 39]. Furthermore they can be used as auxiliary data structure to speed up Ziv-Lempel based compression schemes [12].

Jones [39] studied dynamic Huffman codes based on splay trees. A Huffman code implicitly maintains a code tree. Associated with each leaf in the tree is a symbol of the given alphabet $\Sigma = \{x_1, \ldots, x_n\}$. The code for symbol x_i can be read by following the path from the root of the tree to the leaf containing x_i. Each left branch on the path corresponds to a 0, and each right branch corresponds to a 1. A dynamic Huffman code is obtained by splaying the code tree at certain nodes each time symbol x_i had to be encoded. Note that a Huffman code stores the information at the leaves of the tree, with the internal nodes being empty. Therefore, we may not execute regular splaying in which an accessed leaf would become an internal node, i.e. the root, of the tree. Jones presented a variant of splaying in which the set of leaves remains the same during the operation. He evaluated the algorithm experimentally and showed that the code achieves a very good compression on image data. On text and object files, the codes were not as good, in particular they performed worse than a dynamic Huffman code developed by Vitter [75].

Grinberg *et al.* [33] studied alphabetic codes based on splay trees. Consider an alphabet $\Sigma = \{x_1, \ldots, x_n\}$ in which there is an alphabetic order among the symbols $x_1, \ldots, x_n$. In an alphabetic code, the code words for the symbols have to preserve this alphabetic order. As before, a code tree is maintained. In the algorithm proposed by Grinberg *et al.*, whenever a symbol x_i had to be coded, the code tree is splayed at the parent of the leaf holding x_i.

Grinberg *et al.* analyzed the compression achieved by this scheme. Let S be an arbitrary string of length m and let m_i be the number of occurrences of symbol x_i in S. Furthermore, let $m_{min} = \min\{m_i | 1 \leq i \leq n\}$. We denote by $A_{SP}(S)$ the average number of bits to encode one symbol in S.

Theorem 47. *For any input sequence* S,

$$A_{SP}(S) \leq 2 + 3H(S) + \frac{n}{m}\log(\frac{m}{m_{min}}),$$

where $H(S) = \sum_{i=1}^{n} \frac{m_i}{m}\log(\frac{m}{m_i})$.

Grinberg *et al.* also investigated alphabetic codes based on semisplaying, see Section 3.5. Let $A_{SSP}(S)$ denote the average number of bits needed by semisplaying to encode one symbol in S.

Theorem 48. *For any input sequence S,*

$$A_{SSP}(S) \leq 2 + 2H(S) + \frac{n}{m}\log(\frac{m}{m_{min}}),$$

where $H(S) = \sum_{i=1}^{n} \frac{m_i}{m} \log(\frac{m}{m_i})$.

Thus semisplaying achieves a slightly better performance than splaying.

References

1. N. Abramson. *Information Theory and Coding.* McGraw-Hill, New York, 1983.
2. G.M. Adel'son-Vel'skii and E.M. Landis. An algorithm for the organization of information. *Soviet Math. Dokl.*, 3:1259–1262, 1962.
3. S. Albers. Unpublished result.
4. S. Albers. Improved randomized on-line algorithms for the list update problem. In *Proc. of the 6th Annual ACM-SIAM Symposium on Discrete Algorithms*, pages 412–419, 1995.
5. S. Albers and M. Mitzenmacher. Average case analyses of list update algorithms, with applications to data compression. In *Proc. of the 23rd International Colloquium on Automata, Languages and Programming, Springer Lecture Notes in Computer Science, Volume 1099*, pages 514–525, 1996.
6. S. Albers, B. von Stengel, and R. Werchner. A combined BIT and TIMESTAMP algorithm for the list update problem. *Information Processing Letters*, 56:135–139, 1995.
7. B. Allen and I. Munro. Self-organizing binary search trees. *Journal of the ACM*, 25(4):526–535, October 1978.
8. E.J. Anderson, P. Nash, and R.R. Weber. A counterexample to a conjecture on optimal list ordering. *Journal on Applied Probability*, 19:730–732, 1982.
9. C.R. Aragon and R.G. Seidel. Randomized search trees. In *Proc. 30th Symp. on Foundations of Computer Science*, pages 540–545, 1989.
10. R. Bachrach and R. El-Yaniv. Online list accessing algorithms and their applications: Recent empirical evidence. In *Proc. of the 8th Annual ACM-SIAM Symposium on Discrete Algorithms*, pages 53–62, 1997.
11. J. Bell and G. Gupta. Evaluation of self-adjusting binary search tree techniques. *Software—Practice & Experience*, 23(4):369–382, April 1993.
12. T. Bell and D. Kulp. Longest-match string searching for Ziv-Lempel compression. *Software- Practice and Experience*, 23(7):757–771, July 1993.
13. S. Ben-David, A. Borodin, R.M. Karp, G. Tardos, and A. Wigderson. On the power of randomization in on-line algorithms. *Algorithmica*, 11:2–14, 1994.
14. J.L. Bentley, K.L. Clarkson, and D.B. Levine. Fast linear expected-time algorithms for computing maxima and convex hulls. In *Proc. of the 1st Annual ACM-SIAM Symposium on Discrete Algorithms*, pages 179–187, 1990.
15. J.L. Bentley and C.C. McGeoch. Amortized analyses of self-organizing sequential search heuristics. *Communication of the ACM*, 28:404–411, 1985.
16. J.L. Bentley, D.S. Sleator, R.E. Tarjan, and V.K. Wei. A locally adaptive data compression scheme. *Communication of the ACM*, 29:320–330, 1986.
17. J.R. Bitner. Heuristics that dynamically organize data structures. *SIAM Journal on Computing*, 8:82–110, 1979.

18. M. Burrows and D.J. Wheeler. A block-sorting lossless data compression algorithm. Technical Report 124, DEC SRC, 1994.
19. P.J. Burville and J.F.C. Kingman. On a model for storage and search. *Journal on Applied Probability*, 10:697–701, 1973.
20. R. Chaudhuri and H. Hoft. Splaying a search tree in preorder takes linear time. *SIGACT News*, 24(2):88–93, Spring 1993.
21. R.P. Cheetham, B.J. Oommen, and D.T.H. Ng. Adaptive structuring of binary search trees using conditional rotations. *IEEE Transactions on Knowledge & Data Engineering*, 5(4):695–704, 1993.
22. F.R.K. Chung, D.J. Hajela, and P.D. Seymour. Self-organizing sequential search and hilbert's inequality. In *Proc. 17th Annual Symposium on the Theory of Computing*, pages 217–223, 1985.
23. D. Cohen and M.L. Fredman. Weighted binary trees for concurrent searching. *Journal of Algorithms*, 20(1):87–112, January 1996.
24. R. Cole. On the dynamic finger conjecture for splay trees. Part 2: Finger searching. Technical Report 472, Courant Institute, NYU, 1989.
25. R. Cole. On the dynamic finger conjecture for splay trees. In *Proc. Symp. on Theory of Computing (STOC)*, pages 8–17, 1990.
26. R. Cole, B. Mishra, J. Schmidt, and A. Siegel. On the dynamic finger conjecture for splay trees. Part 1: Splay sorting $\log n$-block sequences. Technical Report 471, Courant Institute, NYU, 1989.
27. T. Cormen, C. Leiserson, and R. Rivest. *Introduction to Algorithms*. McGraw-Hill, New York, NY, 1990.
28. C.A. Crane. Linear lists and priority queues as balanced binary trees. Technical Report STAN-CS-72-259, Dept. of Computer Science, Stanford University, 1972.
29. R. El-Yaniv. There are infinitely many competitive-optimal online list accessing algorithms. Manuscript, May 1996.
30. P. Elias. Universal codeword sets and the representation of the integers. *IEEE Transactions on Information Theory*, 21:194–203, 1975.
31. M.J. Golin. Phd thesis. Technical Report CS-TR-266-90, Department of Computer Science, Princeton University, 1990.
32. G.H. Gonnet, J.I. Munro, and H. Suwanda. Towards self-organizing linear search. In *Proc. 19th Annual IEEE Symposium on Foundations of Computer Science*, pages 169–174, 1979.
33. D. Grinberg, S. Rajagopalan, R. Venkatesan, and V.K. Wei. Splay trees for data compression. In *Proc. of the 6th Annual ACM-SIAM Symposium on Discrete Algorithms*, pages 522–530, 1995.
34. G.H. Hardy, J.E. Littlewood, and G. Polya. *Inequalities*. Cambridge University Press, Cambridge, England, 1967.
35. W.J. Hendricks. An extension of a theorem concerning an intersting Markov chain. *Journal on Applied Probability*, 10:886–890, 1973.
36. J.H. Hester and D.S. Hirschberg. Self-organizing linear search. *ACM Computing Surveys*, 17:295–312, 1985.
37. S. Irani. Two results on the list update problem. *Information Processing Letters*, 38:301–306, 1991.
38. S. Irani. Corrected version of the SPLIT algorithm. Manscript, January 1996.
39. D.W. Jones. Application of splay trees to data compression. *Communications of the ACM*, 31(8):996–1007, August 1988.
40. G. Schay Jr. and F.W. Dauer. A probabilistic model of a self-organizing file system. *SIAM Journal on Applied Mathematics*, 15:874–888, 1967.

41. Y.C. Kan and S.M. Ross. Optimal list orders under partial memory constraints. *Journal on Applied Probability*, 17:1004–1015, 1980.
42. R. Karp and P. Raghavan. From a personal communication cited in [61].
43. W.F. Klostermeyer. Optimizing searching with self-adjusting trees. *Journal of Information & Optimization Sciences*, 13(1):85–95, January 1992.
44. D.E. Knuth. Optimum binary search trees. *Acta Informatica*, pages 14–25, 1971.
45. D.E. Knuth. *The Art of Computer Programming, Sorting and Searching*, volume 3. Addison-Wesley, Reading, MA, 1973.
46. K. Kulik II and D. Wood. A note on some tree similarity measures. *Information Processing Letters*, 15:39–42, 1982.
47. K. Lam, M.K. Sui, and C.T. Yu. A generalized counter scheme. *Theoretical Computer Science*, 16:271–278, 1981.
48. J.M. Lucas. The rotation graph of binary trees is Hamiltonian. *Journal of Algorithms*, 8(4):503–535, December 1987.
49. J.M. Lucas. Arbitrary splitting in splay trees. Technical Report DCS-TR-234, Rutgers University, 1988.
50. J.M. Lucas. Canonical forms for competitive binary search tree algorithms. Technical Report DCS-TR-250, Rutgers University, 1988.
51. F. Luccio and L. Pagli. On the upper bound on the rotation distance of binary trees. *Information Processing Letters*, 31(2):57–60, April 1989.
52. E. Makinen. On the rotation distance of binary trees. *Information Processing Letters*, 26(5):271–272, January 1988.
53. M.S. Manasse, L.A. McGeoch, and D.D. Sleator. Competitive algorithms for online problems. In *Proc. 20th Annual ACM Symposium on Theory of Computing*, pages 322–33, 1988.
54. C. Martel. Self-adjusting multi-way search trees. *Information Processing Letters*, 38(3):135–141, May 1991.
55. J. McCabe. On serial files with relocatable records. *Operations Research*, 12:609–618, 1965.
56. K. Mehlhorn. Nearly optimal binary search trees. *Acta Informatica*, 5:287–295, 1975.
57. K. Mehlhorn. Dynamic binary search. *SIAM Journal on Computing*, 8(2):175–198, 1979.
58. K. Mehlhorn. *Data Structures and Algorithms*. Springer-Verlag, New York, 1984. (3 volumes).
59. G. Port and A. Moffat. A fast algorithm for melding splay trees. In *Proceedings Workshop on Algorithms and Data Structures (WADS '89)*, pages 450–459, Berlin, West Germany, 1989. Springer-Verlag.
60. N. Reingold and J. Westbrook. Optimum off-line algorithms for the list update problem. Technical Report YALEU/DCS/TR-805, Yale University, 1990.
61. N. Reingold, J. Westbrook, and D.D. Sleator. Randomized competitive algorithms for the list update problem. *Algorithmica*, 11:15–32, 1994.
62. R. Rivest. On self-organizing sequential search heuristics. *Communication of the ACM*, 19:63–67, 1976.
63. M. Sherk. Self-adjusting k-ary search trees. *Journal of Algorithms*, 19(1):25–44, July 1995.
64. D.D. Sleator and R.E. Tarjan. Amortized efficiency of list update and paging rules. *Communication of the ACM*, 28:202–208, 1985.
65. D.D. Sleator and R.E. Tarjan. Self-adjusting binary search trees. *Journal of the ACM*, 32:652–686, 1985.

66. D.D. Sleator, R.E. Tarjan, and W.P. Thurston. Rotation distance, triangulations, and hyperbolic geometry. In *Proc. 18th Symp. on Theory of Computing (STOC)*, pages 122–135, 1986.
67. R. Sundar. Twists, turns, cascades, deque conjecture, and scanning theorem. In *Proc. 30th Symp. on Foundations of Computer Science (FOCS)*, pages 555–559, 1989.
68. R. Sundar. Twists, turns, cascades, deque conjecture, and scanning theorem. Technical Report 427, Courant Institute, New York University, January 1989.
69. R.E. Tarjan. *Data Structures and Network Algorithms.* Society for Industrial and Applied Mathematics, Philadelphia, PA., 1983.
70. R.E. Tarjan. Amortized computational complexity. *SIAM Journal on Algebraic and Discrete Methods*, 6:306–318, 1985.
71. R.E. Tarjan. Sequential access in splay trees takes linear time. *Combinatorica*, 5(4):367–378, 1985.
72. B. Teia. A lower bound for randomized list update algorithms. *Information Processing Letters*, 47:5–9, 1993.
73. A. Tenenbaum. Simulations of dynamic sequential search algorithms. *Communication of the ACM*, 21:790–79, 1978.
74. A.M. Tenenbaum and R.M. Nemes. Two spectra of self-organizing sequential search. *SIAM Journal on Computing*, 11:557–566, 1982.
75. J.S. Vitter. Two papers on dynamic Huffman codes. Technical Report CS-85-13, Brown University Computer Science, Providence. R.I., Revised December 1986.
76. R. Wilber. Lower bounds for accessing binary search trees with rotations. *SIAM Journal on Computing*, 18(1):56–67, February 1989.
77. I.H. Witten and T. Bell. The Calgary/Canterbury text compression corpus. Anonymous ftp from ftp.cpsc.ucalgary.ca /pub/text.compression/corpus/ text.compression.corpus.tar.Z.

3

Competitive Analysis of Paging

SANDY IRANI

1 Introduction

This chapter surveys the competitive analysis of paging. We present proofs showing tight bounds for the competitive ratio achievable by any deterministic or randomized on-line algorithm. We then go on to discuss variations and refinements of the competitive ratio and the insights they give into the paging problem. Finally, we discuss variations of on-line paging to which competitive analysis has been applied.

The paging problem has inspired several decades of theoretical and applied research and has now become a classical problem in computer science. This is due to the fact that managing a two level store of memory has long been, and continues to be, a fundamentally important problem in computing systems. The paging problem has also been one of the cornerstones in the development of the area of on-line algorithms. Starting with the seminal work of Sleator and Tarjan which initiated the recent interest in the competitive analysis of on-line algorithms, the paging problem has motivated the development of many important innovations in this area.

1.1 Definitions

Consider a two-level store consisting of a fast memory (the *cache*) that can hold k pages, and a slow memory that can store n pages. The n pages in slow memory represent virtual memory pages. A *paging algorithm* is presented with a sequence of requests to virtual memory pages. If the page requested is in fast memory (a *hit*), no cost is incurred; but if not (a *fault*), the algorithm must bring it into the fast memory at unit cost. The algorithm must decide which of the k pages currently in fast memory to evict in order to make room for the newly requested page.

If we knew the future, the decision would be clear. It has long been known that the optimal off-line algorithm, called MIN, is the algorithm which always evicts the page whose next request is furthest in the future [4]. Unfortunately, in practice paging decisions are made without knowledge of future requests. Typically, a paging algorithm must be *on-line*, meaning that it must make the decision of which page to evict without knowing which pages will be requested in the future.

How do we evaluate the performance of such an on-line paging algorithm? A traditional worst-case analysis of paging is completely uninformative, since any paging algorithm can be made to fault on every single request by an adversary which always requests the most recently discarded page. Hence, from the worst-case point of view, all on-line paging algorithms are equivalent. However, in practice some algorithms perform much better than others and a theoretical analysis should reflect this difference.

Alternatively, one could employ average-case analysis. The problem here is that one must postulate a statistical model for the input. It is difficult to come up with a fixed probability distribution that captures realistic instances since the patterns of access tend to change dynamically with time and across applications. Nonetheless, several of the early analyses of paging algorithms were done assuming such fixed probability distributions.

Motivated by these observations, Sleator and Tarjan proposed the idea of *competitive analysis.* [1] In competitive analysis, the performance of the on-line algorithm is compared to the performance of the optimal off-line algorithm. Let $\mathtt{cost}_A(\sigma)$ be the cost incurred by an on-line algorithm A on the input sequence σ. In the case of paging, σ is a sequence of page requests, and $\mathtt{cost}_{k,A}(\sigma)$ is the number of page faults incurred by algorithm A on the sequence σ when the fast memory can hold k virtual memory pages. Let OPT be the optimal off-line algorithm, and let $\mathtt{cost}_{k,OPT}(\sigma)$ be the cost incurred by the optimal off-line algorithm on input σ when the fast memory can hold k virtual memory pages.

We say that the on-line algorithm A is c-competitive if there exists a constant b such that on every request sequence σ,

$$\mathtt{cost}_{k,A}(\sigma) \leq c \cdot \mathtt{cost}_{k,OPT}(\sigma) + b.$$

In some sense $\mathtt{cost}_{k,OPT}(\sigma)$ measures the inherent difficulty of σ, and we only ask an on-line algorithm to perform well relative to the difficulty of the input. The *competitive ratio* of the algorithm A, denoted $c_{k,A}$ is the infimum over c such that A is c-competitive. An algorithm is said to be *strongly competitive* if it achieves the best possible competitive ratio for a problem. The competitive ratio has become the standard measuring stick for on-line algorithms in recent years and has been used by the vast majority of recent work in on-line algorithms.

[1] Competitive analysis was implicit in early work on bin-packing in the sixties.

Notice that we are not placing any computational restrictions on the algorithm – we are simply measuring, from an information theoretic point of view, what kind of solution quality can be obtained given the fact that the decisions have to be made with partial information.

2 Deterministic algorithms

Sleator and Tarjan give tight bounds on the best competitive ratio which can be achieved by any deterministic on-line paging algorithm [27]. They show that two commonly used paging algorithms achieve a competitive ratio of k. These algorithms are First-In-First-Out (FIFO) which on a fault evicts the page that was placed in the fast memory least recently and Least-Recently-Used (LRU) which on a fault evicts the page that was used least recently. They then show a lower bound of k for the competitive ratio achievable by any on-line algorithm, thus establishing the optimality of LRU and FIFO by the competitive measure.

Below is the proof for the lower bound. Lower bounds for the competitive ratio are often proven using an adversary-style argument, as is common in a more traditional worst-case algorithmic analysis. The idea is that the algorithm plays against an adversary who concocts the worst-case scenario for the algorithm. In competitive analysis, however, the adversary has two tasks. First, it must devise a costly input sequence for the algorithm. Then it must service that sequence, showing an upper bound on the optimal cost for that sequence. The argument below is an example of this type of lower bound argument.

Theorem 1. *[27] If A is any deterministic on-line paging algorithm, then $c_{k,A} \geq k$.*

Proof. We assume that A and OPT both start with the same set of pages in the fast memory. The adversary restricts its request sequence to a set of $k+1$ pages: the k pages initially residing in fast memory and one other page. The adversary always requests the page that is outside of A's fast memory. This process can be continued for an arbitrary number of requests, resulting in an arbitrarily long sequence σ on which A faults on every request.

We must now show that $\mathtt{cost}_{k,OPT}(\sigma) \leq \lceil |\sigma|/k \rceil$. At each fault, the adversary adopts the following strategy: evict the page whose first request occurs farthest in the future. Suppose a page x is evicted by OPT. The next fault occurs the next time x is requested. The adversary is guaranteed that all the other pages in the adversary's fast memory will be requested before x is requested again. There will be at least $k-1$ pages requested between any two faults, so the adversary faults at most on every k^{th} request. □

Sleator and Tarjan actually proved a generalization of this result by allowing the on-line and off-line algorithms to have different memory capacities. Let k_{on} be the number of pages which can fit into the on-line algorithm's fast memory. Let k_{opt} be the number of pages which can fit into the optimal algorithm's fast memory. They show tight bounds of $k_{on}/(k_{on} - k_{opt} + 1)$ for the competitive ratio of the optimal on-line algorithm as long as $k_{on} \geq k_{opt}$. If $k_{on} < k_{opt}$, the competitive ratio is not bounded.

Now we turn to proving upper bounds for deterministic on-line algorithms. Instead of showing the upper bound for LRU directly, we show an upper bound for a general class of algorithms called *marking* algorithms. This class was formally defined by Karlin, Manasse, Sleator, and Rudolph [19] and includes LRU.

A marking algorithm proceeds in phases. At the beginning of a phase all the nodes are unmarked. Whenever a page is requested, it is marked. On a fault, the marking algorithm evicts an unmarked page (chosen by a rule specified by the algorithm), and brings in the requested page. A phase ends just before the first fault after every page in the fast memory is marked (equivalently, a phase ends just before the request to the $k+1^{st}$ distinct page requested in the phase). At this point all the nodes become unmarked and a new phase begins. Marking algorithms and the notion of phases are key concepts which continually reappear in the study of the competitive analysis of paging.

Note that the phases are completely determined by the sequence and not by the choice of which unmarked page the algorithm evicts. The intuition behind the marking algorithms is that an adversary can force any deterministic on-line paging algorithm to fault on every request. Given this fact, the algorithm can only hope to pick pages to evict so that if the adversary always picks a page outside the algorithm's fast memory, his cost will also increase. This idea is made more explicit in the proof below.

Theorem 2. *[19] Any marking algorithm is k-competitive.*

Proof. The proof is based on the simple observation that the optimal algorithm must incur at least one fault in a phase. To see this, divide the sequence in segments that start on the second request of a phase and end with the first request of the next phase. The claim is that any algorithm must fault at least once in a segment. At the beginning of a segment, the algorithm has the most recent request in its fast memory (this is just the first request of the phase). If it does not fault during the remainder of the phase, then it must have all k pages requested during the phase residing its fast memory. The first request of the next phase (the last request of the segment) is, by definition, to a page not in this set. Thus, the optimal algorithm must fault.

Meanwhile, the marking algorithm will incur only k faults in a phase. This follows from the fact that there are exactly k distinct pages requested in one phase. Furthermore, once a page has been requested, it becomes marked and will not be evicted for the remainder of the phase. Thus, a marking algorithm will never fault more than once on a given page during a phase.

□

Another way to view this principle is that, in some sense, a marking algorithm uses the recent past to predict the future. All of the marked pages have been requested more recently than any of the unmarked pages. On the assumption that pages that have been recently requested will be more likely to be requested again (i.e. the sequence exhibits locality of reference), a marking algorithm will not evict any marked page. Thus, an on-line algorithm can take advantage of any local repetitions in the input sequence. An algorithm which can do this will be favored by a competitive analysis for the following reason. The off-line algorithm will have a lower cost on sequences the exhibit locality of reference and will tend to have a higher cost on sequences in which many different pages are requested in turn. Since the on-line algorithm is evaluated in comparison to the off-line algorithm, it must fare well on those sequences for which the off-line algorithm has a low cost. These are exactly the sequences in which exhibit locality of reference. *Least-Recently-Used* (which is also a marking algorithm) uses the recent history of requests to predict the future even more explicitly.

3 Randomized algorithms

Some care must be taken in discussing competitive analysis of randomized on-line algorithms since there is no unique notion of an adversary. The issue which distinguishes different adversaries is the extent to which they know the outcomes of random choices made by the algorithm and how they themselves service the sequence they generate. A discussion of the different types of adversaries is beyond the scope of this article, so we will suffice it to say that our discussion will always assume an *oblivious* adversary.[2] Such an adversary must choose the entire request sequence σ, without any knowledge either of the outcome of the coin tosses or of the specific actions taken as a result of the coin tosses. However, the oblivious adversary does know the algorithm itself including the probability distribution of actions taken for a given input. Formally, we say that a randomized on-line algorithm A is c-competitive if there exists a constant b such that on every request sequence σ,

$$E[\mathtt{cost}_{k,A}(\sigma)] \leq c \cdot \mathtt{cost}_{k,OPT}(\sigma) + b.$$

(See [5] for a more detailed discussion of the different adversaries.)

4 Upper bounds

4.1 Memoryless randomized paging

We begin our discussion of randomized paging algorithms with the simplest randomized algorithm which we call RANDOM: on a fault, evict a random page from the cache. This is a memoryless algorithm: the only information used in making replacement decisions is which pages are currently in the cache.

It turns out that this particular use of randomization does not yield a lower competitive ratio than that achieved by the deterministic algorithms we have

[2] The lower bound of k for the competitive ratio of any deterministic paging algorithm shown in Section 2 can easily be extended to apply to randomized algorithms against an adaptive on-line adversary. Thus, no advantage can be gained in using randomization against an adaptive adversary.

discussed, even against an oblivious adversary: this algorithm is k-competitive against any oblivious adversary. The lower bound is a special case of the following theorem proved by Raghavan and Snir.

Theorem 3. *[25] Any memoryless paging algorithm A has competitive ratio at least k against an oblivious adversary.*

4.2 Memory versus randomness

Raghavan and Snir also make the interesting observation that RANDOM uses $\lceil \log k \rceil$ random bits at every fault while FIFO is a deterministic algorithm which can be implemented using $\lceil \log k \rceil$ bits of memory. Note that $\lceil \log k \rceil$ bits are always necessary at every fault in order to specify which page to evict. To implement FIFO using $\lceil \log k \rceil$ bits of memory, a counter is kept which always has a value in the range $\{0, 1, \ldots, k-1\}$. The pages are stored as a circular array in memory and the counter points to the next page to be evicted. When the next fault occurs, the requested page replaces the page in the location specified by the counter. Then the counter is incremented mod k.

This suggests a whole family of algorithms which are hybrids of RANDOM and FIFO. The algorithm A^i uses i bits of memory and $j = \lceil \log k \rceil - i$ random bits on each fault. Let $I = 2^i$ and $J = k/I$. The pages in memory are organized in a $I \times J$ matrix. The i bits specify a row of the matrix. On a fault, A^i evicts a page chosen randomly from the row pointed to by the i-bit counter. The counter is then incremented mod i. Raghavan and Snir prove the following theorem, demonstrating the trade-off between random bits and memory:

Theorem 4. *[25] The competitive ratio of A^i is k.*

4.3 The randomized marking algorithm

Although no memoryless randomized algorithm has competitiveness below k against oblivious adversaries, there do exist non-memoryless randomized algorithms which beat the deterministic bound. Fiat, Karp, Luby, McGeogh, Sleator and Young [14] were able to show that a randomized marking algorithm has competitive ratio $2H_k$ against an oblivious adversary, where H_k is the kth harmonic number. The algorithm, called RMA (for Randomized Marking Algorithm), is the marking algorithm which on a fault, evicts an unmarked page chosen uniformly at random from the set of unmarked pages.

Theorem 5. *[14] The randomized marking algorithm RMA has competitive ratio $2H_k$ against any oblivious adversary, where H_k is the kth harmonic number.*

Proof. Assume that OPT and RMA start with the same cache contents. As before we divide the sequence σ into phases. The ith phase ends immediately before the $k+1^{st}$ distinct page is requested in the phase. We will analyze the cost of both OPT and RMA phase by phase. Note that once a page is marked, it is not evicted from the cache for the remainder of the phase. Therefore, if we denote the set of pages requested in phase i by P_i, then at the end of a phase i, the contents of RMA's cache is exactly P_i. Furthermore, RMA will not fault twice on the same page within a phase. Thus, we need only account for faults incurred on the first request to any given page in a phase.

Let m_i be the number of *new* requests in phase i, (i.e. the number of pages requested in phase i that were not requested in phase $i-1$). A page requested in the first phase is also new if it is not one of the pages initially in the cache. Since any new page is not in RMA's cache at the beginning of a phase, RMA must fault once on every new page requested. Now we must analyze the expected number of faults on requests to *old* pages (i.e., pages which are not new). What is the probability that RMA faults on the jth old page requested? Let's suppose that just before the jth old page is requested, there have been ℓ new pages requested so far in the phase. It is easy to show by induction that at this time there are exactly ℓ pages in P_{i-1} that are not in the cache. Furthermore, these are distributed uniformly at random among the $k-(j+\ell-1)$ unmarked pages in P_{i-1}. Since the adversary has fixed the request in advance, the probability that it is not in the cache is exactly $\ell/k-(j+\ell-1)$. Since ℓ is always at most m_i, the probability of a fault on the request to the jth old page is at most

$$\frac{m_i}{k-(j+m_i-1)}.$$

Therefore, the expected cost of the marking algorithm in the ith phase is

$$\begin{aligned} E[\mathtt{cost}_{RMA_i}(\sigma)] &\le m_i + \sum_{1\le j\le k-m_i} \frac{m_i}{k-j-m_i+1} \\ &\le m_i H_k. \end{aligned}$$

Summing up over all phases, we get that the expected cost for the algorithm over the entire sequence is

$$E[\mathtt{cost}_{RMA}(\sigma)] \le H_k \sum_i m_i.$$

Now we must prove a lower bound for the optimal cost. We claim that

$$\mathtt{cost}_{OPT}(\sigma) \ge \sum_i \frac{m_i}{2}.$$

Consider the $(i-1)^{st}$ and i^{th} phases. The number of distinct pages requested in both phases is $k+m_i$. Since OPT has only k pages in the cache at the beginning of the $(i-1)^{st}$ phase, it must incur at least m_i faults during the two phases. Applying this argument to every pair of adjacent phases, we have that $\mathtt{cost}_{OPT}(\sigma) \ge \sum_i m_{2i}$ and $\mathtt{cost}_{OPT}(\sigma) \ge \sum_i m_{2i+1}$. Therefore OPT has cost at least the average of these, i.e. $\mathtt{cost}_{OPT}(\sigma) \ge \sum_i m_i/2$. Thus, $E[\mathtt{cost}_{RMA}(\sigma)] \le 2H_k\mathtt{cost}_{OPT}(\sigma)$. □

A much more complicated algorithm was shown to be H_k competitive by McGeoch and Sleator[23]. Recently, Achlioptas, Chrobak and Noga have shown a simpler algorithm which is also H_k-competitive [1]. Their algorithm uses $O(k^2 \log k)$ bits of memory and $O(k^2)$ time per request. The algorithm is based on a very elegant characterization of the *work function* for the paging problem which was developed by Koutsoupias and Papadimitriou in [21]. Achlioptas, Chrobak and Noga also prove that the competitive ratio of RMA is exactly $2H_k - 1$.

4.4 Lower bounds

A common technique for proving lower bounds on the competitive ratio of a randomized on-line algorithm against oblivious adversaries is to examine the performance of deterministic algorithms on inputs from a given probability distribution $\mathcal{P}$ over request sequences σ.

Definition 6. Let $\mathcal{P}$ be a probability distribution on request sequences σ. An algorithm A is *c-competitive against* $\mathcal{P}$ if there exists a constant, b, such that

$$E_{\mathcal{P}}(\mathtt{cost}_A(\sigma)) \leq c \cdot E_{\mathcal{P}}(\mathtt{cost}_{OPT}(\sigma)) + b$$

Definition 7. Let $c_A^{\mathcal{P}}$ denote the infimum over all c such that A is c-competitive against $\mathcal{P}$, and let c_R denote the competitive ratio of a randomized algorithm R against an oblivious adversary.

The basic theorem that allows us to prove lower bounds for randomized on-line algorithms is the following.

Theorem 8.

$$\inf_R c_R = \sup_{\mathcal{P}} \inf_A c_A^{\mathcal{P}}.$$

In other words, the competitive ratio of the best randomized algorithm is equal to the competitive ratio of the best deterministic algorithm, A, on inputs generated from the "worst" probability distribution. The proof of this theorem follows from the minimax theorem of game theory and was first observed by Yao [29] in the context of ordinary complexity theory, and by Borodin, Linial and Saks [7] in the context of on-line algorithms.

We now illustrate the use of Theorem 8 for proving lower bounds on c_R by showing a lower bound for randomized paging algorithms against an oblivious adversary. The bound comes close to matching the upper bound shown in Section 4.3. The following theorem is due to Fiat, Karp, Luby, McGeoch, Sleator and Young [14].

Theorem 9. *Let R be any randomized paging algorithm. If the number of pages is greater than or equal to $k+1$, where k is the size of the cache, the competitive ratio of R against any oblivious adversary is greater than or equal to H_k.*

Proof. We will find our lower bound on c_R by exhibiting a probability distribution $\mathcal{P}$ for which $c_A^{\mathcal{P}} \geq H_k$ for all deterministic algorithms A. Let S be a set of $k+1$ pages which include the k pages initially in the cache. Take $\mathcal{P}$ to be the uniform distribution on the $k+1$ pages in S. That is, a sequence σ of m requests is generated by independently selecting each request at random from S. Clearly, the expected performance of any deterministic algorithm on inputs generated from this distribution is

$$E_{\mathcal{P}}[\texttt{cost}_A(\sigma)] = \frac{m}{k+1} \tag{1}$$

for $|\sigma| = m$. To see this, note that the probability that a given requested page is not in the cache is $\frac{1}{k+1}$. We now need an upper bound on the expected performance of OPT.

Once again, we divide the sequence of page requests into non-overlapping phases such that each phase contains maximal runs of requests to at most k distinct pages. As we have seen, if there are r distinct phases then the optimal algorithm can service the sequence with at most $r+1$ faults. At the beginning of a phase, OPT replaces the one page currently in the cache that will not be requested in the phase. Therefore, if $N(m)$ is the random variable which is the number of phases in a sequence σ of length m (generated from $\mathcal{P}$), then the expected off-line cost satisfies

$$E_{\mathcal{P}}(\texttt{cost}_{OPT}) \leq E(N(m) + 1). \tag{2}$$

Since the durations of successive phases, are independent, identically distributed random variables, we have by the elementary renewal theorem [3] that

$$\lim_{m\to\infty} \frac{m}{E(N(m))} = E(X_i),$$

where $E(X_i)$ is the expected length of the ith phase.

The expected length of a phase, $E(X_i)$ is easily seen to be $(k+1)H_k$ (this is a so-called "coupon collectors problem".)

Therefore, we have that

$$\begin{aligned}
\lim_{m\to\infty} \frac{E_{\mathcal{P}}(\texttt{cost}_A)}{E_{\mathcal{P}}(\texttt{cost}_{OPT})} &\geq \frac{m/k+1}{m/E(X_i)} \\
&\geq \frac{E(X_i)}{k+1} \\
&= \frac{(k+1)\cdot H_k}{k+1} \\
&= H_k.
\end{aligned}$$

Applying Theorem 8 yields a lower bound of H_k on c_R. □

[3] For a detailed treatment of renewal theory, see for example Feller's book [12] or almost any book on probability theory and stochastic processes.

5 Variations on competitive analysis

It would appear at this point that the case has been closed on the competitive analysis of paging since tight bounds for both deterministic and randomized algorithms have been determined. Interestingly, however, these results have been instrumental in illustrating some of the deficiencies with competitive analysis, thus opening up new lines of research in addressing these shortcomings. In fact, a survey of the competitive analysis of paging is in some ways a survey of the refinements of competitive analysis.

Some of the reservations which have been raised about the competitive analysis of paging are its inability to discern between LRU and FIFO (algorithms whose performances differ markedly in practice), and the fact that the theoretical competitiveness of LRU is much larger than what is observed in practice. Also unsettling is fact that in the standard competitive model, a fixed amount of lookahead does not give an on-line algorithm any advantage. If an algorithm is allowed to see l requests into the future, the adversary can choose to repeat each request l times (called the "l-stuttered version" in [21]), thus taking away any advantage in having the lookahead. Meanwhile, the cost for the optimal algorithm does not change. It seems intuitive that in practice, an on-line algorithm would benefit from some limited lookahead, but this is not reflected in the model.

Another feature missing from the model is that it is based on the assumption that the on-line algorithm has absolutely no knowledge whatsoever about the upcoming sequence. However, this may be an unnecessarily pessimistic assumption. It is possible that by preprocessing the program generating the request sequence or by observing past behavior of the program, one can glean some useful information about future requests. One would like to use knowledge of a program's access pattern to improve paging performance.

Most of the refinements of competitive analysis which we discuss attempt to address some or all of these issues.

5.1 Access graphs

The reason for the practical success of LRU has long been known: most programs exhibit *locality of reference*. This means that if a page is referenced, it is more likely to be referenced in the near future (temporal locality) and pages near it in memory are more likely to be referenced in the near future (spatial locality). Indeed, a two-level store is only useful if request sequences are *not* arbitrary.

Motivated by these observations, Borodin, Irani, Raghavan and Schieber [6] introduce a graph theoretic model which restricts page requests so that they conform to a notion of locality of reference. An *access graph* $G = (V, E)$ for a program is a graph that has a vertex for each page that the program can reference. Locality of reference is imposed by the edge relation – the pages that can be referenced after a page p are the neighbors of p in G or p itself. Thus, a request sequence σ must be a walk on G. The specific walk that is generated is determined only at execution time since it depends on the input given to the program. The definition of competitiveness remains the same as before, except for this restriction on the request sequences. Let $c_{A,k}(G)$ denote the competitiveness of an on-line algorithm A on access graph G with k pages of fast memory. We denote by $c_k(G)$ the infimum of $c_{A,k}(G)$ over on-line algorithms A with k pages of fast memory. Thus $c_k(G)$ is the best that any on-line algorithm can do on the access graph G.

An access graph may be either directed or undirected. An undirected access graph might be a suitable model when the page reference patterns are governed by the data structures used by the program. For example, if a program performs operations on a tree data structure, and the mapping of the tree nodes to pages of virtual memory represents a contraction of a tree, then the appropriate access graph might be a tree. For a program doing picture processing or matrix computations, the access graph is likely to resemble a mesh. Alternatively, if we were to completely ignore data, and focus only on the flow of control inherent in the structure of the program, a directed access graph might be a more suitable model.

Typical questions of interest in this model are: (1) How do LRU and FIFO compare on different access graphs? (2) Given the access graph model of a program, what is the performance of LRU on that program (i.e., what is $c_{LRU,k}(G)$)? (3) Given an access graph G, what is $c_k(G)$? can a good paging algorithm be tailor-made given an access graph G? (4) Is there a "universal" algorithm whose competitiveness is close to $c_k(G)$ on every access graph? (5) What is the power of randomization in the access graph model?

Borodin *et al.* give an in-depth study of LRU. The main results determine $c_{LRU,k}(G)$ for every G and k to within a factor of two (plus additive constant). The technique uses combinatorial properties of small subgraphs of G involving the number of articulation nodes in each subgraph. Another useful way of viewing their tight bounds for LRU is as a characterization of "bad" access graphs for LRU. A refinement of the analysis shows that LRU is optimal among on-line algorithms for the important special case when G is a tree. In a recent paper by Chrobak and Noga [11] it was proven that the competitive ratio of LRU is at most that of FIFO on every access graph. The result was proven by obtaining a tighter characterization of LRU's performance.

Borodin *et al.* devise an extremely simple and natural universal algorithm (FAR) and prove the following theorem

Theorem 10. *[6] For any undirected G and k, $c_{FAR,k}(G) \leq 2 + 4c_k(G)\lceil \log 2k \rceil$.*

This was later improved by Irani, Karlin and Phillips who show

Theorem 11. *[18] For any undirected G and k, $c_{FAR,k}(G) = O(c_k(G))$.*

FAR is a marking algorithm which evicts the unmarked page whose distance in the access graph to to a marked page is maximum. The intuition behind FAR is as follows: it is known that the optimal (off-line) paging algorithm on any sequence is to evict the page which corresponds to the node whose next request occurs furthest in the future. FAR attempts to approximate this behavior by vacating a node that is far from the set of marked pages, and thus likely to be requested far in the future.

The class of *structured program graphs* defined in [6] are directed access graphs that represent a subset of the stream of instruction references made by a structured program. Borodin *et al.* analyze a simple generalization of FAR, called 2FAR, that is optimal for structured program graphs in which all strongly connected components have at most $k+1$ nodes [6]. Irani *et al.* introduce a variant of 2FAR, called EVEN, and prove the following theorem:

Theorem 12. *[18] The algorithm EVEN is strongly competitive on the class of structured program graphs. In other words, for any structured program graph G,* $c_{k,EVEN}(G) = O(c_k(G))$.

In a third paper on the access graph model, Fiat and Karlin show a simple randomized universal algorithm with competitive ratio $O(c^R(G))$ for every undirected access graph G. $c^R(G)$ is the best possible competitive ratio achievable by any randomized algorithm on the access graph G.

Fiat and Karlin also consider the multiple pointer case motivated by the observation that in practice there may be multiple flows of control through the virtual memory pages. The different paths through an access graph may represent a multiprogramming environment or operations on multiple data structures within a single application program. If there are m pointers in the access graph, then the next request is to one of the current locations of the pointers or one of their neighbors in the access graph. Denote by $c_m(G)$ the best competitive ratio of any on-line paging algorithm operating on input sequences generated by m pointers walking on access graph G. Fiat and Karlin show an algorithm whose competitive ratio for any m on every access graph G is $O(c_m(G))$. For the $m=1$ case, their proof reduces to a simpler proof of the fact that the competitive ratio of FAR is $O(c(G))$.

Fiat and Rosen in [16] consider several variations of FAR and evaluate then empirically in relation to FAR and LRU. Their algorithms are truly on-line in that they build the access graph on the fly instead of as a preprocessing step. The edges in the graph have weights which are decreased whenever the edge is traversed. They also add a notion of "forgetfulness" in that the edge weights are periodically increased so that edges which have not been recently traversed in the sequence will have a higher relative weight. This allows for the graph to adjust dynamically to the particular phase of the program execution. Their algorithms empirically outperform LRU on most of their traces which are generated from a variety of applications. They also often outperform the static version of FAR which was originally proposed with the access graph model.

A subsequent paper [15] by Fiat and Mendel gives strongly competitive algorithms in the access graph model without knowing the access graph in advance.

Open Problem 13. One of the major drawbacks of the universal algorithms for access graphs is that they require a considerable amount of time to determine which page to evict on a fault. Is there an efficient way to perform some of this computation in a preprocessing step instead of as the program is executing? One idea is to embed paging instructions in the code of the program telling the operating system to evict a certain page or switch to a particular page replacement policy. Is there an efficient algorithm which can preprocess a program and determine a set of effective instructions to insert into the code of the program?

5.2 Probabilistic analysis

Markov paging. Karlin, Phillips, and Raghavan [20] take the access graph idea one step farther by not only providing the on-line algorithm with the graph but also probabilities along the edges of the graph: they analyze paging algorithms where the request sequence is generated by a markov chain, such that each node in the markov chain is a virtual memory page. Since the distribution is completely known to the on-line algorithm, the goal is simply to minimize the fault rate (expected number of faults per request) instead of minimizing the fault rate in comparison to the off-line algorithm. Although the optimal on-line algorithm (which minimizes the fault rate) is available to the on-line algorithm in information theoretic terms, it is in general, not efficient to compute. Thus, the goal is to approximate this algorithm. Karlin *et al.* first prove that many seemingly reasonable algorithms do not have a fault rate that it is within a constant of the best on-line algorithm. They then show a somewhat more complicated algorithm which does achieve this goal.

General distributions. Lund, Phillips and Reingold consider probabilistic analysis of paging in the context of designing interfaces between IP networks and connection-oriented networks [22]. Although their motivation is different from page replacement policies, the abstract formulation of the problem turns out to be the same. Lund *et al.* consider more general distributions over input sequences than those generated by markov chains. Fortunately, their algorithm requires only limited information about the request sequence. At each point in the request sequence, the algorithm needs to know for every pair of pages in memory (p, q), the probability that the next request to p occurs before the next request to q. We will denote this probability by $w(p, q)$. They prove that for any set of k pages in memory at any point in time, there is a *dominating distribution* over the pages in memory. A dominating distribution has the property that if p is chosen according the distribution, then for each page q in fast memory,

$$E[w(p,q)] \leq \frac{1}{2}.$$

They use this fact to show that if the evicted page is always chosen according to a dominating distribution, then the expected fault rate is at most a factor of 4 times the cost of the optimal on-line algorithm.

Open Problem 14. Lund, Phillips and Reingold note that the dominating distribution over k pages can be computed by solving a linear program over k

variables. Suppose that the distribution over input sequences is generated by a markov chain. Thus, for a given set of k pages, the dominating distribution is static. On every fault, the contents of the fast memory is altered by one page. Is there a way to compute the new dominating distribution from the previous one more efficiently than recomputing the entire linear programming problem?

5.3 Diffuse adversaries

In the same spirit of restricting the input sequence to the on-line algorithm, Koutsoupias and Papadimitriou define *the diffuse adversary model* where the input is generated by a distribution [21]. Although the on-line algorithm does not know the exact distribution, it does know that it is chosen from a class Δ of distributions. The adversary can pick the worst $\mathcal{D} \in \Delta$ for an on-line algorithm A. Once this choice is made, the cost of A on input σ is compared to the optimal off-line cost on σ, where σ is picked according to $\mathcal{D}$. The competitive ratio of A is

$$\max_{\mathcal{D} \in \Delta} \frac{E_{\sigma \in \mathcal{D}}[\mathtt{cost}_A(\sigma)]}{E_{\sigma \in \mathcal{D}}[\mathtt{cost}(\sigma)]}.$$

Naturally, the more restricted Δ is, the more information the on-line algorithm has. Notice that when Δ is the set of all distributions, this is just the usual competitive ratio.

Koutsoupias and Papadimitriou illustrate their new measure by defining the class of distributions Δ_ϵ which is the set of all distributions such that for any possible prefix ρ and any page a, the probability that the next request is a given that the sequence seen so far is ρ, is at most ϵ. The smallest ϵ can be is $1/n$, where n is the total number of pages, in which case the only distribution in the class is the one which generates every sequence uniformly at random. The largest ϵ can be is 1 in which case every distribution is contained in Δ_ϵ. They prove that for any ϵ, LRU is the the algorithm which achieves the optimal competitive ratio against a diffuse adversary which chooses from the class Δ_ϵ. Furthermore, they show that for $k = 2$, the optimal competitive ratio is in between $1 + \sqrt{\epsilon}/2$ and $1 + 2\sqrt{\epsilon}$. For larger k, they obtain a description of (although not a closed form for) the optimal competitive ratio as a function of ϵ.

5.4 Comparative ratio

In another variation of the competitive ratio, called the *comparative ratio*, Koutsoupias and Papadimitriou pit the on-line algorithm against less powerful adversaries than the optimal off-line algorithm [21]. Consider two classes of algorithms $\mathcal{A}$ and $\mathcal{B}$. Typically, $\mathcal{A} \subseteq \mathcal{B}$. $\mathcal{B}$ may have a more powerful information regime or more computational resources. The comparative ratio of $\mathcal{A}$ and $\mathcal{B}$ is

$$c(\mathcal{A}, \mathcal{B}) = \max_{B \in \mathcal{B}} \min_{A \in \mathcal{A}} \max_{\sigma} \frac{\mathtt{cost}_A(\sigma)}{\mathtt{cost}_B(\sigma)}.$$

The comparative ratio is probably best viewed as a two player game. Player $\mathcal{B}$ (typically the adversary) picks an algorithm $B \in \mathcal{B}$. Player $\mathcal{A}$ (typically the on-line algorithm) picks an algorithm $A \in \mathcal{A}$. Then player B picks the input σ so as to maximize the ratio of A's cost on σ to B's cost on σ. Note that if $\mathcal{B}$ is the set of all algorithms (off-line and on-line), and $\mathcal{A}$ is the set of all on-line algorithms, then $c(\mathcal{A}, \mathcal{B})$ is simply the best competitive ratio for the given problem.

One appealing feature of comparative analysis is that it allows one to address the anomaly observed in competitive analysis that a fixed amount of lookahead does not give any advantage. The idea is that if an on-line algorithm competes against a class of algorithms with limited lookahead instead of an omniscient algorithm, it does gain some advantage with additional lookahead. The authors show that $c(\mathcal{L}_0, \mathcal{L}_l) = l + 1$ for paging, where $\mathcal{L}_i$ is the class of all on-line algorithms with lookahead i.

Open Problem 15. It would be nice to be able to say that an on-line algorithm which plays against an adversary with limited lookahead does better with more lookahead (i.e., for $i < j < l$, $c(\mathcal{L}_i, \mathcal{L}_l) > c(\mathcal{L}_j, \mathcal{L}_l)$). What are tight bounds for $c(\mathcal{L}_i, \mathcal{L}_l)$ as a function of i and l?

5.5 Loose competitiveness

Young's definition of *loose competitiveness* [31] is based on the observation that lower bounds for on-line algorithms often use an adversary which carefully tailors the request sequence to the particular "hardware configuration" of the algorithm. In the case of paging, the hardware configuration is simply k, the number of pages that can fit in fast memory. Thus, we are evaluating the algorithm on its most damaging request sequence, designed to take advantage of its specific circumstances. In evaluating the loose competitiveness of an algorithm, we require that the adversary concoct a sequence which is bad for an algorithm under most circumstances (i.e., most values of k). The idea was first introduced in [31], however, the following definition is taken from [32] which has generalized and improved bounds.

Definition 16. Algorithm A is (ϵ, δ)-loosely c-competitive if for any request sequence σ, and any n, at least $(1 - \delta)n$ of the values of $k \in \{1, 2, \ldots, n\}$ satisfy

$$\mathtt{cost}_{A,k}(\sigma) \leq \max\{c \cdot \mathtt{cost}_{OPT}(\sigma), \epsilon|\sigma|\}.$$

There are actually two differences between this definition and traditional competitiveness. The first is that the algorithm only has to be competitive on most values of k. The second is that the algorithm does not have to be competitive if its fault rate is small (at most ϵ times the length of the sequence).

Using loose-competitiveness, the competitive ratio decreases dramatically:

Theorem 17. *[32] Every* $\left(\frac{k_{on}}{k_{on}-k_{off}+1}\right)$*-competitive algorithm is* (ϵ, δ)*-loosely competitive for any* $0 < \epsilon, \delta < 1$ *and*

$$c = e\frac{1}{\delta}\lceil \ln \frac{1}{\epsilon} \rceil,$$

where k_{on} is the number of pages that can fit in the on-line algorithm's fast memory and k_{off} is the number of pages that can fit in the optimal off-line algorithm's fast memory.

Theorem 18. *[32] Every $O\left(\ln \frac{k_{on}}{k_{on}-k_{off}}\right)$-competitive algorithm is (ϵ, δ)-loosely competitive for any $0 < \epsilon, \delta < 1$ and*

$$c = O(1 + \ln \frac{1}{\delta} + \ln \ln \frac{1}{\epsilon}).$$

These theorems were actually proven for a generalization of the paging problem discussed in Section 6.

Open Problem 19. The idea of loose competitiveness has the potential to shed light on many other problems in the area of on-line algorithms. In particular, for many scheduling problems, lower bounds are proven by concocting job arrival sequences which are bad for the particular number of processors (or more generally, hardware configuration) available to the algorithm. Are these lower bounds still attainable under the model of loose competitiveness?

5.6 Measuring total memory access time

Torng [28] adds a new twist to the paging problem by making the argument that the correct cost for paging algorithms should be total memory access time and not simply the number of page faults. This adds an extra parameter to the paging problem which is the additional time to access a page in slow memory versus the time to access a page in fast memory. That is, if the access time for fast memory is 1, then the access time for slow memory is $1 + s$. If an algorithm A incurs f faults in a sequence of n requests, then the cost of A on that sequence is

$$\mathtt{cost}_A = (n - f) \cdot 1 + f \cdot (s + 1).$$

Using this cost model, Torng re-examines the competitive ratio of paging algorithms. Thus, the goal still is to bound $\mathtt{cost}_{k,A}(\sigma)/\mathtt{cost}_{k,OPT}(\sigma)$, but $\mathtt{cost}_{k,A}$ and $\mathtt{cost}_{k,OPT}$ are defined to be the total memory access time instead of the number of faults. Note that the previous analysis assumes that $s = \infty$ so that hits do not contribute to memory access time. Using this model, Torng generalizes the previous bounds for the standard algorithms.

Theorem 20. *[28] Any marking algorithm achieves a competitive ratio of*

$$\frac{k(s+1)}{k+s} \approx \min\{k, s+1\}.$$

Theorem 21. *[28] The randomized marking algorithm achieves a competitive ratio of* $\min\{2H_s, 2H_k\}$ *against an oblivious adversary.*

Torng also proves a lower bound showing that Theorem 20 is tight. There is also a very natural notion of locality of reference with this model. Define $L(\sigma, k)$ to be the average phase length in the request sequence σ with k slots of fast memory. Phases here are defined in an identical manner as they were in the definition of marking algorithms. Naturally, the larger $L(\sigma, k)$ is, the more locality of reference exhibited by the sequence. For sequences with large values for $L(\sigma, k)$, some improvement can be gained.

Theorem 22. *[28] If the adversary is restricted to sequences where $L(\sigma, k) > as$, then any marking algorithm achieves a competitive ratio of $(1 + \frac{k-1}{a+1})$.*

Theorem 23. *[28] If the adversary is restricted to sequences where $L(\sigma, k) > kas$, where $a < 1/2$, then the competitive ratio of the Randomized Marking Algorithm is at most $2+2(H_k - H_{2ak}) \approx 2(1+\ln \frac{1}{2a})$. If $a \geq 1/2$, then the competitive ratio is $(1 + \frac{1}{2a})$.*

Another appealing feature of this model is that allowing the on-line algorithm some lookahead does give it some advantage. For example, the l-stuttered version of a sequence gives both the on-line algorithm and off-line algorithm additional hits which contributes an equal additive cost to the on-line and off-line algorithms and serves to reduce the competitive ratio.

Consider the following natural generalization of LRU with lookahead: on a fault, evict the page not in the lookahead buffer which was accessed the least recently. If all pages in the fast memory are in the lookahead buffer, then evict the page whose next reference is farthest in the future. We denote this version of LRU with lookahead l by $LRU(l)$. $LRU(l)$ proves difficult to analyze in Torng's model, so he considers a simple variant called $LRU(l, k)$ which behaves exactly like $LRU(l)$, except that it restricts its lookahead to the requests in the current phase.

Theorem 24. *[28] The competitive ratio of $LRU(l, k)$ is at most*

$$\min\left\{2 + \sqrt{\frac{2ks}{l}}, s+1\right\},$$

when $l < sk$. If $l \geq sk$, the competitive ratio is at most 2.

5.7 Lookahead

Various refinements of the competitive ratio which we have seen so far have resulted in a model in which some limited amount of lookahead does give an on-line algorithm some advantage. The issue of lookahead has also been addressed in a more direct manner by simply changing the definition of lookahead l.

Strong lookahead. Albers defines the notion of *strong lookahead* l which means that the algorithm can see the minimal prefix of the remaining sequence which contains requests to l distinct pages [2]. She also considers the algorithm $LRU(l)$ defined in Section 5.6 and proves the following theorem:

Theorem 25. *[2] The competitive ratio of $LRU(l)$ under the strong lookahead model is exactly $k - l$ when $l \leq k - 2$.*

The following theorem establishes that this upper bound is tight:

Theorem 26. *[2] No deterministic algorithm can obtain a competitive ratio of less than $k - l$ with strong lookahead l when $l \leq k - 2$.*

Albers also examines a variant of her model in which requests arrive in blocks of l distinct requests. The motivation behind this model is that in practice a paging algorithm will be allowed some amount of lookahead because the arrival of requests is bursty. This is more likely to result in batch arrivals than a steady stream in which there is always a backlog of l unserved requests. The algorithm $LRU(l) - blocked$ behaves exactly like $LRU(l)$ except that it limits its lookahead to the requests in the current block. Albers proves the following theorem.

Theorem 27. *[2] LRU(l)-blocked is $(k - l + 1)$-competitive when $l \leq k - 2$.*

Albers also shows that a variant of the Randomized Marking Algorithm is $(2H_{k-l})$-competitive with strong lookahead l. She proves that this bound is tight to within a factor of 2.

Resource bounded lookahead. Young defines the notion of *resource-bounded lookahead* l in which the algorithm can see the maximal prefix of future requests for which the algorithm must incur l faults [30]. Young proves a generalization of the traditional competitive ratio by allowing the on-line algorithm to potentially have more fast memory than the off-line algorithm. Let k_{on} be the number of pages that can fit into the on-line algorithm's fast memory. Let k_{off} be the number of pages that can fit into the off-line algorithm's fast memory. Let $a = k_{on} - k_{off}$.

Theorem 28. *[30] There is a marking-like algorithm which achieves a competitive ratio of at most*

$$\max\left\{2\frac{k_{on} + a + l}{a + l + 1}, 2\right\},$$

with resource-bounded lookahead l.

Interestingly, the above result shows a direct trade-off between extra memory and lookahead. Young shows a lower bound of $\frac{k_{on}+a+l}{a+l+1}$ for the competitive ratio of any on-line deterministic algorithm with resource bounded lookahead. He then shows a variant of Randomized Marking which achieves a competitive ratio of

$2(\ln \frac{k}{l} + 1)$ and gives a lower bound of $\ln \frac{k+l}{l} - \ln \ln \frac{k+l}{l} - \frac{2}{l}$ for any randomized paging algorithm with resource bounded lookahead l.

Breslauer improves Young's upper bound by showing that $LRU(l)$ achieves a competitive ratio of exactly $\frac{k_{on}+a+l}{a+l+1}$ with resource bounded lookahead l [8]. He also defines an alternate definition of lookahead which he calls *natural lookahead.* In this model, the algorithm is allowed to see the maximal prefix of future requests which contain l distinct pages not in the algorithm's fast memory. Breslauer proves that $LRU(l)$ achieves a competitive ratio of $\frac{k_{on}+a+l}{a+l+1}$ and that this is the best competitive ratio that can be achieved under his model of lookahead.

6 Variations on the paging problem

This chapter has mainly addressed algorithms for the basic on-line page replacement problem and different measures which can be used to evaluate them. There have also been many variations on the page replacement problem itself which have been examined using competitive analysis. In this section, we give a very brief survey of some of those problems.

6.1 Weighted caching

In some cases, the cost of bringing a page into memory may vary, even though the number of different pages which the cache can hold remains fixed. That is, each page has a fixed weight which denotes the cost of bringing that page into memory. Since this problem is a generalization of paging, the lower bound of k on the competitive ratio of any deterministic paging algorithm holds for weighted caching. The first matching upper bound was proven by Chrobak, Karloff, Payne and Vishwanathan in [10] who proved that an extension of FIFO called BALANCE is k-competitive. Neal Young, then showed that a whole class of algorithms called Greedy-Dual are $k_{on}/(k_{on} - k_{off} + 1)$-competitive [31], where k_{on} and k_{off} are the sizes of the on-line and off-line caches respectively. This class of algorithms includes BALANCE as well as a generalization of LRU.

6.2 Multi-size pages

The paging problem in which pages have varying sizes has been examined by Irani in [17]. This problem is motivated by cache management for World Wide Web documents where documents held in the cache can vary dramatically in size, depending largely on the type of information they contain (e.g. text, video, audio). Again, since this is a generalization of paging, the lower bound of k for the competitive ratio of any deterministic on-line algorithm still applies. It is relatively straight-forward to prove that the natural extension of LRU is k-competitive in this case. k is the ratio of the size of the cache to the size of the smallest page.

It has also been proven that a version of Neal Young's Greedy Dual algorithm which is a generalization of LRU is k-competitive for the multi-size weighted cache problem where pages have varying sizes as well as a weight associated with each page [9]. Weighted caching is especially pertinent in the multi-size paging problem since the time to cache a web document can vary greatly depending on the location of the document and network traffic. Subsequently, Young proved that the entire range of algorithms covered by Greedy-Dual are k-competitive [32]. In fact, he proved they are $k_{on}/(k_{on} - k_{off} + 1)$-competitive [31], where k_{on} and k_{off} are the sizes of the on-line and off-line caches respectively. The proof is a generalization and simplification of the proof in [31]. Young also has bounds for the loose competitiveness which can be achieved for this problem which improve, generalize and simplify the results from [31]. (See Section 5.5 for a definition of loose competitiveness).

Interestingly, when the pages have varying sizes, the optimal algorithm is not nearly as straight-forward as in the uniform-size case. Irani considers two cost models. In the first, (the FAULT model), the cost to bring a page into the cache is constant, regardless of its size. In the second, (the BIT model), the cost to bring a page into memory is proportional to its size. Off-line algorithms for both cost models are shown which obtain approximation factors of $O(\log k)$. Randomized on-line algorithms for both cost models are shown which are $O(\log^2 k)$-competitive. In addition, if the input sequence is generated by a known distribution, algorithms for both cost models are given whose expected cost is within a factor of $O(\log^2 k)$ of any other on-line algorithm.

6.3 Paging for shared caches

Consider the following problem. The page replacement algorithm is given a sequence of requests which is an interleaving of p different request sequences that are known in advance. The on-line aspect of the problem is that the paging algorithm does not know how the request sequences will be interleaved until it sees each request of the sequence in an on-line manner. The problem models the situation where there are p applications which are simultaneously sharing a cache. Each application uses knowledge about its own request sequence to enable the paging algorithm to make optimal paging decisions about its own request sequence. However, it is unknown how fast each application will progress in its request sequence relative to the other applications. The problem was originally studied by Cao, Felten and Li who showed that a competitive ratio of $2p+2$ can be achieved, where p is the number of processes [24]. Barve, Grove and Vitter use randomization to improve this bound to $2H_{p-1}+2$, where H_p is the pth harmonic number [26]. They also show a lower bound of H_{p-1} for the competitive ratio of any randomized on-line algorithm.

Other models of multi-threaded paging are addressed in Feuerstein [13] and Alborzi *et al.* [3].

References

1. D. Achlioptas, M. Chrobak, and J. Noga. Competitive analysis of randomized paging algorithms. In *Proc. 4th European Symposium on Algorithms*, LNCS, pages 419–430. Springer, 1996.

2. S. Albers. The influence of lookahead in competitive paging algorithms. In *First Annual European Symposium on Algorithms*, pages 1–12, 1993.
3. H. Alborzi, E. Torng, P. Uthaisombut, and S.Wagner. The k-client problem. In *Proc. of the 8th Annual ACM-SIAM Symposium on Discrete Algorithms*, pages 73–82, 1997.
4. L.A. Belady. A study of replacement algorithms for virtual storage computers. *IBM Systems Journal*, 5:78–101, 1966.
5. S. Ben-David, A. Borodin, R. Karp, G. Tardos, and A. Widgerson. On the power of randomization in on-line algorithms. In *Proc. 22nd Symposium on Theory of Algorithms*, pages 379–386, 1990.
6. A. Borodin, S. Irani, P. Raghavan, and B. Schieber. Competitive paging with locality of reference. In *Proc. 23rd ACM Symposium on Theory of Computing*, pages 249–259, 1991.
7. A. Borodin, N. Linial, and M. Saks. An optimal online algorithm for metrical task systems. In *Proc. 19th Annual ACM Symposium on Theory of Computing*, pages 373–382, 1987.
8. D. Breslauer. On competitive on-line paging with lookahead. In *13th Annual Symposium on Theoretical Aspects of Computer Science.*, pages 593–603, 1996.
9. P. Cao and S. Irani. Cost-aware WWW proxy caching algorithms. Technical report, 1343, Dept. of Computer Sciences, University of Wisconsin-Madison, May 1997. A shorter version appears in the 2nd Web Caching Workshop, Boulder, Colorado, June 1997.
10. M. Chrobak, H. Karloff, T. H. Payne, and S. Vishwanathan. New results on server problems. *SIAM Journal on Discrete Mathematics*, 4:172–181, 1991. Also in Proceedings of the 1st Annual ACM-SIAM Symposium on Discrete Algorithms, San Francisco, 1990, pp. 291-300.
11. M. Chrobak and J. Noga. LRU is better than FIFO. In *Proc. of 9th ACM-SIAM Symposium on Discrete Algorithms*, pages 78–81, 1998.
12. W. Feller. *An introduction to probability theory and its applications.* John Wiley and Sons, 1950.
13. E. Feuerstein. *On-line pagine of Structured Data and Multi-threaded Paging.* PhD thesis, University of Rome, 1995.
14. A. Fiat, R. Karp, M. Luby, L. A. McGeoch, D. Sleator, and N.E. Young. Competitive paging algorithms. *Journal of Algorithms*, 12:685–699, 1991.
15. A. Fiat and M. Mendel. Truly online paging with locality of reference. In *Proc. for the 38th Annual Symposium on Foundations of Computer Science*, pages 326–335, 1997.
16. A. Fiat and Z. Rosen. Experimental studies of access graph based heuristics: Beating the LRU standard? In *Proc. of 8th ACM-SIAM Symposium on Discrete Algorithms*, pages 63–72, 1997.
17. S. Irani. Page replacement with multi-size pages and applications to web caching. In *Proc. 29th ACM Symposium on the Theory of Computing*, pages 701–710, 1997.
18. S. Irani, A. Karlin, and S. Phillips. Strongly competitive algorithms for paging with locality of reference. In *3rd Annual ACM-SIAM Symposium on Discrete Algorithms*, pages 228–236, 1992.
19. A. Karlin, M. Manasse, L. Rudolph, and D. Sleator. Competitive snoopy caching. *Algorithmica*, 3(1):79–119, 1988.
20. A. Karlin, S. Phillips, and P. Raghavan. Markov paging. In *Proc. 33rd IEEE Symposium on Foundations of Computer Science*, pages 208–217, 1992.

21. E. Koutsoupias and C. Papadimitriou. Beyond competitive analysis. In *Proc. 25th Symposium on Foundations of Computer Science*, pages 394–400, 1994.
22. C. Lund, S. Phillips, and N. Reingold. Ip over connection-oriented networks and distributional paging. In *35th IEEE Symposium on Foundations of Computer Science*, pages 424–435, 1994.
23. L. McGeoch and D. Sleator. A strongly competitive randomized paging algorithm. *J. Algorithms*, 6:816–825, 1991.
24. K. Li P. Cao, E.W. Felten. Application-controlled file caching policies. In *Proc. for the Summer USENIX Conference*, 1994.
25. P. Raghavan and M. Snir. Memory versus randomization in online algorithms. In *16th International Colloquium on Automata, Languages, and Programming, Lecture Notes in Computer Science vol. 372*, pages 687–703. Springer-Verlag, 1989.
26. J.S. Vitter R.D. Barve, E.F. Grove. Application-controlled paging for a shared cache. In *Proc. for the 36th Annual Symposium on Foundations of Computer Science*, pages 204–213, 1995.
27. D. Sleator and R. E. Tarjan. Amortized efficiency of list update and paging rules. *Communications of the ACM*, 28:202–208, 1985.
28. E. Torng. A unified analysis of paging and caching. In *36th IEEE Symposium on Foundations of Computer Science*, pages 194–203, 1995.
29. A.C. Yao. Probabilistic computations: Towards a unified measure of complexity. In *Proc. 12th ACM Symposium on Theory of Computing*, 1980.
30. N. Young. Dual-guided on-line weighted caching and matching algorithms. Technical Report Tech. Report CS-TR-348-91, Comp. Sci. Dept., Princeton University, 1991.
31. N. Young. The k-server dual and loose competitiveness for paging. In *Proc. of 2nd ACM-SIAM Symposium on Discrete Algorithms*, pages 241–250, 1991.
32. N. Young. Online file caching. In *Proc. of 9th ACM-SIAM Symposium on Discrete Algorithms*, pages 82–86, 1998.

4

Metrical Task Systems, the Server Problem and the Work Function Algorithm

MAREK CHROBAK
LAWRENCE L. LARMORE

1 Introduction

This expository chapter covers on-line algorithms for task systems and the server problem. We concentrate on results related to the *work function algorithm* (WFA), especially its competitive analysis that uses the *pseudocost* approach.

After defining WFA and pseudocost, we show how to estimate the competitiveness of WFA in terms of pseudocost. Using this approach, we present the proof that WFA is $(2n-1)$-competitive for n-state metrical task systems and $(n-1)$-competitive for n-state metrical service systems (also called forcing task systems). We also show examples of metrical service systems in which WFA performs worse than an optimally competitive algorithm.

We then turn our attention to the k-server problem. We present proofs of the following results: that WFA is $(2k-1)$-competitive for k-servers, that it is 2-competitive for two servers, and k-competitive for metric spaces with at most $k+2$ points.

2 Metrical task systems

Bob's ice cream machine can produce two types of ice cream: vanilla (V) or chocolate (C). Only one type can be produced at any given time, and changing the mode from one to another costs \$1. In mode V the machine produces a gallon of vanilla ice cream at a cost of \$1. In mode C, it produces a gallon of chocolate ice cream at a cost of \$2. Whenever Bob receives an order for a gallon of ice cream (v or c), he can either use his machine (switching the mode if necessary) or produce the ice cream without using the machine at a cost of \$2 per gallon for vanilla and \$4 per gallon for chocolate. The table below summarizes the cost function:

mode/order	vanilla	chocolate
V	1	4
C	2	2

When Bob should change the mode of the machine?

Bob's objective is to minimize his cost. If he knew all future orders, he could compute the optimal way of serving the orders using dynamic programming. In reality, however, Bob must make mode changes based only on information about past orders. He cannot always make optimal decisions, for whatever his decision is, there is a sequence of future orders that will make this decision suboptimal.

Since computing the optimal decision sequence is impossible in such an on-line setting, it is important to formulate exactly what Bob's objective is. One objective would be to minimize expected cost. Bob could assume that the orders arrive according to a certain probability distribution. He could estimate this distribution using the frequencies of past orders. For example, each order might be c with probability 0.25 and v with probability 0.75.

What can Bob do, however, if it is not feasible to define a meaningful probability distribution on the orders? For example, the frequencies could be time-dependent. The distributional approach, as described above, is not very useful in such situations.

Another approach, called *competitive analysis*, has attracted a substantial amount of interest. The idea of competitive analysis is to compare the cost of the given on-line algorithm $\mathcal{A}$ to the optimal cost. If $\mathcal{A}$'s cost is always at most c times the optimal cost plus an additive constant independent of the data sequence, then we say that $\mathcal{A}$ is c-competitive. Thus the asymptotic ratio of $\mathcal{A}$'s cost to the optimal cost does not exceed c. The smallest c for which $\mathcal{A}$ is c-competitive is called the *competitive ratio* of $\mathcal{A}$, or simply the *competitiveness* of $\mathcal{A}$.

In this chapter, we use the competitive approach to analyze problems similar to the Ice Cream Problem described above. Borodin, Linial and Saks [4] proposed a model for such on-line problems called *metrical task systems*. In such a system a player faces a sequence of tasks that must be served on-line. Each task can be served from a number of available states, but the cost of service depends on the state. There are also costs associated with state changes, and we assume that these costs constitute a metric on the state set.

Formally, a *metrical task system* (MTS) is a pair $\mathcal{S} = (M, \mathcal{T})$, where M is a finite metric space whose elements are called *states*, and $\mathcal{T}$ is a finite set of functions $\tau : M \to \mathbf{R}^+$ called *tasks*. For notational convenience, if $x, y \in M$, we denote the distance from x to y simply as xy.

Suppose we are given an initial state $x_0 \in M$ and a sequence of tasks $\bar{\tau} = \tau_1 \ldots \tau_n$. Any sequence $\bar{x} = x_0, x_1, \ldots, x_m$, where $x_i \in M$, is called a *schedule*, or a *service schedule* for $x_0, \bar{\tau}$. We define

$$cost(x_0, \bar{\tau}, \bar{x}) = \sum_{i=1}^{m} [x_{i-1}x_i + \tau_i(x_i)],$$

$$opt(x_0, \bar{\tau}) = \min_{\bar{x}} \{cost(x_0, \bar{\tau}, \bar{x})\}.$$

An *on-line strategy* for $\mathcal{S}$ is a function $\mathcal{A} : M \times \mathcal{T}^* \to M$. Given an initial state $x_0 \in M$ and a task sequence $\bar{\tau}$, $\mathcal{A}$ will service the sequence $\bar{\tau}$, ending up in state $\mathcal{A}(x_0, \bar{\tau})$. Let $\bar{\tau} = \tau_1 \ldots \tau_m$, and let $x_i = \mathcal{A}(x_0, \tau_1 \ldots \tau_i)$ for each i. Then the sequence $\bar{x} = x_0, \ldots, x_m$ is called the *schedule produced by* $\mathcal{A}$, and the *cost of* $\mathcal{A}$ *on* $x_0, \bar{\tau}$ is the cost of this schedule: $cost_{\mathcal{A}}(x_0, \bar{\tau}) = cost(x_0, \bar{\tau}, \bar{x})$. Strategy $\mathcal{A}$ is called *c-competitive with initial function* $\alpha : M \to \mathbf{R}$ if for each $x_0 \in M$ and $\bar{\tau} \in \mathcal{T}^*$,

$$cost_{\mathcal{A}}(x_0, \bar{\tau}) \leq c \cdot opt(x_0, \bar{\tau}) + \alpha(x_0).$$

The smallest c for which $\mathcal{A}$ is c-competitive is called the *competitive ratio* of $\mathcal{A}$.

We will be dealing with several types of metrical task systems. An MTS $\mathcal{S} = (M, \mathcal{T})$ is called *unrestricted* if $\mathcal{T}$ is the set of all nonnegative real functions on M. An MTS is called *uniform* if all distances in M are 1.

Another type of task system is where each task can be performed at no cost, but only in certain states. One can think of such tasks as *forcing* tasks, since any algorithm is forced to move to a state belonging to a designated set. Such systems were introduced in [19] and called *forcing task systems*. They were also independently introduced in [9, 10], under the name *metrical service systems*. Formally, a metrical service system (MSS) is a pair $\mathcal{S} = (M, \mathcal{R})$, where M is a finite metric space and $\mathcal{R}$ is a family of non-empty subsets of M called *requests*. If the set $\rho \in \mathcal{R}$ is requested, the algorithm is forced to move the server to some point in ρ to serve the request. An MSS $\mathcal{S}$ is called *unrestricted* if $\mathcal{R}$ consists of all non-empty subsets of M.

Note that a metrical service system $(M, \mathcal{R})$ reduces to the task system $(M, \mathcal{T})$, where $\mathcal{T} = \{\tau_\rho : \rho \in \mathcal{R}\}$ where $\tau_\rho(x) = 2\min_{y \in \rho} xy$, because service of the request τ_ρ from x can be emulated by moving to the nearest point in the set ρ and then back to x.

2.1 Work functions in a task system

Let $\mathcal{S} = (M, \mathcal{T})$ be a metrical task system. Let x_0 be an initial state and $\bar{\tau} = \tau_1 \ldots \tau_m$ a sequence of tasks. Define $\omega : M \to \mathbf{R}$, the *work function* associated with x_0 and $\bar{\tau}$, as follows: for each $x \in M$, $\omega(x)$ is the minimum cost of servicing $\bar{\tau}$, starting from x_0 and ending at x. More formally,

$$\omega(x) = \min_{\bar{x}} \{cost(x_0, \bar{\tau}, \bar{x}) + x_m x\},$$

where the minimum is over all schedules $\bar{x} = x_0, x_1, \ldots, x_m$ for $x_0, \bar{\tau}$. The following observation is immediate from the definition:

Observation 1 *The optimal cost for servicing $\bar{\tau}$ from x_0 is the minimum value of the work function ω associated with x_0 and $\bar{\tau}$,* i.e., $opt(x_0, \bar{\tau}) = \min_{x \in M} \omega(x)$.

As we show later, a work function encompasses all information that an on-line algorithm needs to make decisions at each step. Thus it is not necessary to keep information about the past request sequence. In fact, it is not necessary to think of a work function as being associated with any request sequence at all – instead, one can think of a work function as an abstract function that represents past obligations of an optimal algorithm. Work functions associated with request sequences, in the way described above, will be called *reachable*.

Properties of work functions. Note that any reachable work function satisfies the 1-Lipschitz condition, namely that $\omega(x) - \omega(y) \leq xy$ for all $x, y \in M$. We say that a work function ω is *supported* by a set $K \subseteq M$ if, for all $x \in M$, there exists $y \in K$ such that $\omega(x) = \omega(y) + xy$. We define the *support* of ω to be the smallest set which supports ω. We define a *cone* to be a work function supported by a single point. We denote by χ_x the cone which is zero on x and is supported by $\{x\}$. If x_0 is the initial state, then χ_{x_0} is the initial work function. If $X \subseteq M$, we denote by χ_X the *generalized cone* on X, the work function which is zero on X and is supported by X.

Updating the work function. Let ω be any work function and $\tau \in \mathcal{T}$. We define $\omega \wedge \tau$, also a work function, as follows. For any $x \in M$

$$\omega \wedge \tau(x) = \min_{y \in M} \{\omega(y) + \tau(y) + yx\}.$$

We call "$\wedge$" the *update operator*. We sometimes use the notation ω^τ for $\omega \wedge \tau$. If $\bar{\tau} = \tau_1 \ldots \tau_n$ is a sequence of tasks, we write variously $\omega^{\bar{\tau}}$ or $\omega \wedge \bar{\tau}$ to denote the iterated update $\omega \wedge \tau_1 \wedge \ldots \wedge \tau_m$.

Lemma 1. *If $\omega \wedge \tau(x) = \omega(y) + \tau(y) + yx$ then $\omega \wedge \tau(y) = \omega(y) + \tau(y)$.*

Proof. For all z, $\omega(y) + \tau(y) + yx \leq \omega(z) + \tau(z) + zx$. Thus $\omega(y) + \tau(y) \leq \omega(z) + \tau(z) + zx - yx \leq \omega(z) + \tau(z) + yz$, and the lemma follows. □

Corollary 2. *If x is in the support of $\omega \wedge \tau$ then $\omega \wedge \tau(x) = \omega(x) + \tau(x)$.*

Proof. Suppose $\omega \wedge \tau(x) = \omega(y) + \tau(y) + yx$, where $y \neq x$. By Lemma 1, $\omega \wedge \tau(x) = \omega \wedge \tau(y) + yx$, contradicting the hypothesis that x is in the support of $\omega \wedge \tau$. □

As we mentioned earlier, the work function captures all relevant information about the past. This means, in particular, that no algorithm for optimizing costs can perform better than an algorithm whose only information about the past is the work function. To state this formally, we define a *work-function based strategy* to be a function $\mathcal{A}$, that to a given state x, current work function ω, and new task τ assigns the state $\mathcal{A}(x, \omega, \tau) \in M$ to which $\mathcal{A}$ moves in order to service the task τ. It is sufficient to define $\mathcal{A}(\omega, x, \tau)$ only for reachable work functions ω.

Theorem 3. *If $\mathcal{A}$ is any on-line algorithm for a metrical task system $\mathcal{S} = (M, \mathcal{T})$, and if $\mathcal{A}$ is c-competitive with initial function $\alpha : M \to \mathbf{R}$, then there is a work-function based strategy $\mathcal{B}$ which is also c-competitive with initial function α.*

Proof. For each $x_0 \in M$ and each $\bar{\tau} \in \mathcal{T}^*$, define $L(x_0, \bar{\tau}) = cost_{\mathcal{A}}(x_0, \bar{\tau}) - c \cdot opt(x_0, \bar{\tau})$. Note that $L(x_0, \bar{\tau}) \leq \alpha(x_0)$ by the definition of competitiveness. For all x and ω let $F(x, \omega)$ be the set of all pairs $(x_0, \bar{\tau})$ for which $\mathcal{A}(x_0, \bar{\tau}) = x$ and $\chi_{x_0} \wedge \bar{\tau} = \omega$. If $F(x, \omega) \neq \emptyset$ then we say that the pair (x, ω) is *$\mathcal{A}$-reachable*. For each $\mathcal{A}$-reachable pair (x, ω) define

$$\Phi(x, \omega) = \inf_{(x_0, \bar{\tau}) \in F(x, \omega)} \sup_{\bar{\sigma} \in \mathcal{T}^*} \{L(x_0, \bar{\tau}\bar{\sigma}) - L(x_0, \bar{\tau})\}.$$

Note that $\Phi(x, \omega) \geq 0$ because $\bar{\sigma}$ may be chosen to be the empty sequence, and $\Phi(x_0, \chi_{x_0}) \leq \alpha(x_0)$ because $\bar{\tau}$ may be chosen to be the empty sequence. If (x, ω) is $\mathcal{A}$-reachable and $\mu = \omega \wedge \tau$, then we claim that there is a y for which

$$xy + \tau(y) + \Phi(y, \mu) - \Phi(x, \omega) \leq c \cdot (\min \mu - \min \omega). \tag{1}$$

To prove the claim, for each $\epsilon > 0$ choose $(x_0, \bar{\tau}) \in F(x, \omega)$ for which

$$\begin{aligned} \Phi(x, \omega) + \epsilon &\geq \sup_{\bar{\sigma}} L(x_0, \bar{\tau}\bar{\sigma}) - L(x_0, \bar{\tau}) \\ &\geq \sup_{\bar{\sigma}} L(x_0, \bar{\tau}\tau\bar{\sigma}) - L(x_0, \bar{\tau}). \end{aligned}$$

Let $y_\epsilon = \mathcal{A}(x_0, \bar{\tau}\tau)$. Pick $y \in \bigcap_{\delta > 0} \{y_\epsilon\}_{\epsilon < \delta}$, which is non-empty since M is finite. Since $\Phi(y, \mu) \leq \sup_{\bar{\sigma}} L(x_0, \bar{\tau}\tau\bar{\sigma}) - L(x_0, \bar{\tau}\tau)$, we have

$$\begin{aligned} xy + \tau(y) + \Phi(y, \mu) - \Phi(x, \omega) &\leq xy + \tau(y) - L(x_0, \bar{\tau}\tau) + L(x_0, \bar{\tau}) + \epsilon \\ &= c \cdot (\min \mu - \min \omega) + \epsilon \end{aligned}$$

for arbitrarily small choices of ϵ. The result follows.

For each $\mathcal{A}$-reachable pair (x, ω), we now define $\mathcal{B}(x, \omega, \tau)$ to be any choice of $y \in M$ for which $xy + \tau(y) + \Phi(y, \omega \wedge \tau)$ is minimized. For given x_0 and $\bar{\tau}$, inequality (1) implies, by summation over all requests in $\bar{\tau}$, that

$$cost_{\mathcal{B}}(x_0, \bar{\tau}) \leq c \cdot \min(\chi_{x_0} \wedge \bar{\tau}) + \Phi(x_0, \chi_{x_0}) \leq c \cdot opt(x_0, \bar{\tau}) + \alpha(x_0),$$

completing the proof. □

Intersection property of metrical service systems. Let $\mathcal{S}$ be a metrical service system. We say that an on-line algorithm for $\mathcal{S}$ is *lazy* if it never changes state if the current state is in the request set.

Lemma 4. *Given any on-line algorithm $\mathcal{A}$ for an* MSS $\mathcal{S} = (M, \mathcal{R})$, *there exists a lazy on-line algorithm $\mathcal{B}$ for $\mathcal{S}$ whose cost is never greater than $\mathcal{A}$'s.*

Proof. We define $\mathcal{B}(x_0, \bar{\rho})$ for any request sequence recursively on the length of $\bar{\rho}$.

$$\mathcal{B}(x_0, \epsilon) = x_0$$

$$\mathcal{B}(x_0, \bar{\rho}\rho) = \begin{cases} \mathcal{B}(x_0, \bar{\rho}) & \text{if } \mathcal{B}(x_0, \bar{\rho}) \in \rho \\ \mathcal{A}(x_0, \bar{\rho}\rho) & \text{otherwise} \end{cases}$$

We say that a request $\rho \in \mathcal{R}$ is *futile* if the state of $\mathcal{B}$ is already in ρ. $\mathcal{B}$ works just like $\mathcal{A}$, except that it does not change state if a request is futile. At the next non-futile request, it immediately moves to $\mathcal{A}$'s position. By the triangle inequality, $\mathcal{B}$'s cost for any request sequence is not greater than $\mathcal{A}$'s. □

Lemma 5. *If $\mathcal{S} = (M, \mathcal{R})$, then, without loss of generality, $\mathcal{R}$ is closed under non-empty intersection. More specifically, if $\mathcal{R}'$ is the set of all nonempty intersections of the sets in $\mathcal{R}$, then $\mathcal{S}$ has a c-competitive on-line algorithm if and only if $\mathcal{S}' = (M, \mathcal{R}')$ has a c-competitive on-line algorithm.*

Proof. We prove the following statement, which by induction implies the lemma: Suppose $\rho_1, \rho_2 \in \mathcal{R}$, and $\rho_1 \cap \rho_2 \notin \mathcal{R}$. Let $\mathcal{S}' = (M, \mathcal{R}')$, where $\mathcal{R}' = \mathcal{R} \cup \{\rho_1 \cap \rho_2\}$ Then there is a c-competitive on-line algorithm for $\mathcal{S}$ if and only if there is a c-competitive on-line algorithm for $\mathcal{S}'$.

The ($\Leftarrow$) direction is trivial. We prove ($\Rightarrow$). Suppose that $\mathcal{A}$ is an on-line algorithm for $\mathcal{S}$. By Lemma 4, we may assume that $\mathcal{A}$ is lazy. We construct an on-line algorithm, $\mathcal{A}'$, for $\mathcal{S}'$ whose competitiveness is no larger than that of $\mathcal{A}$. Let $\mathcal{G}$ be the adversary for $\mathcal{S}$ and $\mathcal{G}'$ be the adversary for $\mathcal{S}'$. Our method is to show that a player, which we call White, who plays the role of both $\mathcal{A}'$ and $\mathcal{G}$ against an opponent, which we call Black, who plays the role of both $\mathcal{A}$ and $\mathcal{G}'$ can force c', the competitiveness of $\mathcal{S}'$, to be no larger than c, the competitiveness of $\mathcal{S}$, for that instance.

By Lemma 4, we can assume that $\mathcal{A}$ is lazy. We shall also assume that $\mathcal{G}'$ does not make multiple consecutive identical requests, since $\mathcal{A}'$ can ignore all but the first one, simply remaining in the same state at no cost.

The game consists of a number of *phases*, each consisting of various moves by Black and White. Let ω_t, $cost_t$, x_t denote, respectively, the work function in $\mathcal{S}$, the cost of $\mathcal{A}$ and the state of $\mathcal{A}$ after t phases. Symbols ω'_t, $cost'_t$, x'_t denote the corresponding quantities for $\mathcal{S}'$ and $\mathcal{A}'$.

We claim that (I1) $\omega_t = \omega'_t$, and (I2) $cost'_t + x_t x'_t \leq cost_t$, for all t. The lemma follows immediately from (I1), (I2) and from Observation 1.

Both invariants (I1), (I2) hold trivially at the beginning. We now show that they are preserved by the playing of just one phase. Let D and d be the diameter and smallest non-zero distances of M, respectively, and let $N \geq D/d+1$ be an integer. At phase t, Black, playing the role of $\mathcal{G}'$, makes a request $\rho' \in \mathcal{R}'$. If $\rho' \in \mathcal{R}$, then White, playing the role of $\mathcal{G}$, requests ρ'. Playing the role of $\mathcal{A}$, Black services the request by moving to some state x_t, and playing the role of $\mathcal{A}'$, White chooses $x'_t = x_t$. If, on the other hand, $\rho' = \rho_1 \cap \rho_2$, then White, playing the role of $\mathcal{G}$, makes $2N$ requests, namely $(\rho_1\rho_2)^N$. In response, Black, playing the role of $\mathcal{A}$, must service those requests with a sequence of length $2N$, ending in a state x_t. If $x_t \in \rho' = \rho_1 \cap \rho_2$, then White, playing the role of $\mathcal{A}'$, chooses $x'_t = x_t$. Otherwise, White chooses $x'_t \in \rho_1 \cap \rho_2$ arbitrarily.

We now verify that first invariant is preserved. The trivial case is $\rho' \in \mathcal{A}$. Then $\omega'_t = \omega'_{t-1}\wedge\rho' = \omega_{t-1}\wedge\rho' = \omega_t$. Suppose now that $\rho' = \rho_1 \cap \rho_2$. Since ω_t agrees with ω_{t-1} on ρ', and since ω'_t is the maximum 1-Lipschitz function that agrees with ω'_{t-1} on ρ', we immediately have $\omega_t \leq \omega'_t$. Conversely, suppose that $\omega_t(u) < \omega'_t(u)$ for some $u \in M$. Then $\omega'_t(u) \leq \omega'_{t-1}(v) + vu$ for all $v \in \rho'$, and $\omega_t(u) = \omega_{t-1}(w) + wv_1 + v_1v_2 + \ldots v_{2N-1}v_{2N} + v_{2N}u$ for some $v_1 \ldots v_{2N}$, where $v_i \in \rho_1$ if i is odd, and $v_i \in \rho_2$ if i is even. If any one of the v_i lies in ρ', then by the triangle inequality and the 1-Lipschitz property, $\omega_t(u) \leq \omega'_{t-1}(v_i) + v_iu \leq \omega_t(u)$, contradiction. Otherwise, $v_{i-1}v_i \geq d$ for all $1 < i \leq 2N$, which implies that $\omega'_t(u) \geq \omega_{t-1}(w) + (2N-1)d > \omega_{t-1}(w) + 2D \geq \omega_{t-1}(w) + wv + vu \geq \omega_{t-1}(v) + vu = \omega_t(u)$, which contradicts the hypothesis for this case.

We now prove (I2). If $x_t = x'_t$, then $cost'_t = cost'_{t-1} + x'_{t-1}x'_t \leq cost_{t-1} - x_{t-1}x'_{t-1} + x'_{t-1}x'_t \leq cost_{t-1} + x_{t-1}x'_t = cost_t$. Otherwise, *i.e.*, if $x_t \neq x'_t$, then $\rho' = \rho_1 \cap \rho_2$. Let $v_1, v_2, \ldots v_{2N} = x_t$ be the sequence of states of $\mathcal{A}$ which service $(\rho_1\rho_2)^N$. If any $v_i \in \rho'$, then $v_j = v_i = x_t$ for all $j \geq i$, since $\mathcal{A}$ is lazy. This would imply that $x'_t = x_t$ by the definition of $\mathcal{A}'$, which is a contradiction. Thus $v_i \in \{\rho_1 - \rho_2\}$ if i is odd, and $v_i \in \{\rho_2 - \rho_1\}$ if i is even, which implies that $\mathcal{A}$ pays at least $(2N-1)d \geq 2D$ during the phase.

Since, $\mathcal{G}'$ did not request ρ' at phase $t-1$, we have $x'_{t-1} = x_{t-1}$. Thus, $cost'_t = cost'_{t-1} + x'_{t-1}x'_t \leq cost'_{t-1} + D \leq cost_{t-1} + D \leq cost_t - D \leq cost_t - x_tx'_t$. □

2.2 The competitive analysis of the ice cream problem

In this subsection we show how we can use work functions to analyze the Ice Cream Problem defined at the beginning of this section.

In the analysis of specific problem instances, it is frequently more convenient to use *offset functions*, instead of work functions. An offset function is simply a work function whose minimum is zero. One advantage of using offset functions is that for finite problems in which distances and costs are integers, the number of reachable offset functions is finite. Offset functions are updated just like work functions, except that at each step the constant $\min(\omega\wedge\tau)$ is subtracted from $\omega\wedge\tau$ in order to make the minimum zero. This constant is considered the optimal cost associated with this step (and thus "charged" to the adversary). Figure 1 shows offset functions for the Ice Cream Problem. Each offset function ω is represented by the pair of numbers $\omega(V), \omega(C)$. Arcs depict the transitions corresponding to the update operator, and are labeled with the tasks and optimal costs associated with these transitions.

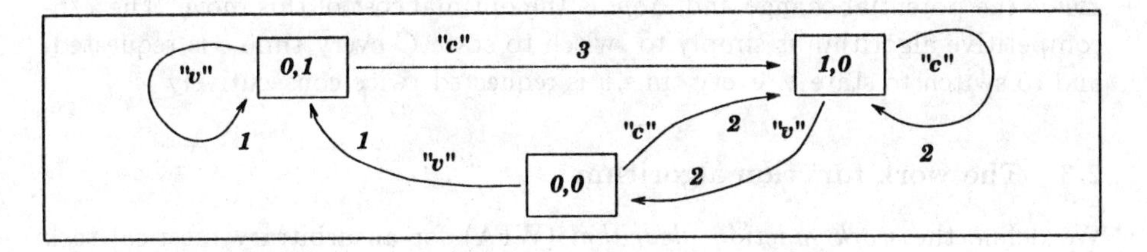

Fig. 1. Offset functions in the Ice Cream Problem.

Figure 2 shows the complete game. Each state is a pair consisting of an offset function and the algorithm state. Arcs leaving each state correspond to tasks and are labeled with the optimal costs. Open dots represent intermediate game states, where a task has been requested but not serviced. The on-line algorithm moves are represented by arcs leaving these intermediate states, and are labeled with their costs. Certain obviously bad moves, *e.g.*, changing state from V to C when v has been requested, are not shown.

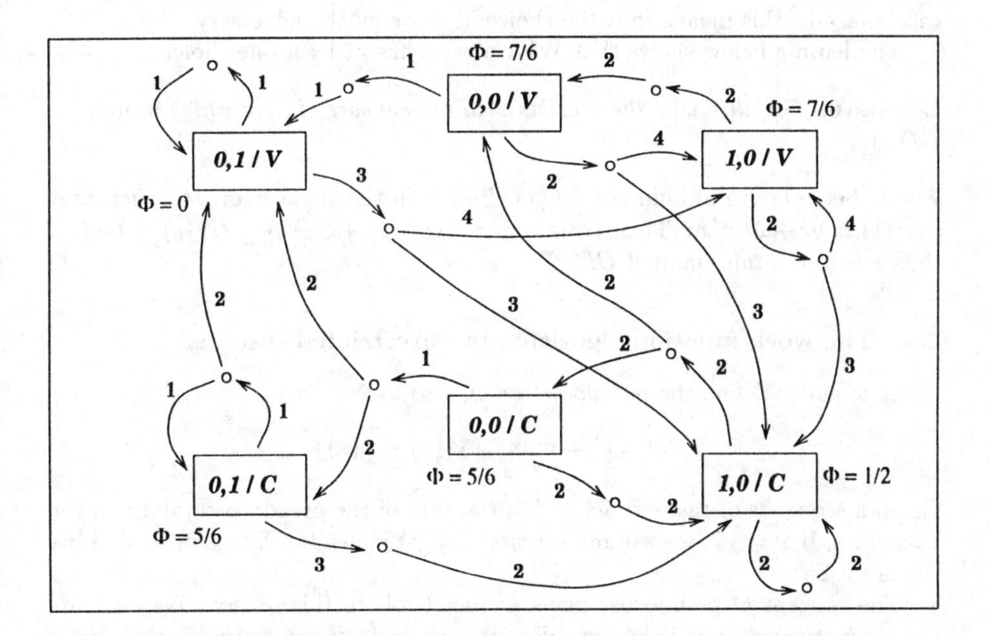

Fig. 2. The complete game graph in the Ice Cream Problem.

The figure also proves that there is a 7/6-competitive algorithm for this problem. The values of potential Φ are given at every game state in the figure. We leave it as an exercise to the reader to verify that for each pair (ω, x) and task τ, it is possible to pick y such that $xy + \tau(y) + \Delta\Phi \leq (7/6)\Delta opt$, where $\Delta\Phi$ is the potential change and Δopt is the optimal cost of this move. The 7/6-competitive algorithm is simply to switch to state C every time c is requested, and to switch to state V every time v is requested twice consecutively.

2.3 The work function algorithm

We define the *work function algorithm* (WFA) for an arbitrary metrical task system $\mathcal{S} = (M, \mathcal{T})$. As we will show, WFA is optimally competitive for some problems and it has been conjectured to be optimal for others (although it is known that there are on-line problems for which it is not optimal).

WFA keeps track of the work function at each step. Suppose that the current state of WFA is s, the current work function is ω, and the new task is τ. Then WFA moves to a state x in the support of $\omega \wedge \tau$ that minimizes the value of

$$GF(x) = sx + \omega\wedge\tau(x)$$

Function $GF(\cdot)$ is called the *guide function*. The other arguments of $GF(\cdot)$, namely s, ω, and τ will usually be understood from context. We will write $GF^{\tau}_{s,\omega}(\cdot)$ whenever ambiguity may arise. We also adopt the convention that if there is more than one choice of x satisfying the conditions, the choice is made arbitrarily. In the competitive analysis, as well as other kinds of worst-case analysis, this means that the choice is made by the adversary.

The lemma below shows that WFA always has at least one choice.

Lemma 6. *For all s, ω, τ there exists x in the support of $\omega\wedge\tau$ which minimizes $GF(x)$.*

Proof. Let y be a minimum of $GF(\cdot)$. Pick x in the support of $\omega\wedge\tau$ such that $\omega\wedge\tau(y) = yx + \omega\wedge\tau(x)$. Then $GF(x) = sx - sy - yx + GF(y) \leq GF(y) \leq GF(x)$, thus x is also a minimum of $GF(\cdot)$. □

2.4 The work function algorithm in unrestricted systems

Given ω and τ define the *pseudocost* on ω, τ to be

$$\nabla(\omega, \tau) = \max_x \{\omega\wedge\tau(x) - \omega(x)\}$$

Given a sequence of tasks $\bar{\tau}$, $\nabla(\omega, \bar{\tau})$ is the sum of the pseudocost values on the tasks in $\bar{\tau}$. If $\omega = \chi_x$, we will often write $\nabla(x, \bar{\tau})$ instead of $\nabla(\chi_x, \bar{\tau})$ to simplify the notation.

The concept of pseudocost plays a crucial role in this section. We estimate the competitiveness of WFA, in different task systems, by estimating the ratio between pseudocost and the optimal cost. For this reason, it is convenient to talk about the "competitiveness" of pseudocost. We will say that *pseudocost is c-competitive* in a given task system if there is a function $\alpha : M \to \mathbf{R}$ such that for any initial state x_0 and task sequence $\bar{\tau}$

$$\nabla(x_0, \bar{\tau}) \leq c \cdot opt(x_0, \bar{\tau}) + \alpha(x_0).$$

Lemma 7. *Pseudocost is n-competitive for each task system with n states.*

Proof. The total pseudocost is bounded by the total of all increases of the work function, which is at most $\sum_x \omega(x) \leq n \inf(\omega) + nD$, where ω is the final work function and D is the diameter of the task system, *i.e.*, the maximum distance between states.

A more accurate analysis can be done using a potential argument. Define $\Phi(\omega) = -\sum_x \omega(x)$. We have $-nD \leq \Phi(\chi_x) \leq 0$ and $\Phi(\omega) \geq -n \inf(\omega)$. Let $\mu = \omega \wedge \tau$. Then

$$\nabla(\omega, \tau) + \Phi(\mu) - \Phi(\omega) = \max_x \{\mu(x) - \omega(x)\} - \sum_y [\mu(y) - \omega(y)] \leq 0, \quad (2)$$

Summing (2) over all requests, we obtain $\nabla(x_0, \bar{\tau}) \leq \Phi(\chi_{x_0}) - \Phi(\chi_{x_0} \wedge \bar{\tau}) \leq n \cdot opt(x_0, \bar{\tau})$. □

Lemma 8. *For each metrical task system, start state s_0 and task sequence $\bar{\tau}$, the following inequality holds: $cost_{\text{wfa}}(s_0, \bar{\tau}) \leq 2 \cdot \nabla(s_0, \bar{\tau}) - opt(s_0, \bar{\tau})$.*

Proof. Let the initial state be s_0, the task sequence be $\bar{\tau} = \tau_1, \ldots, \tau_m$, and $s_0, \ldots, s_m$ be the states of WFA. Let $\omega_0 = \chi_{s_0}$ and $\omega_t = \omega_0 \wedge \tau_1 \ldots \tau_t$ for all t. We have

$$\omega_t(s_t) + s_{t-1}s_t = \min_x \{\omega_t(x) - \omega_{t-1}(x)\} \quad (3)$$

$$\omega_t(s_{t-1}) = \min_x \{\omega_t(x) - \omega_{t-1}(x)\} \quad (4)$$

$$\omega_t(s_t) = \omega_{t-1}(s_t) + \tau_t(s_t) \quad (5)$$

Equation (3) follows from the definition of WFA. Equation (4) follows from the fact that ω_t is 1-Lipschitz. Equation (5) follows from Corollary 2. Then

$$\begin{aligned}
& cost_{\text{wfa}}(s_0, \bar{\tau}) \\
& = \sum_{t=1}^{m} \{s_{t-1}s_t + \tau_t(s_t)\} \\
& = \sum_{t=1}^{m} [\omega_t(s_{t-1}) - \omega_t(s_t) + \tau_t(s_t)] \qquad \text{by Equations (3) and (4)} \\
& = \sum_{t=1}^{m} [\omega_t(s_{t-1}) - \omega_{t-1}(s_t)] \qquad \text{by Equation (5)} \\
& = \sum_{t=1}^{m} [\omega_t(s_{t-1}) - \omega_{t-1}(s_{t-1}) + \omega_t(s_t) - \omega_{t-1}(s_t)] + \omega_0(s_0) - \omega_n(s_n) \\
& \leq 2\nabla(\omega_0, \bar{\tau}) - opt(s_0, \bar{\tau})
\end{aligned}$$

completing the proof. □

Lemma 9. *For each metrical service system, each start state s_0 and request sequence $\bar{\rho}$, the following inequality holds:* $cost_{\mathrm{wfa}}(s_0, \bar{\rho}) \leq \nabla(s_0, \bar{\rho}) - opt(s_0, \bar{\rho})$.

Proof. We use notation from the proof of Lemma 9. Equations (3) and (4) hold in this case also. Then

$$\begin{aligned} cost_{\mathrm{wfa}}(s_0, \bar{\rho}) &= \sum_{t=1}^{m} s_{t-1}s_t \\ &= \sum_{t=1}^{m} [\omega_t(s_{t-1}) - \omega_t(s_t)] \qquad \text{by Equations (3) and (4)} \\ &= \sum_{t=1}^{m} [\omega_t(s_{t-1}) - \omega_{t-1}(s_{t-1})] + \omega_0(s_0) - \omega_n(s_n) \\ &\leq \nabla(\omega_0, \bar{\rho}) - opt(s_0, \bar{\rho}) \end{aligned}$$

completing the proof. □

Theorem 10. (a) WFA *is* $(n-1)$*-competitive for each metrical service system with n states.* (b) WFA *is* $(2n-1)$*-competitive for each metrical task system with n states.*

Proof. Both (a) and (b) follow from the previous lemmas. For example, to prove (a), we have

$$cost_{\mathrm{wfa}}(s_0, \bar{\rho}) \leq \nabla(s_0, \bar{\rho}) - opt(s_0, \bar{\rho}) \ \leq \ (n-1) \cdot opt(s_0, \bar{\rho}),$$

by Lemma 9. The proof of (b) is similar. □

Theorem 11. (a) *No on-line algorithm for an n-state unrestricted service system is better than* $(n-1)$*-competitive.* (b) *No on-line algorithm for an n-state unrestricted task systems is better than* $(2n-1)$*-competitive.*

Proof. We show (a) first. Let $\mathcal{S} = (M, \mathcal{R})$ be an arbitrary service system with n states, and $\mathcal{A}$ an on-line algorithm for $\mathcal{S}$. Let also D be the diameter of the system.

The adversary strategy is to request a sequence $\bar{\rho}$ such that if at the ith step $\mathcal{A}$ is in state s_i then the next request is $M - s_i$. We need to show that $cost_{\mathcal{A}}(s_0, \bar{\rho})$ is at least $n-1$ times as great, asymptotically, as $opt(s_0, \bar{\rho})$ for this request sequence.

Let s_0 be the initial state. We define $n-1$ distinct schedules, $\mathcal{A}_j$ for $j = 1, \ldots n-1$. Before the first request, the $\mathcal{A}_j$ all move to different states x_j, covering all states except s_0. The invariant is that after each step, $\mathcal{A}$ and the $\mathcal{A}_j$ together cover every state. At each request, $\mathcal{A}$ changes state from s_{t-1} to s_t. Then whichever of the $\mathcal{A}_j$ which is at s_t moves to s_{t-1}, preserving the invariant.

Note that at each step the cost of $\mathcal{A}$ is equal to the sum of the costs of all $\mathcal{A}_j$. Thus

$$cost_{\mathcal{A}}(s_0, \bar{\tau}) = \sum_j cost(s_0, \bar{\tau}, \mathcal{A}_j) - \sum_j s_0 x_j,$$

thus there exists j for which $cost_{\mathcal{A}}(s_0, \bar{\tau}) \geq (n-1)cost(s_0, \bar{\tau}, \mathcal{A}_j) - (n-1)D$. Finally, for any arbitrarily large integer B, the adversary can force $\mathcal{A}$ to pay at least B by ensuring that the length of $\bar{\rho}$ is at least B/d, where d is the smallest positive distance between states. This completes the proof.

(b) The proof is similar, we just need to change the adversary strategy slightly and add more schedules. Let $\delta_y(y) = 1$ and $\delta_y(x) = 0$ for $x \neq y$. The adversary will only use requests of the form $\epsilon\delta_y$ for small $\epsilon > 0$.

The adversary strategy is to request $\epsilon_t \delta_{s_{t-1}}$ at step t, where ϵ_t will be determined later. Let $\mathcal{A}_j$ be the schedules defined as in the previous case, and let $\mathcal{B}_x$ be the schedule which moves immediately to state x before the first request and stays there. The invariant is that each state is covered by precisely two of the $2n$ schedules, $\mathcal{A}$, the $\mathcal{A}_j$, and the $\mathcal{B}_x$. If $s_t = s_{t-1}$, none of the $\mathcal{A}_j$ move. Otherwise, that $\mathcal{A}_j$ whose state is s_t changes state to s_{t-1}.

Let $\epsilon_t = 2^{-i}$, where i is the number of times $\mathcal{A}$ has changed state before step t. Let the *excess* cost at each step be the total cost of all $(2n-1)$ adversary schedules at that step, minus the cost of $\mathcal{A}$ for that step. The excess is zero if $\mathcal{A}$ does not move, and is 2^{-i+1} at $\mathcal{A}$'s i^{th} move. Thus, the total excess over the entire request sequence does not exceed 2. If $\mathcal{C}$ is the one of the adversary schedules of minimum cost, then $cost_{\mathcal{A}}(x_0, \bar{\tau}) \geq (2n-1)cost_{\mathcal{C}}(x_0, \bar{\tau}) - 2$.

To prove that $\mathcal{A}$ cannot be less than $(2n-1)$-competitive, it remains only to show that for an arbitrary large integer B, $cost_{\mathcal{A}}(x_0, \bar{\tau}) \geq B$ if the length of $\bar{\tau}$ is sufficiently long. Let d be the smallest non-zero distance in M. Define $\ell = \lceil B/d \rceil$ and let $\bar{\tau}$ have length $\ell + \ell B 2^{\ell}$. Let i be the number of times $\mathcal{A}$ changes state. If $i \geq \ell$, then $\mathcal{A}$ pays at least $\ell d \geq B$. On the other hand, if $i \leq \ell$, the $\mathcal{A}$ stays in the same state at least $B2^{\ell}$ times, and pays at least $2^{-\ell}$ each of those times, incurring a total cost of at least B. □

Borodin, Linial and Saks [4] were the first to prove that the optimal competitive constant for each n-state unrestricted task system is $2n-1$. The idea of the work function algorithm has been considered, independently, by a number of researchers, including H. Karloff, M. Chrobak and L. L. Larmore, L. McGeoch and D. Sleator, and quite possibly others as well. Chrobak and Larmore [10] investigated basic properties of WFA, introduced the concept of pseudocost, and proved the relationship between the pseudocost and the competitiveness of the WFA (for the server problem). That WFA is $(2n-1)$-competitive was proven by Moti Ricklin (unpublished result), without using the idea of pseudocost. The proof given above has been inspired by his approach. The lower bound proofs are based on the elegant averaging idea from [19].

WFA is not optimally competitive in arbitrary task systems (see the next section). An obvious extension of WFA is to modify the guide function (2) to

$GF_\alpha(x) = sx + \alpha \cdot \omega\wedge\tau(x)$, for some constant α. For $\alpha = 3$ this gives a 9-competitive strategy for an MSS in which each request is a 2-point set, and the constant 9 is optimal [9]. The 2-point request problem is closely related to layered graph traversal [1, 12, 13, 20, 21] and to the so-called Cow Problem. Applications of such a modified WFA to searching k-layered graphs have been investigated by Burley (see, for example, [5]).

We have shown exact competitiveness bounds for unrestricted metrical task systems and service systems with a given number of states and arbitrary tasks. This leads to the following problem: Suppose that we are given a task system with at most m requests. How does the competitiveness constant depend on m? The technique used to prove Theorem 10 can be modified to show that if $m \geq n$, the competitive constant can be as high as $2n - 1$, so this problem is open only for smaller values of m.

2.5 Uniform service systems

Recall that a uniform MSS is a pair $S = (M, \mathcal{R})$ where M is a uniform space and $\mathcal{R}$ is the set of admissible requests. Without loss of generality, we assume that $\mathcal{R}$ is closed under intersection and that it contains all singletons. Note that every work function is of the form $\chi_X + a$ where $X \in \mathcal{R}$.

Given $X \in \mathcal{R}$ and two points $x, y \in X$ define $\delta_X(x, y)$ as the largest m for which there exist request sets $X_1, \ldots, X_m$ such that

$$\{y\} = X_1 \subset X_2 \subset \ldots \subset X_m \subset X$$

and $x \notin X_m$. By definition, $\delta_X(x, x) = 0$. If $X = M$ we will simply write $\delta(x, y)$ instead of $\delta_M(x, y)$. Construct a weighted directed graph G_S whose vertices are the points in M and the weight of arc (x, y) is $\delta(x, y)$. If P is a path, then $\delta(P)$ is the sum of all $\delta(x, y)$ for edges $(x, y) \in P$.

Theorem 12. *The competitiveness of* WFA *in S is equal to*

$$c_{\mathsf{wfa}} = \max_C \frac{\delta(C)}{|C|},$$

where the maximum is over all directed cycles C in the graph G_S constructed above.

Proof. We first construct a function ϕ such that

$$\phi(y) - \phi(x) + \delta(x, y) \leq c_{\mathsf{wfa}} \tag{6}$$

for all edges (x, y) of G_S. Let $\phi(x) = \max\{\delta(P) - c_{\mathsf{wfa}}|P| : P \text{ is a path starting from } x\}$. Since $\delta(C) - c_{\mathsf{wfa}}|C| \leq 0$ for any cycle C, we may maximize over only simple paths. Thus $\phi(x)$ is well-defined because there are only finitely many simple paths. Inequality (6) follows immediately.

Now, given a work function $\omega = \chi_X + a$ and $x \in X$, we define a potential $\Phi_x(\omega)$ if the work function is ω and our server is at x, as follows:

$$\Phi_x(\omega) = \max_{y \in X} \{\delta_X(x, y) + \phi(y)\} - c_{\mathsf{wfa}} a.$$

Let $Y \in \mathcal{R}$ be the next request, $\mu = \omega^Y$, and suppose that WFA moves the server from x to y. We need only consider two cases, namely $x \notin Y \subset X$, and $Y \cap X = \emptyset$.

Suppose $x \notin Y \subset X$. Then $\mu = \chi_Y + a$. Choose $z \in Y$ such that $\Phi_y(\mu) = \delta_Y(y, z) + \phi(z) - c_{\mathsf{wfa}} a$. Then

$$\begin{aligned} cost_{\mathsf{wfa}}(\omega, x, Y) + \Phi_y(\mu) &= 1 + \delta_Y(y, z) + \phi(z) - c_{\mathsf{wfa}} a \\ &\le \delta_X(x, z) + \phi(z) - c_{\mathsf{wfa}} a \\ &\le \Phi_x(\omega) \end{aligned}$$

Suppose now that $Y \cap X = \emptyset$. Then $\mu = \chi_Y + a + 1$. Choose $z \in Y$ such that $\Phi_y(\mu) = \delta_Y(y, z) + \phi(z) - c_{\mathsf{wfa}}(a + 1)$. Then

$$\begin{aligned} cost_{\mathsf{wfa}}(\omega, x, Y) + \Phi_y(\mu) &\le 1 + \delta_Y(y, z) + \phi(z) - c_{\mathsf{wfa}}(a + 1) \\ &\le \delta(x, z) + \phi(z) - c_{\mathsf{wfa}}(a + 1) \\ &\le \phi(x) - c_{\mathsf{wfa}} a \\ &\le \Phi_x(\omega), \end{aligned}$$

completing the proof. □

That WFA can be much worse than an optimally competitive algorithm can be seen from the following example: Let $\mathcal{S} = (M, \mathcal{R})$, where $M = \{1, 2, \ldots, n\}$, with the uniform metric, and let $\mathcal{R}$ contain requests $\{1, \ldots, i\}$ for all i, and a singleton request $\{n\}$. Then $c_{\mathsf{wfa}} = n/2$, because the adversary can insist that each request be served with the largest number possible. But $\mathcal{S}$ has a 1-competitive algorithm, namely to serve every request $\{1, \ldots, i\}$ with 1.

Uniform ϵ-task systems. An ϵ-task system is a system with a distinguished family $\mathcal{R}$ of subsets of M, where for each $\rho \in \mathcal{R}$ and $\epsilon > 0$ we have a task τ of the following form: $\tau(x) = \epsilon$ for $x \in \rho$ and $\tau(x) = 0$ elsewhere. Problem: Determine the optimal strategy for a given uniform ϵ-task system. Note that this is a natural extension of a forcing task system, which can be thought of as a uniform ϵ-task system with $\epsilon = \infty$.

3 The k-server problem

The k-server problem is defined as follows: we are given k servers that reside and can move in a metric space M. At every time step, a request $r \in M$ is read, and in order to satisfy the request, we must move one server to the request point. Our cost is the total distance traveled by the servers. The choice of server at each step must be made on-line, *i.e.*, without knowledge of future requests, and therefore (except for some degenerate situations) no algorithm can achieve the optimal cost on each request sequence. Our goal is to design a strategy with a small competitive ratio.

The k-server problem was introduced by Manasse, McGeoch and Sleator [18, 19]. They proved that k is a lower bound on the competitive ratio for any deterministic algorithm for the problem in any metric space with at least $k+1$ points. They also gave an algorithm for the 2-server problem which is 2-competitive, and thus optimal, for any metric space, and a k-competitive algorithm if the metric space has exactly $k+1$ points. The general case was left open.

The problem has attracted a great deal of attention. Chrobak *et al.* [6] presented a k-competitive algorithm for k servers when M is a line. This was later generalized by Chrobak and Larmore [8] to trees. A different 2-competitive algorithm for arbitrary spaces was given by Chrobak and Larmore in [7]. Berman *et al.* [3] proved that there is a c-competitive algorithm for $k = 3$ for some very large c. Chrobak and Larmore [11] gave an 11-competitive algorithm for $k = 3$. The question whether there is a competitive algorithm for any k was settled by Fiat, Rabani and Ravid [14], who showed an $O(2^k)$-competitive algorithm for each k. A slightly better result was given in [15] and later in [2].

Several authors have independently proposed the work function algorithm (WFA), the same algorithm described in the Section 2, for the server problem. WFA was investigated by Chrobak and Larmore [10], who suggested that competitive analysis of WFA can be done by estimating the pseudocost. They also proved that WFA is 2-competitive for 2 servers. More recently, Koutsoupias and Papadimitriou [16] proved that WFA is $(2k-1)$-competitive for all k. They also proved that WFA is k-competitive on metric spaces with $k+2$ points [17]. We will present these results in this section.

The k-server problem is a special case of the MSS problem. Define a *configuration* to be a k-tuple of points of M. Let $M' = \Lambda^k M$ be the set of all such configurations, a metric space with the minimum-matching metric. For each $r \in M$ define ρ_r to be the set of the configurations that contain r, and let $\mathcal{R} = \{\rho_r\}_{r \in M}$. Then the k-server problem on M is simply the metrical service system $(M', \mathcal{R})$.

A lower bound of k on the competitive constant of any algorithm for the server problem can be derived from Theorem 11. If M is any space with $k+1$ points, then M is identical to the subset M'' of M' consisting of the k-tuples in which all points are distinct. Without loss of generality, a k-server algorithm never moves two servers to the same point, so it will only use configurations in M''. Below, for the sake of completeness, we give a direct proof of this lower bound.

Theorem 13. *Let M be an arbitrary metric space with at least $k+1$ points, and $\mathcal{A}$ a k-server on-line algorithm in M. Then the competitiveness of $\mathcal{A}$ is at least k.*

Proof. Assume M has $k+1$ points. Let $\mathcal{A}$ be any algorithm. Let the initial configuration be $S = M - x_0$. The request sequence is the standard "cruel" sequence; at each step, the adversary requests the one point where $\mathcal{A}$ has no server. Let us call this request sequence ϱ.

We shall define k distinct service schedules for ϱ, which we call $\sigma_1, \ldots, \sigma_k$. The sum of the costs for all σ_i to service ϱ equals the cost of $\mathcal{A}$, plus a constant that does not depend on the length of ϱ.

Let $S = \{x_1 \ldots x_k\}$. Initially, σ_i moves a server from x_i to x_0. Henceforth, the following invariant is maintained: for each $x \in M$, either $\mathcal{A}$ has no server at x or one of the σ_i has no server at x.

Suppose at a given step the request point is r. Suppose that $\mathcal{A}$ services that request by moving a server from x to r. Each of the σ_i has a server on r, so each serves the request at no cost. There is exactly one of the σ_i which has no server on x. That one σ_i then moves its server from r to x, preserving the invariant. The cost of that move equals the cost of $\mathcal{A}$ for that time step. We thus have

$$\sum_{i=1}^{k} cost(x_0, \varrho, \sigma_i) = \sum_{i=1}^{k} x_i x_0 + cost_{\mathcal{A}}(x_0, \varrho)$$

We immediately conclude that the cost of $\mathcal{A}$ to service ϱ is, asymptotically, at least k times the minimum cost of the σ_i. $\square$

3.1 The work function algorithm

The work function algorithm (WFA) can be applied to the server problem, in which case it can be stated as follows: given a current work function ω, a server configuration S, and a request r, move the server from a point x such that the configuration $S' = S + r - x$ minimizes the quantity

$$SS' + \omega(S').$$

When we discuss the server problem, we use the notation $\omega \wedge r$ to denote $\omega \wedge \rho_r$. Given a work function ω and request r, recall that we define the pseudocost on r as

$$\nabla(\omega, r) = \sup_X \{\omega \wedge r(X) - \omega(X)\}$$

Since the k-server problem is a special case of a metrical service system, we conclude that:

Corollary 14. *If the pseudocost is $(k+1)$-competitive, then WFA is k-competitive.*

If M has $k+1$ points, then the relationship between M and M'' (defined earlier in this section) and Theorem 10, part (a), imply the following.

Theorem 15. *If $|M| = k + 1$ then WFA is k-competitive in M.*

3.2 Quasiconvexity

A function $\omega : \Lambda^k M \to \mathbf{R}$ is called *quasiconvex* if for all configurations X, Y there is a bijection $f : X \to Y$ with the following property:

(qc) If $A \subseteq X$ and $B = X - A$ then

$$\omega(X) + \omega(Y) \geq \omega(A \cup f(B)) + \omega(f(A) \cup B)$$

Recall that we define a *reachable work function* to be a function of the form $\chi \wedge \varrho$ where χ is a cone and ϱ is a request sequence.

Lemma 16. *If ω is a reachable work function, then ω is quasiconvex. Furthermore, if f is the bijection satisfying condition (qc), then we can insist that f is fixed on $X \cap Y$, that is $f(z) = z$ for $z \in X \cap Y$.*

Proof. The proof is by induction on the number of requests. If ω is a cone, $\omega = \chi_C$, and X, Y are two given configurations, then let $g : X \to C$ and $h : Y \to C$ be the minimum matching bijections. Define $f = h^{-1}g$. That f satisfies (qc) follows directly from the definition of the minimum matching metric.

Let us now assume that ω satisfies the lemma, and $\mu = \omega \wedge r$. We will first prove that μ is quasiconvex. Pick two configurations X, Y. There are $x \in X$ and $y \in Y$ such that

$$\mu(X) = \omega(X - x + r) + xr$$
$$\mu(Y) = \omega(Y - y + r) + yr$$

By induction, there is a bijection $f : (X - x + r) \to (Y - y + r)$ that satisfies (qc) and that is fixed on $(X - x + r) \cap (Y - y + r)$. In particular, $f(r) = r$. Define $g : X \to Y$ by $g(z) = f(z)$ for $z \neq x$ and $g(x) = y$. We will prove that g satisfies (qc). Choose an arbitrary $A \subseteq X$ and let $B = X - A$. Without loss of generality, $x \in A$. Then $A - x + r \subseteq X - x + r$, $(X - x + r) - (A - x + r) = B$, and thus

$$\begin{aligned}\mu(X) + \mu(Y) &= \omega(X - x + r) + \omega(Y - y + r) + xr + yr \\ &\geq \omega(\,(A - x + r) \cup f(B)\,) + \omega(\,f(A - x + r) \cup B\,) + xr + yr \\ &= \omega(\,(A - x + r) \cup g(B)\,) + \omega(\,(g(A) + r - y) \cup B\,) + xr + yr \\ &\geq \mu(A \cup g(B)) + \mu(g(A) \cup B),\end{aligned}$$

proving that μ is indeed quasiconvex.

Note that g is not necessarily fixed on $X \cap Y$. Thus we still need to show that we can transform g into g' that satisfies (qc) for μ and is fixed on $X \cap Y$.

Suppose that there is $u \in X \cap Y$ such that $g(u) = v \neq u$. We show that there is another bijection $h : X \to Y$ that is fixed on u. Pick $p = g^{-1}(u) \in X$, and define $h(u) = u$, $h(p) = v$, and $h(z) = z$ for $z \in X - \{p, u\}$. Let $A \subseteq X$ and $B = X - A$. Assume, without loss of generality, that $p \in A$.

We need to show that $\mu(X)+\mu(Y) \geq \mu(A \cup h(B)) + \mu(h(A) \cup B)$. If $u \in A$, then $h(A) = g(A)$ and $h(B) = g(B)$, and we are done. So suppose now $u \notin A$. Then

$$\begin{aligned} \mu(X)+\mu(Y) &\geq \mu((A+u) \cup g(B-u)) + \mu(g(A+u) \cup (B-u)) \\ &= \mu(A \cup h(B)) + \mu(h(A) \cup B), \end{aligned}$$

and h has one more fixed point in $X \cap Y$ than g. By repeating this process, we obtain g' that is fixed on $X \cap Y$. □

3.3 Maximizers

A *1-dual* of a work function ω is a function $\hat{\omega} : M \to \mathbf{R}$ defined by

$$\hat{\omega}(x) = \max_A \{xA - \omega(A)\}$$

where $xA = \sum_{a \in A} xa$. We call A an *(ω, x)-maximizer* if it maximizes the expression above, that is $\hat{\omega}(x) = xA - \omega(A)$.

Lemma 17. *If $\omega^r = \omega$, then there is a (ω, x)-maximizer A such that $r \in A$.*

Proof. Let B be an (ω, x)-maximizer. Then $\omega(B) = \omega(B - b + r) + br$ for some $b \in B$. Let $A = B - b + r$. Then $Ax - \omega(A) = Bx - bx + rx - \omega(B) + br \geq Bx - \omega(B)$, hence A is also a (ω, x)-maximizer. □

Lemma 18. [The Duality Lemma] *If ω is a quasiconvex work function, $\mu = \omega \wedge r$, and A is a (ω, r)-maximizer, then*

(a) A is a (μ, r)-maximizer,
(b) The pseudocost is maximized on A, that is $\nabla(\omega, r) = \mu(A) - \omega(A)$.

Proof. (a) Let B be a (μ, r)-maximizer. By Lemma 17, we may assume that $r \in B$. Using quasiconvexity, pick a such that $\omega(B) + \omega(A) \geq \omega(B - r + a) + \omega(A - a + r)$. Since A is a (ω, r)-maximizer, $r(B - r + a) - \omega(B - r + a) \leq rA - \omega(A)$. Since μ is 1-Lipschitz, $\mu(A) \leq ra + \mu(A - a + r)$. Then we have

$$\begin{aligned} rB - \mu(B) &= rB - \omega(B) \\ &\leq rB + \omega(A) - \omega(B - r + a) - \omega(A - a + r) \\ &= r(B - r + a) - ra + \omega(A) - \omega(B - r + a) - \mu(A - a + r) \\ &\leq rA - \mu(A), \end{aligned}$$

implying that A is a (μ, r)-maximizer.

(b) Pick a such that $\mu(A) = \omega(A + r - a) + ar$. Let B be an arbitrary configuration, and using quasiconvexity choose b such that $\omega(A + r - a) + \omega(B) \geq$

$\omega(A{+}b{-}a){+}\omega(B{+}r{-}b)$. Since A is a (ω, r)-maximizer, $r(A{+}b{-}a){-}\omega(A{+}b{-}a) \leq rA - \omega(A)$. Then

$$\begin{aligned}\mu(A) + \omega(B) &= \omega(A + r - a) + ar + \omega(B) \\ &\geq \omega(A + b - a) + \omega(B + r - b) + ar \\ &\geq \omega(A) + br + \omega(B + r - b) \\ &\geq \omega(A) + \mu(B),\end{aligned}$$

completing the proof. □

Corollary 19. *If ω is a quasiconvex function and $\mu = \omega \wedge r$, then $\nabla(\omega, r) = \hat{\omega}(r) - \hat{\mu}(r)$.*

Proof. Let A be a (ω, r)-maximizer. Then $\hat{\omega}(r) = rA - \omega(A)$. By Lemma 18 we also have $\hat{\mu}(r) = rA - \mu(A)$ and $\nabla(\omega, r) = \mu(A) - \omega(A)$, implying that $\nabla(\omega, r) = \hat{\omega}(r) - \hat{\mu}(r)$. □

3.4 Proof of $(2k - 1)$-competitiveness for $k \geq 3$

The proof of $(2k - 1)$-competitiveness given in this chapter uses Corollary 14 and a potential-based argument. Let ω be a work function. The potential for ω is defined by

$$\begin{aligned}\Psi_\omega(X) &= \sum_{x \in X} \hat{\omega}(x) - k\omega(X) \\ \Psi_\omega &= \max_X \Psi_\omega(X)\end{aligned}$$

Lemma 20. *If ω is a work function with last request r then $\Psi_\omega(X)$ is maximized when $X \ni r$.*

Proof. Note that $\hat{\omega}$ is k-Lipschitz: $\hat{\omega}(x) - \hat{\omega}(y) \leq k \cdot xy$ for all $x, y \in M$. Suppose that $\omega(X) = \omega(X + r - y) + ry$. Then

$$\begin{aligned}\Psi_\omega(X) &= \sum_{x \in X - y} \hat{\omega}(x) + (\hat{\omega}(y) - k \cdot ry) - k\omega(X + r - y) \\ &\leq \sum_{x \in X + r - y} \hat{\omega}(x) - k\omega(X + r - y) \\ &= \Psi_\omega(X + r - y),\end{aligned}$$

completing the proof. □

Theorem 21. WFA *is $(2k - 1)$-competitive for k servers.*

Proof. It is sufficient to prove that the pseudocost is $2k$-competitive. Given ω, r and $\mu = \omega \wedge r$, by Lemma 20 we can pick $X \ni r$ for which $\Psi_\mu = \Psi_\mu(X)$. By Corollary 19, we have

$$\begin{aligned} \nabla(\omega, r) + \Psi_\mu &= \sum_{x \in X - r} \hat{\mu}(x) + \hat{\omega}(r) - k\omega(X) \\ &\leq \sum_{x \in X} \hat{\omega}(x) - k\omega(X) \\ &\leq \Psi_\omega. \end{aligned}$$

Given an initial work function χ_X and request sequence ϱ, we can assume that $\chi_X \wedge \varrho = \chi_Y + a$ for some Y and a. For otherwise, we could continue making requests on the configuration Y that minimizes $\chi_X \wedge \varrho(Y)$, increasing the pseudocost but not the optimal cost. By summation over the requests in ϱ we obtain $\nabla(\chi, \varrho) \leq \Psi_{\chi_X} - \Psi_{\chi_Y} = (2k - 1)a$. □

3.5 Competitiveness proof for 2 servers

We now prove that WFA is 2-competitive for 2 servers. Given ω, introduce a function:

$$\bar{\omega}(x) = \max_{ab} \{ab - \omega(xa) - \omega(xb)\}$$

We define the following potential:

$$\begin{aligned} \Psi_\omega(x) &= \hat{\omega}(x) + \bar{\omega}(x) \\ \Psi_\omega &= \max_x \Psi_\omega(x) \end{aligned}$$

Lemma 22. *If ω is a work function with last request r, then the potential $\Psi_\omega(x)$ is maximized for $x = r$.*

Proof. By Lemma 17, there exist points a, b, c such that

$$\hat{\omega}(x) + \bar{\omega}(x) = xa + xr - \omega(ar) + bc - \omega(bx) - \omega(cx)$$

Now

$$\omega(bx) = \min \left\{ \begin{matrix} \omega(rx) + br \\ \omega(br) + rx \end{matrix} \right\} \quad \text{and} \quad \omega(cx) = \min \left\{ \begin{matrix} \omega(rx) + cr \\ \omega(cr) + rx \end{matrix} \right\}$$

making four cases overall, although by symmetry only three need be considered. If $\omega(bx) = \omega(br) + rx$ and $\omega(cx) = \omega(cr) + rx$, then

$$\begin{aligned} \hat{\omega}(x) + \bar{\omega}(x) &= ax + rx - \omega(ar) + bc - \omega(br) - rx - \omega(cr) - rx \\ &\leq [ar - \omega(ar)] + [bc - \omega(br) - \omega(cr)] \\ &\leq \hat{\omega}(r) + \bar{\omega}(r) \end{aligned}$$

If $\omega(bx) = \omega(rx) + br$ and $\omega(cx) = \omega(cr) + rx$, then

$$\begin{aligned}\hat{\omega}(x) + \bar{\omega}(x) &= ax + rx - \omega(ar) + bc - \omega(rx) - br - \omega(cr) - rx \\ &\leq [cr - \omega(cr)] + [ax - \omega(ar) - \omega(rx)] \\ &\leq \hat{\omega}(r) + \bar{\omega}(r)\end{aligned}$$

If $\omega(bx) = \omega(rx) + br$ and If $\omega(cx) = \omega(rx) + cr$, then

$$\begin{aligned}\hat{\omega}(x) + \bar{\omega}(x) &= ax + rx - \omega(ar) + bc - \omega(rx) - br - \omega(rx) - cr \\ &\leq [rx - \omega(rx)] + [ax - \omega(rx) - \omega(ar)] \\ &\leq \hat{\omega}(r) + \bar{\omega}(r)\end{aligned}$$

This completes the proof. □

Theorem 23. WFA *is 2-competitive for 2 servers.*

Proof. Fix ω and r. Let $\mu = \omega \wedge r$. Then, using Lemma 22 and Corollary 19 we have

$$\nabla(\omega, r) + \Psi_\mu = \hat{\omega}(r) + \bar{\mu}(r) \;\leq\; \hat{\omega}(r) + \bar{\omega}(r) \;\leq\; \Psi_\omega.$$

2-competitiveness follows by summation over the request sequence, as in the proof of Theorem 21. □

3.6 Proof for $k+2$ points

Let M be a metric space with $k+2$ points. Introduce a function $\sigma_\omega : M^2 \to \mathbf{R}$ defined as follows: if $x, y \in M$, then

$$\sigma_\omega(x, y) = -\omega(M - x - y) - xy,$$

We will view (M, σ_ω) as a weighted graph, in which (x, y) is assigned weight $\sigma_\omega(x, y)$.

Theorem 24. *If M has $k+2$ points then* WFA *is k-competitive in M.*

Proof. Given a work function ω, define Ψ_ω to be the weight of the maximum-weight spanning tree in the weighted graph (M, σ_ω). Since a spanning tree in (M, σ_ω) has exactly $k+1$ edges, Ψ_ω has exactly $k+1$ negative ω-terms.

Fix ω, r and $\mu = \omega \wedge r$. Let A be a (ω, r)-maximizer. Without loss of generality, we can assume that $r \notin A$. By Lemma 18 we know that A is also a (μ, r)-maximizer and that $\nabla(\omega, r) = \mu(A) - \omega(A)$. Let $x = M - A - r$. Since A is a (μ, r)-maximizer, we have

$$\sigma_\mu(r, x) = -\mu(A) - xr \;=\; \max_{b \in M - r} \{-\mu(M - r - b) - br\},$$

i.e., that the edge (r, x) has the maximum weight of any edge incident to r, and hence must belong to a maximum spanning tree T of (M, σ_μ). Then

$$\begin{aligned}\nabla(\omega, r) + \Psi_\mu &= \mu(A) - \omega(A) + \sum_{(u,v)\in T} \sigma_\mu(u, v) \\ &= \sigma_\omega(x, r) + \sum_{(u,v)\in T-(x,r)} \sigma_\mu(u, v) \\ &\leq \Psi_\omega.\end{aligned}$$

The theorem follows by summation over the request sequence, as in Theorem 21. □

Acknowledgements

Marek Chrobak acknowledges support by NSF grant CCR-9503498. Lawrence L. Larmore acknowledges support by NSF grant CCR-9503441. Part of this work was done while Lawrence L. Larmore was visiting Institut Informatik V, Universität Bonn, Germany.

References

1. R. Baeza-Yates, J. Culberson, and G. Rawlins. Searching in the plane. *Information and Computation*, 106(2):234–252, 1993. Preliminary version in Proc. 1st Scandinavian Workshop on Algorithm Theory, Lecture Notes in Computer Science 318, Springer-Verlag, Berlin, 1988, 176-189. Also Tech. Report CS-87-68, University of Waterloo, Department of Computer Science, October, 1987.
2. Y. Bartal and E. Grove. The harmonic k-server algorithm is competitive. *To appear in* Journal of ACM, 1994.
3. P. Berman, H. Karloff, and G. Tardos. A competitive algorithm for three servers. In *Proceedings of the 1st Annual ACM-SIAM Symposium on Discrete Algorithms*, pages 280–290, 1990.
4. A. Borodin, N. Linial, and M. Saks. An optimal online algorithm for metrical task systems. In *Proc. 19th Annual ACM Symposium on Theory of Computing*, pages 373–382, 1987.
5. W. Burley. Traversing layered graphs using the work function algorithm. Technical Report CS93-319, Department of Computer Science and Engineering, University of California at San Diego, 1993.
6. M. Chrobak, H. Karloff, T. H. Payne, and S. Vishwanathan. New results on server problems. *SIAM Journal on Discrete Mathematics*, 4:172–181, 1991. Also in Proceedings of the 1st Annual ACM-SIAM Symposium on Discrete Algorithms, San Francisco, 1990, pp. 291-300.
7. M. Chrobak and L. L. Larmore. A new approach to the server problem. *SIAM Journal on Discrete Mathematics*, 4:323–328, 1991.
8. M. Chrobak and L. L. Larmore. An optimal online algorithm for k servers on trees. *SIAM Journal on Computing*, 20:144–148, 1991.
9. M. Chrobak and L. L. Larmore. Metrical service systems: Deterministic strategies. Technical Report UCR-CS-93-1, Department of Computer Science, University of California at Riverside, 1992. Submitted for publication in a journal.

10. M. Chrobak and L. L. Larmore. The server problem and on-line games. In *DIMACS Series in Discrete Mathematics and Theoretical Computer Science*, volume 7, pages 11–64, 1992.
11. M. Chrobak and L. L. Larmore. Generosity helps or an 11-competitive algorithm for three servers. *Journal of Algorithms*, 16:234–263, 1994. Also in Proceedings of ACM/SIAM Symposium on Discrete Algorithms, 1992, 196-202.
12. X. Deng and C. Papadimitriou. Exploring an unknown graph. In *Proc. 31st IEEE Symp. Foundations of Computer Science*, pages 355–361, 1990.
13. A. Fiat, D. Foster, H. Karloff, Y. Rabani, Y. Ravid, and S. Vishwanathan. Competitive algorithms for layered graph traversal. In *Proc. 32nd IEEE Symposium on Foundations of Computer Science*, pages 288–297, 1991.
14. A. Fiat, Y. Rabani, and Y. Ravid. Competitive k-server algorithms. In *Proc. 22nd IEEE Symposium on Foundations of Computer Science*, pages 454–463, 1990.
15. E. Grove. The harmonic k-server algorithm is competitive. In *Proc. 23rd ACM Symposium on Theory of Computing*, pages 260–266, 1991.
16. E. Koutsoupias and C. Papadimitriou. On the k-server conjecture. In *Proc. 25th Symposium on Theory of Computing*, pages 507–511, 1994.
17. E. Koutsoupias and C. Papadimitriou. The 2-evader problem. *Information Processing Letters*, 57:249–252, 1996.
18. M. Manasse, L. A. McGeoch, and D. Sleator. Competitive algorithms for online problems. In *Proc. 20th Annual ACM Symposium on Theory of Computing*, pages 322–333, 1988.
19. M. Manasse, L. A. McGeoch, and D. Sleator. Competitive algorithms for server problems. *Journal of Algorithms*, 11:208–230, 1990.
20. C. H. Papadimitriou and M. Yannakakis. Shortest paths without a map. In *16th International Colloquium on Automata, Languages, and Programming, Lecture Notes in Computer Science vol. 372*, pages 610–620. Springer-Verlag, 1989.
21. H. Ramesh. On traversing layered graphs on-line. In *Proc. 4th Annual ACM-SIAM Symp. on Discrete Algorithms*, pages 412–421, 1993.

5

Distributed Paging

YAIR BARTAL

1 Introduction

Many modern information services know no national boundaries. The widespread use of the Internet and Internet-related applications such as the World Wide Web is growing fantastically on an annual basis.

This chapter deals with *distributed data management* problems. Such problems may arise as a memory management problem for a globally addressed shared memory in a multiprocessor system as well as in a distributed network of processors where data files are kept in different sites and may be accessed for information retrieval by dispersed users and applications. In this context, a file may be a conventional single file, a system database, fragments of a database, or any combination of these.

When a processor wishes to access a file it must send a request to a processor holding the file and the desired information is transmitted back. The communication cost incurred thereby is proportional to the distance between the corresponding processors.

Files can be replicated and allocated so as to reduce communication costs. Such transactions incur however a high communication cost proportional to the file size times the communication distance. In any case, consistency of the data must be preserved. The main issue in distributed data management is the design of a dynamic allocation of file copies in a network in order to achieve low communication costs.

Distributed file assignment problems have been extensively studied in management science, engineering, and computer systems. The 1981 survey paper by Dowdy and Foster [16], compares studies on different models, and cites close to a hundred references. The 1990 paper by Gavish and Sheng [19], gives a survey of recent research on dynamic file assignment problems, where the locations of file copies may be changed over time. All of these models rely heavily on prior knowledge regarding potential usage patterns of the system databases.

This chapter describes work concerning with the *competitive analysis* [25] of distributed data management problems. This approach makes no prior assumptions on the input sequence, and provides a worst-case guarantee on the performance of the algorithms.

The following are the main problems we survey:

- The *file migration* problem [14] allows only one copy of each file to be kept in the network.
- The *file allocation* problem [11] deals with the more general case where files may be replicated and deleted in response to a sequence of read and write requests.
- The *distributed paging* problem [11] further generalizes the model to deal with memory capacity limitations at the network processors.

The *file allocation* problem [11, 2] is the fundamental case where a *single* file resides among the network nodes. Copies of the file may be stored in the local storage of some subset of processors. Copies may be replicated and discarded over time as to optimize communication costs, but multiple copies must be kept consistent and at least one copy must be stored somewhere in the network at all times. An on-line file allocation algorithm must minimize communication costs, over arbitrary sequences of read and write requests issued at different locations over the network.

The file allocation problem can be viewed as the generalization of two other basic problems due to Black and Sleator [14]. The *file replication* problem where only read requests are issued, and the *file migration* problem where only one copy of the file may be kept in the network, which admits with write-only file allocation.

The file allocation solutions can be used only if every processor has enough memory to accommodate all files. If this is true, then every file can be handled separately. On the other hand, if this is not true, then the replication of a file into a processor's local memory may require the migration of another file copy onto some other processor.

The *distributed paging* problem is the simultaneous solution of many individual file allocation problems, subject to the constraints of the local memory capacities.

1.1 Network model

A *network* is modeled by an undirected weighted graph $G = (V, E)$, where processors are represented by vertices, and edge weights the length or cost of a bi-directional link between the two corresponding adjacent processors. Processors communicate only by exchanging messages. The weighted graph need not obey the triangle inequality, but a natural *metric space* can be defined where the points are processors and distance between two points $p, q \in V$ is equal to the length of the shortest path between the processors in the weighted graph, and denoted $d(p, q)$. We use the terms network, weighted graph, and metric space as called for by discussion, but they refer to the same underlying interconnection network.

Let $|V| = n$ be the number of processors in the network. Let Δ denote the diameter of the network, i.e., the maximal distance between any two pair of processors, assuming the minimal distance is 1.

2 File migration

2.1 Problem definition

The file migration problem consists of a single copy of a file, F, held at some processor in a network of processors. A sequence of file accesses are issued at different processors over time. Each access to the file costs the distance between the requesting processor and the processor holding the file copy. In addition to serving the access requests, an algorithm is allowed at any time to migrate the file from its current location to another processor at a cost equal to the distance between them times the file size, D.

2.2 Lower bound

Black and Sleator [14] used a result of Karlin et al. [21] to get a lower bound of 3 for deterministic data migration algorithms.

The following theorem is a modification that gives a lower bound on the competitive ratio of any file migration algorithm in any network topology against adaptive on-line adversaries.

Theorem 1. *Let G be any network over a set of at least two processors. The competitive ratio of any randomized on-line file migration algorithm for G against adaptive on-line adversaries is at least 3.*

Proof. Let two different processors in the network be p and q, and assume the distance between p and q is 1.

Assume the on-line algorithm holds a copy of the file at processor p. We define 3 different adaptive on-line adversaries as follows:

- The *p-adversary*: holds a single copy of the file at p.
- The *q-adversary*: holds a single copy of the file at q.
- The *jumping-adversary*: holds a single copy of the file at a processor not holding a copy of the file by the on-line algorithm (if such exists and otherwise the adversary's configuration remains unchanged) ;I.e., at processor q.

Now a write request is generated at q.

The cost for this request for both the q-adversary and the jumping-adversary is 0. The cost for the request for the p-adversary as well as the on-line algorithm is 1.

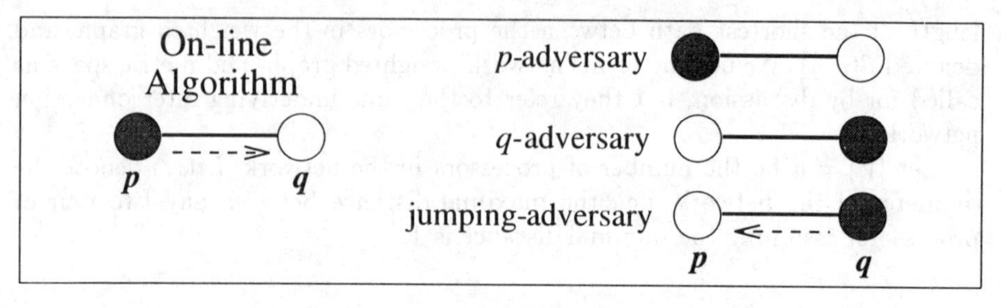

Fig. 1. Lower bound of 3 for file migration on 2 processors.

Consider next the cost incurred for configuration changes. The p-adversary and the q-adversary never change their configuration. If the on-line algorithm performs migration the jumping-adversary may change its configuration thereby incurring a cost of D.

We therefore conclude that over the entire sequence of events the on-line cost is equal to the cost of the 3 adversaries up to an additive term, implying the lower bound. ■

2.3 A simple randomized algorithm

Westbrook [26] gave a simple randomized algorithm that achieves the optimal competitive ratio of 3 in any network topology against adaptive on-line adversaries.

The Coin-Flipping Algorithm.
Given an access request issued at v move the file to v with probability $1/2D$.

Theorem 2. *The Coin-Flipping algorithm is 3-competitive against an adaptive on-line adversary.*

Proof. Define the potential function $\Phi = 3D \cdot d(a,b)$ where a and b are the adversary's and the on-line's file locations respectively. Clearly, Φ is nonnegative. When the adversary migrates the file the potential function changes by at most 3 times its cost.

Now consider an access request at v. The expected cost incurred by the algorithm is

$$\mathrm{E}(\mathrm{Cost}_{\mathrm{CF}}) = d(b,v) + \frac{1}{2D} \cdot D \cdot d(b,v)$$

The expected change in the potential is

$$\mathrm{E}(\Delta\Phi) = \frac{1}{2D} \cdot 3D \cdot (d(a,v) - d(a,b)).$$

Therefore
$\mathrm{E}(\mathrm{Cost}_{\mathrm{CF}}) + \mathrm{E}(\Delta\Phi) = \frac{3}{2}(d(a,v) + d(b,v) - d(a,b)) \leq 3 \cdot d(a,v) = 3 \cdot \mathrm{Cost}_{\mathrm{Adv}}$.
■

2.4 Deterministic algorithms

Black and Sleator [14] give optimal deterministic 3-competitive algorithms for the uniform network and for trees.

In contrast to the competitive ratio of 3 against adaptive on-line adversaries, Chrobak et. al. [15] show a network with a lower bound greater than 3 for deterministic algorithms, specifically $85/27 \approx 3.148$.

Awerbuch, Bartal and Fiat [2] gave the first deterministic file migration algorithm called Move-To-Min (MTM) with competitive ratio 7 for all network topologies. The best currently known deterministic file migration algorithm is due to Bartal, Charikar, and Indyk [10], called Move-To-Local-Min (MTLM) which has competitive ratio ~ 4.086.

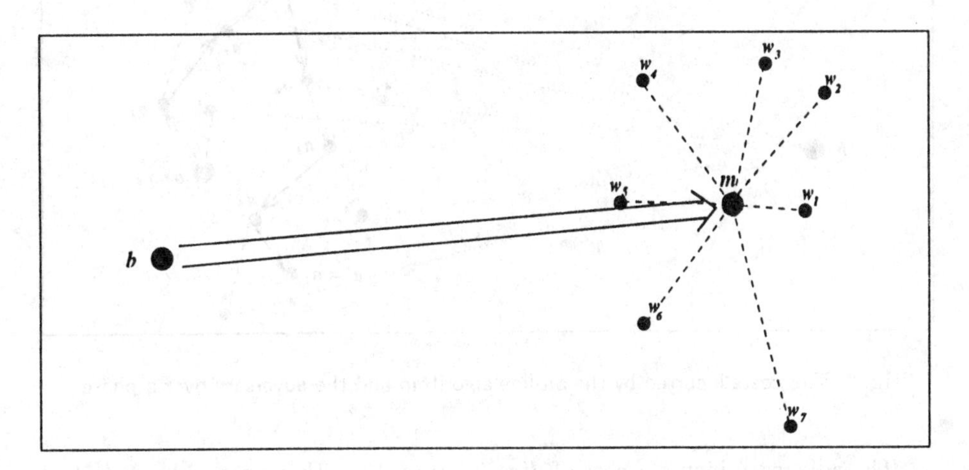

Fig. 2. Algorithm Move-To-Min (MTM). In the example D=7. m minimizes the sum of distances to the D access locations.

We present here the proof of the simpler Move-To-Min algorithm.

Algorithm **Move-To-Min** (MTM).

The algorithm divides the request sequence into phases. Each phase consists of D consecutive accesses to the file at processors $v_1, v_2, \ldots, v_D$. During a phase the algorithm doesn't move the copy of the file. At the end of a phase migrate the copy to a processor m in the network such that the sum of distances from m to the v_i's is minimized.

Theorem 3. *Algorithm Move-To-Min is 7-competitive for file migration on arbitrary network topologies.*

Proof. We analyze the performance of the algorithm in a phase. Let $a = a_0$ denote the position of the adversary copy at the beginning of a phase, and $a_i, 1 \leq i \leq D$ denote its position after the i'th request of the phase. Also let b denote the position of the on-line copy at the beginning of the phase.

During the phase MTM incurs a cost of $d(b, v_i)$ on the i'th request. At the end of the phase its cost for migrating the file is $D \cdot d(b, m)$. Thus its total cost for the phase is

$$\begin{aligned}\mathrm{Cost}_{\mathsf{MTM}} &= \sum_i d(b, v_i) + D \cdot d(b, m) \\ &\leq \sum_i d(a, v_i) + D \cdot d(b, a) + D \cdot d(b, m).\end{aligned}$$

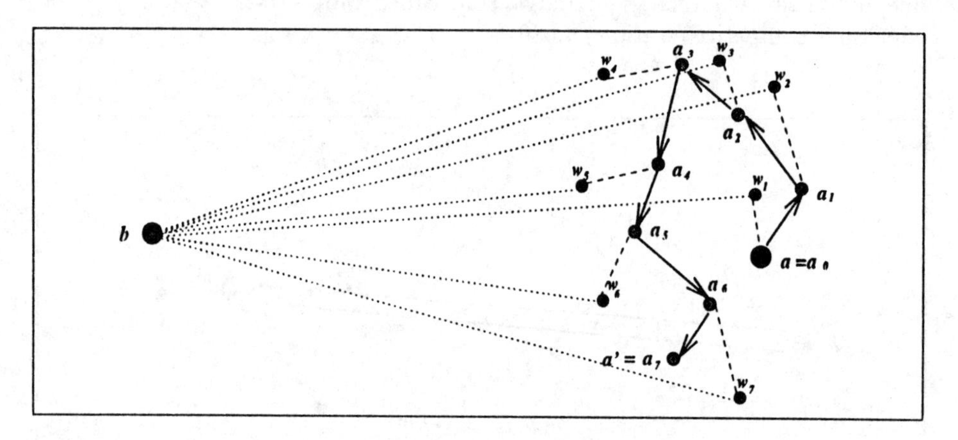

Fig. 3. The costs incurred by the on-line algorithm and the adversary over a phase.

The adversary incurs a cost of $d(a_{i-1}, v_i)$ on the i'th request, plus a cost of $D \cdot d(a_{i-1}, a_i)$ for its migration after the request. Thus the adversary cost during the entire phase is

$$\mathrm{Cost}_{\mathsf{Adv}} = D \cdot \sum_i d(a_{i-1}, a_i) + \sum_i d(a_{i-1}, v_i).$$

Using the triangle inequality we get that for any $0 \leq \ell \leq D$:

$$\begin{aligned}\mathrm{Cost}_{\mathsf{Adv}} &= D \cdot \sum_i d(a_{i-1}, a_i) + \sum_i d(a_{i-1}, v_i) \\ &\geq \sum_i \{d(a_\ell, a_{i-1}) + d(a_{i-1}, v_i)\} \\ &\geq \sum_i d(a_\ell, v_i).\end{aligned}$$

Thus we obtain

$$D \cdot d(a_\ell, m) \leq \sum_i d(a_\ell, v_i) + \sum_i d(m, v_i) \leq 2 \sum_i d(a_\ell, v_i) \leq 2\mathrm{Cost}_{\mathsf{Adv}}.$$

We use the potential function $\Phi = 2D \cdot d(a, b)$. We show that over a phase it increases by at most 7 times the adversary cost and decreases by at least the on-line cost. Let $a' = a_D$.

$$\begin{aligned}
\Delta\Phi &= 2D \cdot (d(a', m) - d(a, b)) \\
&\leq D(2d(a', m) + d(a, m) - d(b, m) - d(a, b)) \\
&= 2D \cdot d(a', m) + D \cdot d(a, m) \\
&+ \sum_i d(a, v_i) - \sum_i d(a, v_i) - D \cdot d(b, m) - D \cdot d(a, b) \\
&\leq 7\mathrm{Cost}_{\mathsf{Adv}} - \mathrm{Cost}_{\mathsf{MTM}}.
\end{aligned}$$

■

The Move-To-Local-Min (MTLM) algorithm of [10] also operates in phases. The length of a phase is $L = cD$. At the end of the phase, the page is migrated to the node x which minimizes the function $f(x) = \sum_{i=1}^{L} d(x, v_i) + \delta D \cdot d(b, x)$, where $c \approx 1.841$ and $\delta \approx 0.648$.

The first term in the minimizer function $f(x)$ ensures that the page is moved to a node in the network which is close to where the activity is taking place (reflected by the requests in the last phase). The second term reflects the cost of moving the page to the new node, weighted by parameter δ. This ensures that the cost of making this move is not too high. This additional term is similar to, and in fact inspired by the additional term in the minimizer function used by the Work Function Algorithm for the k-server problem and metrical task systems. The competitive ratio of Move-To-Local-Min is approximately 4.086.

2.5 Randomization against an oblivious adversary

Chrobak et al. [15] studied randomized page migration algorithms against an oblivious adversary. They show that the optimal randomized competitive ratio for a network consisting of two nodes connected by an edge is $2 + \frac{1}{2D}$. They then use the algorithm for an edge to obtain a $(2 + \frac{1}{2D})$-competitive algorithm for trees. Lund et al. [22] as similar result for uniform networks. All of these algorithms are work-function based, i.e., on the optimal costs of serving the requests and ending at different configurations.

Westbrook [26] gives randomized algorithms for arbitrary network topologies. The algorithm operates in phases. The length of a phase is drawn out of a fixed probability distribution. The file is not moved during a phase and at the end of a phase it is migrated to the location of the last request. The competitive ratio given by the best choice of a probability distribution approaches $1 + \phi \approx 2.618$ as D goes to infinity.

2.6 *k*-page migration

The *k-page migration* problem is a generalization of the standard file migration problem where there are k copies of a file residing in the network instead of just one. An access request initiated in one of the processors costs the distance to the nearest page copy. Each of the copies may also be migrated at the a cost of D (the size of the page) times the distance traveled. As before the problem is to minimize the total access and migration costs.

The k-page migration problem can be viewed as a special case of a more general framework of "*relaxed task systems*" which are problems that can be formulated as a relaxed version of some other problem where configuration changes are forced. In the case of k-page migration this would be the k-server problem.

Awerbuch, Azar, and Bartal [1] give a randomized transformation from k-server algorithms into k-page migration algorithms with competitive ratio $O(k)$ against an adaptive on-line adversary. Bartal, Charikar and Indyk [10] give a deterministic transformation yielding competitive ratio $O(k^2)$.

[10] also give a $(2k+1)$ lower bound on the competitive ratio for the problem in any network topology. A matching upper bound is given for the uniform network and a $(2k+1)(1+\frac{1}{D})$ competitive ratio for trees.

3 File allocation

3.1 Problem definition

The *file allocation* problem [11] assumes that data is organized in indivisible blocks such as files (or pages), consisting of D data units.

Let P denote the set of processors. Initially, a subset $Q \subseteq P$ of processors is each assigned a copy of the file. The algorithm receives a sequence of requests initiated by processors in P. Each request is either a *read* request or a *write* request. A read request at processor r is served by the closest processor p holding a copy of the file. The cost associated with this transmission is the distance between p and r. In response to a *write* request initiated at processor w, the algorithm must transmit an update to all currently held copies of the file – the subset $Q \subseteq P$. It pays a cost equal to the minimum Steiner tree spanning $Q \cup \{w\}$.

In between requests, the algorithm may re-arrange the copies of the database. A processor may delete the copy it is holding, unless it is the last copy in the network, at no cost. The file may also be *replicated* from a processor p, which holds a copy, to a subset $Q' \subset P$. The cost of replicating is equal to D times the minimum Steiner tree spanning $Q' \cup \{p\}$. D represents the ratio between the size of the entire file and the size of the minimal data unit being read or updated. A new current subset Q of processors holding copies of the file is determined as a result of delete and replicate steps. A combination of a replicate step from a processor p to a processor q, followed by a delete at p, is sometimes called a *migration* step. The subset Q is called the configuration of the algorithm.

3.2 Lower bounds

A lower bound of 3 in any network topology follows from the lower bounds for file migration described in the previous section by restricting the request sequence to write requests.

Bartal, Fiat and Rabani [11] give a lower bound of $\Omega(\log n)$ for a specific family of networks based on a lower bound for the on-line Steiner tree problem [20]. This lower bound holds even against an oblivious adversary.

3.3 Specific network topologies

[11] give an optimal 3-competitive, deterministic file allocation algorithm for uniform networks. This also gives an alternative algorithm for file migration when restricting to write requests only.

[11] also give a memoryless 3-competitive algorithm for continuous tree. [22] give a 3-competitive algorithm for trees based on work-functions.

In what follows we describe the optimal algorithm for uniform networks. Let P denote the set of processors in the network.

Algorithm **Count.**

Count is defined for each processor $p \in P$ separately. It maintains a counter c, and performs the following algorithm. We say that Count is *waiting*, if there is a single copy of the file and the processor holding the file is performing step 4 of the algorithm. Initially, set $c := 0$. If p holds a copy of the file, begin at step 4.

1. While $c < D$, if a *read* is initiated by p, or if a *write* is initiated by p, and Count is waiting, increase c by 1.
2. *Replicate* a copy of the file to p.
3. While $c > 0$, if a *write* is initiated by any other processor, decrease c by 1.
4. If p holds the last copy of the file, wait until it is replicated by some other processor.
5. *Delete* the copy held by p.
6. Repeat from step 1.

Theorem 4. *Algorithm* Count *is 3-competitive for uniform networks.*

Proof. Fix a processor p. One iteration of steps 1–6 at p is named a *phase.* Note that if Count is waiting then it is executing step 4 in the single processor holding a copy of the file, and it is executing step 1 in all the other processors. Count's cost is charged on individual processors as follows:

1. A processor initiating a read is charged the cost of the read.
2. If Count is waiting, a processor initiating a write is charged the cost of the write.
3. If Count is not waiting, and a write is initiated, the cost of 1 is charged at each processor holding a copy, except for the initiating processor. Note that the sum of costs charged here is exactly the cost of that write.
4. The cost D of replicating is charged at the processor receiving the copy.

The adversary's cost is charged the same, except that a replication is not charged. Rather, it registers a debit of D at the processor receiving the copy. That debit is paid (and a cost of D is charged) when the copy is deleted. Debits are initially set to 0 for processors *not* holding copies and to D for processors that initially do hold a copy. Note that the charging of the adversary's cost minus the sum of initial debits is a lower bound on its actual cost.

At the beginning Count is waiting after all but one copy are deleted, so that no cost is incurred. Now, during a phase of a processor p, Count's cost charged to p is at most $3D$. Steps 1 and 2 cause a charge of D each. Step 3 causes a charge of D. The total cost of Count is the sum of costs over all phases of all processors. There can be at most n partial phases (which are not over).

The adversary's cost during a full phase (note that the duration of a phase is determined by Count) is at least D. If the adversary ever deletes a copy from the processor during a phase, it is charged D. Otherwise, it either holds a copy at that processor when Count begins step 3 (and therefore not waiting), so it pays D during that step; or, it does not hold a copy at the end of step 1, and since it could not delete during that step, it must have been charged at least D for the requests of step 1. The reason is that during step 1 the processor initiated a total of D requests, counting read requests and write requests initiated while Count was waiting. ■

3.4 Arbitrary networks

Randomized algorithm [11] give a randomized algorithm for the file allocation problem on all networks, which is competitive against an adaptive on-line adversary.

This is done by a reduction to the on-line Steiner tree problem. Let G be an arbitrary network. Let A be a strictly (no additive term) c-competitive *Steiner tree* algorithm on G. We use A can to obtain a competitive randomized file allocation algorithm on G. Assume that the initial configuration consists of only one copy of the file.

Algorithm **Steiner Based** (SB).
Algorithm SB simulates a version of the Steiner tree algorithm A starting with p as the initial configuration. At all times, the set of processors in which SB keeps copies of the file is equal to the set of processors covered by A's Steiner tree.

Upon receiving a *read* request initiated at node r, the algorithm serves it, and then with probability $1/D$ feeds A with a new request at vertex r. In response A computes a new Steiner tree T' in place of its previous tree T. SB *replicates* new copies of the file at the processors corresponding to the vertices that A added to its tree.

Upon receiving a *write* request initiated at node w, the algorithm serves it, and with probability $1/\alpha D$ *deletes* all copies of the file, leaving only one copy at the processor closest to w, and then migrates the file to w, initializing a new version of A starting at vertex w as its initial configuration.

SB achieves best performance for $\alpha = \sqrt{3}$.

Theorem 5. *If A is a strictly c-competitive Steiner tree algorithm against adaptive on-line adversaries on a network G, then* SB *is a* $(2+\sqrt{3})c$*-competitive algorithm for the file allocation problem on G against adaptive on-line adversaries.*

In particular by using the greedy Steiner algorithm [20] we get an $O(\log n)$ competitive ratio.

Deterministic algorithm [2] give a deterministic $O(\log n)$-competitive file allocation algorithm for arbitrary networks.

The algorithm partitions the request sequence into phases. Each phase, except perhaps for the last one, contains exactly D successive *write* requests.

Within each phase, the algorithm deals only with *read* requests.

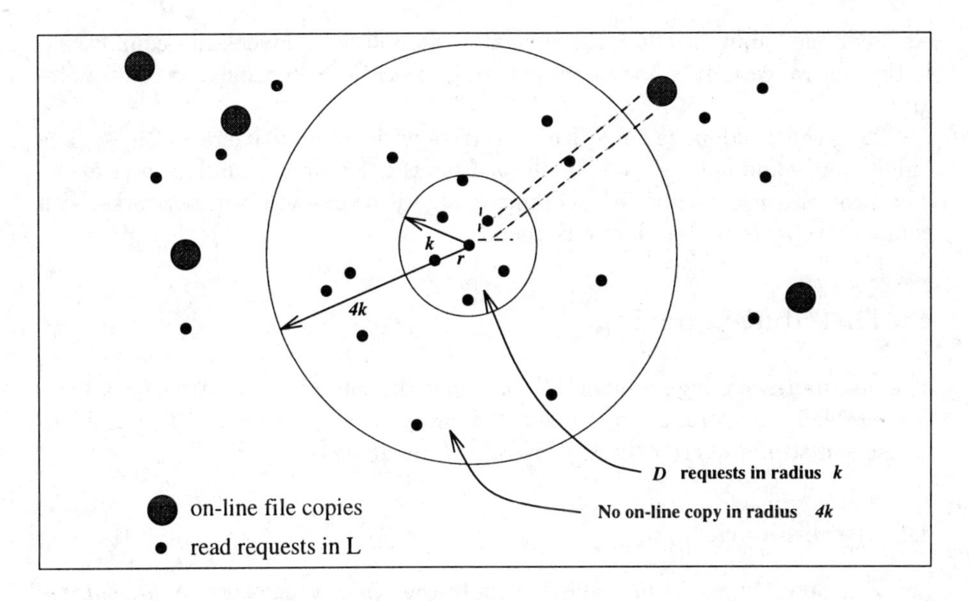

Fig. 4. The file allocation algorithm (FA): dealing with read requests. r is the last read request. ($\lambda = 4$.)

Dealing with read requests. At all times the algorithm maintains a partial list, L, of previous read requests in the current phase.

Upon receiving a new *read* request initiated by processor r, add r to L.

If L consists of more than D *read* requests, then consider the smallest k-neighborhood of r containing D requests.

If the algorithm does not hold a copy of the file at distance less than λk from r, *replicate* a copy of the file to r from the closest processor to r holding a copy, and remove the D requests from L.

Dealing with write requests. When a phase ends, let $w_1, w_2, \ldots w_D$ be the locations of the D *writes* initiated during the phase. Let m be a processor such that the sum of distances from m to the w_i's (i.e., $\sum_i d(m, w_i)$) is minimized. Now, *replicate* a copy of the file to m from the closest processor to it holding a copy, and then *delete* all copies of the file except the one at m. The read requests list, L, is initialized to be empty.

Theorem 6. *Algorithm* FA *is* $O(\min\{\log n, \log(\Delta)\})$*-competitive for file allocation on arbitrary network topologies.*

3.5 Randomization against an oblivious adversary

As mentioned above randomization against an oblivious adversaries cannot help in the general case. It is however interesting to study for specific network topologies.

[22] give a randomized algorithm for trees with competitive ratio $2 + \frac{1}{D}$. The only lower bound is $2 + \frac{1}{2D}$ which follows from the the lower bound for migration. The problem of obtaining a randomized algorithm for uniform networks with competitive ratio better than 3 is open.

4 Distributed paging

The *distributed paging* problem [11]. paging is the solution of multiple file allocation problems, constrained by the local memory of the processors. (The problem is also sometimes referred to as *constrained file allocation*).

4.1 Preliminaries

Let $\mathcal{F}$ denote the set of files and $\mathcal{P}$ denote the set of processors. A *file copy* is an ordered pair $< F, p >$ where $F \in \mathcal{F}$ denotes the file and $p \in \mathcal{P}$ denotes the processor.

Reads and writes are issued at some processor for a file. Requests are of the form *read*(p, F) and *write*(p, F), for $F \in \mathcal{F}$ and $p \in \mathcal{P}$.

We assume that all files are of the same size. Let $m = \sum_p k_p$, the total number of files that can be stored in the network, and $k = \max_p k_p$, the maximal number of files that can be stored in any one processor.

4.2 Deterministic algorithms

[11] give a lower bound of $2m - 1$ on the competitive ratio of any deterministic algorithm for the distributed paging problem.

They also describe the Distributed-Flush-When-Full algorithm (DFWF) which is $3m$ for uniform networks.

The algorithm uses the following terminology. We say a processor p is *free* if it holds less than k_p different files. A copy of a file is called *single* if there are no other copies of that file currently in the network.

The algorithm works in phases. Copies of files can be either marked or unmarked. At the beginning of a phase, all counters are zero and all copies are unmarked. Throughout, an unmarked copy is single, a marked copy may be not single.

Algorithm **Distributed-Flush-When-Full** (DFWF).

The algorithm is defined for each processor p separately. Every processor maintains a counter c_F for every file F. Initially, or as a result of a *restart* operation, all counters are set to zero and all markings are erased. Arbitrarily, copies of files are deleted until there is exactly one copy of every file somewhere in the network.

Every processor p follows the following procedure simultaneously for all files F:

1. While $c_F < D$, if a *read*(p, F) request is initiated at p, or if a *write*(p, F) request is initiated at p and F is unmarked, increase c_F by 1, if p does not contain a copy of F.
2. (a) If p is free, *replicate* F to p and mark it. If F was unmarked, *delete* the unmarked copy.
   (b) Otherwise, if all file copies in p are marked then restart.
   (c) Otherwise, choose S to be an arbitrary unmarked copy in p.
      i. If F is unmarked, switch between S and F, and mark F in p.
      ii. Otherwise, if some free processor q is available, *dump* S to q, and *replicate* a copy of F to p, mark this copy.
      iii. Otherwise, restart.
3. While $c_F > 0$, if a *write*(p, F) request is initiated by any other processor, decrease c_F by 1.
4. Restart.

Theorem 7. *Algorithm DFWF is* $3m$*-competitive for distributed paging on uniform networks.*

4.3 Randomization against an oblivious adversary

Awerbuch, Bartal and Fiat [3] study randomized algorithms against an oblivious adversary for the distributed paging problem.

They prove the following lower bound:

Theorem 8. *Let the* $f > k$ *be the number of different files in the network. Let* $\tilde{S}(r)$ *be the maximal* r*-server lower bound achievable for any metric space over* $r+1$ *points. The competitive ratio of any randomized distributed paging algorithm for* $\mathcal{N}$ *against oblivious adversaries is* $\Omega(\max\{\tilde{S}(\min\{m+1-f, n-1\}), \log k\})$.

In particular it follows from a result of [17] that $S(\mathcal{U}, m+1-f) = \Omega(\log(m-f))$. Hence,

Corollary 9. *The competitive ratio of any randomized distributed paging algorithm is* $\Omega(\max\{\log(m-f), \log k\})$ *for uniform networks.*

[3] also give an $O(\max\{\log(m-f), \log k\})$ competitive algorithm for uniform networks called HEAT & DUMP.

The algorithm is quite complicated and therefore we concentrate on the read-only distributed paging case and assume the file size $D = 1$. Thus the problem considered in the rest of this section is the basic problem where given a request for a file in some processor the file must be replicated into that processor. The algorithm must decide which other file to drop off that processor's local memory and where to migrate ("dump") a copy of that file if the processor holds the last copy of the file.

The HEAT & DUMP algorithm Algorithm HEAT & DUMP is modularly composed of two conflicting strategies.

The first is some *uniprocessor paging strategy* run within each processor's local memory. The paging strategy must deal with the memory constraints of the specific processor and decide which files should be erased to make place for other more popular files.

The paging strategy though may sometimes decide upon erasing the last single copy of some file. We then invoke the *global memory management strategy* which deals with the management of the entire aggregate memory in the network. The global memory management strategy is responsible of making room for keeping a copy of the file that must be removed according to the uniprocessor paging strategy. For that purpose it decides upon erasing a copy of some file, currently having multiple copies.

It follows that the global memory management strategy, may interfere with the local paging algorithms by forcing the deletion of some file required by the paging strategy. Therefore the actual files configuration in every processor somewhat differs the configuration it would have maintained if it was running the uniprocessor paging strategy without interference.

In what follows, we describe the HEAT & DUMP algorithm.

Following the completion of the t'th event in the input sequence, we say that the time is t. At all such times, associated with every processor p is a local paging algorithm $\mathcal{L}_p$.

The *virtual configuration* associated with p is the configuration of the local paging algorithm associated with the processor. The list of files associated with the local processor, had it been a uniprocessor under control of the local paging algorithm, is denoted by $\mathcal{V}(p,t)$. We call $\mathcal{V}(p,t)$ the "virtual configuration" with the understanding that all other configuration data required by the local paging algorithm is also stored.

In reality, the "real" list of files stored at the processor, $\mathcal{R}(p,t)$ contains some other files, not in $\mathcal{V}(p,t)$. These files were "dumped" into the processor p by the global strategy. The *real configuration* may also contain some empty file *slots*, in place of missing "virtual" files.

A *file copy* is an ordered pair $< F, p >$ where $F \in \mathcal{F}$ denotes the file and $p \in \mathcal{P}$ denotes the processor. We associate labels, *hot*, *cold*, and *dumped*, with file copies, a file copy must have exactly one of these labels. The sets of hot,

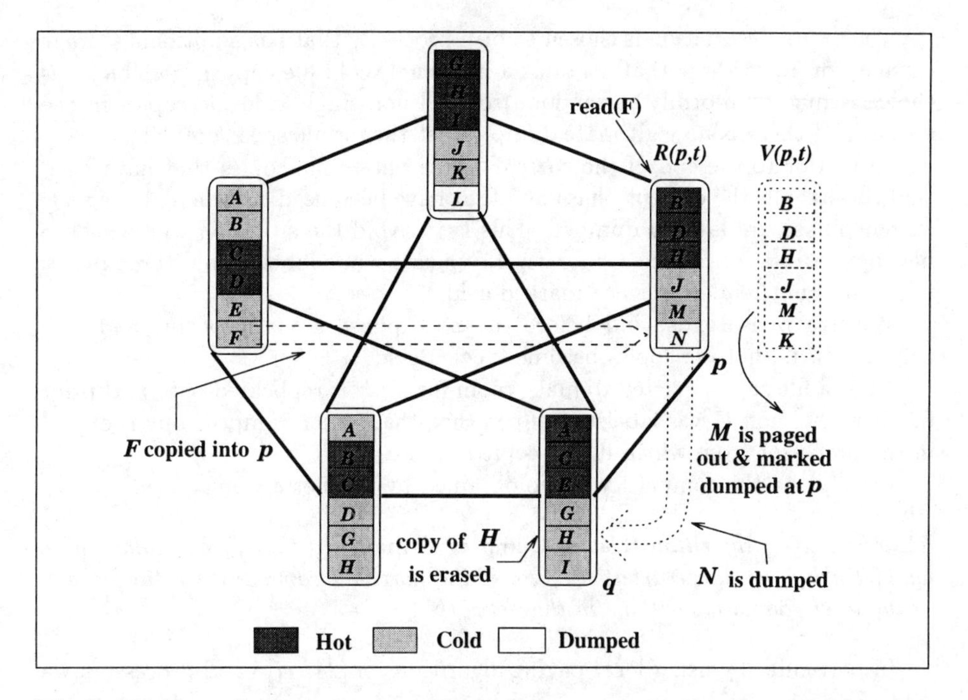

Fig. 5. The HEAT & DUMP algorithm. File copies are marked hot, cold and dumped. F is requested at processor p at time t. M is paged out from the virtual configuration $\mathcal{V}(p,t)$, and is marked dumped at p. p now contains two file copies marked dumped, so an arbitrary dumped file copy (N) is dumped out of p to q overwriting H, a non-single file copy marked cold chosen at random amongst all such candidates in the network.

cold, and dumped files in processor p at time t are denoted $\mathcal{H}(p,t)$, $\mathcal{C}(p,t)$, and $\mathcal{D}(p,t)$ respectively. Note that $\mathcal{R}(p,t) = \mathcal{H}(p,t) \cup \mathcal{C}(p,t) \cup \mathcal{D}(p,t)$.

A file is called *single* if only one copy of the file resides in the network. If more than one copy of the file resides in the network, then the file is called *non-single*.

We may also say that a file is hot, if there exists a hot copy of the file in the network. We say that a file is cold or dumped if all copies of the file are cold or dumped, respectively.

Initially, all files are single and are labeled dumped.

When a request for file F appears at some processor p, the simulated local paging algorithm $\mathcal{L}_p$ is fed with a request for F. This results in erasing some file from the virtual configuration $\mathcal{V}(p,t)$. Files that are ejected from the virtual configuration in a processor are either erased from the processor (if non-single) or become labeled dumped. This assures that p either contains some empty slots or that it contains some files marked dumped.

When a file is ejected from the processor to make room for another file, we say that the file has been *dumped*. The general strategy is to avoid dumps, if possible, and if impossible, to dump out of the processor a dumped file.

The dump destination is chosen to be a processor that has additional storage capacity or a processor that contains a non-single cold file copy, if possible, this choice is made uniformly at random from all non-single cold file copies in the network. If there is no legitimate dump target then a phase is completed.

Just prior to the end of the current phase phase, file copies that have been cold throughout the current phase and that have been used elsewhere during the current phase are labeled dumped. This is to avoid the situation where such a file copy resides in a processor for several phases while labeled cold throughout, unless it is not requested since marked cold.

When a phase ends, and before the next phase actually begins and new activity starts, hot file copies become labeled cold.

After a file slot is evicted to make room for F, F is replicated onto p. If prior to the replication F was labeled dumped then the former dumped copy is erased in the processor from which it was copied.

Let $c(p)$ be the competitive ratio obtained by $\mathcal{L}_p$. Let $c = \max_p c(p)$.

Theorem 10. *Algorithm* Heat & Dump *is* $O(\max\{\log(m-f), c\})$*-competitive against oblivious adversaries, where c is the maximal competitive ratio for any of the local paging algorithms in the network.*

In particular by using local paging algorithms of [24, 17] in all processors we get:

Corollary 11. *Algorithm* Heat & Dump *is* $O(\max\{\log(m-f), \log k\})$*-competitive against oblivious adversaries.*

4.4 Algorithms for arbitrary networks

Awerbuch, Bartal and Fiat [4] deal with the distributed paging problem on arbitrary networks. The solution they present holds against some weak adversary as will be explained in the sequel. However they develop a general framework for reducing the problem to the read-only case. This is done by letting the read-only distributed paging algorithm define a set of active and in-active processors for each of the files. Then a modification of the file allocation strategies that can deal with a dynamically changing network are are with these sets of active processors to deal with the write requests.

The goal of giving an algorithm for distributed paging seems to be very difficult. Moreover the best competitive ratio possible for deterministic algorithms is $\Omega(m)$.

This leads [4] to defining a weaker but reasonable adversary, against which they can obtain an algorithm for read-only distributed paging with polylogarithmic competitive ratio.

The model they use was suggested in [25, 21, 23, 27] for uniprocessor paging, where the adversary has a smaller cache size than the on-line algorithm. This can also be viewed as a solution for the problem where the cache size may vary (cf. [27]), but the algorithm should pay for enlarging its cache size.

More precisely, we measure the performance of an on-line distributed paging algorithm by two measures: one is the ratio between its cost for serving the request sequence (including it's communication cost in the distributed setting.) and that of the adversary, and the second is the ratio between its cache size and that of the adversary's in every processor.

Definition 12. Given a distributed paging algorithm, DP, and any adversary, Adv, let k_p denote the on-line memory capacity of processor p, and let h_p denote the adversary's memory capacity for processor p.

Let c and s be constants at least 1. If for any adversary, Adv such that for all p, $k_p \geq s \cdot h_p$, the algorithm is c-competitive against Adv, then we say the algorithm is (c, s)-competitive.

In this terminology, [25, 21] give deterministic (2,2)-competitive uniprocessor paging algorithms.

Thus, in the uniprocessor case the assumption of some small advantage in memory for the on-line algorithm reduces the competitive ratio dramatically. The situation is similar for the distributed paging problem.

[4] give a read-only distributed paging algorithm called HIERARCHICAL PAGING.

The major idea is to use uni-processor paging on "areas of activity". These areas are defined using the sparse partitions of [8]. After sufficiently many accesses to a file have been issued by processors within such an "area of activity", a copy of the file is fetched into this area. For every such "area of activity" we simulate a uni-processor paging process that determines what files are to be ejected/erased to make space for new files.

Theorem 13. *The* HIERARCHICAL PAGING *algorithm is* (polylog(m,Δ),polylog(m,Δ))*-competitive.*

4.5 Constrained file migration

It is only natural to consider the migration version of the distributed paging problem where only one copy of each file is allowed to be kept in the network. (This problem is also sometimes referred to as *constrained file migration*.)

This has been first consider by Albers and Koga [5] who give $\Theta(k)$ and $\Theta(\log k)$ deterministic and randomized, upper and lower bounds for the problem on uniform networks, respectively.

Bartal [9] gives a deterministic lower bound of $\Omega(m)$, a randomized lower bound of $\Omega(\log m)$ for the problem in some network topology, and an $O(\log m \log^2 n)$ randomized upper bound for arbitrary network topologies.

The upper bound is based on a general method of probabilistically approximating metric spaces by metric spaces from a set of "nice" metric spaces called HST's.

Definition 14. An *r-hierarchically well-separated tree* (r-HST) is defined as a rooted weighted tree with the following properties:

- The edge weight from any node to each of its children is the same.
- The edge weights along any path from the root to a leaf are decreasing by a factor of at least r.

The result of [9] can be summarized as follows:

Theorem 15. *Say there is a c-competitive algorithm on r-HST's. There exists a randomized algorithm with competitive ratio* $O(rc \log n \log_r n)$ *for arbitrary graphs against an oblivious adversary.*

The result for constrained file migration on arbitrary networks follows from the following theorem

Theorem 16. *Given a network represented by a* r-HST, T, *for some* $r \geq 2$ *then there exists a randomized* $O(\log m)$ *competitive algorithm for the constrained file migration problem on* G.

Proof Sketch. We can concentrate on the case $D = 1$ (as follows from the general reductions in [1, 10]). Apply a copy of the marking paging algorithm [17] for each subtree as follows. Given a request for a file invoke the algorithm at the topmost level containing the request, where the file is missing, and then repeat until the file arrives at the requesting processor. Now consider a marking algorithm for a particular subtree. Its cache size is at most m. The virtual configuration of the algorithm contains some files that may actually be outside its subtree. Since a file in the virtual configuration has been requested inside the subtree in either the current or the previous marking phases, this may happen only if the file has been later requested outside the subtree. Thus these two requests have an inherent cost proportional to the cost of migrating the file into the subtree. The claim follows since the expected cost on requests for files not in the virtual configuration of the algorithm is bounded by $O(\log m)$ times the optimal cost. ■

Corollary 17. *Given a network* G, *there exists a randomized* $O(\log m \log^2 n)$ *competitive algorithm for the constrained file migration problem on* G.

5 Discussion

The fundamental distributed paging problem is still open for arbitrary network topologies. This seems to be an excellent problem for further study.

Obvious generalizations of the problems discussed herein are to deal with network congestion and delay issues rather than simple communication costs. Distributed algorithms for data management problems are discussed in [11, 13, 2, 4, 9]. These algorithms make decisions at the network processors based on the local information gathered via communication. However these results assume serialization of the requests, thus issues of concurrency must be addressed. The current file allocation models also assume that all file copies must be updated at every moment. This is a common practical demand. However perhaps there are more efficient solutions if file copies are allowed to be maintained in some clever partially inconsistent manner.

Modern distributed systems allow not only migrating data along the network, but also the migration of processes. The *process allocation* problem can be defined as the dynamic re-allocation of processes as to minimize the sum of communication costs between them (see [18]).

Perhaps the most promising avenue of research is to attempt to implement some or all of the ideas presented here in real systems. Here, there are additional practical considerations and constraints that have to be addressed. It seems that in practice the issues of distributed control over centralized control are less important and that it is probably our centralized algorithms that are of greater interest. A key practical consideration is how to co-exist with existing networks and protocols.

In many cases, competitive analysis and its variants yield algorithms that seem to be natural for the problem and may prove to have a real advantage in practice over currently used heuristics, and in some cases may even provide the only complete solution for the problem.

For example, the concepts presented herein seem to be highly suitable for implementation in large network caches.

Acknowledgements

This work was partially supported by the Rothschild Postdoctoral fellowship and by the National Science Foundation operating grants CCR-9304722 and NCR-9416101.

References

1. B. Awerbuch, Y. Azar, and Y. Bartal. On-line Generalized Steiner Problem. In *Proc. of the 7th Ann. ACM-SIAM Symp. on Discrete Algorithms*, pages 68–74, January 1996.
2. B. Awerbuch, Y. Bartal, and A. Fiat. Competitive Distributed File Allocation. In *Proc. 25th ACM Symp. on Theory of Computing*, pages 164–173, May 1993.
3. B. Awerbuch, Y. Bartal, and A. Fiat. Heat & Dump: Randomized competitive distributed paging. In *Proc. 34rd IEEE Symp. on Foundations of Computer Science.* IEEE, pages 22–31,November 1993.
4. B. Awerbuch, Y. Bartal, and A. Fiat. Distributed Paging for General Networks. In *Proc. of the 7th Ann. ACM-SIAM Symp. on Discrete Algorithms*, pages 574–583, January 1996.
5. S. Albers and H. Koga. New On-line Algorithms for the Page Replication Problem. In *Proc. of the 4th Scandinavian Workshop on Algorithmic Theory*, Aarhus, Denmark, July 1994.
6. S. Albers and H. Koga. Page Migration with Limited Local Memory Capacity. In *Proc. of the 4th Workshop on Algorithms and Data Structures*, , August 1995.

7. B. Awerbuch, S. Kutten, and D. Peleg. Competitive Distributed Job Scheduling. In *Proc. of the 24th Ann. ACM Symp. on Theory of Computing*, pages 571–580, May 1992.
8. B. Awerbuch and D. Peleg. Sparse Partitions. In *Proc. of the 31st Ann. Symp. on Foundations of Computer science*, pages 503-513, October 1990.
9. Y. Bartal. Probabilistic Approximation of Metric Spaces and its Algorithmic Applications. In *Proc. of the 37rd Ann. IEEE Symp. on Foundations of Computer Science*, pages 184-193, October 1996.
10. Y. Bartal, M. Charikar and P. Indyk. On Page Migration and Other Relaxed Task Systems. In *Proc. of the 8th Ann. ACM-SIAM Symp. on Discrete Algorithms*, pages 43-52, January 1997.
11. Y. Bartal, A. Fiat, and Y. Rabani. Competitive algorithms for distributed data management. In *Proc. 24th ACM Symp. on Theory of Computing*, pages 39–50, May 1992.
12. A. Borodin, N. Linial, and M. Saks. An Optimal On-Line Algorithm for Metrical Task Systems. In *Proc. of the 19th Ann. ACM Symp on Theory of Computing*, pages 373–382, May 1987.
13. Y. Bartal and A. Rosén. The Distributed k-Server Problem — A Competitive Distributed Translator for k-Server Algorithms. In *Proc. of the 33rd Ann. IEEE Symp. on Foundations of Computer Science*, pages 344-353, October 1992.
14. D.L. Black and D.D. Sleator. Competitive Algorithms for Replication and Migration Problems. Technical Report CMU-CS-89-201, Department of Computer Science, Carnegie-Mellon University, 1989.
15. M. Chrobak, L. Larmore, N. Reingold, and J. Westbrook. Optimal Multiprocessor Migration Algorithms Using Work Functions. In *Proc. of the 4th International Symp. on Algorithms and Computation*. Also *Lecture Notes in Computer Science*, vol. 762, pages 406-415, Hong Kong, 1993, Springer-Verlag.
16. D. Dowdy and D. Foster. Comparative Models of The File Assignment Problem. Computing Surveys, 14(2), June 1982.
17. A. Fiat, R.M. Karp, M. Luby, L.A. McGeoch, D.d. Sleator , and N.E. Young. Competitive Paging Algorithms. Technical Report, Carnegie Mellon University, 1988.
18. A. Fiat, Y. Mansour, A. Rosén, and O. Waarts. Competitive Access Time via Dynamic Storage Rearrangement. In *Proc. of the 36th Ann. IEEE Symp. on Foundations of Computer Science*, pages 392–403, October 1995.
19. B. Gavish and O.R.L. Sheng. Dynamic File Migration in Distributed Computer Systems. In *Communications of the ACM*, 33(2):177-189, 1990.
20. M. Imaze and B.M. Waxman. Dynamic Steiner Tree Problem. In *SIAM Journal on Discrete Mathematics*, 4(3):369-384, August 1991.
21. A.R. Karlin, M.S. Manasse, L. Rudolph, and D.D. Sleator. Competitive Snoopy Caching. In *Algorithmica*, 3(1):79-119, 1988.
22. C. Lund, N. Reingold, J. Westbrook, and D. Yan. On-Line Distributed Data Management. In *Proc. of European Symp. on Algorithms*, 1994.
23. M.S. Manasse, L.A. McGeoch, and D.D. Sleator. Competitive Algorithms for On-Line Problems. In *Proc. of the 20th Ann. ACM Symp. on Theory of Computing*, pages 322–333, May 1988.
24. L.A. McGeoch and D.D. Sleator. A Strongly Competitive Randomized Paging Algorithm. In *Algorithmica*, 1991.
25. D. Sleator and R.E. Tarjan. Amortized Efficiency of List Update and Paging Rules. *Communications of ACM* , 28(2):202–208, 1985.

26. J. Westbrook. Randomized Algorithms for Multiprocess or Page Migration. In *Proc. of DIMACS Workshop on On-Line Algorithms.* American Mathematical Society, February, 1991.
27. N.E. Young. Competitive Paging as Cache Size Varies. In *Proc. 2nd ACM-SIAM Symp. on Discrete Algorithms*, January 1991.

6

Competitive Analysis of Distributed Algorithms

JAMES ASPNES

1 Introduction.

Most applications of competitive analysis have involved on-line problems where a candidate *on-line* algorithm must compete on some input sequence against an optimal *off-line* algorithm that can in effect predict future inputs. Efforts to apply competitive analysis to fault-tolerant distributed algorithms require accounting for not only this *input nondeterminism* but also *system nondeterminism* that arises in distributed systems prone to asynchrony and failures. This chapter surveys recent efforts to adapt competitive analysis to distributed systems, and suggests how these adaptations might in turn be useful in analyzing a wider variety of systems. These include tools for building competitive algorithms by composition, and for obtaining more meaningful competitive ratios by limiting the knowledge of the off-line algorithm.

Like on-line algorithms, distributed algorithms must deal with limited information and unpredictable user and system behavior. Unlike on-line algorithms, in many distributed algorithms the primary source of difficulty is the possibility that components of the underlying system may fail or behave badly. In a distributed system, processes may crash, run at wildly varying speeds, or execute erroneous code; messages may be lost, garbled, or badly delayed. As a consequence, the worst-case performance of many algorithms can be very bad, and may have little correspondence to performance in more typical cases.

It is not surprising that the technique of **competitive analysis** [17] should be useful for taming the excesses of worst-case analysis of distributed algorithms. Of course, new opportunities and complications arise in trying to apply competitive analysis directly to fault-tolerant distributed systems. Distributed systems have a natural split between the inputs coming from the users above and the environment provided by the system below— by exploiting this split it is possible, among other things, to build competitive algorithms by composition. On the other hand, because a distributed system consists of many individual components with limited information, one must be careful in how one defines the powers of the "off-line" algorithm so that the boundaries between these separate components do not become blurred. Work in this area has yielded several useful techniques for carefully controlling how much information the off-line algorithm is allowed to use.

Section 2 discusses distributed systems in general. Section 3 describes competitive analysis in its traditional form. Some variants on traditional competitive measures that are useful in models in which the nondeterminism naturally splits into two categories are described in Section 4. For these *semicompetitive* performance measures, composition of algorithms is possible, as described in Section 5. Examples of applications of these techniques to distributed problems are given in Sections 6 and 7. Finally, Section 8 discusses directions in which this area might profitably be extended.

A caveat: this chapter is concerned primarily with examples in the distributed algorithms literature of applying competitive analysis directly to nondeterminism in the underlying system. No attempt is made to cover the vast body of excellent work on on-line problems, such as distributed paging, load-balancing, routing, mobile user tracking, etc., that arise in networks and other distributed systems.

2 Distributed systems.

There are many forms of distributed systems, and a wide variety of theoretical models used to study them. However, there are two properties that show up in most distributed system models, which distinguish them from uniprocessor or parallel systems. The first and most important property is that a distributed system consists of more than one process, which may represent a real physical CPU, or might correspond to an abstract entity like a process or thread in a timesharing system. The second property is that the multiple processes of a distributed system are poorly coordinated— each runs its own program; the communication channels between them may be slow, expensive, and unreliable; and individual processes may crash, become faulty, or run at varying speeds. It is this poor coordination that distinguishes distributed systems from parallel systems, in which processes typically execute the same program in very close synchrony with each other using a powerful and reliable communication system. The line between parallel and distributed systems is not a sharp one, but a reasonable rule of thumb is that parallel systems are predictable; the same program with the same input running on the same parallel machine should give essentially the same results every time it is run. In contrast, distributed systems are riddled with nondeterminism— distributed algorithms are always at the mercy of an unreliable and sometimes hostile infrastructure, and must strive for robustness and consistency despite it.

Because of this inherent nondeterminism and the difficulty of communication, an individual process in a distributed system is likely to have only an incomplete and often out-of-date picture of the state of the system as a whole. Like an on-line algorithm, it must still make decisions, which may turn out to be suboptimal later, based on this limited information. This property of distributed algorithms gives an on-line flavor to most problems that arise in distributed computing, and suggests that techniques such as competitive analysis that have been useful for dealing with the unpredictable request sequences of on-line problems should be useful for dealing with the unpredictable system behavior of distributed problems. However, it is not immediately obvious how on-line techniques can best be adapted to distributed systems.

The complication is that for many distributed algorithms we can distinguish between two different sources of nondeterminism. The first source is the user-supplied **input** to the algorithm: those distributed algorithms that carry out inherently on-line tasks such as scheduling or load balancing must suffer the whims of unpredictable users just as much as any on-line algorithm running on a single processor. However, these distributed algorithms must also contend with system nondeterminism, often summarized as a **schedule** that specifies when processes can take steps, what order messages are received in, and so forth. This schedule is just as unpredictable (and is often assumed to be just as adversarial) as the input.

3 Competitive analysis.

Competitive analysis [17] compares the performance of a general-purpose algorithm given particular inputs against specialized algorithms optimized for each of those inputs. From an abstract perspective, one considers a class of algorithms $\mathcal{A}$, a class of inputs $\mathcal{I}$, and a **cost function** $\text{cost}(A, I)$ that assigns a non-negative cost to each algorithm A and input I. A particular algorithm A is **k-competitive** if there exists a constant c such that for all inputs I in $\mathcal{I}$ and algorithms A^* in $\mathcal{A}$,

$$\text{cost}(A, I) \leq k \cdot \text{cost}(A^*, I) + c.$$

If $c = 0$, the algorithm is said to be **strictly k-competitive**. The minimum value of k for which an algorithm is k-competitive is the algorithm's **competitive ratio**. If $c = 0$, this quantity is just the maximum over all I of $\text{cost}(A, I) / \min_{A^* \in \mathcal{A}} \text{cost}(A^*, I)$.

The competitive ratio of an algorithm measures its performance on a particular input relative to all other algorithms. Often it is convenient to speak of the optimal algorithm for a particular input. When analyzing on-line algorithms, there is no harm in assuming that the best algorithm is an **off-line algorithm**— one that can predict the input— as the ability to choose an optimal algorithm based on the input is equivalent to using single algorithm with such predictive powers. In contrast, the algorithm being evaluated can be described as the **on-line algorithm** because it does not have such powers.

For distributed algorithms, the assumption that an algorithm can predict the input can have surprising consequences, and it is necessary in some models to limit the class of algorithms $\mathcal{A}$ to exclude full clairvoyance. With such a limitation, it is less misleading to use the neutral term **champion** to refer to the optimal algorithm for a particular input, since this term does not suggest any special knowledge that the champion might have. Similarly, the term **candidate** provides a neutral way to refer to the algorithm that is being evaluated.[1]

3.1 Payoffs instead of costs.

One can also define the competitiveness of an algorithm whose performance is measured in terms of a non-negative payoff function payoff(A, I). The idea here is that for some problems it is more natural to think in terms of how much can be accomplished given a fixed set of resources than how many resources are needed to accomplish a fixed set of tasks. In terms of payoffs, an algorithm is k-competitive if there exists a constant c such that for all I in $\mathcal{I}$ and A^* in A,

$$\text{payoff}(A, I) + c \geq \frac{1}{k}\,\text{payoff}(A^*, I).$$

Note that the competitive ratio is inverted for consistency with the cost version; smaller competitive ratios imply better performance with either measure.

4 Semicompetitive analysis

For distributed algorithms, it often makes sense to assume that the cost function depends on three parameters: the algorithm chosen, an input (which may or may not be common to both the candidate and champion algorithm) and an **environment** (which is always assumed to be common to both the candidate and champion algorithm). A problem is then defined in terms of a set of algorithms $\mathcal{A}$, a set of inputs $\mathcal{I}$, and a set of environments $\mathcal{E}$, together with a cost function cost(A, I, E) that assigns a non-negative cost to each algorithm A, input I, and environment E. From an abstract perspective it is not necessary to worry about how the nondeterminism in a system is split into the input and the environment, though in practice it may strongly affect how a problem looks.

One effect of such split nondeterminism is that it allows what we will call **semicompetitive analysis**, in which a candidate algorithm is measured against an optimal champion running with the same environment but not necessarily the same input. The goal is still to be able to measure the performance of a particular general-purpose candidate algorithm relative to the best specialized algorithms in some class. However, there are now several possible measures depending on how the inputs are chosen for the candidate and champion algorithms.

[1] These terms were suggested by [2].

Common Inputs. If one assumes that the input and environment are the same for both candidate and champion, one has the traditional definition of k-competitiveness. An algorithm A is **traditionally k-competitive** if there exists a constant c such that for all A^* in $\mathcal{A}$, I in $\mathcal{I}$, and E in $\mathcal{E}$,

$$\text{cost}(A, I, E) \leq k \cdot \text{cost}(A^*, I, E) + c.$$

Worst-Case vs. Best-Case Inputs. A more useful measure arises if one assumes that the inputs are not shared between the candidate and the champion. Let us define an algorithm A in a split-nondeterminism framework as **k-competitive** if there exists a constant c such that for all A^* in $\mathcal{A}$, I and I^* in $\mathcal{I}$, and E in $\mathcal{E}$,

$$\text{cost}(A, I, E) \leq k \cdot \text{cost}(A^*, I^*, E) + c.$$

In effect, this definition assumes that the candidate and champion share the same environment, but that the candidate faces a worst-case input and the champion a best-case input. The assumption that the candidate and champion do not have the same inputs— the only difference between this definition and traditional k-competitiveness— has far-reaching consequences. In particular it allows the modular construction of competitive algorithms, as described below in Section 5.

The definition of k-competitiveness given above is based on the "throughput model" of [5]. The use of the term "k-competitiveness" for this measure is justified by several practical considerations. First, the semicompetitive definition is stronger than the traditional definition. Not only is a k-competitive algorithm a traditionally k-competitive algorithm (immediate from inspection of the quantifiers), but if one also assumes that the set of inputs consists of a single point or that the input has no effect on the cost of an algorithm, the new definition reduces to the traditional definition. Second, in many respects the new measure captures the intuitive notion of competitiveness at least as well as the traditional measure in those situations where one can reasonably assume that only part of the nondeterminism affecting an algorithm is likely to be shared between it and an optimal algorithm (for example, when considering a subroutine whose input, in the form of procedure calls, is supplied by a higher-level candidate algorithm). Finally, there is little chance of confusion between the new and old definitions, as the new definition applies only in the case of split nondeterminism, and the presence of such a split will usually be obvious from context.

Worst-Case vs. Worst-Case Inputs. Yet another useful measure is obtained if it is assumed that both the candidate and champion algorithms face worst-case inputs. Define an algorithm A to be **k-optimal** if there exists a constant c such that for all A^* in $\mathcal{A}$, I in $\mathcal{I}$, and E in $\mathcal{E}$, there exists an input I^* in $\mathcal{I}$ such that

$$\text{cost}(A, I, E) \leq k \cdot \text{cost}(A^*, I^*, E) + c.$$

The difference between this definition and the previous one lies only in the choice of quantifiers. In effect, the input I^* given to the champion algorithm A^*

is a worst-case input, chosen to maximize the champion's cost. Consequently, k-optimality is a weaker notion than k-competitiveness; however, it may still be a useful measure in contexts where k-competitiveness tells us little about the actual performance of an algorithm. An example is given in Section 7.

The definition of k-optimality given above was suggested by [5] by analogy to their semicompetitive definition of k-competitiveness. It also corresponds very closely to an earlier measure used by Patt-Shamir and Rajsbaum in their work on clock synchronization [15].

4.1 Relations between the measures

There is a nice relationship between semicompetitive analysis and traditional competitive analysis:

Theorem 1. *If an algorithm is k-competitive, then it is traditionally k-competitive. If an algorithm is traditionally k-competitive, it is k-optimal.*

Proof. Immediate from the definitions. □

In particular, if one has an upper bound on the competitive ratio of an algorithm (in the split-nondeterminism sense), one immediately gets an upper bound on the traditional competitive ratio of an algorithm. Similarly, a lower bound on optimality implies a lower bound on competitiveness. Thus even if one is not interested in the semicompetitive measures directly, they may still provide a useful tool for bounding the traditional competitive ratio of an algorithm.

4.2 Payoffs instead of costs

Just as one can define the competitiveness of an algorithm in the traditional sense in terms of payoffs instead of costs, one can do the same for the semicompetitive measures. Let us assume as above that we have a class of algorithms $\mathcal{A}$ against which the candidate algorithm will compete, a set of inputs $\mathcal{I}$ and a set of environments $\mathcal{E}$. Suppose further that we have a payoff function assigning a non-negative value $\text{payoff}(A, I, E)$ to each A in $\mathcal{A}$, I in $\mathcal{I}$, and E in $\mathcal{E}$.

An algorithm is traditionally k-competitive if there exists a constant c such that for all A^* in $\mathcal{A}$, I in $\mathcal{I}$, and E in $\mathcal{E}$,

$$\text{payoff}(A, I, E) + c \geq \frac{1}{k} \cdot \text{payoff}(A^*, I, E).$$

An algorithm is k-competitive if there exists a constant c such that for all A^* in $\mathcal{A}$, I and I^* in $\mathcal{I}$, and E in $\mathcal{E}$,

$$\text{payoff}(A, I, E) + c \geq \frac{1}{k} \cdot \text{payoff}(A^*, I^*, E). \tag{1}$$

Finally, an algorithm is k-optimal if there exists a constant c such that for all A^* in $\mathcal{A}$, I in $\mathcal{I}$, and E in $\mathcal{E}$, there exists an input I^* in $\mathcal{I}$ such that

$$\text{payoff}(A, I, E) + c \geq \frac{1}{k} \cdot \text{payoff}(A^*, I^*, E).$$

It is not difficult to see that k-competitiveness implies traditional k-competitiveness which in turn implies k-optimality in the payoff model just as in the cost model.

5 Modularity

One of the fundamental tools in traditional algorithm design is the ability to construct algorithms by composition. However, competitive analysis in general appears to forbid such modular constructions of competitive algorithms. If A is an algorithm that uses a subroutine B, the fact that B is competitive says nothing at all about A's competitiveness, since A must compete against algorithms that do not use B. This lack of modularity impedes the development of practical competitive algorithms.

Fortunately, by treating the input to a subroutine separately from its environment, it is possible to recover the ability to compose algorithm while retaining the advantages of competitive analysis. The key is the notion of **relative competitiveness** defined by [5]. Relative competitiveness is a measure of how well an algorithm A uses a competitive subroutine B, which takes into account not only how large A's cost is relative to B but also whether or not the decision to use B was a good idea in the first place. To do so, it considers the relative performance of three distinct executions: an execution of the combined algorithm $A \circ B$; an execution of an optimal A^* (which may or may not use a subroutine corresponding to B); and an execution of an optimal B^*. This approach requires operating within a semicompetitive framework, as described in Section 4, in which the input and environment of an algorithm are treated separately. The details of the definition of relative competitiveness are given below in Section 5.1.

The remarkable property of relative competitiveness is that it acts like a traditional worst-case performance measure when composing algorithms together. Glossing over some small technical details, if an algorithm A is l-competitive relative to a subroutine B, and B is itself k-competitive, then the combined algorithm $A \circ B$ is kl-competitive, even when compared against algorithms that do not use B or anything like B. This is exactly the same multiplicative effect that one gets with traditional worst-case analysis, where a parent algorithm that calls a subroutine l times at a cost of k units per call pays kl total cost. Except that here we are looking at competitive ratios.

The details of how one can compose competitive algorithms in this fashion are given below in Section 5.2.

5.1 Relative competitiveness

Formally, it is assumed that we have two sets of inputs and environments. Algorithm A takes an input from $\mathcal{I}_A$ and an environment from $\mathcal{E}_A$; similarly, B takes an input from $\mathcal{I}_B$ and an environment from $\mathcal{E}_B$. In the composite algorithm $A \circ B$ the input to B is provided by its parent routine A— one can think of this input as the procedure calls given to B. Similarly, the environment of A is provided by its subroutine B— one can think of this environment as the return values from these procedure calls. The input to A and the environment of B are provided by the adversary.

In the executions of A^* and B^* the situation is not quite symmetric. The difference arises from the fact that A^* and B^* are completely independent algorithms; and while B^* is in effect an optimal version of B, A^* is not an optimal version of A but instead is an optimal version of $A \circ B$. The effect of this difference is that while B^* takes an input from $\mathcal{I}_B$ and an environment from $\mathcal{E}_B$, just as B does, A^* takes an input from $\mathcal{I}_A$ but its environment comes from $\mathcal{E}_B$.

To complete the picture, it is necessary to have two cost measures. The first, cost_{AB}, measures the cost of algorithms that see inputs from $\mathcal{I}_A$ and environments from $\mathcal{E}_B$; it will be applied to both the composite algorithm $A \circ B$ and the champion A^*. The second, cost_B, measures the cost of algorithms that see inputs from $\mathcal{I}_B$ and environments from $\mathcal{E}_B$; it will be applied to B (as a component of $A \circ B$) and B^*. To avoid difficulties with division by zero, it is convenient to require that cost_B always be positive.

The definition is as follows. Given $\mathcal{A}_A$, $\mathcal{A}_B$, $\mathcal{I}_A$, $\mathcal{I}_B$, $\mathcal{E}_A$, $\mathcal{E}_B$, cost_{AB}, and cost_B, an algorithm A is l-competitive relative to an algorithm B if there exists a constant c such that for all E in $\mathcal{E}_B$,

$$\max_{I_A \in \mathcal{I}_A} \frac{\text{cost}_{AB}(A \circ B, I_A, E) - c}{\text{cost}_B(B, I_B, E)} \leq l \cdot \frac{\min_{A^* \in \mathcal{A}_A, I_A^* \in \mathcal{I}_A} \text{cost}_{AB}(A^*, I_A^*, E)}{\min_{B^* \in \mathcal{A}_B, I_B^* \in \mathcal{I}_B} \text{cost}_B(B^*, I_B^*, E)}. \quad (2)$$

Note that the input I_B to B is the input generated by A when run in interaction with B. If the constant c is zero, then A is **strictly** l-competitive relative to B.

It is sometimes possible to show that an algorithm A is l-competitive relative to *any* algorithm in a class of algorithms $\mathcal{A}_B$. In this case we we write that A is l-competitive relative to $\mathcal{A}_B$. The competitiveness of an algorithm relative to the class of all correct subroutines for solving a particular problem is often a more useful measure than its competitiveness relative to a particular subroutine— it implies that one can substitute a more efficient subroutine for a less efficient one and get an improvement in performance.

The payoff version of relative competitiveness is defined analogously as follows. Given $\mathcal{A}_A$, $\mathcal{A}_B$, $\mathcal{I}_A$, $\mathcal{I}_B$, $\mathcal{E}_A$, $\mathcal{E}_B$, payoff_{AB}, and payoff_B, an algorithm A is l-competitive relative to an algorithm B if there exists a constant c such that for all E in $\mathcal{E}_B$,

$$\max_{I_A \in \mathcal{I}_A} \frac{\text{payoff}_{AB}(A \circ B, I_A, E) + c}{\text{payoff}_B(B, I_B, E)} \geq \frac{1}{l} \cdot \frac{\max_{A^* \in \mathcal{A}_A, I_A^* \in \mathcal{I}_A} \text{payoff}_{AB}(A^*, I_A^*, E)}{\max_{B^* \in \mathcal{A}_B, I_B^* \in \mathcal{I}_B} \text{payoff}_B(B^*, I_B^*, E)}.$$

Note that the sign of the additive constant c has changed, in order to be consistent with the definition of k-competitiveness. Again, it is required that payoff_B is never zero; however, in the payoff model it is in fact possible to drop this requirement without causing too many difficulties (for details see [5]).

It is also possible to define relative optimality for both the cost and payoff models. For relative optimality, the inputs I_A^* and I_B^* are chosen to maximize the costs (minimize the payoffs) of A^* and B^*; there are no other differences.

5.2 The composition theorem

The usefulness of relative competitiveness is captured in Composition Theorem of [5]. The theorem as given here is taken more-or-less directly from [5], where it was first stated in a slightly less general form. The version given here has been adapted slightly to fit into the more general framework used in this chapter. To simplify the statement of the theorem, let us assume throughout this section that $\mathcal{A}_A$, $\mathcal{A}_B$, $\mathcal{I}_A$, $\mathcal{I}_B$, $\mathcal{E}_A$, $\mathcal{E}_B$ and either cost_{AB} and cost_B or payoff_{AB} and payoff_B are fixed.

Theorem 2 (Composition Theorem). *Let A be l-competitive relative to B, B k-competitive, and suppose that there exists a non-negative constant c such that for all E in $\mathcal{E}_E$, either*

$$\min_{I_A^* \in \mathcal{I}_A, A^* \in \mathcal{A}_A} \text{cost}_{AB}(A^*, I_A^*, E) \geq c \min_{I_B^* \in \mathcal{I}_B, B^* \in \mathcal{A}_B} \text{cost}_B(B^*, I_B^*, E) \tag{3}$$

or

$$\max_{I_A^* \in \mathcal{I}_A, A^* \in \mathcal{A}_A} \text{payoff}_{AB}(A^*, I_A^*, E) \leq c \max_{I_B^* \in \mathcal{I}_B, B^* \in \mathcal{A}_B} \text{payoff}_B(B^*, I_B^*, E) \tag{4}$$

(as appropriate). Then $A \circ B$ is kl-competitive.

Proof. The proof given here is adapted from [5].

Let us consider only the case of costs; the case of payoffs is nearly identical and in any case has been covered elsewhere [5]. The proof involves only simple algebraic manipulation, but it is instructive to see where the technical condition (3) is needed.

To avoid entanglement in a thicket of superfluous parameters let us abbreviate the cost of each algorithm X with the appropriate input and environment as $C(X)$, so that $\text{cost}_{AB}(A \circ B, I_A, E)$ becomes $C(A \circ B)$, $\text{cost}_B(B, I_B, E)$ becomes $C(B)$, $\text{cost}_{AB}(A^*, I_A^*, E)$ becomes $C(A^*)$, and $\text{cost}_B(B^*, I_B^*, E)$ becomes $C(B^*)$.

Fix E. The goal is to show that $C(A \circ B) \leq kl \cdot C(A^*) + c_{AB}$, where C_{AB} does not depend on E. We may assume without loss of generality that A^* and its input I_A^* are chosen to minimize $C(A^*)$, and that the input I_A is chosen to maximize $C(A)$. So in particular $C(A^*$ is equal to the left-hand side of inequality (3) and the numerator of the right-hand side of inequality (2). Similarly, choose B^* and I_B^* to minimize $C(B^*)$.

Thus (2) can be rewritten more compactly as

$$\frac{C(A \circ B) - c_{AB}}{C(B)} \leq l \cdot \frac{C(A^*)}{C(B^*)},$$

where c_{AB} is the constant from (2). Multiplying out the denominators gives

$$(C(A \circ B) - c_{AB})\, C(B^*) \leq l \cdot C(A^*)C(B). \tag{5}$$

Similarly, the k-competitiveness of B means that

$$C(B) \leq k \cdot C(B^*) + c_B, \tag{6}$$

where c_B is an appropriate constant.

Plugging (6) into the right-hand side of (5) gives

$$\begin{aligned}(C(A \circ B) - c_{AB})\, C(B^*) &\leq l \cdot C(A^* \cdot (k \cdot C(B^*) + c_B) \\ &= kl \cdot C(A^*)C(B^*) + l \cdot C(A^*)c_B.\end{aligned}$$

Dividing both sides by $C(B^*)$ and moving c_{AB} gives

$$C(A \circ B) \leq kl \cdot C(A^*) + c_{AB} + lc_B \frac{C(A^*)}{C(B^*)}. \tag{7}$$

The last term is bounded by a constant if (3) holds; thus $A \circ B$ is kl-competitive. □

Much of the complexity of the theorem, including the technical conditions (3) and (4), is solely a result of having to deal with the additive constant in the k-competitiveness of B. If one examines the last steps of the proof, it is evident that the situation is much simpler if B is strictly competitive.

Corollary 3. *Let A be l-competitive relative to B and let B be strictly k-competitive. Then $A \circ B$ is kl-competitive. If in addition A is strictly competitive relative to B, $A \circ B$ is strictly kl-competitive.*

Proof. Consider the proof of Theorem 2. Since B is strictly competitive, c_B in (7) is zero, and the technical condition (3) is not needed. If A is strictly competitive, then c_{AB} is zero as well, implying $A \circ B$ is strictly competitive. A similar argument shows that the corollary also holds in the payoff model. □

Note that the Composition Theorem and its corollary can be applied transitively: if A is competitive relative to $B \circ C$, B is competitive relative to C, and C is competitive, then A is competitive (assuming the appropriate technical conditions hold).

6 Example: The wait-free shared memory model

In this section we describe some recent approaches to applying competitive analysis to problems in the **wait-free shared memory model** [13]. In this model, a collection of n processes communicate only indirectly through a set of single-writer atomic registers. A protocol for carrying out some task or sequence of tasks is **wait-free** if each process can finish its current task regardless of the relative speeds of the other processes. Timing is under the control of an adversary scheduler that is usually modeled as a function that chooses, based on the current state of the system, which process will execute the next operation. This adversary is under no restrictions to be fair; it can, for example, simulate up to $n-1$ process crashes simply by choosing never to schedule those processes again.

The wait-free shared memory model is a natural target for competitive analysis. The first reason is that many of the algorithms that are known for this model pay for their high resilience (tolerating up to $n-1$ crashes and arbitrary asynchrony) with high worst-case costs. Thus there is some hope that competitive analysis might be useful for showing that in less sever cases these algorithms perform better. The second reason is that the absence of restrictions on the adversary scheduler makes the mathematical structure of the model very clean. Thus it is a good jumping off point for a more general study of the applicability of competitive analysis to the fault-tolerant aspects of distributed algorithms. Finally, a third reason is that under the assumption of single-writer registers, there is a natural problem (the *collect problem*) that appears implicitly or explicitly in most wait-free shared-memory algorithms. By studying the competitive properties of this problem, we can learn about the competitive properties of a wide variety of algorithms.

Sections 6.1 and 6.2 describe the collect problem and solution to it without regard to issues of competitiveness. Section 6.3 discusses why competitiveness is a useful tool for analyzing the performance of collect algorithms. The following sections describe how it has been applied to such algorithms and the consequences of doing so.

Some of the material in these sections is adapted from [2], [5], and [4].

6.1 Collects: A fundamental problem

When a process starts a task in the wait-free model, it has no means of knowing what has happened in the system since it last woke up. Thus to solve almost all non-trivial problems a process must be able to carry out a **collect**, an operation in which it learns some piece of information from each of the other processes.

The collect problem was first abstracted by Saks, Shavit, and Woll [16]. The essential idea is that each process owns a single-writer multi-reader atomic register, and would like to be able carry out write operations on its own register and **collect operations** in which it learns the values in all the registers. The naive method for performing a collect is simply to read all of the registers directly; however, this requires at least $n-1$ read operations (n if we assume that a process does not remember the value in its own register). By cooperating with other processes, it is sometimes possible to reduce this cost by sharing the work of reading all the registers.

There are several versions of the collect problem. The simplest is the "one-shot" collect, in which the contents of the registers are fixed and each process performs only one collect. The one-shot version of the problem will not be discussed much in this chapter, but it is worth noting some of the connections between one-shot collects and other problems in distributed computing. Without the single-writer restriction, the one-shot collect would be isomorphic to the problem of Certified Write-All, in which n processes have to write to n locations while individually being able to certify that all writes have occurred (indeed, the collect algorithm of Ajtai et al. [2] is largely based on a Certified Write-All algorithm of Anderson and Woll [3]). One can also think of the collect problem as an asynchronous version of the well-known **gossip problem** [12], in which n persons wish to distribute n rumors among themselves with a minimum number of telephone calls; however, in the gossip problem, which persons communicate at each time is fixed in advance by the designer of the algorithm; with an adversary controlling timing this ceases to be possible.

The more useful version of the problem, and the one that we shall consider here, is the **repeated collect** problem. This corresponds exactly to simulating the naive algorithm in which a process reads all n registers to perform a collect. An algorithm for repeated collects must provide for each process a *write* procedure that updates its register and a *collect procedure* that returns a vector of values for all of the registers. Each of these procedures may be called repeatedly. The values returned by the collect procedure must satisfy a safety property called **regularity** or **freshness**: any value that I see as part of the result of a collect must be **fresh** in the sense that it was either present in the appropriate register when my collect started or it was written to that register while my collect was in progress.

The repeated collect problem appears at the heart of a wide variety of shared-memory distributed algorithms (an extensive list is given in [2]). What makes it appealing as a target for competitive analysis is that the worst-case performance of any repeated collect algorithm is never better than that of the naive algorithm. The reason for this is the strong assumptions about scheduling. In the wait-free shared-memory model, a process cannot tell how long it may have been asleep between two operations. Thus when a process starts a collect, it has no way of knowing whether the register values have changed since its previous collect. In addition, the adversary can arrange that each process does its collects alone by running only one process at a time. If the adversary makes these choices, any algorithm must read all of the $n-1$ registers owned by the other processes to satisfy the safety property.

But there is hope: this lower bound applies to any algorithm that satisfies the safety property— including algorithms that are chosen to be optimal for the timing pattern of a particular execution. Thus it is plausible that a competitive approach would be useful.

6.2 A randomized algorithm for repeated collects

Before jumping into the question of how one would apply competitive analysis to collects, let us illustrate some of the issues by considering a particular collect algorithm, the "Follow the Bodies" algorithm of [4].

As is often the case in distributed computing, one must be very precise about what assumptions one makes about the powers of the adversary. The Follow-the-

Bodies algorithm assumes that a process can generate a random value and write it out as a single atomic operation. This assumption appears frequently in early work on consensus; it is the "weak model" of Abrahamson [1] and was used in the consensus paper of Chor, Israeli, and Li [11]. In general, the weak model in its various incarnations permits much better algorithms (e.g., [8, 10]) for such problems as consensus than the best known algorithms in the more fashionable "strong model" (in which the adversary can stop a process in between generating a random value and writing it out). The assumption that the adversary cannot see coin-flips before they are written is justified by an assumption that in a real system failures, page faults, and similar disastrous forms of asynchrony are likely to be affected by *where* each process is reading and writing values but not by *what* values are being read or written. For this reason the adversary in the weak model is sometimes called *content-oblivious*.

Even with a content-oblivious adversary the collect problem is still difficult. Nothing prevents the adversary from stopping a process between reading new information from another process's register and writing that information to its own register. Similarly the adversary can stop a process between making a random choice of which register to *read* and the actual read operation. (This rule corresponds to an assumption that not all reads are equal; some might involve cache misses, network delays, and so forth.)

This power of the adversary turns out to be quite important. Aspnes and Hurwood show that an extremely simple algorithm works in the restricted case where the adversary cannot stop a read selectively depending on its target. Each process reads registers randomly until it has all the information it needs. However, the adversary that *can* stop selected reads can defeat this simple algorithm simple algorithm by choosing one of the registers to be a "poison pill": any process that attempts to read this register will be halted. Since on average only one out of every n reads would attempt to read the poisonous register, close to n^2 reads would be made before the adversary would be forced to let some process actually swallow the poison pill.

In order to avoid this problem, in the Follow-the-Bodies algorithm a process leaves a note saying where it is going before attempting to read a register.[2] Poison pills can thus be detected easily by the trail of corpses leading to them. The distance that a process will pursue this trail will be $\lambda \ln n$, where λ is constant chosen to guarantee that the process reaches its target with high probability.

Figure 1 depicts one pass through the loop of the resulting algorithm. The description assumes that each process stores in its output register both the set of values S it has collected so far and its *successor*, the process it selected to read from most recently.

For the moment we have left out the termination conditions for the loop, as they may depend on whether one is trying to build a one-shot collect or a repeated collect (which requires some additional machinery to detect fresh register values).

---

[2] It is here that the assumption that one can flip a coin and write the outcome atomically is used.

- Set p to be a random process, and write out p as our successor.
- Repeat $\lambda \ln n$ times:
  - Read (S', p') from the register of p.
  - Set S to be the union of S and S'.
  - Set p to p'.
  - Write out the new S and p.

Fig. 1. The Follow-the-Bodies Algorithm (One Pass Only)

The Follow-the-Bodies algorithm has the property that it spreads information through the processes' registers rapidly regardless of the behavior of the adversary. In [4] it is shown that:

Theorem 4. *Suppose that in some starting configuration the register of each process p contains the information K_p and that the successor fields are set arbitrarily. Fix $\lambda \geq 9$, and count the number of operations carried out by each process until its register contains the union over all p of K_p, and write W for the sum of these counts.*

$$\Pr W \geq 37\lambda^2 n \ln^3 n \leq \frac{2}{n^{\lambda-5}}.$$

Here the rapidity of the spread of information is measured in terms of total work. Naturally, the adversary can delay any updates to a particular process's register for an arbitrarily long time by putting that process to sleep; what the theorem guarantees is that (a) when the process wakes up, it will get the information it needs soon (it is likely that by then the first process it looks at will have it); and (b) the torpor of any individual process or group of processes cannot increase the total work of spreading information among their speedier comrades.

The proof of the theorem is a bit involved, and the interested reader should consult [4] for details. In the hope of making the workings of the algorithm more clear we mention only that the essence of the proof is to show that after each phase consisting of $O(n \ln n)$ passes through the loop, the size of the smallest set of processes that collectively know all of the information is cut in half. After $O(\ln n)$ of these phases some process knows everything, and it is not long before the other processes read its register and learn everything as well. (The extra log factor comes from the need to read $\lambda \ln n$ registers during each pass through the loop).

In [4] it is shown how to convert this rumor-spreading algorithm to a repeated-collect algorithm by adding a simple timestamp scheme. Upon starting a collect a process writes out a new timestamp. Timestamps spread through the process's registers in parallel with register values. When a process reads a value *directly* from its original register, it tags that value by the most recent timestamp it has from each of the other processes. Thus if a process sees a value tagged with its own most recent timestamp, it can be sure that that value was present in the registers after the process started its most recent collect, i.e. that the value is fresh.

The resulting algorithm is depicted in Figure 2. Here, S tracks the set of values (together with their tags) known to the process. The array T lists each process's most recent timestamps. Both S, T, and the current successor are periodically written to the process's output register.

- Choose a new timestamp τ and set our entry in T to τ.
- While some values are unknown:
  - Set p to be a random process, write out p as our successor and T as our list of known timestamps.
  - Repeat $\lambda \ln n$ times:
    - Read the register of p. Set S to be the union of S and the values field. Update T to include the most recent timestamps for each process. Set p to the successor field.
    - Write out the new S and T.
- Return S.

Fig. 2. Repeated Collect Algorithm

The performance of this algorithm is characterized by its **collective latency** [2], an upper bound on the total amount of work needed to complete all collects in progress at some time t:[3]

Theorem 5. *Fix a starting time t. Fix $\lambda \geq 9$. Each process carries out a certain number of steps between t and the time at which it completes the collect it was working on at time t. Let W be the sum over all processes of these numbers. Then*

$$\Pr W \geq 74\lambda^2 n \ln^3 n \leq \frac{4}{n^{\lambda-5}}.$$

Proof. Divide the steps contributing to W into two classes: (i) steps taken by processes that do not yet know timestamps corresponding to all of the collects in progress at time t; and (ii) steps taken by processes that know all n of these timestamps. To bound the number of steps in class (i), observe that the behavior of the algorithm in spreading the timestamps during these steps is equivalent to the behavior of the non-timestamped algorithm. Similarly, steps in class (ii) correspond to an execution of the non-timestamped algorithm when we consider the spread of values tagged by all n current timestamps. Thus the total time for both classes of steps is bounded by twice the bound from Theorem 4, except for a case whose probability is at most twice the probability from Theorem 4. □

[3] The "collective" part of "collective latency" refers not to the fact that the algorithm is doing collect operations but instead to the fact that the latency it is measuring is a property of the group of processes as a whole.

6.3 Competitive analysis and collects

Theorem 5 shows that the Follow-the-Bodies algorithm improves in one respect on the naive collect algorithm. For the naive algorithm, the collects in progress at any given time may take n^2 operations to finish, while with high probability Follow-the-Bodies finishes them in $O(n \log^3 n)$ operations. Unfortunately, this improvement does not translate into better performance according to traditional worst-case measures.

As noted above, if a process is run in isolation it must execute at least $n - 1$ operations to carry out a collect, no matter what algorithm it runs, as it can learn fresh values only by reading them directly. Under these conditions the naive algorithm is optimal. Even if one considers the amortized cost of a large number of collects carried out by different processes, the cost per collect will still be $\Omega(n)$ if no two processes are carrying out collects simultaneously. Yet in those situations where the processes have the opportunity to cooperate, Theorem 5 implies that they can do so successfully, combing their efforts so that in especially good cases (where a linear number of processes are running at once) the expected amortized cost of a single collect drops to $O(\log^3 n)$.

It is clear that algorithms like Follow-the-Bodies improve on the naive collect. But how can one quantify this improvement in a way that is meaningful outside the limited context of collect algorithms? Ideally, one would like to have a measure that recognizes that distributed algorithms may be run in contexts where many or few processes participate, where some processes are fast and some are slow, and where the behavior of processes may vary wildly from one moment to the next. Yet no simple parameter describing an execution (such as the number of active processes) can hope to encompass such detail.

Competitive analysis can. By using the performance of an optimal, specialized algorithm as a benchmark, competitive analysis provides an objective measure of the difficulty of the environment (in this case, the timing pattern of an asynchronous system) in which a general-purpose algorithm operates. To be competitive, the algorithm must not only perform well in worst-case environments; its performance must adapt to the ease or difficulty of whatever environment it faces. Fortunately, it is possible to show that many collect algorithms have this property.

Some complications arise. In order to use competitive analysis, it is necessary to specify precisely what is included in the environment— intuitively, on what playing field the candidate and champion algorithms will be compared. An additional issue arises because assuming that the champion has complete knowledge of the environment (as the hypothetical "off-line" algorithm has when computing the competitive ratio of an on-line algorithm) allows it to implicitly communicate information from one process to another at no cost that a real distributed algorithm would have to do work to convey. Neither difficulty is impossible to overcome. In the following sections we describe two measures of competitive performance for distributed algorithms that do so.

6.4 A traditional approach: Latency competitiveness

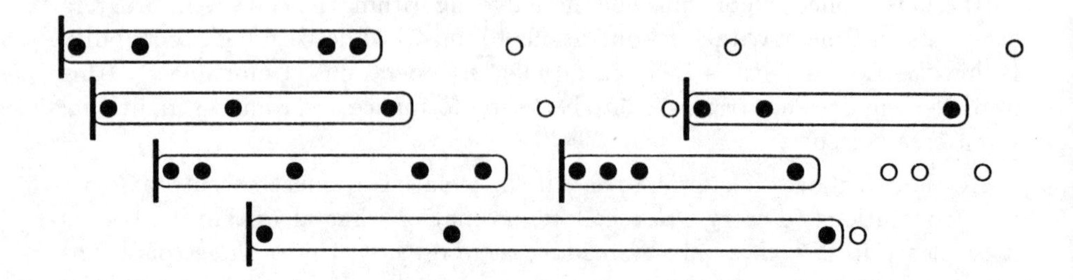

Fig. 3. Latency model. New high-level operations (ovals) start at times specified by the adversary (vertical bars). Adversary also specifies timing of low-level operations (small circles). Cost to algorithm is number of low-level operations actually performed (filled circles).

The competitive latency model of Ajtai et al. [2] uses competitive analysis in its traditional form, in which all nondeterminism in a system is under the control of the adversary and is shared between both the candidate and champion algorithms. In the context of the repeated collect problem, it is assumed that the adversary controls the execution of an algorithm by generating (possibly in response to the algorithm's behavior) a schedule that specifies when collects start and when each process is allowed to take a step (see Figure 3). A process halts when it finishes a collect; it is not charged for opportunities to take a step in between finishing one collect and starting another (intuitively, we imagine that it is off doing something else). The **competitive latency** of a candidate algorithm is the least constant k, if any, that guarantees that the expected total number of operations carried out by the candidate on a given schedule σ is at most k times the cost of an optimal distributed algorithm running on the same schedule (modulo an additive constant).

In terms of the general definition of competitive analysis from Section 3, $\mathcal{E}$ consists of all schedules as defined above. The set of algorithms $\mathcal{A}$ consists of all distributed algorithms that satisfy a correctness condition. The cost measure is just the number of operations required by a particular algorithm. For the collect problem, this correctness condition requires that an algorithm guarantee that the values returned are fresh regardless of the schedule. In particular, while processes in a champion algorithm are allowed to use their implicit knowledge of the schedule to optimize their choices of what registers to read and how to cooperate with one another, they cannot use this knowledge to avoid verifying the freshness of the values they return.

There is a general principle involved here that points out one of the difficulties of applying competitive analysis directly to distributed algorithms. The difficulty is that if one is not careful, one can permit the processes in the champion algorithm to communicate with each other at zero cost by virtue of their common knowledge of the schedule. For example: consider a schedule in which no process ever writes to a register. If the champion is chosen after the schedule is fixed, as occurs whenever we have an off-line adversary, then it can be assumed that every process in the champion implicitly knows the schedule. In particular, each process in the champion can observe that no processes are writing to the registers and thus that the register values never change. So when asked to perform a collect, such a process can return a vector consisting of the initial values of the registers at zero cost. Needless to say, a situation in which the champion pays zero cost is not likely to yield a very informative competitive ratio.

What happened in this example is that if one assumes that the champion processes all have access to the schedule, each champion process can then compute the entire state of the system at any time. In effect, one assumes that the champion algorithm is a global-control algorithm rather than a distributed algorithm. It is not surprising that against a global-control algorithm it is difficult to be competitive, especially for a problem such as collect in which the main difficulty lies in transmitting information from one process to another.

On the other hand, notice that the champion algorithm in this case is not correct for all schedules; any schedule in which any register value changes at some point will cause later collects to return incorrect values. Limiting the class of champion algorithms to those whose correctness does not depend on operating with a specific schedule restores the distributed character of the champion algorithm. Yet while such a limit it prevents pathologies like zero-cost collects, it still allows the champion to be "lucky", for example by having all the processes choose to read from the one process that happens in a particular execution to have already obtained all of the register values. Thus the reduction in the champion's power is limited only to excluding miraculous behavior, but not the surprising cleverness typical of off-line algorithms.

With this condition, Ajtai et al. show that if an algorithm has a maximum collective latency of L at all times, then its competitive ratio in the latency model is at most $L/n + 1$. Though this result is stated only for deterministic algorithms, as observed in [4] it can be made to apply equally well to randomized algorithms given a bound L on the expected collective latency.

The essential idea of the proof in [2] of the relationship between collective latency and competitive latency is to divide an execution into segments and show that for each such segment, the candidate algorithm carries out at most $L + n$ operations and the champion carries out at least n operations. The lower bound on the champion is guaranteed by choosing the boundaries between the segments so that in each segment processes whose collects start in the segment are given at least n chances to carry out an operation; either they carry out these n operations (because their collects have not finished yet) or at least n read operations must have been executed to complete these collects (because otherwise the values returned cannot be guaranteed to be fresh). For an algorithm with expected collective latency L, the cost per segment is at most $L + n$; n for the n steps during the segment, plus at most an expected L operations to complete any collects still in progress at the end of the segment. By summing over all segments this gives a ratio of $L/n + 1$.

For the Follow-the-Bodies algorithm, the expected value of L is easily seen to be $O(n \log^3 n)$ (from Theorem 5). It follows that:

Theorem 6. *The competitive latency of the Follow-the-Bodies algorithm is* $O(\log^3 n)$.

It is worth noting that this result is very strong; it holds even against an *adaptive off-line* adversary [9], which is allowed to choose the champion algorithm after seeing a complete execution of the candidate. In contrast, the best known lower bound is $\Omega(\log n)$ [2].

It is still open whether or not an equally good deterministic algorithm exists. The best known deterministic algorithm, from [2], has a competitive latency of $O(n^{1/2} \log^2 n)$.

6.5 A semicompetitive approach: Throughput competitiveness

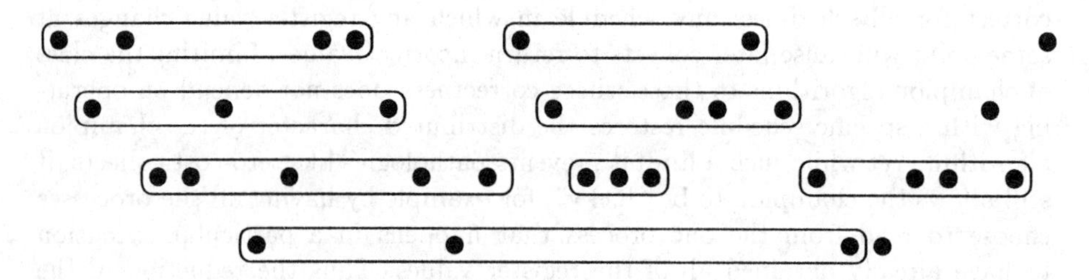

Fig. 4. Throughput model. New high-level operations (ovals) start as soon as previous operations end. Adversary controls only timing of low-level operations (filled circles). Payoff to algorithm is number of high-level operations completed.

A semicompetitive approach of the sort discussed in Section 4 is used in the **competitive throughput** model of Aspnes and Waarts [5]. In this model, the adversary no longer controls the starting time of collects; instead, both the candidate and the champion try to complete as many collects as possible in the time available (see Figure 4). It also distinguishes between the environment, which takes the form of a schedule that now specifies only when low-level operations such as reads and writes occur; and the input, which takes the form of a specification of what sequence of high-level operations each process is to perform but not when it must perform them. (For the collect problem the input is generally not very interesting, since the collect algorithm can only perform one kind of high-level operation.)

Formally, the model can be viewed as using the semicompetitive definition of competitiveness from Section 4. The set of algorithms $\mathcal{A}$ consists of all correct distributed algorithms; as in the preceding section, correctness must be defined independently of any single schedule to avoid pathologies. The set of environments $\mathcal{E}$ consists of all schedules defining the timing of low-level operations. The set of inputs $\mathcal{I}$ consists of all sequences of requests to perform high-level tasks. The competitive throughput model is a payoff model; competitiveness is defined in terms of a payoff function payoff(A, I, E) that measures *how many* of the high-level tasks in I can be completed by A given the schedule E. The **throughput competitiveness** of an algorithm A is the least value k for which the payoff of A is always within a factor of k of the payoff of an optimal A^* (modulo an additive constant).

The motivation for adopting a semicompetitive measure like throughput competitiveness is that it allows competitive algorithms to be constructed modularly as described in Section 5. This is particularly important for collect algorithms since collect appears as a subroutine in such a large number of other algorithms. If one can show that these algorithms are throughput competitive relative to collect, then the only thing needed to make them throughput competitive in their own right is to plug in a throughput-competitive collect subroutine.

In order to do this, a small technical complication must first be dealt with. No interesting algorithm performs only collects; for information to pass from one process to another somebody must be writing to the registers. So instead of using a subroutines that performs only collects one uses a subroutine that supports two operations: the collect operation as described above, and a **write-collect**, consisting of a write operation followed by a collect. The intent is to simulate an object that provides both writes and collects. However, because throughput competitiveness assumes a worst-case choice of operations for the candidate to perform, but a best-case choice for the champion, it is important that all operations supported by the subroutine have roughly the same granularity. Otherwise it becomes possible for the candidate to be forced to carry out only expensive operations (like collects) while the champion carries out only cheap operations (like writes).[4] Fortunately, in practice in almost all algorithms that use writes and collects, each process performs at most one write in between each pair of collects, and so one can treat such algorithms as using write-collect and collect operations. Furthermore, since all known cooperative collect algorithms start each collect with a low-level write operation (in order to write out a new timestamp), it costs nothing to modify such an algorithm to perform write-collects as well as collects.

[4] In principle, there are other ways to avoid this difficulty. For example, one could define the payoff function so that completed collect operations produced more value than completed write operations; but then it might be difficult to show that an algorithm that performed both writes and collects to achieve some higher goal is in fact competitive relative to this mixed-payoff subroutine. Alternatively, one could use the notion of k-optimality (where the champion also faces a worst-case input) instead of k-competitiveness. However, k-optimality is weaker than traditional k-competitiveness, so results about the performance of higher-level algorithms obtained using the k-optimality of the collect subroutine would not immediately imply competitiveness of the higher-level algorithms in the traditional sense. Still, there are not yet very many examples in the literature of the use of these techniques, so it remains to be seen which approach is likely to be most useful in the long run.

It turns out that the same bound on collective latency— the total number of operations required to complete all collects in progress at any given time— used to show the latency competitiveness of a collect algorithm can also be used to show throughput competitiveness. A second condition is also needed; there must be a bound on the **private latency**— the number of operations required to complete any *single* collect. This second condition can easily be guaranteed by dovetailing a collect algorithm that guarantees low collective latency with the naive algorithm that guarantees low private latency by having a process read all n registers itself.

With this modification, and the earlier modification to add write-collect operations, the Follow-the-Bodies algorithm from Section 6.2 has an collective latency of $O(n \log^3 n)$ (on average) and a private latency of at most $2n$. A general bound on throughput competitiveness based on these two parameters is given in [5]; plugging in the values for Follow-the-Bodies gives:

Theorem 7. *The competitive throughput of the Follow-the-Bodies algorithm is* $O(n^{1/2} \log^{3/2} n)$.

The proof of this bound is quite complex and it would be difficult even to try to summarize it here. Its overall structure is similar to the technique for bounding latency competitiveness described in Section 6.4, in that it proceeds by dividing an execution into segments and proving a lower bound on the candidate's payoff and an upper bound on the champion's payoff for each segment. The mechanism used for this division is a potential function taking into account many fine details of the execution. The interested reader can find the complete argument in the full version of [5].

It might be surprising that the competitive ratio for essentially the same algorithm is much larger in the throughput model than in the latency model. The reason for the difference is the increased restrictions placed on the champion by the latency model. Because the champion can only start collects when the candidate does, it is not possible for the champion to take advantage of a particularly well-structured part of the schedule to finish many collects using its superior knowledge while the candidate wanders around aimlessly. Instead, the most the champion can hope to do is finish quickly the same collects that the candidate will finish more slowly— but the candidate will not finish these collects too much more slowly if it uses an algorithm with low collective latency. By contrast, in the throughput model it is possible to construct schedules in which the champion has a huge advantage. An example (taken from [5]) is given in the following section.

6.6 Lower bound on throughput competitiveness of collect

To illustrate some of the difficulties that arise in the throughput model, consider the following theorem and its proof:

Theorem 8. *No cooperative collect protocol has a throughput competitiveness less than $\Omega(\sqrt{n})$.*

Proof. (This proof is taken directly from [5].)

Because of the additive term in the definition of competitive ratio, it is not enough to exhibit a single schedule on which a given collect algorithm fails. Instead, we will describe a randomized adversary strategy for producing schedules of arbitrary length on which the ratio between the number of collects performed by a given candidate algorithm and the number of collects performed by an optimal champion tends to $\Theta\sqrt{n}$.

The essential idea is that we will build up an arbitrarily-long schedule out of phases. In each phase, most of the work will be done by a randomly-chosen "patsy". Profiting from the patsy's labors, in the champion algorithm, will be $\sqrt{n}$ "active" processes. These same processes, in the candidate algorithm, will not benefit from the patsy's work, since we will terminate a phase as soon as any active process discovers the patsy.

Let us be more specific. The adversary fixes the $\sqrt{n}$ active processes at the start of the schedule, and chooses a new patsy uniformly at random from the set of all non-active processes at the start of each phase. Each phase consists of one or more rounds. In each round, first each active process takes one step, then the patsy takes $n + \sqrt{n} + 1$ steps, and finally each active process takes a second step. If, during any round, some active process reads the patsy's register or learns all n register values, the adversary cleans up by running each active process and the patsy in an arbitrary order until each finishes its current collect. A new phase then starts with a new patsy, chosen at random independently of previous choices.

In the champion algorithm, each active process writes a timestamp to its register at the start of each round. The patsy then reads these timestamps ($\sqrt{n}$ steps), reads all the registers (n steps), and writes a summary of their contents, together with the timestamps, to its own output register (1 step). Finally, each active process reads the patsy's register. The result is that in each round, the champion algorithm completes $\sqrt{n} + 1$ collects. To simplify the analysis, we will assume that the champion algorithm does no work at all during the clean-up stage and the end of a phase. Under this assumption the champion completes $k(\sqrt{n} + 1)$ collects during a k-round phase.

In the candidate algorithm, each active process completes exactly one collect during each phase. Since no active process reads the patsy's register until the last round of the phase, the patsy cannot use any values obtained by the active processes prior to the last round. Thus the patsy completes at most one collect for every n operations that it executes prior to the last round, plus at most two additional collects during the last round and the clean-up stage. Since the patsy executes $n + \sqrt{n} + 1$ operations per round, the total number of collects it completes during a k-round phase is thus at most $k(1 + \frac{1}{\sqrt{n}} + \frac{1}{n}) + 2$, and the total number of collects completed by all processes during a k-round phase is at most $k(1 + \frac{1}{\sqrt{n}} + \frac{1}{n}) + 2 + \sqrt{n}$.

It remains only to determine the expected number of rounds in a phase. A phase can end in two ways: either some process finds the patsy, or some process learns all n values. Since the patsy is chosen at random, finding it requires an expected $(n - \sqrt{n} + 1)/2$ reads and at most $n - \sqrt{n}$ reads. Learning all n values requires n reads of the registers. Thus finding the patsy is a better strategy for the active processes, and since they together execute $2\sqrt{n}$ operations per round, they can find it in an expected $(\sqrt{n} - 1)/4$ rounds.

Plugging this quantity in for k gives an expected $(n-1)/4$ collects per phase for the champion algorithm and at most an expected $\frac{\sqrt{n}-1/n}{4} + 2 + \sqrt{n} \leq 2\sqrt{n}$ collects per phase for the candidate. In the limit as the number of phases goes to infinity, the ratio between the payoff to the candidate and the payoff to the champion goes to the ratio of these expected values, giving a lower bound on the competitive ratio of $\Omega(\sqrt{n})$. □

6.7 Competitive collect as a subroutine

The importance of collect is that it appears as a subroutine in many wait-free distributed algorithms. The value of showing that a particular collect algorithm is competitive in the throughput model is that one can then apply the Composition Theorem (Theorem 2) to show that many of these algorithms are themselves competitive when they are modified to use a competitive collect. Several examples of this approach are given in [5]; one simple case is reproduced below, in order to make more concrete how one might actually show that an algorithm is competitive relative to some class of subroutines.

A **snapshot** algorithm, like a collect algorithm, simulates an array of n single-writer registers. It supports a **scan-update** operation that writes a value to one of the registers (this part is the "update") and then returns a snapshot, which is a vector of values for all of the registers (this part is the "scan"). Unlike a collect algorithm, which must satisfy only the requirement that all values returned by a collect are at least as recent as the start of that collect, the snapshot algorithm must guarantee that different snapshots appear to be instantaneous. In practice, this means that process P cannot be shown an older value than process Q in one register and a newer value in a different register.

There are many known wait-free algorithms for snapshot (a list is given in [5]). The best algorithm currently known in terms of the asymptotic work required for a single snapshot is the algorithm of Attiya and Rachman [6], in which each process uses $O(\log n)$ alternating writes and collects in order to complete a single scan-update operation. (The reason why a scan-update requires several rounds of collects is that the processes must negotiate with each other to ensure that the snapshots they compute are consistent.) If the collects are implemented using the naive algorithm that simply reads all n registers directly, this bound translates into a cost of $O(n \log n)$ primitive operations per collect in the worst case. However, by plugging in a competitive collect algorithm, it is possible to improve on this bound in many executions.

In order to do so it is first necessary to argue that the Attiya-Rachman algorithm is competitive relative to a subroutine providing write-collect and collect operations as described in Section 6.5. To remain consistent with the definition given in Section 5.1, which requires nonzero payoffs for the subroutine, we will consider only schedules in which the candidate collect algorithm used as subroutine in Attiya-Rachman completes at least one collect.[5] We would like to show that for a suitable choice of l, for any such schedule E, that there exists a constant c such that

$$\max_{I_A \in \mathcal{I}_A} \frac{\mathrm{payoff}_{AB}(A \circ B, I_A, E) + c}{\mathrm{payoff}_B(B, I_B, E)} \geq \frac{1}{l} \cdot \frac{\max_{A^* \in \mathcal{A}_A, I_A^* \in \mathcal{I}_A} \mathrm{payoff}_{AB}(A^*, I_A^*, E)}{\max_{B^* \in \mathcal{A}_B, I_B^* \in \mathcal{I}_B} \mathrm{payoff}_B(B^*, I_B^*, E)} \tag{8}$$

holds, where A is the Attiya-Rachman snapshot, B is an arbitrary collect algorithm, and $\mathcal{I}_A$, $\mathcal{I}_B$, etc., are defined appropriately.

It is easy to get an upper bound of one on the ratio on the right-hand side. Because a single scan-update operation can be used to simulate a single write-collect or collect, the payoff (i.e., number of scan-updates completed) of an optimal snapshot algorithm cannot exceed the payoff (number of collects and/or write-collects completed) of an optimal collect algorithm. For the left-hand side, the Attiya-Rachman algorithm completes one scan-update for every $O(\log n)$ write-collects done by any single process. So if c is set to n (to account for partially completed scan-updates), then

$$\frac{\mathrm{payoff}\, AB(A \circ B) + n}{\mathrm{payoff}(B)} \geq \frac{1}{O(\log n)}.$$

It follows that (8) holds for $l = O(\log n)$.

To use Theorem 2, it is also necessary to show that the technical condition (4) holds. But this just say that the maximum number of scan-updates that can be completed in a given schedule is at most a constant times the maximum number of write-collects. As observed above, this constant is one, and so Theorem 2 implies that plugging Follow-the-Bodies ($O(n^{1/2} \log^{3/2} n)$-competitive) into Attiya-Rachman ($O(\log n)$-competitive relative to collect) gives an $O(n^{1/2} \log^{5/2} n)$-competitive snapshot algorithm.

Observe that in the above argument, essentially no properties of the Attiya-Rachman snapshot algorithm were used except (a) the fact that scan-update is a "stronger" operation than either collect or write-collect, and (b) the fact that Attiya-Rachman uses $O(\log n)$ write-collects per scan-update. It is not unusual to be able to treat a parent algorithm as a black box in this way in traditional worst-case analysis— but this is one of the first examples of being able to do the same thing while doing competitive analysis. In addition, since almost any operation is stronger than write-collect in the wait-free model, a similar argument works for a wide variety of algorithms that use collects. Some additional examples are given in [5].

Because cooperative collect has a throughput competitiveness of $\Omega(\sqrt{n})$, using it directly as a subroutine can only give algorithms whose throughput competitiveness is also $\Omega(\sqrt{n})$. However, in a recent paper [7] Aumann has shown

[5] As noted in Section 5.1, for payoff models like the throughput model this requirement is not strictly necessary.

that the fundamental *consensus problem*, in which n processes must agree on a bit despite failures, can be solved using single-writer registers with $O(\log^4 n)$ competitiveness in both the latency and throughput models. The algorithm uses a weak version of cooperative collect in which freshness is not considered, thus avoiding the lower bound described in Section 6.6.

7 Example: Guessing your location in a metric space

So far we have seen examples of analyzing distributed problems using both the traditional definition of competitiveness and the semicompetitive definition of competitiveness from Section 4. For some problems neither approach is appropriate, and instead the notion of k-optimality gives more information about the actual merits of an algorithm.

One such problem can be described simply as guessing one's location in a metric space based on locally available information. Imagine that you have been placed by an adversary somewhere in a metric space (say, the surface of the Earth). You have the ability to observe the local landscape, and you would like to be able to make as close a guess as possible to your actual location based on what you see. In some locations (at the base of the Eiffel Tower; in a store that displays its wide selection of Global Positioning System boxes), the best possible guess is likely to be quite accurate. In others (somewhere in the Gobi desert; inside a sealed cardboard box), the best possible guess is likely to be wildly off. Yet it is clear that some algorithms for making such guesses are better than others.

How can one measure the performance of an algorithm for this problem? A traditional competitive approach might look something like this: the actual location x is the environment that is shared between the candidate and champion. Both candidate and champion are shown some view v, a function $f(x)$ of the location. The cost of a guess y is just the distance between y and x.

Unfortunately, in this framework, the optimal champion always guesses x correctly— at a cost of zero— while the candidate must choose among the many possible y for which $f(y) = f(x)$. Unless the candidate is lucky enough also to hit x exactly (which the adversary can prevent if it is only moderately clever), the competitive ratio of any algorithm is infinite.

No improvement comes from moving to a semicompetitive approach. Suppose that we assume that it is the observed view v that is shared between candidate and champion, but that the actual locations x and x^* may be any two points for which $f(x) = f(x^*) = v$. More formally, let us use the semicompetitive definition of k-competitiveness with the shared environment being the view and the differing inputs being the actual locations of the candidate and champion. Unfortunately, once again the champion (which is given a best-case input) guesses its location exactly, and unless the subset of the metric space that produces the appropriate view is very small indeed, the candidate once again has an infinite competitive ratio.

The solution is to use k-optimality. As above, let us assume that the the view v is shared between both candidate and champion, the actual locations x and x^* of the candidate and champion may be distinct; but in this case, suppose that *both* candidate and champion are given worst-case locations. Now the cost to the champion of choosing y^* will be the distance between y^* and the most distant x^* for which $f(x^*) = v$. If the candidate algorithm chooses any point y for which $f(y) = v$, then from the triangle inequality $d(y, x) \leq d(y, y^*) + d(y^*, x^*) \leq 2d(y^*, x^*)$, and so any candidate algorithm that is smart enough makes a guess consistent with what it sees will be 2-optimal. Particularly clever algorithms (say, those which always guess a *center* of the set of points consistent with the observed data) can be as good as 1-optimal. Though the range of optimality ratios is not large, it does give us a way to distinguish exceptionally stupid algorithms (worse than 2-optimal) from plausible algorithms (2-optimal) from good algorithms (1-optimal).

This example has been abstracted almost to the point of triviality, but it does illustrate issues that have been studied in the context of real distributed systems. The example is inspired by work on clock synchronization by Patt-Shamir and Rajsbaum [15]; in this work, the goal of a clock synchronization algorithm is to make a good estimate of the offsets between different clocks in a distributed system based on data piggybacked on the messages sent by some underlying algorithm. For some communication patterns (e.g., ones in which no messages are exchanged), it is impossible to make a very accurate estimate—but the possibility of using traditional competitive analysis is foreclosed by the ability of an off-line algorithm to guess the correct offsets exactly. Their solution was to use a notion of 1-optimality, which the definition in Section 4 generalizes.

Of course, how one actually computes a good estimate of the offsets between clocks in a distributed system based on limited information is a much more complicated problem than simply picking a good point somewhere in the middle of the appropriate metric space. The interested reader will find algorithms for solving several interesting variants of the problem in [15].

8 Conclusions

The preceding sections have concentrated on how the techniques of competitive analysis have been useful for analyzing fault-tolerant distributed algorithms, where the primary source of difficulty is not unpredictable inputs but unpredictable behavior in the underlying system. This approach is still quite new, and there are many questions that have yet to be considered. Not only may it be possible to apply competitive analysis to a wider variety of distributed models, but the adaptations needed to fit competitive analysis to the models described above may be useful for analyzing problems that do not necessarily involve distributed systems. There are three main areas of research suggested by this work.

Extensions to other models of distributed computation. The models described in Section 6 depend on a number of details of the wait-free shared-memory model.

This model is a good testbed for new measures, since the lack of restraints on the behavior of the adversary makes the model relatively simple from a formal perspective; however, in many cases other models are more realistic. These include models in which there are much tighter controls on the timing of events (for example, those in which synchrony or fairness conditions hold or in which a small upper bound exists on the number of failures), models with different communications primitives, and models in which not all processes can be trusted to carry out the algorithm correctly. Very little has been done to study how competitive analysis can be used to analyze how algorithms respond to hostile system behavior in these models.

Non-distributed applications of relative competitiveness. Relative competitiveness and the composition theorem were defined in [5] to analyze a class of problems that arise in a particular model of distributed systems. However, as the authors of [5] point out, when viewed abstractly there is nothing about relative competitiveness that depends on having a distributed system. The notion of relative competitiveness opens up the possibility of constructing competitive algorithms for a wide variety of problems by composition: building them up one subroutine at a time, with the competitiveness of each subroutine, proved separately, contributing to the competitiveness of the whole. With this ability, in addition to looking for good superproblems, like k-servers or Metrical Task Systems, that subsume many on-line problems, it may be fruitful to look for good subproblems whose solutions can be used as subroutines in many relative-competitive algorithms.

Limiting the clairvoyance of the off-line algorithm. Competitive analysis is not very useful for "lady or the tiger" problems (like the example in Section 7) in which an off-line algorithm can always guess correctly information that no on-line algorithm could ever hope to obtain. In many such cases it is known from real-world constraints that no algorithm of any kind could deduce the correct information. When this is true, it makes sense to limit the knowledge of the off-line algorithm to what might reasonably be available in the real-world situation being modeled. However, traditional competitive analysis provides no mechanism for doing so. [6] Here, a semicompetitive approach using k-optimality

[6] The technique of **comparative analysis** [14] might seem to be the right answer to these problems. Comparative analysis measures the value of information by comparing the best algorithms in two "information regimes" (e.g., paging algorithms with l-lookahead versus paging algorithms with no lookahead). A complication, from the point of view of someone trying to construct algorithms, is that the definition of comparative analysis assumes that the algorithm in the weaker class is chosen only after the algorithm in the stronger class is known. This does not prevent the stronger algorithm from knowing at birth bizarrely detailed properties of the request sequence (for example, a paging algorithm might know that if page 127 is requested first, the remainder of the request sequence includes twice as many even as odd-numbered pages); it just means that some hypothetical weaker algorithm may know these properties too. Thus comparative analysis is the opposite of what we want:

may be useful. Information that the champion algorithm is allowed to use on can be made part of the environment shared with the candidate. Information that the champion algorithm is not allowed to use can be made part of the input that is assumed to be worst-case for both champion and candidate.

Acknowledgments

Some of the material in this chapter has been adapted from [2], [5], and [4]. None of this work would have been possible without the efforts of my co-authors Miklos Ajtai, Cynthia Dwork, Will Hurwood, and Orli Waarts. This work was partially supported by NSF grants CCR-9410228 and CCR-9415410.

References

1. K. Abrahamson. On achieving consensus using a shared memory. In *Proceedings of the Seventh ACM SIGACT-SIGOPS Symposium on Principles of Distributed Computing*, pages 291–302, August 1988.
2. M. Ajtai, J. Aspnes, C. Dwork, and O. Waarts. A theory of competitive analysis for distributed algorithms. In *Proc. 35th Symp. of Foundations of Computer Science*, pages 401–411, 1994.
3. R. Anderson and H. Woll. Wait-free parallel algorithms for the union-find problem. In *Proc. 23rd ACM Symposium on Theory of Computing*, pages 370–380, 1991.
4. J. Aspnes and W. Hurwood. Spreading rumors rapidly despite an adversary. In *Proc. 15th ACM Symposium on Principles of Distributed Computing*, pages 143–151, 1996.
5. J. Aspnes and O. Waarts. Modular competitiveness for distributed algorithms. In *Proc. 28th ACM Symposium on the Theory of Computing*, pages 237–246, 1996.
6. H. Attiya and O. Rachman. Atomic snapshots in $o(n \log n)$ operations. In *Proc. 12th ACM Symposium on Principles of Distributed Computing*, pages 29–40, 1993.
7. Y. Aumann. Efficient asynchronous consensus with the weak adversary scheduler. In *Proc. 16th ACM Symposium on Principles of Distributed Computing*, 1997.
8. Y. Aumann and M.A. Bender. Efficient asynchronous consensus with the value-oblivious adversary scheduler. In *Proc. 23rd International Colloquium on Automata, Languages, and Programming*, 1996.
9. S. Ben-David, A. Borodin, R. Karp, G. Tardos, and A. Widgerson. On the power of randomization in on-line algorithms. In *Proc. 22nd Symposium on Theory of Algorithms*, pages 379–386, 1990.
10. T. Chandra. Polylog randomized wait-free consensus. In *Proc. 15th ACM Symposium on Principles of Distributed Computing*, 1996.
11. B. Chor, A. Israeli, and M. Li. Wait–free consensus using asynchronous hardware. *SIAM J. Comput.*, 23(4):701–712, August 1994. Preliminary version appears in *Proceedings of the 6th ACM SIGACT-SIGOPS Symposium on Principles of Distributed Computing*, pages 86-97, 1987.

---

instead of taking away the implausible knowledge of the off-line algorithm, it merely adds to the implausible knowledge of its on-line competitor.

12. S. Even and B. Monien. On the number of rounds needed to disseminate information. In *Proc. 1st ACM Symposium on Parallel Algorithms and Architectures*, 1989.
13. M. Herlihy. Wait-free synchronization. *ACM Transactions on Programming Languages and Systems*, 13(1):124–149, January 1991.
14. E. Koutsoupias and C. Papadimitriou. Beyond competitive analysis. In *Proc. 25th Symposium on Foundations of Computer Science*, pages 394–400, 1994.
15. B. Patt-Shamir and S. Rajsbaum. A theory of clock synchronization. In *Proc. 26th ACM Symposium on the Theory of Computing*, pages 810–819, 1994.
16. M. Saks, N. Shavit, and H. Woll. Optimal time randomized consensus — making resilient algorithms fast in practice. In *Proceedings of the Second Annual ACM-SIAM Symposium on Discrete Algorithms*, pages 351–362, 1991.
17. D. Sleator and R. E. Tarjan. Amortized efficiency of list update and paging rules. *Communications of the ACM*, 28:202–208, 1985.

7

On-line Packing and Covering Problems

János Csirik
Gerhard J. Woeginger

1 Introduction

This chapter deals with a couple of problems that are all related to the classical bin packing problem. In this problem, one is given a list L of items $\langle a_1, a_2, \ldots, a_n \rangle$, each item $a_i \in (0, 1]$, and the goal is to find a packing of these items into a minimum number of unit-capacity bins. This packing problem is one of the basic problems in Theoretical Computer Science and Combinatorial Optimization. It arises in many real world applications: e.g. in the (real-world) packing of (real-world) bins, in memory allocation in paged computer systems, in assigning newspaper articles to newspaper pages, in loading trucks, in packet routing in communication networks, in assigning commercials to station breaks on television, in cutting-stock problems, etc. The bin packing problem is well-known to be NP-hard [54], and thus research has concentrated on the study and development of approximation algorithms.

A packing algorithm is called *on-line* if it packs every item a_i solely on the basis of the sizes of the items a_j, $1 \leq j \leq i$, i.e., without any information on subsequent items. The decisions of the algorithm are irrevocable; packed items cannot be repacked at later times. The bin packing problem holds a special place, both in the history of approximation algorithms and in the history of on-line algorithms. Essentially, bin packing was the first problem that was carefully investigated in these two directions. Moreover, it was in the area of bin packing where for the first time the quality of an on-line algorithm was measured against the optimal off-line strategy (Johnson [71]), and it was in the area of bin packing where for the first time an adversary argument was used for deriving lower bound results (Yao [111]).

We survey *worst case* results and *average case* results for the classical bin packing problem, for generalizations to higher dimensions, for variants with several bin types, and for bin covering problems. We mention another survey by Coffman, Garey and Johnson [23] that covers the one dimensional topics. This chapter is structured as follows: The worst case results are contained in Sections 3–7, and the average case results in Sections 8–10. First, Section 2 collects the basic definitions on worst case ratios and expected ratios of approximation algorithms. Section 3 concentrates on the classical one dimensional bin packing problem. We describe old, new and very new algorithms and discuss lower bound techniques. Section 4 deals with higher dimensional generalizations of the bin packing problems where one has to pack vectors or boxes into boxes or strips. Section 5 treats dynamic bin packing (where the items do not only arrive but also leave after some time) and variable sized bin packing (where bins of different sizes are available). Section 6 deals with the one dimensional bin covering problem: Here the goal is to pack a list of items into a maximum number of bins such that the contents of each bin is at least one. Section 7 deals with some variants of on-line packing that are only close to being on-line: here we have some limited lookahead to some of the future items or earlier packed items may be rearranged in some limited way. Section 8 deals with average case results for the classical bin packing problem, Section 9 with average case results for higher dimensional generalizations, and Section 10 closes with average case results for covering problems.

2 Basic definitions

This section summarizes the essential definition for the worst case analysis and for the average case analysis of bin packing problems. For a list $L = \langle a_1, a_2, \ldots, a_n \rangle$ of items and an approximation algorithm A, we denote by $\mathrm{OPT}(L)$ and $A(L)$, respectively, the number of bins of size 1 used by an optimum algorithm and the number of bins used by algorithm A to pack the input list L. The *absolute worst case ratio* of A, denoted by R_A, is given by

$$R_A = \sup_L \{A(L)/\mathrm{OPT}(L)\}.$$

The *asymptotic worst case ratio* R_A^∞ of algorithm A is defined by

$$R_A^n = \max\{A(L)/\mathrm{OPT}(L) \mid \mathrm{OPT}(L) = n\},$$

$$R_A^\infty = \limsup_{n\to\infty} R_A^n.$$

The experiences of the last 25 years suggest that the asymptotic worst case ratio is the more reasonable measure of performance for the quality of a bin packing algorithm: It is robust against anomalies with a small number of bins in the optimum packing, and it also allows the packing algorithm more freedom while packing the first few bins. Throughout this chapter, we will often simply write *worst case ratio* instead of 'asymptotic worst case ratio'. In some applications, an a priori upper bound α $(0 < \alpha \leq 1)$ on the size of all items is known (intuitively, it is clear that small items are easier to pack than large items). To measure the performance of algorithms on lists with small items, the *parametric worst case ratio* $R_A^\infty(\alpha)$ is introduced by

$$R_A^n(\alpha) = \max\{A(L)/\mathrm{OPT}(L) \mid \mathrm{OPT}(L) = n \text{ and all } a_i \in L \text{ are } \leq \alpha\},$$

$$R_A^\infty(\alpha) = \limsup_{n\to\infty} R_A^n(\alpha).$$

In the *average case analysis* of bin packing problems, a standard convention is to assume that all items in the input list L have independent, identically distributed sizes. We denote a list L with n random items by $L_n = (X_1, X_2, \ldots, X_n)$, and we write $X_i \sim F$ to denote that the common distribution is F. With this convention, almost everything related to lists becomes a random variable: $\mathrm{OPT}(L_n)$, $A(L_n)$, $R_A(L_n) = A(L_n)/\mathrm{OPT}(L_n)$, and $s(L_n)$ (the total item size). We will also investigate $W_A(L_n) = A(L_n) - s(L_n)$ which is the empty (= wasted) space when packing list L_n according to algorithm A. For measuring the quality of the average case behaviour of A, we define

$$R_A^n(F) = E[R_A(L_n)] = E\left[\frac{A(L_n)}{\mathrm{OPT}(L_n)}\right]$$

and

$$W_A^n(F) = E[A(L_n) - s(L_n)]$$

for lists L_n with item size distribution F. The *asymptotic expected ratio* of A under F is given by

$$R_A^\infty(F) = \lim_{n\to\infty} R_A^n(F).$$

The *uniform distribution* on $[a, b]$ with $0 \leq a < b \leq 1$ will be denoted by $U(a, b)$. In this chapter, we will mainly deal with average case results for such uniform distributions.

Next, let $t_1 = 2$ and $t_{i+1} = t_i(t_i - 1) + 1$ for $i \geq 1$. This sequence starts with $\langle 2, 3, 7, 43, 1807, 3263443, \ldots \rangle$ and its growth is doubly exponential. The sequence turned out to be essential in the analysis of on-line bin packing algorithms, and also in proving lower bound results for on-line bin packing. Moreover, the closely related number

$$h_\infty = \sum_{i=1}^{\infty} \frac{1}{t_i - 1} = 1 + \frac{1}{2} + \frac{1}{6} + \frac{1}{42} + \frac{1}{1806} + \ldots \approx 1.69103$$

will appear many times in our statements.

3 Worst case analysis for one dimensional bin packing

This section is divided into four subsections: Subsection 3.1 deals with the 'classical' packing algorithms like NEXT FIT, FIRST FIT, BEST FIT, and WORST FIT. These are the algorithms that were introduced and analyzed in the beginning of the 1970s. Subsection 3.2 considers *bounded-space* algorithms. Such algorithms manage to pack with only a small number of simultaneously open bins. Subsection 3.3 describes the algorithms that currently yield the best worst case performance of all known on-line algorithms. Finally, Subsection 3.4 investigates lower bounds on the asymptotic worst case ratio of one dimensional bin packing algorithms.

3.1 Classical algorithms

The simplest, oldest and fastest on-line algorithm for one dimensional bin packing is the NEXT FIT (NF): NF always packs a newly arriving item a_i into a unique, so-called *active* bin. In case item a_i does not fit into the active bin, the current active bin is closed (and never used again) and an empty bin is opened and becomes the next active bin. Clearly, the running time of NF is linear in the number of items packed. It is easy to show that $\mathrm{NF}(L) \leq 2\mathrm{OPT}(L)$ for all lists L. Moreover, the list $L = \langle \frac{1}{2}, \varepsilon, \frac{1}{2}, \varepsilon, \frac{1}{2}, \varepsilon, \ldots \rangle$ consisting of n pairs $\frac{1}{2}$, ε where $0 < \varepsilon < \frac{1}{n}$ demonstrates that this bound is tight. Hence, $R^\infty_{NF} = 2$ holds.

Theorem 1. (Johnson [71, 72]) *For* $1/2 \leq \alpha \leq 1$, $R^\infty_{NF}(\alpha) = 2$ *and for* $\alpha < 1/2$, $R^\infty_{NF}(\alpha) = 1/(1-\alpha)$.

The worst case example above illustrates that NEXT FIT sometimes wastes a lot of space, since it does not reuse the empty space in earlier packed bins. This naturally leads to the FIRST FIT (FF) algorithm: Here, during the packing we never close active bins. When packing some new item a_i, FF puts it into the lowest indexed bin into which it will fit. A new bin is started only in the case where a_i will not fit into any non-empty bin. The $O(n \log n)$ running time of FF is slower than the $O(n)$ running time of NEXT FIT. However, the worst case ratio of FF indeed yields an improvement over NEXT FIT.

Theorem 2. (Johnson, Demers, Ullman, Garey, Graham [73]) *Let* $m \in \mathbb{N}$ *with* $\frac{1}{m+1} < \alpha \leq \frac{1}{m}$. *Then for the FIRST FIT algorithm,*

- $R^\infty_{FF}(\alpha) = 17/10$ *holds for* $m = 1$, *and*
- $R^\infty_{FF}(\alpha) = (m+1)/m$ *for* $m \geq 2$.

The proof of the 17/10-result is based on a so-called *weighting function* $w : [0,1] \to [0, 8/5]$ which assigns to the size a_i of every item a corresponding weight $w(a_i)$. Intuitively, this weight $w(a_i)$ measures the space that item a_i uses when packed according to FF (where space equals the size of item plus some wasted space). For $w(L) = \sum_{i=1}^n w(a_i)$, it then can be shown that (i) $\mathrm{FF}(L) \leq w(L) + 2$ and that (ii) $w(L) \leq (17/10)\mathrm{OPT}(L)$ holds.

Algorithm	Time	R_A^∞	$R_A^\infty(\frac{1}{2})$	$R_A^\infty(\frac{1}{3})$	$R_A^\infty(\frac{1}{4})$
NEXT FIT	$\Theta(n)$	2.0	2.0	1.5	1.33...
WORST FIT	$\Theta(n \log n)$	2.0	2.0	1.5	1.33...
ALMOST WORST FIT	$\Theta(n \log n)$	1.7	1.5	1.33...	1.25
FIRST FIT	$\Theta(n \log n)$	1.7	1.5	1.33...	1.25
BEST FIT	$\Theta(n \log n)$	1.7	1.5	1.33...	1.25

Table 1. Asymptotic worst case ratios for the classical algorithms.

At first sight, the packing rule of FF ('use the lowest indexed bin') seems to be rather primitive. Obvious variants for this packing rule yield the BEST FIT (BF), WORST FIT (WF) and ALMOST WORST FIT (AWF): BEST FIT behaves like FIRST FIT, except that the new item a_i is placed in the bin into which it will fit with the smallest gap left over (ties are broken arbitrarily). WORST FIT puts a_i into the non-empty bin with the largest gap (ties are broken arbitrarily), starting a new bin only if this largest gap is not big enough. ALMOST WORST FIT tries to put a_i into the non-empty bin with second largest gap first, and in case a_i does not fit AWF behaves like WORST FIT.

All these three algorithms belong to the class of so-called ANY FIT algorithms: An ANY FIT (AF) algorithm *never* puts an item a_i into an empty bin unless the item does not FIT into ANY partially filled bin. Similarly, an ALMOST ANY FIT (AFF) algorithm is an ANY FIT algorithm that never puts an item into a partially filled bin with lowest level, unless there is more than one bin with lowest level or unless that bin is the only one that has enough room. The following surprising result indicates a borderline for all algorithms defined till now.

Theorem 3. (Johnson [72]) *For any* $0 < \alpha \leq 1$,

- *For every AF algorithm A,* $R_A^\infty(\alpha) \geq R_{FF}^\infty(\alpha)$ *holds.*
- *For every AAF algorithm A,* $R_A^\infty(\alpha) = R_{FF}^\infty(\alpha)$ *holds.*

Table 1 states the worst case ratios for the introduced algorithms together with some of their parametric worst case ratios. Note that ALMOST WORST FIT – somewhat surprisingly – performs much better than WORST FIT does.

3.2 Bounded-space algorithms

An on-line bin packing algorithm is said to use *k-bounded-space* if for each item a_i, the choice of bins into which it may be packed is restricted to a set of k or fewer *active* bins, where each bin becomes active when it receives its first item; once a bin is declared *inactive* (or *closed*), it can never become active again. Since for $k = 1$, the only reasonable k-bounded-space algorithm is the NEXT FIT with worst case ratio 2, we assume from now on that $k \geq 2$.

The bounded-space restriction models situations in which bins are exported once they are packed. Consider the problem of packing trucks at a loading dock that has positions for only k trucks. If the next item to be packed does not fit into any of the trucks currently backed up to the dock, we will need a new truck. If there are already k trucks present, one of them will have to drive away to make room (and presumably to start making its deliveries). Alternatively, consider a communication channel in which information moves in large fixed-size blocks. If these blocks are filled with smaller packets of various sizes that must be assigned to blocks as they arrive at the entrance to the channel, we have an on-line bin packing problem. If the buffer for the channel input has only room for k blocks, we also are subject to the k-bounded memory restriction.

The most natural bounded-space bin packing algorithms are defined via simple *packing* rules for items and *closing* rules for bins: A new item may always be packed into the lowest indexed bin (as in the FIRST FIT) or into the bin with the smallest remaining gap (as in the BEST FIT). If the new item does not fit into any active bin, some active bin has to be closed; in this case one may always choose the lowest indexed bin (the FIRST bin) or the fullest bin (the BEST bin). This leads to four algorithms AFF_k, AFB_k, ABB_k, and ABF_k that are classified according to these rules: Here A stands for *Algorithm*, the second letter denotes the packing rule (Best fit or First fit), the third letter denotes the closing rule (Best bin or First bin), and k is an upper bound on the number of active bins.

- Algorithm AFF_k. This algorithm was already introduced in 1974 by Johnson [72] under the name NEXT-k FIT (NF_k). Just fifteen years later, Csirik and Imreh [37] discovered a sequence of worst case examples that imply $R^\infty_{NF_k} \geq 1.7 + \frac{3}{10(k-1)}$. Mao [90] proved a matching upper bound on $R^\infty_{NF_k}$.
- Algorithm AFB_k. Zhang [112] showed that for this algorithm, the analysis of Mao [90] can be adapted to prove the same worst case bound as for AFF_k. Thus, $R^\infty_{AFB_k} = 1.7 + \frac{3}{10(k-1)}$.
- Algorithm ABF_k. Mao [89] proved that $R^\infty_{ABF_k} = 1.7 + \frac{3}{10k}$.
- Algorithm ABB_k. This algorithm was investigated by Csirik and Johnson [38] under the name k-BOUNDED BEST FIT (BBF_k). In some sense, BBF_k is the most clever of the four algorithms since it uses the better packing and the better closing rule. Csirik and Johnson confirmed in a rather sophisticated proof that best is indeed better than first: $R^\infty_{BBF_k} = 17/10$ always holds, independently of the value of k.

Summarizing, the worst case guarantees of all four algorithms tend to 1.7 as the number k of active bins tends to infinity. The big surprise is that BBF_2 performs as well as FIRST FIT and BEST FIT, while using only *two* active bins. However, since all four algorithms fulfill a kind of ANY FIT constraint, they can not beat FIRST FIT. To find an algorithm with $R^\infty_A < 1.7$, it thus is necessary to start new bins even if the current item fits into one of the active bins.

In this spirit, C.C. Lee and D.T. Lee [83] introduced the HARMONIC algorithm H_k. This algorithm is based on a nonuniform partition of the interval

(0, 1] into k subintervals where the partitioning points are $1/2, 1/3, \ldots, 1/k$. To each of these subintervals there always corresponds a single active bin and only items belonging to this subinterval are packed into this corresponding bin. If a new item arrives that does not fit into its corresponding bin, this bin is closed and a new bin is used.

Theorem 4. (Lee and Lee [83]) *As k tends to infinity, $R^{\infty}_{H_k}$ tends to the number $h_\infty \approx 1.69103$.*

It can be shown that algorithm H_7 (with at most seven active bins) has a worst case ratio of 1.6944 and thus beats FIRST FIT. In fact, there is another bounded-space algorithm SH_6 that beats FIRST FIT while using only *six* active bins: It behaves very similar to H_7, but uses only a single active bin for the items in (1/6, 1/4]. This algorithm belongs to the class of SIMPLIFIED HARMONIC (SH_k) algorithms introduced by Woeginger [109]. The SH_k are modifications of the H_k that use a more complicated interval structure.

k	NF_k	ABF_k	BBF_k	H_k	SH_k	Champion
2	2.00000	1.85000	1.70000	2.00000	2.00000	BBF
3	1.85000	1.80000	1.70000	1.75000	1.75000	BBF
4	1.80000	1.77500	1.70000	1.71429	1.72222	BBF
5	1.77500	1.76000	1.70000	1.70000	1.70000	BBF, H, SH
6	1.76000	1.75000	1.70000	1.70000	1.69444	SH
7	1.75000	1.74286	1.70000	1.69444	1.69388	SH
8	1.74286	1.73750	1.70000	1.69388	1.69106	SH
9	1.73750	1.73333	1.70000	1.69345	1.69104	SH
∞	1.70000	1.70000	1.70000	1.69103	1.69103	H, SH

Table 2. Asymptotic worst case ratios for bounded-space algorithms, rounded to five decimal places.

A summary of the worst case ratios of the bounded-space algorithms for some small values of k is given in Table 2. In the column for H_k, the values for $k \leq 7$ are the exact worst case ratios and the values for $k \in \{8, 9\}$ are upper bounds on the worst case ratio. The tight results for $k \in \{4, 5\}$ are due to van Vliet [108], the other results are from Lee and Lee [83]. Note that the worst case ratios of all five algorithms always remain above h_∞. In fact, for bounded-space algorithms it is not possible to do better:

Theorem 5. (Lee and Lee [83]) *Every bounded-space on-line bin packing algorithm A fulfills $R^{\infty}_{A} \geq h_\infty$.*

The worst case ratios of H_k and SH_k approach h_∞ in the limit. However, none of the known bounded-space algorithms reaches the worst case ratio h_∞ while using a *finite* number of active bins. Deciding whether this is possible at all remains an open problem. The parametric worst case ratio of H_k was analyzed by Galambos [46] (see Table 3 for some of these results).

3.3 The current champion

The first on-line algorithm for bin packing with $R_A^\infty < h_\infty$ was the REVISED FIRST FIT (RFF) designed by Yao [111] in 1980. Both, the definition and the analysis of RFF are rather involved: RFF is essentially based on FIRST FIT, but similarly as the H_k it uses separate bins for items from the intervals $(0, 1/3]$, $(1/3, 2/5]$, $(2/5, 1/2]$, and $(1/2, 1]$, respectively. Moreover, every sixth item from the interval $(1/3, 2/5]$ receives a special treatment. Yao proved that RFF has $R_{RFF}^\infty = 5/3$.

In the same paper in which they introduced the harmonic algorithms, Lee and Lee [83] described the REFINED HARMONIC (RH) algorithm with worst case ratio $R_{RH}^\infty = 373/228 \approx 1.639$. RH is based on H_{20}, but the intervals $(1/3, 1/2]$ and $(1/2, 1]$ both are broken into two corresponding subintervals. As a next step, Ramanan, Brown, Lee and Lee [94] investigated another modification of the harmonic algorithm which they called MODIFIED HARMONIC (MH). MH is a refinement of the REFINED HARMONIC, uses a more complicated interval structure, and fulfills $R_{MH}^\infty \approx 1.616$. Ramanan et al [94] also prove that their underlying basic approach can never yield a ratio $R_A^\infty < 1.583$.

The current champion algorithm is the HARMONIC+1 of Richey [101] introduced in 1991. The definition of HARMONIC+1 uses more than 70 intervals. Its design and analysis was performed with the help of several large linear programs. Its worst case ratio is less than 1.5888.

3.4 Lower bounds

Theorem 5 yields a lower bound on the performance of bounded-space bin packing algorithms. To derive lower bounds on the performance of *general* on-line bin packing algorithms, one uses ideas from everyday life: Assume that you have to pack your suitcase. Then you will usually start with first packing the large items and afterwards you will try to fit in the smaller items. Therefore, if one wants to make some algorithm A to behave poorly, it is natural to confront it first with a set of small items. In case A packs these small items very tight, it will not be able to pack the larger items that may arrive later. In case A leaves lots of room for large items while packing the small items, the large items will not arrive. In either case, the produced packing will be very bad.

Yao [111] was the first to translate this idea into a more mathematical formulation: He first confronted the algorithm with a list consisting of n items of size $1/7+\varepsilon$ (small items), then with n items of size $1/3+\varepsilon$ (medium size items), and finally with n items of size $1/2+\varepsilon$ (large items). Yao showed that for this list, every on-line algorithm must be worse than the optimum by a factor of $3/2$, either after the first or after the second or after the third sublist. Hence, every on-line bin packing algorithm fulfills $R_A^\infty \geq 1.5$. Just a short time later, Brown

[15] and Liang [87] simultaneously and independently from each other generalized this lower bound to 1.53635. They used lists with items of sizes $1/1807+\varepsilon$, $1/43+\varepsilon$, $1/7+\varepsilon$, $1/3+\varepsilon$ and $1/2+\varepsilon$, respectively. Note that these item sizes correspond to the numbers t_i, $1 \leq i \leq 5$, introduced in Section 2 for the definition of h_∞. The analysis in [87] is based on some hand-woven inequalities and takes some effort to read. It took more than ten years until van Vliet [106, 107] was able to give a tight analysis of the Brown/Liang construction. He did this by designing a very elegant linear programming formulation.

Theorem 6. (Van Vliet [106, 107]) *For any on-line bin packing algorithm A, the bound* $R_A^\infty \geq 1.5401$ *holds.*

Note that the gap between this lower bound and the upper bound of 1.5888 described in Section 3.3 is less than 0.05. Galambos [45] and Galambos and Frenk [48] simplified and extended the Brown/Liang construction to the parametric case. Combining these results with the linear programming formulation of van Vliet [106] yields the lower bounds given in Table 3. For comparison, corresponding upper bounds for some algorithms are also given.

α	1	1/2	1/3	1/4	1/5
Lower bound on $R_A^\infty(\alpha)$	1.540	1.389	1.291	1.229	1.188
Current champion	1.588	1.423	1.302	1.234	1.191
$R_{H_k}^\infty(\alpha)$	1.691	1.423	1.302	1.234	1.191
$R_{FF}^\infty(\alpha)$	1.700	1.500	1.333	1.250	1.200

Table 3. Lower bounds for the parametric case.

Chandra [18] argued that all these lower bounds can be carried over to *randomized* on-line bin packing algorithms (algorithms that are allowed to flip coins and act according to the outcome of these flips) even in the case of oblivious adversaries.

4 Worst case analysis for higher dimensional packing

The one dimensional bin packing problem has been generalized to higher dimensions in several ways. First, there is the generalization to *vectors* as treated in Subsection 4.1: Instead of packing numbers, one has to pack vectors subject to the constraint that in every bin the sum of all vectors is at most one in every coordinate. Subsection 4.2 deals with the generalization to *boxes*: Here one has to pack geometric boxes into unit cubes. The boxes must not be rotated and different boxes must not overlap. Subsection 4.3 deals with another geometric generalization called *strip packing*: The items are two dimensional objects that have to be packed into a vertical strip so as to minimize the total height of the strip needed.

4.1 Vector packing

Instead of each item being a single number, in the d-dimensional vector packing problem each item is a d-dimensional vector $a_i = (v_1(a_i), \ldots, v_d(a_i))$, where $0 \leq v_j(a_i) \leq 1$ holds for $1 \leq j \leq d$. The goal is to pack all items into the minimum number of bins in such a way that in every bin the sum of all vectors is at most one in every coordinate. The vector packing problem arises as a crucial subproblem in scheduling with resource constraints (cf. Garey, Graham, Johnson and Yao [53]).

Kou and Markowsky [79] call an approximation algorithm A for d-dimensional vector packing *reasonable*, if A never yields packings in which the contents of two non-empty bins could be combined into a single bin (this obviously corresponds to the ANY FIT algorithms described in Section 3.1 for the one dimensional problem). Kou and Markowsky show that any reasonable vector packing algorithm A obeys the bound $R_A^\infty \leq d+1$. We note that the obvious generalization of FIRST FIT to d-dimensional vector packing is reasonable and hence obeys the above-mentioned (not very impressive) bound. In 1976, Garey, Graham, Johnson and Yao [53] performed an exact analysis of the d-dimensional FIRST FIT and derived a slight improvement over $d+1$:

Theorem 7. (Garey, Graham, Johnson, Yao [53]) *The generalization of FIRST FIT to d-dimensional vector packing has worst case ratio* $R_{FF}^\infty = d + 7/10$.

Note that this theorem reduces to the familiar 17/10 result in the one dimensional case for the standard FIRST FIT (cf. Theorem 2). Until today, these ancient results provide the best known worst case bounds for d-dimensional on-line vector packing. Especially, the question whether $R_A^\infty < d$ is possible for some on-line approximation algorithm A remains open (in fact, it even remains open for *off-line* approximation algorithms!).

Now let us turn to lower bounds on the worst case ratios of on-line algorithms for vector packing. Yao [111] proved that in a reasonable model of computation, any off-line approximation algorithms with $o(n \log n)$ running time must have a worst case ratio $R_A^\infty \geq d$. Since on-line algorithms form a subclass of the off-line algorithms, this result indicates that no fast and simple on-line vector packing algorithm can break the barrier of d. Next, we observe that the 1.54 lower bound for one dimensional bin packing (cf. Theorem 6) trivially carries over to vector packing in $d \geq 2$ dimensions: We just replace every item a_i in the one dimensional construction by a vector $(a_i, a_i, \ldots, a_i)$. Slight improvements on this observation were made by Galambos, Kellerer and Woeginger [49] and then in turn were slightly improved by Blitz, van Vliet and Woeginger [14]. The exact values of these lower bounds for dimensions $2 \leq d \leq 8$ are depicted in Table 4. The underlying technique for proving all these lower bounds is essentially the same as for the one dimensional case: One designs an appropriate list L of items (L usually consists of a small number of homogeneous sublists), and then derives a number of combinatorial properties for the packing patterns of prefixes of L. The goal is to show that *every* algorithm has to perform badly on one of these prefixes. Hence, these lower bounds are tedious to derive, but nevertheless are

d	2	3	4	5	6	7	8	∞
Lower bound [49]	1.6707	1.7509	1.8003	1.8334	1.8572	1.8750	1.8889	2.000
Lower bound [14]	1.6712	1.7584	1.8326	1.8729	1.8980	1.9149	1.9271	2.000

Table 4. Lower bounds for d-dimensional on-line vector packing, rounded to four decimal places.

not impressive: as the dimension d tends to infinity, the bounds tend to 2. The main open problem in on-line d-dimensional vector packing consists in narrowing the wide gap between 2 and $d + 7/10$.

4.2 Box packing

Box packing is a purely geometric way of generalizing the one dimensional bin packing problem to d dimensions: Similarly as in the vector packing problem, each item a_i is described by a d-dimensional vector $a_i = (v_1(a_i), \ldots, v_d(a_i))$, with $0 \leq v_j(a_i) \leq 1$ for $1 \leq j \leq d$. However in this version of the problem these vectors determine the dimensions of a d-dimensional box, a d-dimensional solid geometric object. The goal is to pack all items into the minimum number of bins (which now are d-dimensional unit cubes) in such a way that (i) each item is entirely contained inside its bin with all sides parallel to the sides of the bin, (ii) no two items in a bin overlap and (iii) the orientation of any item is the same as the orientation of the bin (in other words, the items must not be rotated).

Most of the approximation algorithms developed for d-dimensional box packing are off-line algorithms (e.g. Chung, Garey, and Johnson [20]). The first on-line algorithm for 2-dimensional box packing is due to Coppersmith and Raghavan [29]. The main idea of this algorithm is to first round up the larger dimension of a newly arrived item to some number in the set $\{2^{-k}, 2^{-k}/3 | k \geq 0\}$ and then to pack it according to a classification scheme, in a similar way as the HARMONIC algorithm. Since the rounding just wastes a constant factor and since the packing just wastes another constant factor, it can be shown that this yields an on-line approximation algorithm CRA with worst case ratio $R^{\infty}_{CRA} \leq 3.25$. Csirik, Frenk and Labbé [34] took a closer look at the packing scheme of the CRA algorithm. They showed that if one packs items belonging to the same class according to the NEXT FIT rule, then $R^{\infty}_{CRA} = 3.25$. However, in case one packs these items according to the FIRST FIT rule then the worst case ratio drops to $R^{\infty}_{CRA} \leq 3.06$. As a next result, Csirik and van Vliet [41] combined a rounding procedure that is more complex than that of CRA with a clever generalization of the harmonic algorithm to d dimensions. The resulting algorithm is the current champion for all dimensions $d \geq 2$.

Theorem 8. (Csirik and van Vliet [41]) *For d-dimensional on-line box packing, there exists an on-line algorithm with worst ratio bounded by* $h^d_{\infty} \approx 1.69^d$.

The situation for lower bounds for the d-dimensional box packing problem is similarly disappointing as it is in the case of d-dimensional vector packing: The first improvement over the trivial 1.54 bound induced by the one dimensional case was done by Galambos [47]. He provides a lower bound of 1.6 for the 2-dimensional box packing problem. Galambos and van Vliet [50] apply a refinement of the same proof technique to derive a better bound of 1.808. In his Ph.D. thesis [107], van Vliet extends his construction to establish a lower bound of 1.851 at the cost of having to work with *very* complex packing patterns. Currently, the best lower bound known for 2-dimensional box packing is 1.907 which is due to Blitz, van Vliet and Woeginger [14]. For 3-dimensional box packing, [14] provides a lower bound of 2.111. Any straightforward generalization of these lower bound methods to $d \geq 4$ dimensions will yield values that are smaller than 3. Since the current champions for d-dimensional box packing have worst case ratios that are exponential in d, there is ample space for improvements.

Lassak and Zhang [82] and Januszewski and Lassak [69] investigate a somewhat different type of on-line box packing. Their goal is to pack a set of boxes into a *single* unit bin. Among other results, they prove that in two dimensions every set of boxes with total area at most 0.0903 can be packed on-line into the unit square. Moreover, every set of squares with total area at most 0.3125 can be on-line packed into the unit square. For $d \in \{3,4\}$, they show that every set of cubes with total volume at most $\frac{3}{2}(\frac{1}{2})^d$ can be packed into the d-dimensional unit cube. For $d \geq 5$, the volume bound is $2(\frac{1}{2})^d$. The result for $d \geq 5$ is as good as the best possible corresponding off-line result of Meir and Moser [91]: Two cubes of edge length $\frac{1}{2} + \varepsilon$ cannot be packed together.

4.3 Strip packing and TETRIS

In $d = 2$ dimensions, there is another reasonable version of higher dimensional bin packing besides box packing. Again, all items are boxes with given heights and widths. The goal is to pack them into a vertical strip of width W so as to minimize the total height of the strip needed. Without loss of generality, we may assume that $W = 1$ and that the widths of all items are at most one (this can be reached by scaling all these numbers by an appropriate factor). The boxes must be packed orthogonally; in general, rotations are forbidden. This 2-dimensional version of bin packing is called *strip packing*.

This problem has many applications in VLSI-design or stock-cutting. The application to stock-cutting is as follows. In a variety of industrial settings, the raw material comes in rolls (rolls of cloth, rolls of paper, rolls of sheet material). From these rolls we want to cut patterns (for clothes, labels, boxes) or merely smaller and shorter rolls. In the simplest case we can view the objects we wish to cut from the rolls as being or approximating boxes. We minimize our wastage if we minimize the amount of the roll (i.e., the strip length) used. Some form of orthogonality may be justified, since in many applications the cutting is done by blades parallel or perpendicular to the length of the material, and the material may have a bias that dictates the orientation of the boxes.

Similarly as for d-dimensional box packing, most of the algorithms developed for strip packing are off-line algorithms. Typically, they preorder the items by decreasing height or by decreasing width (see e.g. the papers by Sleator [105], Baker, Coffman and Rivest [10], and Baker, Brown and Katseff [7]). The only on-line algorithms analyzed from a worst case point of view for this problem are the *shelf algorithms* introduced by Baker and Schwarz [11]. Before describing these approximation algorithms, we stress the fact that they assume that the heights of all boxes are also bounded by 1.

Note that in case all the boxes have the same height, the 2-dimensional problem collapses to the classical one dimensional problem: in an optimal packing, the bins may be placed in rows or 'shelfs'. Each shelf in the packing then corresponds to a bin and the height of the strip packing corresponds to the number of bins used. The basic idea of the shelf algorithms is to classify the arriving boxes according to their heights. The packing is then constructed as a sequence of shelfs, and boxes of similar heights are packed into the same shelfs. The main differences between distinct shelf algorithms consist in the way how boxes within the same class are packed. More precisely, each shelf algorithm takes a parameter $0 < r \leq 1$, which is a measure of how much wasted space can occur. All shelfs have heights r^j, $j \geq 0$. Whenever a box (h_i, w_i) is to be packed, one first determines the value of j for which $r^{j+1} < h_i \leq r^j$ holds. If there is already a shelf of height r^j in the packing, the NEXT FIT SHELF algorithm checks whether the box fits into the current active one and in case it fits, it packs the box there. Otherwise, the box is placed into a newly opened shelf of height r^j which becomes the currently active one for that height. The FIRST FIT SHELF algorithm behaves analogously, but tries to fit the new box into the *first* shelf of corresponding height into which it will fit.

Theorem 9. (Baker and Schwarz [11]) *For the shelf packing algorithms NEXT FIT SHELF and FIRST FIT SHELF with parameter* $0 < r \leq 1$, $R^\infty_{NFS} = 2/r$ *and* $R^\infty_{FFS} = 1.7/r$ *holds. For* $r = 0.622$, *the* absolute *worst case ratio of FFS obeys* $R_{FFS} \leq 6.9863$.

Note that these worst case ratios approach the corresponding worst case ratios of NEXT FIT and of FIRST FIT as r approaches 1. In an analogous way, Csirik and Woeginger [42] transform the HARMONIC algorithm into the corresponding HARMONIC SHELF algorithm. The worst case ratios of the HARMONIC SHELF algorithm can be made arbitrarily close to h_∞. Improving on h_∞ in on-line strip packing is probably difficult: Csirik and Woeginger [42] argue that no shelf algorithm can yield a worst case ratio better than h_∞. Hence, a completely different approach will be necessary.

Lower bounds on the worst case ratio of any on-line strip packing algorithm are investigated by Baker, Brown and Katseff [8]. Among many other results for algorithms with preorderings, they prove that the *absolute* worst case ratio of any on-line strip packing algorithm A must fulfill $R_A \geq 2$. For the *asymptotic* worst case ratio, the 1.54 lower bound can be carried over from the one dimensional case to strip packing.

Li and Cheng [86] investigate the 3-dimensional version of strip packing: Here the strip is a column with a unit square cross section and the items are 3-dimensional boxes. [86] proposes and analyzes nine on-line generalizations of the 2-dimensional shelf algorithms. Each of these algorithms uses a separate packing heuristic in each dimension. The overall worst case ratio of such an algorithm equals the product of the worst case ratios of the individual heuristics used independently in different dimensions. By choosing certain parameters appropriately, one arrives at an algorithm with worst case ratio 2.897.

Up to now we assumed that every newly arrived box can be placed without problems into any free space of the size of the rectangle. Azar and Epstein [5] consider a variant that also takes into consideration the way how the box is moved to its place. In their model, a box arrives from the top of the strip and has to be moved continuously around only in the free space until it reaches its place, and afterwards cannot be moved again. As in classical strip packing, the goal is to minimize the total height of the strip needed. For the case where rotations are not allowed, [5] prove that no on-line algorithm can have a finite worst case guarantee. The argument uses boxes that become arbitrarily thin. However, if every item has a width of at least β then there exists an on-line packing algorithm with worst case ratio $O(\log(1/\beta))$, and no worst case ratio better than $\Omega(\sqrt{\log(1/\beta)})$ is possible. For the case where rotations by ninety degrees are allowed, Azar and Epstein construct an on-line algorithm with worst case ratio 4.

We close this section with a short discussion of the computer game TETRIS that also is a kind of on-line strip packing problem. Designed by the Soviet mathematician Alexey Pazhitnov in the late eighties, TETRIS won a record number of software awards in 1989. In this game, the board is ten units wide and twenty units tall; when the game starts the board is empty. Then tetrominoes (= groups of four connected cells, each cell covering exactly one grid square) appear at the top of the board and fall down. When a tetromino reaches a point where it can fall no further without two cells overlapping, it remains in that spot and the next tetromino appears at the top of the board. The player uses rotations and horizontal translations to guide the falling items. When a row is completely filled, the cells in that row are removed from the board. As soon as the packing height exceeds 20, the player loses. In his Master's thesis, Brzustowski [16] developed an adversary strategy that makes every on-line algorithm eventually lose the game. Burgiel [17] showed that even *off-line* algorithms (that know the whole sequence of tetrominoes in advance) can be forced to lose eventually: Just confront them with a sequence that alternates between left-handed and right-handed Z-shaped items. After at most 70.000 items, the player will lose.

5 Dynamic bin packing and variable sized bin packing

In **dynamic bin packing**, the items a_i only stay in the system for some (initially unknown) time period. Hence, there are two possible types of information given to the packer in every step: Either, the packer learns about the arrival of some new item a_i (*insert event*), or the packer is told that now some earlier arrived item leaves the system (*delete event*). At any moment in time, the total size of items currently assigned to any bin must not exceed 1. This problem models memory allocation problems in paged computer systems (cf. Coffman [21]).

For an input list L and an approximation algorithm A, we denote by $A(L)$ the maximum number of bins occupied by A while it is packing L. We shall compare the quality of an approximation algorithm against an optimum packer that is allowed to repack the current packing into an optimum packing each time a new item arrives; $\mathrm{OPT}(L)$ is the maximum number of bins used by such an optimum packer while packing L. In analogy to the asymptotic worst case ratio for classical bin packing, the worst case ratio R_A^{dyn} of algorithm A is defined by

$$R_A^{dyn} = \lim_{k\to\infty} \sup_L \{A(L)/\mathrm{OPT}(L) \mid \mathrm{OPT}(L) = k\}.$$

Coffman, Garey and Johnson [22] investigate a variant of FIRST FIT for dynamic bin packing: Every time a new item arrives, it is packed into the lowest indexed occupied bin that has sufficient available space; if no such bin exists, a new bin is opened. Every time an item a_i is deleted, the available space in the corresponding bin is simply increased by a_i. Coffman et al [22] prove that $2.75 \le R_{FF}^{dyn} \le 2.897$. A slightly modified version FFx of FF that packs every item $> 1/2$ into its own bin and never adds small items to such a large item behaves better: $2.75 \le R_{FFx}^{dyn} \le 2.778$ holds. Moreover, no on-line algorithm for dynamic bin packing can beat a lower bound of 2.5. The situation drastically improves in the parametric version where all item sizes are bounded by $\alpha \le 1/2$: In these cases, worst case ratios of at most 1.788 are possible. Moreover, the lower bounds drop to $1 + \alpha + \alpha^2 + O(\alpha^3)$, and the parametric worst case ratios of the FIRST FIT match these lower bounds up to a term of order $O(\alpha^3)$.

In **variable sized bin packing**, we are given several different types $B_1, \ldots, B_r$ of bins with sizes $1 = s(B_1) > s(B_2) > \ldots > s(B_r)$. There is an infinite supply of bins of each size. The problem is to pack a list of items $a_i \in [0,1]$ into a set of bins with *smallest total size* (observe that for the case where all bins are of size one, this is just the classical one dimensional bin packing problem). For a list L of items and an approximation algorithm A, let $s(A, L)$ denote the total size of bins used by algorithm A. Let $s(\mathrm{OPT}, L)$ denote the total size of bins used in an optimal packing. Then, we measure the quality of algorithm A by

$$R_A^{var} = \lim_{k\to\infty} \sup_L \{s(A,L)/s(\mathrm{OPT},L) \mid \mathrm{OPT}(L) \ge k\}.$$

In the on-line version of variable sized bin packing, every time a new bin is opened, the packing algorithm chooses which bin size to use next.

Friesen and Langston [44] give three approximation algorithms with finite worst case ratios for variable sized bin packing; only one of these three algorithms is on-line. This algorithm always chooses the largest available bin size when a new bin is opened, and otherwise behaves just like NEXT FIT. It is not surprising that it has a worst case ratio of 2, the same as NEXT FIT. In fact, *any* on-line algorithm that always chooses the *largest* available bin size has a worst case ratio of at least 2: Consider an instance with two bin sizes $s(B_1) = 1$ and $s(B_2) = 1/2 + \varepsilon$, and an input list L consisting only of pieces of size $1/2 + \varepsilon$. A similar example demonstrates that on-line algorithms that always choose the *smallest* bin size have worst case ratios of at least 2. Therefore, Kinnersley and Langston [76] design a hybrid strategy that they call FFf. Here the 'FF' stands for FIRST FIT, since FFf uses the packing strategy of FIRST FIT. Moreover, FFf is based on a so-called *filling factor* f, $1/2 \leq f \leq 1$. Suppose that FFf must start a new bin as item a_i arrives. If a_i is a small item with $a_i \leq 1/2$, then FFf starts a new bin of size 1. If a_i is a large item with $a_i > 1/2$, then FFf chooses the smallest bin size in the interval $[a_i, a_i/f]$; in case no such bin size exists, FFf chooses size 1. Intuitively speaking, when FFf packs large items into new bins then it tries to make them full to at least a fraction of f. The worst case ratio of FFf is at most $1.5 + f/2$, and there exist values of f (e.g. $f = 3/5$) for which this bound is tight. Zhang [113] proves that the worst case ratio of FF$\frac{1}{2}$ equals 17/10, thus matching the guarantee of FIRST FIT. Observe that for large items, FF$\frac{1}{2}$ always selects the smallest bin size into which they will fit.

Csirik [31] designs an on-line algorithm for variable sized bin packing that is based on the HARMONIC algorithm (cf. Section 3.2). For every item size a_i, this algorithm computes the most appropriate bin type B_j (i.e., the bin size that in some sense minimizes the expected wasted space for item size a_i). All items that are assigned to the same bin type B_j are then packed with a HARMONIC algorithm into bins of size $s(B_j)$. For any collection of bin types, the worst case guarantee of this algorithm can be made arbitrarily close to $h_\infty \approx 1.691$. Quite surprisingly, the algorithm sometimes performs even better for special collections of bin types:

Corollary 10. (Csirik [31]) *For variable sized one dimensional bin packing with two bin sizes $s(B_1) = 1$ and $s(B_2) = 0.7$, there exists an on-line approximation algorithm A with worst case ratio $R_A^{var} = 1.4$.*

Note that the bound 1.4 is smaller than the 1.5401 lower bound for classical one dimensional bin packing! A partial explanation for this phenomenon goes as follows. Although variable sized bin packing is a much harder problem than classical bin packing, it also gives the on-line packer more freedom for making decisions. Suppose that there were bin types of all sizes between 1/2 and 1. Then the packing of large items with sizes $> 1/2$ can be done without wasting any space, and the small items with sizes $\leq 1/2$ are anyway easy to pack without wasting lots of space.

There remain many open problems in this area. Does there exist an on-line algorithm that has worst case ratio strictly less than h_∞ for all collections of bin types? Which combination of two bin types allows the smallest possible worst case ratio? What can be said about lower bounds depending on the bin sizes for variable sized bin packing?

6 Worst case analysis for bin covering problems

In the *packing* problems treated till now, the goal is to partition a set of items (numbers, vectors, boxes) into the *minimum* number of subsets such that in every subset the total size of the items does not exceed some *upper* bound. In a *covering* problem, the goal is to partition a set of items into the *maximum* number of subsets such that in every subset the total size of the items does not remain under some *lower* bound. Covering problems model a variety of situations encountered in business and in industry, from packing peach slices into tin cans so that each tin can contains at least its advertised net weight, to such complex problems as breaking up monopolies into smaller companies, each of which is large enough to be viable. Since covering problems may be considered to be a kind of inverse or dual version of the packing problem, they are sometimes also called 'dual packing' problems in the literature. This section deals with the one dimensional bin covering problem and with some generalizations to covering by vectors, cubes and boxes.

In the one dimensional bin covering problem, the goal is to pack a list $L = \langle a_1, a_2, \ldots, a_n \rangle$ into a maximum number of bins of size 1 such that the contents of each bin is at least one. Similarly to the definitions introduced in Section 2, we will denote by T_A^∞ the asymptotic worst case ratio of an approximation algorithm for bin covering ($A(L)$ and $\mathrm{OPT}(L)$ denote the number of bins in the packing constructed by algorithm A and in the optimal packing, respectively). It is defined by

$$T_A^n = \min\{A(L)/\mathrm{OPT}(L) \mid \mathrm{OPT}(L) = n\},$$

$$T_A^\infty = \liminf_{n\to\infty} T_A^n.$$

Since we deal with a maximization problem, we have that the larger the worst case ratio of an algorithm, the better the approximation algorithm. The bin covering problem was for the first time investigated in the Ph.D. thesis [3] of Assmann and in the journal article [4] by Assmann, Johnson, Kleitman and Leung. They considered the following adaptation of NEXT FIT, called DUAL NEXT FIT (DNF): DNF always keeps a single active bin. Newly arriving items a_i are packed into the active bin until the active bin is full (i.e., has contents at least one). Then the active bin is closed and another (empty) bin becomes the active bin. It is not too difficult to see that DNF fulfills $T_{DNF}^\infty = \frac{1}{2}$. However, here analogous modifications of FIRST FIT, BEST FIT and HARMONIC *do not* improve this worst case ratio $\frac{1}{2}$ (e.g. modifying FIRST FIT is useless, since after filling a bin, placing further items into it does not make sense). In fact, from the worst case point of view, algorithm DNF is the best possible on-line bin covering algorithm:

Theorem 11. (Csirik, Totik [40]) *Every on-line bin covering algorithm A fulfills* $T_A^\infty \leq \frac{1}{2}$.

Woeginger and Zhang [110] consider bin covering with variable sized bins: There are several different types $B_1, \ldots, B_r$ of bins with sizes $1 = s(B_1) > s(B_2) > \ldots > s(B_r)$; there is an infinite supply of bins of each size. The problem is to cover with a given list of items $a_i \in [0,1]$ a set of bins with *largest total size*. [110] determine for every finite collection of bin sizes the worst case ratio of the best possible on-line covering algorithm. This worst case ratio mainly depends on the largest gap between consecutive bin sizes of the bins with $s(B_j) > 1/2$.

Alon, Azar, Csirik, Epstein, Sevastianov, Vestjens and Woeginger [2] investigate the vector covering problem: Here the input consists of a list L of d-dimensional vectors in $[0,1]^d$, and the goal is to assign L to a maximum number of bins, subject to the constraint that in every bin the sum of all vectors is at least one in every coordinate. For every $d \geq 2$, Alon et al [2] demonstrate the existence of an on-line vector covering algorithm with ratio $1/(2d)$: The main idea is to partition the list of vectors into d sublists in such a way that the i-th sublist ($1 \leq i \leq d$) contains at least a fraction of roughly $1/d$ of the total size of all the i-th coordinates in L. Somewhat surprisingly, this can be done in an on-line fashion. At the same time, the vectors that are assigned to the i-th sublist are used to cover the i-th coordinate of unit bins by applying DUAL NEXT FIT. Moreover, Alon et al [2] show that for $d \geq 2$ every on-line algorithm must fulfill $T_A^\infty \leq 2/(2d+1)$.

There is only a small number of results known for geometric covering by boxes and cubes. In this version of the problem, the boxes (respectively cubes) are allowed to overlap; rotations of the items are *not* allowed. All known results concern on-line covering of a *single* d-dimensional unit cube. Kuperberg [80] developed a beautiful on-line algorithm for covering the unit cube by smaller cubes; this algorithm packs the cubes along a space-filling Peano curve and uses the current position on this curve as a measure of progress. Kuperberg proves that in d-dimensional space, every sequence of cubes whose sum of volumes is greater than 4^d admits an on-line covering of the unit cube. Kuperberg's paper triggered a race which led to more and more elaborate variants. The results of Januszewski and Lassak [66, 68], of Lassak [81] and of Januszewski, Lassak, Rote and Woeginger [70] finally led to the following theorem.

Theorem 12. (Januszewski, Lassak, Rote and Woeginger [70]) *For $d \geq 2$, every sequence of cubes in d-dimensional space whose sum of volumes is greater than $2^d + 3$ admits an on-line covering of the unit cube.*

It is interesting to observe that the estimate in Theorem 12 is almost as good as the best possible off-line estimate: For every $d \geq 2$, the sequence consisting of $2^d - 1$ cubes with edge length $1 - \varepsilon$ can not be used to cover the unit cube (since every corner of the unit cube has to be covered by its own small cube). Consequently, the best off-line estimate must be at least $2^d - 1$. In 1982 Groemer [55] showed that every sequence of cubes in d-dimensional space whose sum of volumes is greater than $2^d - 1$ indeed allows an *off-line* covering of the d-dimensional unit cube. For $d = 2$, Theorem 12 gives an (on-line) upper bound of 7 on the volume. This bound can be improved down to 5.265 (cf. Januszewski and Lassak [68]).

There are also several results on covering the unit cube by axes-parallel boxes (whose edge lengths are bounded by one) or by other convex bodies with bounded diameters (cf. Januszewski and Lassak [65, 67]). We only mention that every sequence of axes-parallel boxes with edge lengths bounded by 1 and total volume at least $2 \cdot 8^{d-1} - 2^{d-1}$ allows an on-line covering of the d-dimensional unit cube.

7 Semi on-line bin packing

This section deals with several variants of bin packing that are 'almost' on-line: Either the input list is presorted, or the packer has some lookahead to some of the future items, or earlier packed items may be rearranged in some limited way.

First, let us assume that the input list is *presorted.* In case the list is in increasing order, this is of no great help to us: the lower bound techniques in Section 3.4 all work with such increasing lists, and hence the 1.5401 lower bound in Theorem 6 applies to this case, too. However, in case the items are sorted by decreasing size, the problem becomes much easier: The NEXT FIT algorithm on a decreasing list is called the NEXT FIT DECREASING (NFD) algorithm, and the FIRST FIT algorithm on a decreasing list is called the FIRST FIT DECREASING (FFD) algorithm. The worst case ratio of NFD equals 1.691 (Baker and Coffman [9]) and the worst case ratio of FFD equals 11/9 (cf. Johnson [72], Baker [6], Csirik [32]). This is quite an improvement over the worst case bounds 2 and 1.7, respectively, for arbitrary lists. On the other hand, Csirik, Galambos and Turán [36] showed that every on-line algorithm for decreasing lists must have a worst case ratio of at least 8/7. Their argument uses two lists L_1 and L_2 where L_1 contains n items of size $1/3 + 2\varepsilon$ and L_2 contains $2n$ items of size $1/3 - \varepsilon$, where $0 < \varepsilon < 1/12$. It can be seen that any on-line algorithm must be at least a factor of 8/7 away from the optimum either if it has to pack list L_1, or if it has to pack the concatenated list $L_1 L_2$.

Next, let us reconsider the k-bounded-space restriction introduced in Section 3.2. According to Theorem 5, each bounded-space on-line algorithm has worst case ratio at least h_∞. Moreover, it is an open problem whether there actually exists a bounded-space algorithm that reaches the worst case ratio h_∞ while using a finite number of active bins. Grove [56] considers a variant of bounded-space bin packing with *bounded lookahead.* Bounded lookahead means that for some fixed constant W, an item a_i need not be packed until one has looked through all the items a_j, $j > i$, for which $\sum_{\ell=i}^{j} a_\ell \leq W$ holds. For every variant with k-bounded space and lookahead W, the h_∞ lower bound from Theorem 5 still applies. However, if k and W are fixed at sufficiently large values, then it can be shown that the worst case ratio of the REVISED WAREHOUSE (RW) algorithm introduced by Grove reaches this lower bound. Algorithm RW is based on a weighting function $w : [0, 1] \to [0, 3/2]$. For each list L of items, the total weight of the items in this list fulfills $w(L) \leq h_\infty \mathrm{OPT}(L)$. If the lookahead W is sufficiently large, then the average weight of the bins closed by RW is at least 1. Hence, $\mathrm{RW}(L) \leq w(L) + k$ holds; this implies $R_{RW}^\infty = h_\infty$.

Galambos and Woeginger [51] introduced yet another variant of bounded-space bin packing where repacking of the currently active bins is allowed. Here repacking means that the packer is allowed to take all items out of the currently active bins and to reassign them before packing the next item. Also for this variant the h_∞ lower bound in Theorem 5 still applies. Similarly as Grove [56], Galambos and Woeginger [51] use a weighting function $w' : [0,1] \to [0,3/2]$ such that for each list L, $w'(L) \leq h_\infty \mathrm{OPT}(L)$ holds. Moreover, function w' has the following property: If one takes three partially filled bins B_1, B_2 and B_3 and repacks their items according to FFD, then afterwards either one bin has become empty or in one bin the total w'-weight of the items is at least 1. The algorithm REP_3 introduced in [51] uses three active bins; every time a new item arrives, it repacks the active bins by FFD, then closes all bins with total w'-weight at least 1, and finally puts the new item into an empty bin. Since the w'-weight of every closed bin is at least 1, $\mathrm{REP}_3(L) \leq w'(L) + 3$ and $R^\infty_{REP_3} = h_\infty$.

Next, let us turn to a variant where the repacking is less restricted (observe that if we allowed completely unlimited repacking then we would end up with an off-line problem). Gambosi, Postiglione and Talamo [52] allow for every new item a small number of *simple* repacking moves. In a simple repacking move, one takes a single item or a bundle of items that currently are assigned to the same bin and transports this item or bundle to another bin. Gambosi et al [52] suggest two algorithms for this model. Their first algorithm uses at most three simple repacking moves per new item, works with a simple classification of the items into four types, and has a worst case ratio $R^\infty_A = 3/2$. Their second algorithm uses up to seven simple repacking moves per new item, works with a more complicated item classification with six types, and has a worst case ratio $R^\infty_A = 4/3$. Hence, both algorithms beat the 1.5401 lower bound for the classical model! Ivković and Lloyd [62, 64] give an even more sophisticated algorithm that achieves a worst case ratio of 5/4 with $O(\log n)$ simple repacking moves per new item and five types of items. Their algorithm is designed to handle the more general dynamic bin packing problem (cf. Section 5). Moving bundles of items is an essential component of the algorithms in [52, 62, 64]. Let us define a *very simple* repacking move to be a simple move that only transports a single item. It can be shown [63] that any algorithm that only makes a bounded number of very simple repacking moves per new item has a worst case ratio of at least 4/3.

8 Average case analysis for one dimensional bin packing

Shortly after the systematic worst case analysis of bin packing algorithms had started, also the first papers on average case analysis of bin packing appeared. Since then a lot of work has been done in this field: several excellent review papers (Coffman, Lueker and Rinnooy Kan [26], and Coffman and Shor [27]) were written, and there are also two books (Hofri [60], and Coffman and Lueker [25]) that partly deal with the probabilistic analysis of bin packing. It turned out that the average case behaviour gives an important supplement to the worst case analysis of approximation algorithms.

This section is divided into two subsections: Subsection 8.1 deals with the average case behaviour of NEXT FIT, HARMONIC, and of variants of these algorithms on uniform distributions. It will turn out that all these algorithms have their asymptotic expected ratios bounded away from 1. Subsection 8.2 treats FIRST FIT and BEST FIT that are asymptotically optimal. Moreover, the expected wasted space for on-line algorithms on uniform distributions is studied in some detail.

8.1 Next Fit, Smart Next Fit, and Harmonic

Let us start with discussing the NEXT FIT algorithm. As a first step, Shapiro [102] performed an analysis of NF for exponential distributions. Then, based on a Markov model approach, Coffman, So, Hofri, and Yao [28] analyzed the behaviour of NF for $U(0,1)$ (recall that $U(0,1)$ is the continuous uniform distribution on $[0,1]$). Their analysis concentrates on the behaviour of the level of the currently last closed bin. They prove that the distribution of this level converges to the distribution x^3, $0 \leq x \leq 1$. Since this distribution has mean $3/4$, we get the following result.

Theorem 13. (Coffman, So, Hofri, Yao [28]) $R^{\infty}_{NF}(U(0,1)) = 4/3$.

Hofri [59] presented an analysis of NF for arbitrary distributions. Such general questions are quite hard to handle, and so he derived sets of equations that implicitly describe the expected number of bins that NF uses for packing n items. For the uniform distribution $U(0,1)$, however, it is easy to read off from his results that

$$E[\mathrm{NF}(L_n)] = \begin{cases} 1, & n = 1 \\ \frac{2}{3}n + \frac{1}{6}, & n \geq 1. \end{cases}$$

Since for $U(0,1)$, $E[\mathrm{OPT}(L_n)]$ grows like $n/2$, this confirms the result stated in Theorem 13.

In order to analyze NF for $U(0,a)$, Karmarkar [74] formulated NF as a Markov process with discrete time steps and continuous state space. He overcomes technical difficulties by a novel technique of converting the differential equation governing the corresponding steady-state distribution into a matrix-differential equation, which then can be solved by standard methods. For $1/2 < a \leq 1$, he gives the closed-form result

$$\rho = \frac{1}{12a^3}(15a^3 - 9a^2 + 3a - 1) + \frac{(1-a)^2}{2\sqrt{2a}} + \tanh\frac{1-a}{\sqrt{2a}}$$

where $E[\mathrm{NF}(L_n)] \sim \rho n$ as $n \to \infty$. These results agree exactly with the computational experiments of Ong, Magazine, and Wee [92]. In particular, Ong et al [92] had observed a strong drop in the efficiency of NF for values of a between 0.8 and 0.9. Karmarkar's work shows that $R^{\infty}_{NF}(U(0,a))$ indeed takes a local maximum at $a = 0.841$.

Ramanan [93] devised a modification of NF that is called SMART NEXT FIT (SNF) and differs from NF in only one respect: In case the new item does not fit into the active bin, the item is put into a newly opened bin, and then the *fuller one* of these two bins is closed and the other one becomes the active bin. Hence, SNF uses 2-bounded space. By applying the method of Karmarkar [74], Ramanan analyzes SNF and derives the asymptotic expected ratios of SNF for $U(0,a)$ with $1/2 \leq a \leq 1$. In particular, $R^{\infty}_{SNF}(U(0,1)) = 1.227$ holds. Hence, SNF is indeed an improvement over the 1.333 bound of NF.

The HARMONIC algorithm is easy to analyze, even for arbitrary distributions. Essentially, the analysis reduces to counting or estimating the expected number of items in every harmonic interval $(1/(j+1), 1/j]$. For the distribution $U(0,1)$ the following holds: if the number k of active bins goes to infinity, then the asymptotic expected ratio of H_k tends to $\pi^2/3 - 2 \approx 1.289$. But even for $k = 3$ active bins, the asymptotic expected ratio of H_3 is about 1.306. These results were derived by many authors independently from each other: e.g. by Csirik and Máté [39], by Hofri and Kahmi [61], or by Lee and Lee [84]. Ramanan and Tsuga [95] investigated the asymptotic expected ratio of the MODIFIED HARMONIC algorithm. They showed that for $U(0,1)$, $R^{\infty}_{MH}(U(0,1)) \leq 1.276$ holds, which is slightly better than the bound for HARMONIC. Moreover, they propose and analyze a simple modification MH* of the MH that brings the asymptotic expected ratio down to 1.189.

Summarizing the results on asymptotic expected ratios for $U(0,1)$ described till now, we get that

$$\begin{pmatrix}\text{MH*}\\1.189\end{pmatrix} < \begin{pmatrix}\text{SNF}\\1.227\end{pmatrix} < \begin{pmatrix}\text{MH}\\1.276\end{pmatrix} < \begin{pmatrix}\text{H}_{1000}\\1.289\end{pmatrix} < \begin{pmatrix}\text{H}_3\\1.306\end{pmatrix} < \begin{pmatrix}\text{NF}\\1.333\end{pmatrix}$$

holds. It is also interesting to note that according to Coffman and Shor [27], each on-line algorithm A that gets along with k-bounded-space must satisfy the inequality

$$W^n_A(U(0,1)) \geq \frac{n}{16(k+1)}.$$

Hence, on average every bounded-space algorithm A wastes a constant fraction of $s(L_n)$, and thus it can not be asymptotically optimal. For $k = 2$, the above expression yields that no on-line algorithm using 2-bounded-space can have an asymptotic expected ratio better than $25/24 \approx 1.041$. SMART NEXT FIT reaches 1.227 with 2-bounded-space. Based on extensive computational experiments, Csirik and Johnson [38] conjecture that their 2-bounded-space BBF_2 algorithm (cf. Section 3.2) has $R^{\infty}_{BBF_2}(U(0,1)) = 1.178$. Hence, there remains quite a gap to be closed between 1.041 and 1.178.

8.2 First Fit, Best Fit, and the problem of wasted space

Although FIRST FIT and BEST FIT are outperformed by the bounded-space algorithms from the worst case point of view, they behave much better than their competitors from the average case of view:

Theorem 14. (Bentley, Johnson, Leighton, McGeoch, McGeoch [13]) $R^{\infty}_{FF}(U(0,1)) = 1$.

Also for BEST FIT, $R^{\infty}_{BF}(U(0,1)) = 1$ holds true. Now immediately the question arises, *how fast* these ratios do tend to 1. This question can be answered via investigating the wasted space. For the FIRST FIT, already Bentley et al [13] showed that $W^n_{FF}(U(0,1))$ grows with $O(n^{\frac{4}{5}})$. Later, Shor [103] gave tighter bounds and proved that the expected wasted space of FF is $\Omega(n^{\frac{2}{3}})$ and $O(n^{\frac{2}{3}}\sqrt{\log n})$. Finally, Coffman, Johnson, Shor and Weber [24] settled the problem completely by showing that $W^n_{FF}(U(0,1))$ is $\Theta(n^{\frac{2}{3}})$, i.e., that Shor's lower bound gives the right order of magnitude. For the BEST FIT algorithm, Shor [103] proved that the expected wasted space $W^n_{BF}(U(0,1))$ is $\Omega(\sqrt{n}\log^{\frac{3}{4}} n)$, and Leighton and Shor [85] proved that it is $O(\sqrt{n}\log^{\frac{3}{4}} n)$. The proofs of most of these results rely on results for certain planar matching problems derived by Ajtai, Komlós and Tusnády [1]. The connection between the bin packing problem and the matching problem was drawn by Karp, Luby and Marchetti-Spaccamela [75]. Similar results were achieved by Rhee and Talagrand [99], [100] for more general distributions.

Summarizing, FF does not waste too much space, and BF wastes even less. What can be said in general on the wasted space of approximation algorithms? Since $W^n_{OPT}(U(0,1)) = \Omega(\sqrt{n})$, even an off-line algorithm cannot do better than $\Omega(\sqrt{n})$. In 1986, Shor [103] gave a very interesting partial answer to this question for on-line algorithms: Consider an on-line algorithm A that receives k items with sizes uniformly distributed according to $U(0,1)$. The value of k is chosen with equal probability from the integers between 1 and n, and A is given no information about the value of k until it has received and packed the last item. Call such an on-line algorithm A an *open* on-line algorithm. Shor proved that each open on-line algorithm must have an expected waste of $\Omega(\sqrt{n\log n})$. Algorithm BF is just a factor of $O(\log^{\frac{1}{4}} n)$ away from this lower bound. In 1991, Shor [104] detected an open on-line algorithm with expected waste $O(\sqrt{n\log n})$, thus matching his lower bound. The algorithm roughly proceeds as follows.

Let us assume for the moment that we a priori know the number k of items that will be packed. Let $p = \sqrt{k/\log k}$. For $1 \le i \le p$, let S_i be the interval $[\frac{i-1}{2p}, \frac{i}{2p}]$ and let L_i be the interval $[1-\frac{i}{2p}, 1-\frac{i-1}{2p}]$. Hence, the union of the intervals L_i covers the large items $> 1/2$ and the union of the intervals S_i covers the small items $\le 1/2$. The bins are divided into $p+3$ classes labeled $\mathcal{B}_0, \mathcal{B}_1, \ldots, \mathcal{B}_{p+2}$. Into any bin, the algorithm will put at most one large and at most one small item. Large items from interval L_i only go into bins of class $\mathcal{B}_{i-1}$ or $\mathcal{B}_i$, and small items from interval S_i only go into bins of class $\mathcal{B}_{i+1}$ or $\mathcal{B}_{i+2}$. This ensures that an item from interval L_i will get paired with an item from interval S_{i-1} or smaller, so that the sizes of the two items always will sum to at most 1. The exact assignment of items to bin classes can be modeled via a *balls in buckets game* for which Shor [104] designs a very sophisticated algorithm. In the end, the overall wasted space is half the expected number of bins with one item in them (since the expected size of an item is 1/2 and since we never put more than two items into a bin), which can be shown to be $O(k/p + p\log p) = O(\sqrt{k\log k})$. To turn this procedure into an open on-line algorithm, one dynamically changes p every time the number of items received so far doubles. Even if one starts over with a new set of bins each time p is updated and leaves the previously half-filled bins forever half-empty, one still gets $O(\sqrt{n \log n})$ expected wasted space.

If an on-line algorithm A is told the number of items in advance, it is easy to reach the best possible bound $W_A^n(U(0,1)) = \Omega(\sqrt{n})$: It simply packs the first $\lfloor k/2 \rfloor$ items into separate bins, and then packs the remaining $\lceil k/2 \rceil$ items by applying BEST FIT (Shor [103]). Also in the case where the total number of items is unknown, but the items are presorted into decreasing order, the best possible bound $\Omega(\sqrt{n})$ can be reached: Based on preliminary work by Frederickson [43], Knödel [77] and Lueker [88] proved independently of each other that in this case the FIRST FIT DECREASING algorithm yields wasted space $\Theta(\sqrt{n})$. These questions were investigated for general distributions by Rhee and Talagrand [96, 97]. A variety of results that yield asymptotically optimal algorithms for various special cases are contained in the papers of Hoffman [57], Knödel [78], Csirik [30], Csirik and Galambos [35], and Chang, Wang, and Kankanhalli [19].

9 Average case analysis for higher dimensional packing

In Section 4, we surveyed worst case results for three types of higher dimensional bin packing problems: For vector packing, for box packing, and for strip packing. For average case analysis, Karp, Luby and Marchetti-Spaccalema [75] investigated the average-case behaviour of the optimal packing for the vector packing and box packing under uniform distribution on $(0,1)^d$. They also gave asymptotically optimal off-line algorithms for these problems. We are not aware of any results for the *on-line vector packing* problem. However, it seems plausible that the d-dimensional variant of FIRST FIT introduced by Garey, Graham, Johnson, and Yao [53] (cf. Theorem 7) is asymptotically optimal if the vectors are independent, identically and uniformly distributed in $[0,1]^d$. For *box packing*, Csirik and van Vliet [41] introduced an on-line algorithm that may be considered to form a higher dimensional generalization of the HARMONIC algorithm (cf. Theorem 8). Indeed, it can be shown that the asymptotic expected ratio of this algorithm equals $(\pi^2/3 - 2)^d$ in d dimensions, if the vectors that describe the side lengths of the boxes are distributed independently, identically and uniformly in $[0,1]^d$. For $d = 1$, this takes us back to the asymptotic expected ratio of the HARMONIC algorithm for one dimensional bin packing (cf. Section 8.1). Chang, Wang and Kankanhalli [19] defined an algorithm HashPacking for (both open and closed) d-dimensional box packing which achieves average wasted space of order $O(n^{(d+1)/(d+2)})$.

In the remainder of this section we will deal with the average case analysis of strip packing (cf. Section 4.3). As before, the strip has width 1. Here we assume that the boxes are given in the form (h_i, w_i) where h_i and w_i are independent variables. Hofri [58] investigates a primitive shelf algorithm in which only one shelf is packed at a time. The boxes are always packed by NEXT FIT onto this single active shelf. The height of the active shelf is not determined until some item fails to fit, whereupon it is fixed at the maximum item height in the shelf and a new shelf is opened above the filled shelf. It is easy to see that the worst case ratio of this algorithm is unbounded: Just consider the list that consists of n sublists with items $(\frac{1}{n}, 1)$ and $(1 - \frac{1}{n}, \frac{1}{n})$. Then the optimum strip packing has height 2, whereas the algorithm constructs a packing of height n. However, the algorithm has a reasonable average case behaviour: If $h_i \sim U(0,1)$ and $w_i \sim U(0,1)$, then the algorithm constructs packings with an expected height $E[H] \sim 0.572n$ and an expected wasted area of $\sim 0.337n$.

Bartholdi, vande Vate and Zhang [12] analyzed a generalization of the NEXT FIT SHELF algorithm described in Section 4.3. In their analysis, they allow arbitrary finite sequences $1 = s_0 > \cdots > s_k > s_{k+1} = 0$ of feasible shelf heights. Remember that for every height s_i, NFS keeps an active shelf into which all boxes with height in $(s_{j+1}, s_j]$ are packed. Whenever a new box does not fit into its shelf, this shelf is closed and a new shelf of height s_j is opened. For the boxes (h_i, w_i), it is assumed that the widths w_i are independent, identically distributed according to $U(0,1)$ and that the heights h_i are independent, identically distributed according to a cumulative distribution F on $[0,1]$. F is assumed to have density function f and a mean μ. Bartholdi et al [12] proved that under these assumptions, the expected height of an NFS packing for n boxes equals

$$E[H] = \frac{2n}{3} \sum_{j=0}^{k} s_j p_j + \frac{1}{6} \sum_{j=0}^{k} s_j \left(1 + np_j(1-p_j)^{n-1} - (1-p_j)^n\right),$$

where for $0 \le j \le k$, $p_j = F(s_j) - F(s_{j+1})$ denotes the probability that the height of an item lies in the interval $(s_{j+1}, s_j]$, so that it is packed onto a shelf of height s_j. From this result one easily derives results on the expected wasted area (which equals $E[H] - \mu/2$):

- For the original NFS as introduced by Baker and Schwarz [11] with shelf heights $s_j = r^j$ and $h_i \sim F = U(0,1)$, the expected wasted area equals ωn with $\omega = \frac{2}{3}(1 + r^{2k+1})/(1+r) - \frac{1}{4}$.
- For $h_i \sim F = U(0,1)$ and shelf heights $s_j = (k-j)/k$, the expected wasted area equals ωn with $\omega = \frac{1}{3}(1 + 1/k) - \frac{1}{4}$.

Moreover, for any given distribution F the expected wasted area is minimized for the shelf heights defined by $s_{k+1} = 0$ and $s_j - s_{j+1} = (F(s_{j+1} - F(s_{j+2}))/f(s_{j+1})$ for $0 \le j \le k$. All these strip packing algorithms have expected waste linear in the number of items. Shor [104] points out that his average case results for the one dimensional case also yield an on-line strip packing algorithm with expected waste $\Theta(\sqrt{n \log n})$ and that this result is best possible.

10 Average case analysis for bin covering

In this section, we discuss some average case results for bin covering problems. Let us start with the one dimensional case. Already in 1983, Assmann [3] proved that for a list L_n with n items $\sim U(0,1)$, $E[\mathrm{OPT}(L_n)] = n/2 - \Omega(n^{1/2})$ holds for n large enough. Several years later, Csirik, Frenk, Galambos and Rinnooy Kan [33] showed that the $\Omega(n^{1/2})$ in this formula is in fact $\Theta(n^{1/2})$. Hence, the behaviour of $E[\mathrm{OPT}(L_n)]$ is completely understood for the uniform distribution. Rhee and Talagrand [98] investigated the optimal solution for general distribution too.

By modeling the DUAL NEXT FIT algorithm (cf. Section 6) as a Markov process with discrete time steps (= the arrival of an item) and continuous state space (= the level of the current bin), Assmann [3] showed that $E[DNF(L_n)]$ grows like $n/e \approx n/2.718$ for the distribution $U(0,1)$. Hence, the asymptotic expected ratio of DNF for $U(0,1)$ equals $2/e \approx 0.735$. It is perhaps surprising that presorting the elements in decreasing order does not help for this problem: In this case the asymptotic expected ratio of DNF becomes worse and drops to $4-\pi^2/3 \approx 0.710$ (see Csirik et al [33]). By using strong techniques from renewal theory, a general result for distributions $U(0,a)$ can be derived.

Theorem 15. (Csirik, Frenk, Galambos, Rinnooy Kan [33]) *If the items in L_n are distributed according to $U(0,a)$, $0 \le a \le 1$, then $E[DNF(L_n)]$ grows like n/τ where*

$$\tau = \sum_{\ell=0}^{\lfloor 1/a \rfloor} (-1)^\ell \frac{1}{\ell!} \left(\frac{1}{a} - \ell\right)^\ell \exp\left(\frac{1}{a} - \ell\right).$$

As a last result for the one dimensional problem we mention that Csirik and Totik [40] designed an asymptotically optimal on-line algorithm for $U(0,1)$.

The only average case results known for higher dimensional covering are for 2-dimensional vector covering. If the vectors are uniformly and independently distributed on $[0,1]^2$, then $\lim_{n\to\infty} E[\mathrm{OPT}(L_n)]/n = 1/2$. The obvious extension DNF_2 of the DUAL NEXT FIT algorithm to 2-dimensional vector covering was analyzed by Csirik et al [33]. They proved that the asymptotic expected ratio of DNF_2 equals $2/\tau \approx 0.633$, where $\tau = 2e - \sum_{t=0}^{\infty} 1/(t!)^2 \approx 3.156$.

Acknowledgements

This work was partially supported by the Start-program Y43-MAT of the Austrian Ministry of Science and by the Project 20u2 of the Austro-Hungarian Action Fund.

References

1. M. Ajtai, J. Komlós, and G. Tusnády. On optimal matching. *Combinatorica*, 4:259–264, 1984.
2. N. Alon any Y. Azar, J. Csirik, L. Epstein, S. V. Sevastianov, A. P. A. Vestjens, and G. J. Woeginger. On-line and off-line approximation algorithms for vector covering problems. In *Proc. 4th European Symposium on Algorithms*, LNCS, pages 406–418. Springer, 1996. To appear in Algorithmica.
3. S. F. Assmann. *Problems in Discrete Applied Mathematics.* PhD thesis, MIT, Cambridge, MA, 1983.
4. S. F. Assmann, D. S. Johnson, D. J. Kleitman, and J. Y. T. Leung. On a dual version of the one-dimensional bin packing problem. *J. Algorithms*, 5:502–525, 1984.

5. Y. Azar and L. Epstein. On two dimensional packing. In *Proc. 5th Scand. Workshop Alg. Theor.*, LNCS, pages 321–332. Springer, 1996.
6. B. S. Baker. A new proof for the First-Fit decreasing bin-packing algorithm. *J. Algorithms*, 6:49–70, 1985.
7. B. S. Baker, D. J. Brown, and H. P. Katseff. A 5/4 algorithm for two-dimensional packing. *J. Algorithms*, 2:348–368, 1981.
8. B. S. Baker, D. J. Brown, and H. P. Katseff. Lower bounds for two-dimensional packing algorithms. *Acta Informatica*, 8:207–225, 1982.
9. B. S. Baker and E. G. Coffman. A tight asymptotic bound for Next-Fit-Decreasing bin-packing. *SIAM J. Alg. Disc. Meth.*, 2:147–152, 1981.
10. B. S. Baker, E. G. Coffman, and R. L. Rivest. Orthogonal packings in two dimensions. *SIAM J. Comput.*, 9:846–855, 1980.
11. B. S. Baker and J. S. Schwartz. Shelf algorithms for two-dimensional packing problems. *SIAM J. Comput.*, 12:508–525, 1983.
12. J. J. Bartholdi, J. H. vande Vate, and J. Zhang. Expected performance of the shelf heuristic for two-dimensional packing. *Oper. Res. Lett.*, 8:11–16, 1989.
13. J. L. Bentley, D. S. Johnson, F. T. Leighton, C. C. McGeoch, and L. A. McGeoch. Some unexpected expected behavior results for bin packing. In *Proc. 16th Annual ACM Symp. Theory of Computing*, pages 279–288, 1984.
14. D. Blitz, A. Van Vliet, and G. J. Woeginger. Lower bounds on the asymptotic worst-case ratio of on-line bin packing algorithms. Unpublished manuscript, 1996.
15. D. J. Brown. A lower bound for on-line one-dimensional bin packing algorithms. Technical Report R-864, Coordinated Sci. Lab., Urbana, Illinois, 1979.
16. J. Brzustowski. Can you win at Tetris? Master's thesis, The University of British Columbia, 1992.
17. H. Burgiel. How to lose at Tetris. Technical report, The Geometry Center, Minneapolis, MN, 1996.
18. B. Chandra. Does randomization help in on-line bin packing? *Inform. Process. Lett.*, 43:15–19, 1992.
19. E. E. Chang, W. Wang, and M. S. Kankanhalli. Multidimensional on-line bin packing: An algorithm and its average-case analysis. *Inform. Process. Lett.*, 48:121–125, 1993.
20. F. R. K. Chung, M. R. Garey, and D. S. Johnson. On packing two-dimensional bins. *SIAM J. Alg. Disc. Meth.*, 3:66–76, 1982.
21. E. G. Coffman. An introduction to combinatorial models of dynamic storage allocation. *SIAM Rev.*, 25:311–325, 1983.
22. E. G. Coffman, M. R. Garey, and D. S. Johnson. Dynamic bin packing. *SIAM J. Comput.*, 12:227–258, 1983.
23. E. G. Coffman, M. R. Garey, and D. S. Johnson. Approximation algorithms for bin packing: a survey. In D. Hochbaum, editor, *Approximation algorithms for NP-hard Problems*, pages 46–93. PWS Publishing Company, 1997.
24. E. G. Coffman, D. S. Johnson, P. W. Shor, and R. R. Weber. Bin packing with discrete item sizes, part ii: average case behaviour of First Fit. Unpublished manuscript, 1996.
25. E. G. Coffman and G. S. Lueker. *Probabilistic analysis of Packing and Partitioning Algorithms.* John Wiley, New York, 1991.
26. E. G. Coffman, G. S. Lueker, and A. H. G. Rinnooy Kan. Asymptotic methods in the probabilistic analysis of sequencing and packing heuristics. *Management Science*, 34:266–290, 1988.

27. E. G. Coffman and P. W. Shor. Packing in two dimensions: asymptotic average-case analysis of algorithms. *Algorithmica*, 9:253–277, 1993.
28. E. G. Coffman, K. So, M. Hofri, and A. C. C. Yao. A stochastic model of bin packing. *Information and Control*, 44:105–115, 1980.
29. D. Coppersmith and P. Raghavan. Multidimensional on-line bin packing: Algorithms and worst case analysis. *Oper. Res. Lett.*, 8:17–20, 1989.
30. J. Csirik. Bin-packing as a random walk: a note on Knoedels paper. *Oper. Res. Lett.*, 5:161–163., 1986.
31. J. Csirik. An on-line algorithm for variable-sized bin packing. *Acta Informatica*, 26:697–709, 1989.
32. J. Csirik. The parametric behaviour of the First Fit decreasing bin-packing algorithm. *J. Algorithms*, 15:1–28, 1993.
33. J. Csirik, J. B. G. Frenk, G. Galambos, and A. H. G. Rinnooy Kan. Probabilistic analysis of algorithms for dual bin packing problems. *J. Algorithms*, 12:189–203, 1991.
34. J. Csirik, J. B. G. Frenk, and M. Labbe. Two dimensional rectangle packing: on line methods and results. *Discr. Appl. Math.*, 45:197–204, 1993.
35. J. Csirik and G. Galambos. An O(n) bin packing algorithm for uniformly distributed data. *Computing*, 36:313–319, 1986.
36. J. Csirik, G. Galambos, and G. Turán. A lower bound on on-line algorithms for decreasing lists. In *Proc. of EURO VI*, 1984.
37. J. Csirik and B. Imreh. On the worst-case performance of the NkF bin packing heuristic. *Acta Cybernetica*, 9:89–105, 1989.
38. J. Csirik and D. S. Johnson. Bounded space on-line bin packing: best is better than first. In *Proc. 2nd ACM-SIAM Symp. Discrete Algorithms*, pages 309–319, 1991.
39. J. Csirik and E. Máté. The probabilistic behaviour of the NFD bin-packing heuristic. *Acta Cybernetica*, 7:241–246, 1986.
40. J. Csirik and V. Totik. On-line algorithms for a dual version of bin packing. *Discr. Appl. Math.*, 21:163–167, 1988.
41. J. Csirik and A. Van Vliet. An on-line algorithm for multidimensional bin packing. *Oper. Res. Lett.*, 13:149–158, 1993.
42. J. Csirik and G. J. Woeginger. Shelf algorithms for on-line strip packing. *Inform. Process. Lett.*, 63:171–175, 1997.
43. G. N. Frederickson. Probabilistic analysis for simple one- and two-dimensional bin packing algorithms. *Inform. Process. Lett.*, 11:156–161, 1980.
44. D. K. Friesen and M. A. Langston. Variable sized bin packing. *SIAM J. Comput.*, 15:222–230, 1986.
45. G. Galambos. Parametric lower bounds for on-line bin packing. *SIAM J. Alg. Disc. Meth.*, 7:362–367, 1986.
46. G. Galambos. Notes on Lee's harmonic fit algorithm. *Annales Univ. Sci. Budapest., Sect. Comp.*, 9:121–126, 1988.
47. G. Galambos. A 1.6 lower bound for the two-dimensional on-line rectangle bin packing. *Acta Cybernetica*, 10:21–24, 1991.
48. G. Galambos and J. B. G. Frenk. A simple proof of Liang's lower bound for on-line bin packing and the extension to the parametric case. *Discr. Appl. Math.*, 41:173–178, 1993.
49. G. Galambos, H. Kellerer, and G. J. Woeginger. A lower bound for on-line vector packing algorithms. *Acta Cybernetica*, 11:23–34, 1994.

50. G. Galambos and A. Van Vliet. Lower bounds for 1-, 2-, and 3-dimensional on-line bin packing algorithms. *Computing*, 52:281–297, 1994.
51. G. Galambos and G. J. Woeginger. Repacking helps in bounded space on-line bin packing. *Computing*, 49:329–338, 1993.
52. G. Gambosi, A. Postiglione, and M. Talamo. New algorithms for on-line bin packing. In *Proc. 1st Italian Conference on Algorithms and Complexity*, 1990.
53. M. R. Garey, R. L. Graham, D. S. Johnson, and A. C. C. Yao. Resource constrained scheduling as generalized bin packing. *J. Comb. Th. Ser. A.*, 21:257–298, 1976.
54. M. R. Garey and D. S. Johnson. *Computers and Intractability (A Guide to the theory of of NP-Completeness.* W. H. Freeman and Company, San Francisco, 1979.
55. H. Groemer. Covering and packing properties of bounded sequences of convex sets. *Mathematika*, 29:18–31, 1982.
56. E. F. Grove. On-line binpacking with lookahead. In *Proc. 6th ACM-SIAM Symp. Discrete Algorithms*, pages 430–436, 1995.
57. U. Hoffman. A class of simple stochastic on-line bin packing algorithms. *Computing*, 29:227–239, 1982.
58. M. Hofri. Two-dimensional packing: Expected performance of simple level algorithms. *Information and Control*, 45:1–17, 1980.
59. M. Hofri. A probabilistic analysis of the Next-Fit bin packing algorithm. *J. Algorithms*, 5:547–556, 1984.
60. M. Hofri. *Probabilistic analysis of algorithms.* Springer, New York, 1987.
61. M. Hofri and S. Kahmi. A stochastic analysis of the NFD bin packing algorithm. *J. Algorithms*, 7:489–509, 1986.
62. Z. Ivkovic and E. Lloyd. Fully dynamic algorithms for bin packing: Being myopic helps. In *Proc. 1st European Symposium on Algorithms*, LNCS, pages 224–235. Springer, 1993.
63. Z. Ivkovic and E. Lloyd. A fundamental restriction on fully dynamic maintenance of bin packing. *Inform. Process. Lett.*, 59:229–232, 1996.
64. Z. Ivkovic and E. Lloyd. Partially dynamic bin packing can be solved within 1+eps in (amortized) polylogaritmic time. *Inform. Process. Lett.*, 63:45–50, 1997.
65. J. Januszewski and M. Lassak. On-line covering by boxes and by covex bodies. *Bull. Pol. Acad. Math.*, 42:69–76, 1994.
66. J. Januszewski and M. Lassak. On-line covering the unit cube by cubes. *Discrete Comput. Geom.*, 12:433–438, 1994.
67. J. Januszewski and M. Lassak. Efficient on-line covering of large cubes by convex bodies of at most unit diameters. *Bull. Pol. Acad. Math.*, 43:305–315, 1995.
68. J. Januszewski and M. Lassak. On-line covering the unit square by squares and the three-dimensional unit cube by cubes. *Demonstratio Math.*, 28:143–149, 1995.
69. J. Januszewski and M. Lassak. On-line packing the unit cube by cubes. unpublished manuscript, University of Bydgoszcz, Poland, 1996.
70. J. Januszewski, M. Lassak, G. Rote, and G. J. Woeginger. On-line q-adic covering by the method of the n-th segment and its application to on-line covering by cubes. *Beitr. Algebra Geom.*, 37:94–100, 1996.
71. D. S. Johnson. *Near-optimal bin packing algorithms.* PhD thesis, MIT, Cambridge, MA, 1973.
72. D. S. Johnson. Fast algorithms for bin packing. *J. Comput. System Sci.*, 8:272–314, 1974.

73. D. S. Johnson, A. Demers, J. D. Ullman, M. R. Garey, and R. L. Graham. Worst-case performance bounds for simple one-dimensional packing algorithms. *SIAM J. Comput.*, 3:256–278, 1974.
74. N. Karmarkar. Probabilistic analysis of some bin packing algorithms. In *Proc. 23rd IEEE Symp. Found. Comp. Sci.*, pages 107–118, 1982.
75. R. M. Karp, M. Luby, and A. Marchetti-Spaccamela. Probabilistic analysis of multi-dimensional binpacking problems. In *Proc. 16th Annual ACM Symp. Theory of Computing*, pages 289–298, 1984.
76. N. G. Kinnersley and M. A. Langston. On-line variable-sized bin packing. *Discr. Appl. Math.*, 22:143–148, 1988.
77. W. Knoedel. A bin packing algorithm with complexity o(nlogn) and performance 1 in the stochastic limit. In J. Gruska and M. Chytil, editors, *Proc. 10th Symp. Math. Found. Comp. Science*, LNCS, pages 369–378. Springer, 1981.
78. W. Knoedel. Ueber das mittlere Verhalten von on-line Packungsalgorithmen. *EIK*, 19:427–433, 1983.
79. L. T. Kou and G. Markowsky. Multidimensional bin packing algorithms. *IBM J. Research and Development*, 21:443–448, 1977.
80. W. Kuperberg. On-line covering a cube by a sequence of cubes. *Discrete Comput. Geom.*, 12:83–90, 1994.
81. M. Lassak. On-line covering a box by cubes. *Beitr. Algebra Geom.*, 36:1–7, 1995.
82. M. Lassak and J. Zhang. An on-line potato-sack theorem. *Discrete Comput. Geom.*, 6:1–7, 1991.
83. C. C. Lee and D. T. Lee. A simple on-line bin packing algorithm. *J. Assoc. Comput. Mach.*, 32:562–572, 1985.
84. C. C. Lee and D. T. Lee. Robust on-line bin packing algorithms. Technical Report 83-03-FC-02, Department of Electrical Engineering and Computer Science, Northwestern University, Evanston, IL, 1987.
85. F. T. Leighton and P. Shor. Tight bounds for minimax grid matching with applications to the average case analysis of algorithms. *Combinatorica*, 9:161–187, 1989.
86. K. Li and K. H. Cheng. Heuristic algorithms for on-line packing in three dimensions. *J. Algorithms*, 13:589–605, 1992.
87. F. M. Liang. A lower bound for on-line bin packing. *Inform. Process. Lett.*, 10:76–79, 1980.
88. G. S. Lueker. An average-case analysis of bin packing with uniformly distributed item sizes. Technical Report 181, Dept. Information and Computer Science, University of California at Irvine, 1982.
89. W. Mao. Best-k-Fit bin packing. *Computing*, 50:265–270, 1993.
90. W. Mao. Tight worst-case performance bounds for Next-k-Fit bin packing. *SIAM J. Comput.*, 22:46–56, 1993.
91. A. Meir and L. Moser. On packing of squares and cubes. *J. Combin. Theory*, 5:126–134, 1968.
92. H. L. Ong, M. J. Magazine, and T. S. Wee. Probabilistic analysis of bin packing heuristics. *Oper. Res.*, 5:983–998, 1984.
93. P. Ramanan. Average case analysis of the Smart Next Fit algorithm. *Inform. Process. Lett.*, 31:221–225, 1989.
94. P. Ramanan, D. J. Brown, C. C. Lee, and D. T. Lee. On-line bin packing in linear time. *J. Algorithms*, 10:305–326, 1989.
95. P. Ramanan and K. Tsuga. Average case analysis of the modified harmonic algorithm. *Algorithmica*, 4:519–533, 1989.

96. W. T. Rhee and M. Talagrand. The complete convergence of best fit decreasing. *SIAM J. Comput.*, 18:909–918, 1989.
97. W. T. Rhee and M. Talagrand. The complete convergence of first fit decreasing. *SIAM J. Comput.*, 18:919–938, 1989.
98. W. T. Rhee and M. Talagrand. Dual bin packing of items of random size. *Math. Progr.*, 58:229–242, 1993.
99. W. T. Rhee and M. Talagrand. On line bin packing with items of random size. *Math. Oper. Res.*, 18:438–445, 1993.
100. W. T. Rhee and M. Talagrand. On line bin packing with items of random sizes — ii. *SIAM J. Comput.*, 22:1252–1256, 1993.
101. M.B. Richey. Improved bounds for harmonic-based bin packing algorithms. *Discr. Appl. Math.*, 34:203–227, 1991.
102. S. D. Shapiro. Performance of heuristic bin packing algorithms with segments of random length. *Inf. and Cont.*, 35:146–148, 1977.
103. P. W. Shor. The average-case analysis of some on-line algorithms for bin packing. *Combinatorica*, 6:179–200, 1986.
104. P. W. Shor. How to pack better than Best-Fit: Tight bounds for average-case on-line bin packing. In *Proc. 32nd IEEE Symp. Found. Comp. Sci.*, pages 752–759, 1991.
105. D. D. K. D. B. Sleator. A 2.5 times optimal algorithm for packing in two dimensions. *Inform. Process. Lett.*, 10:37–40, 1976.
106. A. Van Vliet. An improved lower bound for on-line bin packing algorithms. *Inform. Process. Lett.*, 43:277–284, 1992.
107. A. Van Vliet. *Lower and upper bounds for on-line bin packing and scheduling heuristics.* PhD thesis, Erasmus University, Rotterdam, The Netherlands, 1995.
108. A. Van Vliet. On the asymptotic worst case behaviour of harmonic fit. *J. Algorithms*, 20:113–136, 1996.
109. G. J. Woeginger. Improved space for bounded-space on-line bin packing. *SIAM J. Discr. Math.*, 6:575–581, 1993.
110. G. J. Woeginger and G. Zhang. Optimal on-line algorithms for variable-sized bin covering. Technical Report Woe-022, TU Graz, Austria, 1998.
111. A. C. C. Yao. New algorithms for bin packing. *J. Assoc. Comput. Mach.*, 27:207–227, 1980.
112. G. Zhang. A tight worst-case performance bound for AFB-k. Technical Report 015, Institute of Applied Mathematics, Academia Sinica, Beijing, China, 1994.
113. G. Zhang. Worst-case analysis of the FFH algorithm for on-line variable-sized bin packing. *Computing*, 56:165–172, 1996.

8

On-line Load Balancing

Yossi Azar

1 Introduction

General: The machine load balancing problem is defined as follows: There are n parallel machines and a number of independent tasks (jobs); the tasks arrive at arbitrary times, where each task has an associated *load vector* and duration. A task has to be assigned immediately to exactly one of the machines, thereby increasing the *load* on this machine by the amount specified by the corresponding coordinate of the load vector for the duration of the task. All tasks must be assigned, i.e., no admission control is allowed. The goal is usually to minimize the maximum load, but we also consider other goal functions. We mainly consider non-preemptive load balancing, but in some cases we may allow preemption *i.e.,* reassignments of tasks. All the decisions are made by a centralized controller.

The on-line load balancing problem naturally arises in many applications involving allocation of resources. As a simple concrete example, consider the case where each "machine" represents a communication channel with bounded bandwidth. The problem is to assign each incoming request for bandwidth to one of the channels. Assigning a request to a certain communication channel increases the load on this channel, *i.e.,* increases the percentage of the used bandwidth. The load is increased for the duration associated with the request.

Load vs. Time: There are two independent parameters that characterize the tasks: the first is the duration and second is the type of the load vector. This leads to several load balancing problems. Note that the main difference between scheduling and load-balancing is that in scheduling there is only one axis (duration or load) in load balancing they are two independent axes (duration and load). Thus we may have tasks of high or low load with short or long durations. For example, in storing files on set of servers files that may have various sizes are created or deleted at arbitrary times which may be unrelated to their sizes.

Duration: We classify the duration of tasks as follows: we call tasks which start at arbitrary times but continue forever *permanent*, while tasks that begin and end are said to be *temporary*. The duration of each task may or may not be known upon its arrival. Note that tasks arrive at specific times, depart at specific times and are active for their whole durations. Thus delaying tasks is not allowed. For example, in storing files on servers, once a file is created it must be stored immediately on a server until it is deleted.

Clearly, permanent tasks are an important special case of temporary ones, (departure time is ∞ or very large) and better results can be achieved for them. Also knowing the duration of (temporary) tasks may help in achieving better results compared with the unknown duration case. Note that permanent tasks may be viewed also in the scheduling framework; where "load" corresponds to "execution time" (the durations are ignored). Restating the problem in these terms, our goal is to decrease maximum execution time under the requirement that the arriving tasks are *scheduled immediately*.

Load vector: Formally, each arriving task j has an associated *load vector*, $\mathbf{p}(j) = (p_1(j), p_2(j), \ldots, p_n(j))$ where $p_i(j)$ defines the increase in the load of machine i if we were to assign task j to it. The load vector can be categorized in several classes: identical machines case, related machines case, restricted assignment case and unrelated machines case.

In the *identical* machines case, all the coordinates of a load vector are the same. In the *related* machines case, the ith coordinate of each load vector is equal to $w(j)/v_i$, where the "weight" $w(j)$ depends only on the task j and the "speed" v_i depends only on the machine i. In the *restricted assignment* case each task has a weight and can be assigned to one of a subset of machines. In terms of the load vector the coordinates are either $w(j)$ or ∞. The *unrelated* machines case is the most general case, i.e., $p_i(j)$ are arbitrary non-negative real numbers (∞ may be represented by large M). Clearly, related machines and the restricted assignment are not comparable, but are both special cases of the unrelated machines case. The identical machines case is a special case of the related machines where all the speeds v_i are the same. It is also a special case of restricted assignment where all the coordinates of tasks j are $w(j)$.

The measure: Since the arriving tasks have to be assigned without knowledge of the future tasks, it is natural to evaluate the performance in terms of the *competitive ratio* w.r.t. some performance measure e.g., the maximum load. In our case, the competitive ratio is the supremum, over all possible input sequences, of the ratio between the performance achieved by the on-line algorithm and the performance achieved by the optimal off-line algorithm. The competitive ratio may be constant or depends on the number of machines n (which is usually relatively small). It should not depend on the number of tasks that may be arbitrarily large.

The most popular performance measure is the maximum load, i.e, the maximum over machines and over time of the load. This measure focuses on the worst machine and is the equivalent of the makespan for scheduling problems.

However, the maximum load is not the only reasonable measure. Measures that take into account how well all the machines are balanced are, for example, the L_2 norm or the L_p norm of the load vector.

To emphasize the difference between the maximum load and the L_2 norm we take an example where each task sees a delay in service that is proportional to the number (or total weight) of tasks that are assigned to its machine. Then the traditional load balancing which is to minimize the maximum load corresponds to minimize the *maximum* delay. Minimizing the sum of squares (equivalently, minimizing the L_2 norm) corresponds to minimizing the *average* delay of tasks in the system.

We note that all the definitions and theorems are stated for the maximum load performance measure except in the section where the L_p norm is considered.

Reassignments: We recall that each task must be assigned to some machine for its duration. It turns out that the performance of load balancing algorithms may be significantly improved in some cases if we allow limited amount of reassignments. More specifically, the algorithm can reassign some of the existing tasks.

Observe that if the number of reassignments per task is not limited, it is trivial to maintain optimum load, *i.e.* competitive ratio of 1 by reassigning optimally all the current tasks. However, reassignments are expensive process and should be limited. Thus, we measure the quality of an on-line algorithm by the competitive ratio achieved and the number of reassignments performed during the run.

Virtual circuit routing: Some of the algorithms for virtual circuit routing are extensions of the algorithms for load balancing. We consider the following idealized setting: We are given a network. Requests for virtual circuits arrive on line, where each request specifies source and destination points, and a load vector (the number of coordinates is the number of edges of the network). The routing algorithm has to choose a path from the source to the destination, thereby increasing the load on each edge of the path by the corresponding coordinate of the load vector for the duration of the virtual circuit. The duration may or may not be specified in advance. The goal is to minimize the maximum over all edges of the load or some other function of the load. Reassignments which are called here reroutings may or may not be allowed.

The above problem is called *generalized* virtual circuit routing problem. It is easy to see that the load balancing problem is a special case of generalized virtual circuit routing. Load balancing can be reduced to a generalized virtual circuit routing problem on a 2 vertex network with multiple edges between them. Every edge corresponds to a machine and every arriving task is translated into a request between the two vertices s and t with the same load vector.

In the classical (in contrast to generalized) virtual circuit routing problem we assume that the load vector of request j on edge e is $r(j)/c(e)$ where $r(j)$ is the requested bandwidth and $c(e)$ is the capacity of the edge. Clearly the identical and related machines problems are special cases of virtual circuit routing on the

2 vertices network. There are various models for virtual circuit routing problems (e.g. allowing admission control and maximizing the throughput). For a survey on on-line virtual circuit routing, we refer the reader to the Chapter 11 by Leonardi in this book and to the references listed therein.

2 Definition and basic schemes

2.1 The model

The input sequence consists of task arrival events for permanent tasks and of task arrival and departure events for temporary tasks. Since the state of the system changes only as a result of one of these events, the event numbers can serve as time units, *i.e.*, we can view time as being *discrete*. We say that time t corresponds to the tth event. Initially the time is 0, and time 1 is the time at which the first task arrives. Whenever we speak about the "state of the system at time t" we mean the state of the system *after* the tth event was already handled. In other words, the response to the tth event takes the system from the "state at $t-1$" to the "state at t".

A task j is represented by its "load vector" $\mathbf{p}(j) = (p_1(j), p_2(j), \ldots, p_n(j))$, where $p_i(j) \geq 0$. Let $\ell_i(t-1)$ denotes the load on machine i at time $t-1$. If at time t a task j is assigned to machine i, the load on this machine increases by $p_i(j)$. In other words $\ell_k(t)$ which denotes the load on machine k at time t, *i.e.*, after the tth event is defined as follows:

$$\ell_k(t) = \begin{cases} \ell_k(t-1) + p_k(j) & \text{if } k = i \\ \ell_k(t-1) & \text{otherwise} \end{cases}$$

Similarly, if at time t a task j departs from machine i, the load on this machine decreases by $p_i(j)$.

Let $\sigma = (\sigma_1, \sigma_2, \ldots)$ be a particular sequence of arrivals and departures of tasks. Denote by $\ell_k^*(t)$ the load of the optimal off-line algorithm on machine k at time t. The maximum load achievable by an optimum off-line algorithm is denoted by $OPT(\sigma)$ which is the maximum of $\ell_k^*(t)$ over time and machines. If σ is clear from the context, we will use OPT for $OPT(\sigma)$.

Recall that for *identical* machines, $\forall i, j : p_i(j) = w(j)$. For *related* machines, $\forall i, j : p_i(j) = w(j)/v_i$ where v_i denotes the speed of machine i. For *restricted* assignment $p_i(j)$ is either $w(j)$ or ∞. For *unrelated* machines $p_i(j)$ are arbitrary non-negative real numbers.

2.2 Doubling

Several algorithms are designed as if the value of the optimal algorithm is known. This assumption can be easily eliminated by using simple doubling and losing a factor of 4 in the competitive ratio. More specifically, we define the notion of a *designed performance guarantee* β as follows: the algorithm $A(\Lambda)$ accepts a parameter Λ and never creates load that exceeds $\beta\Lambda$. $A(\Lambda)$ is allowed to return "fail" and to refuse to assign a task if $\Lambda < OPT$ otherwise it has to assign all of the tasks.

The algorithm A works in phases, the difference between phases is the value of Λ assumed by the algorithm. Within a phase the algorithm $A(\Lambda)$ is used to assign tasks, while ignoring all tasks assigned in previous phases. The first phase has $\Lambda = \min_i p_i(1)$, which is the minimum possible load the first (non-zero) task may produce. At the beginning of every subsequent phase the value of Λ doubles. A new phase starts when the algorithm returns "fail". Thus, the last phase will never end.

It is easy to see that this approach sets the competitive factor of A to be larger than the designed performance guarantee by a factor of 4 (a factor of 2 due to the load in all the rest of the phases except the last, and another factor of 2 due to imprecise approximation of OPT by Λ). Thus the competitive ratio is 4β. We note that the factor of 4 can be replaced by $e = 2.7\ldots$ for a restricted class of algorithms that uses randomization (see [26]).

3 Permanent tasks

Tasks which start at arbitrary times but continue forever are called *permanent*. The situation in which all tasks are permanent is classified as permanent tasks. Otherwise, it is classified as temporary tasks. (A permanent task is a special case of a temporary one by assuming ∞ or high departure time.) In the permanent tasks case, task j is assigned at time j. Thus $\ell_k(j)$ corresponds to the load on machine k immediately after task j has been assigned.

We note that for the off-line problems there are polynomial approximation schemes for identical and related machines cases [25, 24] and 2 approximations for restricted assignment and unrelated machines cases [30].

3.1 Identical machines

In the *identical* machines case, all the coordinates of a load vector are the same. This case was first considered by Graham [22, 23] who considered the natural greedy which is assigning the next task to the machine with the current minimum load.

Theorem 1. [22] *The greedy algorithm has a competitive ratio of exactly* $2 - \frac{1}{n}$ *where* n *is the number machines.*

The fact that greedy is not better than $2 - \frac{1}{n}$ is shown by a simple example which is $n(n-1)$ unit size tasks followed by one task of size n. For $n = 2$ and $n = 3$ the competitive ratios are $3/2$ and $5/3$ (respectively) and are optimal. Somewhat better algorithms for small $n \leq 4$ appear in [21, 19]. It took some time until an algorithm whose competitive ratio is strictly below $c < 2$ (for all n) was found [13]. The competitive ratio of this algorithm, which does not always assign the next task to the lowest loaded machine, is $2 - \epsilon$ for a small constant ϵ. The algorithm was modified in [27] and its competitive ratio was improved to 1.945. Recently, Albers [1] designed a 1.923 competitive algorithm and improved the lower bound for large number of machines to 1.852 (the previous lower bound for permanent tasks was 1.8370 [14]). We may also consider randomized algorithms. For example for two machines the competitive ratio (upper and lower bound) is 4/3 [13]. Somewhat better algorithms for small $n \leq 4$ appear in [33]. For large n the best lower bound for randomized algorithms is 1.582 [18, 34] and the best randomized algorithm is just the deterministic one.

It is worthwhile to note that one may consider the case where the value of OPT is known. Then, for two machines the competitive ratio is exactly 4/3 [29]. For any n a deterministic algorithm that is 1.625 competitive is given in [12]. The lower bound for this case is only 4/3.

The identical machines case for permanent tasks is also considered in the on-line scheduling framework. It is also called jobs arriving one by one. A comprehensive survey by Sgall on on-line scheduling appears in Chapter 9 of this book.

Open problem 3.1 *Determine the competitive ratio for deterministic and randomized load balancing algorithm for permanent tasks on identical machines.*

3.2 Related machines

Recall that in the *related* machines case, the ith coordinate of each load vector is equal to $w(j)/v_i$, where the "weight" $w(j)$ depends only on the task j and the "speed" v_i depends only on the machine i. This case was considered in [4] who showed an 8 competitive algorithm for permanent tasks.

Note that the related machines case is a generalization of the identical machines case. However, assigning a task to the machine with minimum load (i.e greedy) results in an algorithm with competitive ratio that is at least the ratio of the fastest to slowest machines speed which maybe unbounded even for two machines. Nevertheless, one may consider the following natural post-greedy algorithm: each task j is assigned upon its arrival to a machine k that minimizes the resulting load *i.e.*, a machine k that minimizes $\ell_k(j-1) + p_k(j)$. It is easy to see that for the identical machines case greedy and post-greedy algorithms are the same. The lower bound of the next theorem was proved in [20] and the upper bound was proved in [4].

Theorem 2. [20, 4] *The post-greedy algorithm has a competitive ratio $\Theta(\log n)$ for related machines.*

A new algorithm is required to achieve constant competitive algorithm for the related machine case. We will use the doubling technique and thus may assume a given parameter Λ, such that $\Lambda \geq OPT$. Roughly speaking, algorithm ASSIGN-R will assign tasks to the slowest machine possible while making sure that the maximum load will not exceed twice Λ. More specifically, we assume that the machines are indexed according to increasing speed. The algorithm assigns a task to the machine i of minimum index such that $\ell_i(j-1) + p_i(j) \leq 2\Lambda$ and it fails if such an index does not exists. The above algorithm may be viewed as an adaptation of the scheduling algorithm [35] to the context of load balancing.

Theorem 3. [4] *If $OPT \leq \Lambda$, then algorithm* ASSIGN-R *never fails. Thus, the load on a machine never exceeds 2Λ. If OPT is unknown in advance the doubling technique for Λ implies that* ASSIGN-R *is 8 competitive.*

Using randomized doubling it is possible to replace the deterministic 8 upper bound by $2e \approx 5.436$ expected value [26]. Recently it was shown by [16] that replacing the doubling by a more refined method improves the deterministic competitive ratio to $3 + \sqrt{8} \approx 5.828$ and the randomized variant to about 4.311. Also the lower bound is 2.438 for deterministic algorithms and 1.837 for randomized ones.

Open problem 3.2 *Determine the competitive ratio for load balancing of permanent tasks on related machines.*

3.3 Restricted assignment

Each task has a weight and can be assigned to one of a subset of admissible machines (which may depend on the task). This case was considered in [11], who described an optimal (up to an additive one) competitive algorithm for permanent tasks. The algorithm *AW* is just a natural greedy algorithm that assigns a task to a machine with minimum load among the admissible machines breaking ties arbitrarily.

Theorem 4. [11] *Algorithm AW achieves a competitive ratio of $\lceil \log_2 n \rceil + 1$. The competitive ratio of any on-line assignment algorithm is at least $\lceil \log_2 n \rceil$.*

It is interesting to realize that randomized algorithms may improve the performance but only by a constant factor. More specifically:

Theorem 5. [11] *The competitive ratio of any randomized on-line assignment algorithm is at least* $\ln(n)$

Using the on-line randomized matching algorithm of [28] the lower bound can be matched for the unit tasks case by an algorithm *AR* defined as follows. Choose a sequence of random permutations $\pi(k)$ of 1 to n for $k \geq 1$. Assign tasks greedy. If the minimum load among the admissible machines for a given task is k then breaks ties by priorities given by $\pi(k)$. The result is summarized in the next theorem.

Theorem 6. [11] *For unit size tasks the (expected) competitive ratio of Algorithm AR is at most* $\ln(n) + 1$ *assuming that the optimal load is* 1.

Open problem 3.3 *Design a randomized algorithm for arbitrary sized tasks for the permanent restricted assignment case that achieves* $\ln(n) + O(1)$ *competitive ratio or show an appropriate lower bound.*

3.4 Unrelated machines

The *general* unrelated machines case for permanent tasks was considered in [4] who described an $O(\log n)$-competitive algorithm.

We note that natural greedy approaches are far from optimal for this case. More specifically consider the post-greedy algorithm in which a task is assigned to a machine that minimizes the resulting load or an algorithm in which a task is assigned to a machine whose increase in load is minimum.

Lemma 7. [4] *The competitive ratios of the above greedy algorithms are exactly n for the unrelated machines case.*

For describing the $O(\log n)$-competitive algorithm we first consider the case where we are given a parameter Λ, such that $\Lambda \geq OPT$. As before, an appropriate value of Λ can be "guessed" using a simple doubling approach, increasing the competitive ratio by at most a factor of 4. We use tilde to denote normalization by Λ, *i.e.* $\tilde{x} = x/\Lambda$.

We use again the notion of *designed performance guarantee.* Let $1 < a < 2$ be any constant and β the designed performance guarantee. The basic step of the algorithm called ASSIGN-U is to assign task j to a machine s such that after the assignment $\sum_{i=1}^{n} a^{\tilde{\ell}_i(j)}$ is as small as possible. More precisely, we compute

$$\text{Increase}_i(j) = a^{\tilde{\ell}_i(j-1)+\tilde{p}_i(j)} - a^{\tilde{\ell}_i(j-1)}$$

and assign the task to a machine s with minimum increase unless $\ell_s(j-1) + p_s(j) > \beta\Lambda$ which results in returning "fail".

Theorem 8. [4] *There exists $\beta = O(\log n)$ such that if $OPT \leq \Lambda$ then algorithm* ASSIGN-U *never fails. Thus, the load on a machine never exceeds $\beta\Lambda$.*

The lower bound for the restricted assignment case implies that the algorithm is optimal (up to a constant factor). Also it is interesting to note that for the restricted assignment case the algorithm becomes the *AW* greedy algorithm described in the previous subsection.

3.5 Virtual circuit routing

Surprisingly, the algorithm for the unrelated machines case can be extended to the more complex case of virtual circuit routing. We are given a (directed or undirected) graph and requests for virtual paths arrive on-line. The jth request is $(s_j, t_j, r(j))$ where s_j, t_j are source and destination points and $r(j)$ is a required bandwidth. If the path assigned to request j uses an edge e then the load $\ell(e)$ is increased by $p_e(j) = r(j)/c(e)$ where $c(e)$ is the capacity of the edge. The goal is to minimize the maximum load over the edges.

As usual we assume that we are given a parameter $\Lambda \geq OPT$. The algorithm assigns a route such that $\sum_{e \in E} a^{\tilde{\ell}_e(j)}$ is as small as possible. More precisely, we compute

$$\text{Increase}_e(j) = a^{\tilde{\ell}_e(j-1)+\tilde{p}_e(j)} - a^{\tilde{\ell}_e(j-1)}$$

and assign the request to the shortest path from s_j to t_j unless some load exceeds $\beta\Lambda$ which results in returning "fail".

Theorem 9. [4] *If $OPT \leq \Lambda$, then there exists $\beta = O(\log n)$ such that the routing algorithm never fails. Thus, the load on a link never exceeds $\beta\Lambda$.*

It is possible to translate the $\Omega(\log n)$ lower bound for restricted assignment to the virtual circuit routing problem on directed graphs. Recently [15] showed that the lower bounds hold also for undirected graphs.

	Competitive ratio
Identical	$2-\epsilon$
Related	$\Theta(1)$
Restricted	$\Theta(\log n)$
Unrelated	$\Theta(\log n)$
Routing	$\Theta(\log n)$

Fig. 1. Summary of competitive ratio for permanent tasks

4 Temporary tasks unknown duration

Tasks that arrive and may also depart are called temporary tasks. Recall that permanent tasks are a special case of temporary tasks. We refer to the case that the duration of a task is unknown at its arrival (in fact until it actually departs) as unknown duration.

4.1 Identical machines

It turns out that the analysis of Graham [22, 23] of the greedy algorithm for permanent tasks also holds for temporary tasks. It is shown in [9] that no algorithm can achieve a better competitive ratio. Thus, the optimal algorithm is greedy which is $(2-\frac{1}{n})$-competitive. Recall that for permanent tasks the competitive ratio is below 1.923 which implies that the temporary tasks case is strictly harder than the permanent tasks case. However, for $n = 2, 3$ the competitive ratio is the same as for permanent tasks. It turns out that randomization does not help much. Specifically, randomized algorithms cannot achieve a competitive ratio which is better than $2 - 2/(n+1)$ [9]. If the sequence is limited to a polynomial size in n then the lower bound is $2-O(\log\log n/\log n) = 2-o(1)$. Also, the tight lower bound for randomized algorithms for $n = 2$ is $3/2$. This is contrast to the $4/3$ randomized competitive ratio for permanent tasks. Hence, randomization cannot help much for temporary tasks and the obvious question is whether it can help at all here.

Open problem 4.1 *Can we get below the $2-1/n$ using randomized algorithms for temporary tasks of unknown durations on identical machines ?*

4.2 Related machines

This case was considered in [10] who showed a 20 competitive algorithm for temporary tasks. Recall that for permanent tasks there is a 5.828 (improvement over the 8) competitive algorithm.

We use again the notion of *designed performance guarantee* and give an algorithm Slow-Fit which guarantees a load of 5Λ given a parameter $\Lambda \geq OPT$. By the simple doubling approach this results in a 20-competitive algorithm. Assume that the machines are indexed according to increasing speed. For a task j that arrives at time t we say j is *assignable* to machine i if $w(j)/v_i \leq \Lambda$ and $\ell_i(t-1) + w(j)/v_i \leq c \cdot \Lambda$. Slow-Fit assign task j to the *assignable* machine of minimum index.

Theorem 10. [10] *Provided $c \geq 5$ and $\Lambda \geq OPT$, the* Slow-Fit *guarantees that every task is assignable. Thus, if OPT is unknown in advance the doubling technique for Λ implies that* Slow-Fit *algorithm is 20-competitive.*

The best lower bound is the following:

Theorem 11. [10] *The competitive factor c of any on-line algorithm for the related machines case satisfies $c \geq 3 - o(1)$.*

Note that the lower bound is valid even when the value of OPT is known to the algorithm (the upper bound is 5 if OPT is known).

Open problem 4.2 *Determine the competitive ratio for load balancing of tasks with unknown durations on related machines.*

4.3 Restricted assignment

Recall that in the restricted assignment case for permanent tasks the competitive ratio of the greedy algorithm is at most $\log n + 2$ and that no algorithm can do better (up to an additive one).

In contrast for the case of temporary tasks with *unknown duration* it is shown in [8, 10] that there is an algorithm with competitive ratio $\Theta(\sqrt{n})$ and that no algorithm can do better. More precisely, the following theorems are proved:

Theorem 12. [8] *The competitive ratio of the greedy on-line assignment algorithm is exactly $\frac{(3n)^{2/3}}{2}(1+o(1))$.*

Theorem 13. [8] *The competitive ratio of any deterministic on-line assignment algorithm is at least* $\lfloor\sqrt{2n}\rfloor$. *For randomized algorithms the lower bound is* $\Omega(n^{1/2})$.

The lower bound is proved using exponential size sequence of requests. It is shown in [32] that the lower bound can also be achieved (up to some constant factor) even on polynomial length sequence.

Next we describe an $O(\sqrt{n})$ competitive algorithm called ROBIN-HOOD. Again we first design an algorithm for the case that we are given a parameter $\Lambda \geq OPT$. A machine g is said to be *rich* at some point in time t if $\ell_g(t) \geq \sqrt{n}\Lambda$, and is said to be *poor* otherwise. A machine may alternate between being rich and poor over time.

If g is rich at t, its *windfall time* at t is the last moment in time it became rich. More precisely, g has windfall time t_0 at t if g is poor at time $t_0 - 1$, and is rich for all times t' $t_0 \leq t' \leq t$.

Algorithm ROBIN-HOOD assigns a new task to some poor machine if possible. Otherwise, it assigns to the machine with the most recent windfall time.

Theorem 14. [10] *The competitive ratio of Algorithm* ROBIN-HOOD *is at most* $2\sqrt{n}+1$.

We can apply doubling to overcome the problem that OPT is unknown in advance. This would result in increasing the competitive ratio by a factor of 4. However, it turns out that we do not need to lose this factor of 4. We do so by maintaining an estimate $L(t)$ and use it instead of Λ. Instead of doubling $L(t)$ it is updated after the arrival of a new task j at time t by setting

$$L(t) = \max\{L(t-1), w_j, \frac{1}{n}(w_j + \sum_g \ell_g(t-1))\}.$$

4.4 Unrelated machines

The only facts that are known for tasks of unknown duration on unrelated machines is that the competitive ratio is at least $\Omega(\sqrt{n})$ (by the restricted assignment lower bound) and at most n (many versions of greedy). The main open problem here is the following:

Open problem 4.3 *Determine the competitive ratio for load balancing of tasks with unknown durations on unrelated machines.*

5 Known duration

It is not known if knowing the durations of tasks helps in the identical and related machines cases. Recall that for identical machines, if the duration are not known then the deterministic and randomized competitive ratios are $2-o(1)$. An interesting open problem is the following:

Open problem 5.1 *Can we get a competitive ratio which is below 2 for identical machines knowing the duration (deterministic or randomized) ?*

Knowing the durations certainly helps in the restricted assignment and unrelated machines cases. Denote by T the ratio of the maximum to minimum duration (the minimum possible task duration is known in advance). Recall the $\Omega(\sqrt{n})$ lower bound on the competitive ratio for the restricted assignment case when the duration of a task is not known upon its arrival [8]. In contrast if the duration is known we have the following:

Theorem 15. [10] *There is an on-line load balancing algorithm for unrelated machines with known tasks duration which is $O(\log nT)$-competitive.*

It is unclear if the $\log T$ is really necessary when the durations are known. Of course, we can ignore the durations and get $O(\sqrt{n})$ competitive algorithm for the restricted assignment (which is better for huge T). The obvious question is whether we can do better.

Open problem 5.2 *Can we get below the $\Theta(\sqrt{n})$ bound for restricted assignment assuming known durations and can we prove lower bounds ?*

Open problem 5.3 *Can we get below the $\Theta(n)$ for unrelated machine case knowing the durations and can we prove lower bounds ?*

	Unknown durations	Known durations	Permanent
Identical	$2 - o(1)$	?	$2 - \epsilon$
Related	$\Theta(1)$	$\Theta(1)$	$\Theta(1)$
Restricted	$\Theta(n^{1/2})$	$O(\log nT)$	$\Theta(\log n)$
Unrelated	?	$O(\log nT)$	$\Theta(\log n)$
Routing	?	$O(\log nT)$	$\Theta(\log n)$

Fig. 2. Summary of competitive ratio for the various models

6 Reassignments

In order to overcome the above large lower bounds for load balancing of tasks with unknown durations, one may allow *task reassignments*.

6.1 Restricted assignment

For unit size tasks (i.e. all coordinates of the load vector are either ∞ or 1), [31] presented an algorithm that achieves $O(\log n)$ competitive ratio with respect to load while making at most a constant amortized number of reassignments per task. Their algorithm belongs to the class of algorithm that does not perform reassignments unless tasks depart. Hence the $\Omega(\log n)$ lower bound for permanent tasks holds for this type of algorithm.

Later, [7] considered the unit size task case. They gave an algorithm that achieves a *constant* competitive ratio while making $O(\log n)$ amortized reassignments per task. However, they required that the optimum load achieved by the off-line algorithm is at least $\log n$. The algorithm has been extended in [36] and that assumption has been removed.

We first describe the algorithm that maintains constant competitive ratio and $O(\log n)$ reassignments per unit size task with the assumption that $OPT \geq \log n$.

As usual we assume that the algorithm has a knowledge of $\Lambda \geq OPT$. The algorithm will maintain the following *stability condition*:

Definition 16. Let j be some task which is currently assigned to machine i. Consider a machine i' which is an eligible assignment for task j (*i.e.* machine i with $p_{i'j} = 1$). We say that the algorithm is in a *stable state* if for any such i and i', we have:

$$\ell_i - \ell_{i'} \leq 2\Lambda / \log n$$

The main idea of the algorithm is to make sure that the above stability condition is satisfied. More precisely the algorithm is described as follows: A new task j is assigned to any eligible machine and when a task departs, it is removed from the machine on which it is currently assigned. If at any moment the stability condition is not satisfied by some task j that is currently assigned to machine i, the algorithm reassigns j to a least loaded machine among the machines that are eligible with respect to j.

Theorem 17. [7] *The assignment algorithm maintains load of at most* 4Λ *with* $O(\log n)$ *reassignments per arrival or departure of a task assuming* $\Lambda \geq OPT$.

As before we eliminate the need to know the optimal load in advance by the doubling technique. This increases the competitive ratio by at most a factor of 4 to be 16.

Observe that, as opposed to the previous algorithm, this algorithm reassigns tasks both as a result of task arrival and departure. As mentioned this is necessary to achieve constant competitive ratio, since the lower bound of [11] implies that an algorithm that does not reassign tasks as a result of task arrivals can not achieve better than $\Omega(\log n)$ competitive ratio.

Next, we describe how to get rid of the assumption that $OPT \geq \log n$ [36]. Regard the problem as a game on a dynamic bipartite graph. On one side are the machines V and on the other side are the tasks U. An edge (u, v) indicates that u can be assigned to v. Edge (u, v) is *matching* if u is assigned to v.

Let $\ell(v)$ denote the load on $v \in V$, i.e., the number of matching edges incident on v. A balancing path is an even-length path sequence of alternating matched and unmatched edges $(v_1, u_1), (u_1, v_2), \ldots, (u_{m-1}, v_m)$ with the property that $\ell(v_i) < \ell(v_1)$ for $1 \leq i \leq m-1$ and $\ell(v_m) < \ell(v_1) - OPT$. A balancing path can be used to reduce the maximum load on $v_1, v_2, \ldots, v_m$ by reassigning u_i to v_{i+1} for $1 \leq i \leq m-1$. The machines are r-balanced if there is no balancing path of length r or less.

The algorithm is described as follows: A new task j is assigned to any eligible machine and when a task departs, it is removed from the machine on which it is currently assigned. If at any moment there is a balancing path of length r or less then re-balance using this path. For $r = \log n$ we have

Theorem 18. [36] *The assignment algorithm above is constant competitive with $O(\log n)$ reassignments per arrival or departure of a task.*

Open problem 6.1 *Is it possible to achieve constant ratio and constant number of reassignments per task.*

6.2 Unrelated machines

The general case *i.e.* load balancing of unknown duration tasks with no restrictions on the load vectors was considered in [7]. They designed a new algorithm that makes $O(\log n)$ reassignments per task and achieves an $O(\log n)$ competitive ratio with respect to the load.

We first assume that the algorithm has a knowledge of $\Lambda \geq OPT$. Let $1 < a < 2$ be a constant. At every instance t, each active task j is assigned to some machine i. Define the *height* $h_i^j(t)$ of task j that is assigned to machine i at time t as follows. It is the sum of $p_i(j)$ for all tasks j' that are currently assigned to i and were last reassigned to machine i before j was last assigned to i. The weight of task j is

$$W_i^j(t) = a^{\tilde{h}_i^j(t)+\tilde{p}_i(j)} - a^{\tilde{h}_i^j(t)}$$

Note that the weight of task j immediately after it is assigned to machine i is:

$$a^{\tilde{\ell}_i(t)} - a^{\tilde{\ell}_i(t)-\tilde{p}_i(j)}$$

where t is the time immediately after the assignment.

From now on we will omit the parameter t. The algorithm maintains the following *stability condition*:

Definition 19. We say that the algorithm is in a *stable state* if for any machine i' we have:

$$W_i^j = a^{\tilde{h}_i^j+\tilde{p}_i(j)} - a^{\tilde{h}_i^j} \leq 2\left(a^{\tilde{\ell}_{i'}+\tilde{p}_{i'}(j)} - a^{\tilde{\ell}_{i'}}\right) .$$

Intuitively, the main idea of the algorithm is to make sure that the above stability condition is satisfied. More precisely the algorithm is described as follows:

A task is assigned upon its arrival to a machine i which minimizes W_i^j. When a task departs it is removed from the machine on which it is currently assigned. If at any moment the stability condition is not satisfied by some task j that is currently assigned to some machine i, the algorithm reassigns j on machine i' that minimizes $W_{i'}^j$.

Observe that the algorithm will never reassign as a result of an arrival of a new task. The stability condition is strong enough to maintain the competitive ratio and weak enough to cause many reassignments.

Theorem 20. [7] *For the unrelated machines problem where the duration of tasks is a-priori unknown, the above assignment algorithm makes $O(\log n)$ reassignments per task and achieves $O(\log n)$ competitive ratio with respect to the load.*

Recall that for restricted assignment (and therefore for unrelated machines) an $\Omega(\log n)$ lower bound was proved on the competitive ratio for the load balancing case where tasks never depart. Observe that the algorithm reassigns tasks only as a result of task departures, and hence can not achieve better than $O(\log n)$ competitive ratio with respect to load.

A natural extension of the algorithm also works for the virtual circuit routing problems [7]. By making $O(\log n)$ reroutings per path, it achieves an $O(\log n)$ competitive ratio with respect to the load.

6.3 Current load

We conclude this subsection by an alternative definition of competitive ratio which requires reassignments to get reasonable results. In the standard definition we compare the maximum on-line load to the maximum off-line load. It was suggested in [36] to compare the current load against the current off-line load. It is easy to see that for permanent tasks the standard definition and the new definition are the same (since the sequence may stop at any time and the on-line and off-line loads are monotonically non-decreasing). However, for temporary tasks it is straightforward to see that if no reroutings are allowed then the lower bound is n even on identical machines. Specifically, n^2 unit tasks appear, after which some machine k must have load at least n. Then, all tasks depart except for those on k. Thus, the current on-line load is n while the current optimal off-line load is 1. Thus, one must allow reassignments to achieve significant results for this model. Algorithms for this purpose appear in [31, 36, 2, 3].

7 L_p norm

In all the previous sections we evaluated the performance of the algorithm by the maximum load. In section we consider the L_p norm measure ($p \geq 1$).

Recall that $\ell_k(t)$ denotes the load on machine k at time t and $\ell_k^*(t)$ denotes the load on machine k at time t of the optimal off-line algorithm. For a given vector $X = (x_1, x_2, \ldots, x_n)$ the L_p norm and L_∞ norm of X are

$$|X|_p = \left(\sum_{1 \leq i \leq n} |x_i|^p \right)^{1/p} \quad \text{and} \quad |X|_\infty = \max_{1 \leq i \leq n} \{|x_i|\}.$$

The L_2 norm is the Euclidean norm, which measures the length of the vector X in Euclidean space. The maximum load measure of an algorithm is the maximum over time of $|\ell(t)|_\infty$. The L_p norm measure for an algorithm A on a sequence σ denoted by $A(\sigma)$ is the maximum over time of $|\ell(t)|_p$. The performance of an algorithm A is the supremum over all sequences of $A(\sigma)/OPT(\sigma)$.

We first consider permanent tasks. It is not hard to show that for identical machines the greedy algorithm (i.e always assigning tasks to the minimum loaded machine) is 2 competitive. In fact, the competitive ratio of greedy is determined in [5]. In particular, for the L_2 norm it is $\sqrt{4/3}$ and no algorithm can achieve a better competitive ratio. Surprisingly, the asymptotic competitive ratio is below $\sqrt{4/3} - \epsilon$.

Next we consider unrelated machines. A natural post-greedy type algorithm for minimizing the L_p norm is to assign a task on a machine to minimize $\sum_i \ell_i^p(j)$. More precisely, when task j arrives we compute weights to the machines,

$$\text{Increase}_i(j) = (\ell_i(j-1) + r_i(j))^p - \ell_i^p(j-1)$$

and assign the task to a machine with minimum increase. Note that for the identical machines and the restricted assignment cases the algorithm is equivalent to the greedy that assigns a task to a least loaded machine.

Theorem 21. [6] *The above algorithm is $1+\sqrt{2}$ competitive with respect to the L_2 norm.*

Theorem 22. [6] *For any constant $p \geq 1$ the above load balancing algorithm is $O(p)$-competitive in the L_p norm. Moreover, any deterministic algorithm must be $\Omega(p)$-competitive even for the restricted assignment case.*

Theorem 23. [17] *For the restricted assignment case with unit jobs the greedy algorithm is approximately 2.01 competitive with respect to the L_2 norm.*

Open problem 7.1 *Design an algorithm for related machine case (permanent tasks) whose competitive ratio in the L_p norm is constant (independent of p).*

It is not quite clear how to define the performance measure for temporary tasks. One possible definition is the maximum over the duration of the L_p norm of the load vector. For the case of known duration one may use a different definition which is the L_p norm of the nT vector of the n machines over the sequence of total length T. For this definition one can achieve a competitive ratio of $O(p)$ (known durations). Not much is known for the unknown duration case.

Open problem 7.2 *Determine the competitive ratio in the L_p norm for tasks with unknown duration for related machines, unrelated machines and for restricted assignment with and without reassignments.*

Acknowledgements

This work was partially supported by the Israel Science Foundation and by the United States-Israel Binational Science Foundation (BSF).

References

1. S. Albers. Better bounds for on-line scheduling. In *Proc. 29th ACM Symp. on Theory of Computing*, pages 130–139, 1997.
2. M. Andrews. Constant factor bounds for on-line load balancing on related machines. Manuscript.
3. M. Andrews, M. Goemans, and L. Zhang. Improved bounds for on-line load balancing. In *COCOON'96*, 1996.
4. J. Aspnes, Y. Azar, A. Fiat, S. Plotkin, and O. Waarts. On-line load balancing with applications to machine scheduling and virtual circuit routing. In *Proc. 25th ACM Symposium on the Theory of Computing*, pages 623–631, 1993.
5. A. Avidor, Y. Azar, and J. Sgall. Ancient and new algorithms for load balancing in the l_p norm. In *Proc. 9th ACM-SIAM Symp. on Discrete Algorithms*, pages 426–435, 1998.
6. B. Awerbuch, Y. Azar, E. Grove, M. Kao, P. Krishnan, and J. Vitter. Load balancing in the l_p norm. In *Proc. 36th IEEE Symp. on Found. of Comp. Science*, pages 383–391, 1995.
7. B. Awerbuch, Y. Azar, S. Plotkin, and O. Waarts. Competitive routing of virtual circuits with unknown duration. In *Proc. 5th ACM-SIAM Symposium on Discrete Algorithms*, pages 321–327, 1994.
8. Y. Azar, A. Broder, and A. Karlin. On-line load balancing. In *Proc. 33rd IEEE Symposium on Foundations of Computer Science*, pages 218–225, 1992.
9. Y. Azar and L. Epstein. On-line load balancing of temporary tasks on identical machines. In *5th Israeli Symposium on Theory of Computing and Systems*, pages 119–125, 1997.
10. Y. Azar, B. Kalyanasundaram, S. Plotkin, K. Pruhs, and O. Waarts. Online load balancing of temporary tasks. In *Workshop on Algorithms and Data Structures*, pages 119–130, 1993.
11. Y. Azar, J. Naor, and R. Rom. The competitiveness of on-line assignments. In *Proc. 3rd ACM-SIAM Symposium on Discrete Algorithms*, pages 203–210, 1992.
12. Y. Azar and O. Regev. On-line bin stretching. Manuscript, 1997.
13. Y. Bartal, A. Fiat, H. Karloff, and R. Vohra. New algorithms for an ancient scheduling problem. *Journal of Computer and System Sciences*, 51:359–366, 1995.
14. Y. Bartal, H. Karloff, and Y. Rabani. A better lower bound for on-line scheduling. *Information Processing Letters*, 50:113–116, 1994.
15. Y. Bartal and S. Leonardi. On-line routing in all-optical networks. In *Proc. 24rd International Colloquium on Automata, Languages, and Programming*, 1997.
16. P. Berman, M. Charikar, and M. Karpinski. A note on on-line load balancing for related machines. In *5th Annual Workshop on Algorithms and Data Structures*, pages 116–125, 1997.

17. R. Boppana and A. Floratos. Load balancing in the euclidean norm. Manuscript, 1997.
18. B. Chen, A. van Vliet, and G. J. Woeginger. Lower bounds for randomized online scheduling. *Information Processing Letters*, 51:219–222, 1994.
19. B. Chen, A. van Vliet, and G. J. Woeginger. New lower and upper bounds for on-line scheduling. *Operations Research Letters*, 16:221–230, 1994.
20. Y. Cho and S. Sahni. Bounds for list schedules on uniform processors. *SIAM Journal on Computing*, 9:91–103, 1988.
21. G. Galambos and G. J. Woeginger. An on-line scheduling heuristic with better worst case ratio than graham's list scheduling. *SIAM J. Computing*, 22:349–355, 1993.
22. R. L. Graham. Bounds for certain multiprocessing anomalies. *Bell System Technical Journal*, 45:1563–1581, 1966.
23. R.L. Graham. Bounds on multiprocessing timing anomalies. *SIAM J. Appl. Math*, 17:263–269, 1969.
24. D. Hochbaum and D. Shmoys. A polynomial approximation scheme for scheduling on uniform processors: Using the dual approximation approach. *SIAM Journal on Computing*, 17:539–551, 1988.
25. Dorit S. Hochbaum and David B. Shmoys. Using dual approximation algorithms for scheduling problems: Theoretical and practical results. *J. of the ACM*, 34(1):144–162, January 1987.
26. P. Indyk, 1996. Personal communication.
27. D. R. Karger, S. J. Phillips, and E. Torng. A better algorithm for an ancient scheduling problem. *J. Algorithms*, 20:132–140, 1996.
28. R. M. Karp, U. V. Vazirani, and V. V. Vazirani. An optimal algorithm for on-line bipartite matching. In *Proceedings of the 22nd Annual ACM Symposium on Theory of Computing*, pages 352–358, Baltimore, Maryland, May 1990.
29. H. Kellerer, V. Kotov, M. G. Speranza, and Zs. Tuza. Semi on-line algorithms for the partition problem. *Operations Research Letters*, 1998. To appear.
30. J.K. Lenstra, D.B. Shmoys, and E. Tardos. Approximation algorithms for scheduling unrelated parallel machines. *Math. Prog.*, 46:259–271, 1990.
31. S. Phillips and J. Westbrook. On-line load balancing and network flow. In *Proc. 25th ACM Symposium on Theory of Computing*, pages 402–411, 1993.
32. S. Plotkin and Y. Ma. An improved lower bound for load balancing of tasks with unknown duration. Manuscript.
33. S. Seiden. Randomized algorithms for that ancient scheduling problem. In *5th annual Workshop on Algorithms and Data Structures*, pages 210–223, 1997.
34. J. Sgall. A lower bound for randomized on-line multiprocessor scheduling. *Information Processing Letters*, 63:51–55, 1997.
35. D. B. Shmoys, J. Wein, and D. P. Williamson. Scheduling parallel machines on-line. *SIAM Journal on Computing*, 24:1313–1331, 1995.
36. J. Westbrook. Load balancing for response time. In *3rd Annual European Symposium on Algorithms*, 1995.

9

On-line Scheduling

JIŘÍ SGALL

1 Introduction

Scheduling has been studied extensively in many varieties and from many viewpoints. Inspired by applications in practical computer systems, it developed into a theoretical area with many interesting results, both positive and negative.

The basic situation we study is the following. We have some sequence of jobs that have to be processed on the machines available to us. In the most basic problem, each job is characterized by its running time and has to be scheduled for that time on one of the machines. In other variants there may be additional restrictions or relaxations specifying which schedules are allowed. We want to schedule the jobs as efficiently as possible, which most often means that the total length of the schedule (the makespan) should be as small as possible, but other objective functions are also considered.

The notion of an on-line algorithm is intended to formalize the realistic scenario, where the algorithm does not have the access to the whole input instance, unlike the off-line algorithms. Instead, it learns the input piece by piece, and has to react to the new requests with only a partial knowledge of the input. Such scheduling algorithms are the topic of this chapter (for a general reference on on-line algorithms see the upcoming book [13].)

Scheduling has continuously been an active research area, reflecting the changes in theoretical computer science. When the theory of NP-completeness was developed, many scheduling problems have been shown to be NP-complete: Garey and Johnson [39] give 18 basic NP-complete scheduling problems; since then many new variants were considered and shown to be NP-complete. After the NP-completeness results, the focus shifted to designing approximation algorithms, often using quite non-trivial techniques and insights. There is extensive literature on these subjects, for recent surveys see e.g. [59, 54].

Many natural heuristics for scheduling are in fact on-line algorithms. Hence when the study of on-line algorithms using competitive analysis became usual, this approach was naturally and quite successfully applied to scheduling.

Organization of the chapter

We define some of the variants of scheduling that have been studied in the on-line setting in Section 2. In Section 3 we discuss the early results on on-line scheduling, focusing on Graham's paper [43]. The three sections 4 to 6 survey the results divided according to the three different on-line paradigms described in Section 2.2. In Section 7 we discuss several papers which study various modification of competitive analysis in which the on-line algorithm is less restricted than in the standard situation.

We keep the exposition as informal as possible, focusing on the intuition behind the results. We include several algorithms and proofs that are reasonably simple and illustrate more general techniques. Each section contains its own subsection of open problems, where we describe the open problems which we feel are the most important ones. In addition, there are of course many cases where we do not know the tight bounds on the competitive ratio, or variants that were not studied at all in the on-line setting. We give the appropriate citations for each result or variant of scheduling; whenever we give the citation in the heading of a section, it means that all the results come from the cited paper(s).

2 Taxonomy of on-line scheduling problems

After giving the general definitions, in Sections 2.2 to 2.5 we survey the features important in every variant of scheduling; these include the on-line character of the problem, the objective function, the use of randomization, release times and precedence constraints. Then we define other features which are used only in a few variants.

2.1 General definitions and preliminaries

The *number of machines* is always denoted m, and n stands for the *number of jobs*. Each job is characterized by its *running time*, which we denote t (it is also often denoted p and called processing time), and perhaps other characteristics as required by each variant of scheduling. The scheduling algorithm is asked to produce a schedule, which means that each job is assigned to one or more machines and one or more time slots, according to the variant of scheduling. Each machine is assigned to a single job at any time, and the processing of a job always takes at least its running time.

All scheduling problems we consider ask for minimization of some objective function (performance measure). The performance of an on-line algorithm is measured by the *competitive ratio* (w.r.t. some objective function). An on-line algorithm is σ-competitive if for each input instance the objective value of the schedule produced by the algorithm is at most σ times larger than the optimal objective value. The competitive ratio may depend on m, but should be independent of n, which reflects the fact that the number of jobs is not known to the on-line algorithm.

The minimal makespan for a given instance is denoted T_{opt}. A machine is *busy* at a given time, if it is assigned to some job, otherwise it is *idle*. The *load* of the machine is the total time during which this machine is busy (i.e., the idle time does not contribute to the load of the machine). A job is *available* at a given time, if it was not scheduled yet and can be scheduled in a time slot(s) starting at that time consistently with the restrictions of the given variant of scheduling (e.g., it is after its release time). In many variants of scheduling we encounter the golden ratio, defined by $\phi = (\sqrt{5}+1)/2 \approx 1.6180$.

While most results work with arbitrary real running times, some make technical restrictions, e.g., the minimal running time is specified or the running times and release times are required to be integral. This is important for example for some results on preemptive scheduling, where we want to divide the time of one machine equally among all jobs. Another possibility how to deal with this problem is to use the common more general definition of competitiveness which includes an additive term, an algorithm is then σ-competitive if its objective value is at most some constant plus σ times the optimal objective value. We do not pay much attention to these restrictions, as they are only technical details.

2.2 Different on-line paradigms

For on-line scheduling the most important classification of the on-line problems is according to which part of the problem is given on-line. There are several very different possibilities.

Scheduling jobs one by one. In this paradigm the jobs are ordered in some list (sequence) and are presented one by one according to this list. Each of them has to be assigned to some machine(s) and time slot(s) before the next jobs are seen, consistently with other restrictions given by the problem. As soon as the job is presented we know all its characteristics, including the running time. We are allowed to assign the jobs to arbitrary time slots (i.e., they can be delayed), thus a job can start running later than the successive jobs in the sequence; however, once we see the successive jobs we cannot change the assignment of the previous jobs.

Unknown running times. Here the main on-line feature is the fact that the running time of a job is unknown until the job finishes; an on-line algorithm only knows whether a job is still running or not. Unlike in the previous paradigm, at any time all currently available jobs are at the disposal of the algorithm; any of them can be started now on any machine(s) or delayed further. Also, if preemptions or restarts are allowed (see Section 2.6), the algorithm can decide to preempt or stop any job which is currently running. The jobs may become available over time according to their release times or precedence constraints (see Section 2.5), but the situation when all jobs are available at the beginning plays an important role in this paradigm, too. If there are other characteristics of a job than its running time, they are known when the job becomes available (such characteristics may include for example the number of parallel processors it uses, which has to be known to guarantee that the job is scheduled legally).

Jobs arrive over time. In this paradigm the algorithm has the same freedom as in the previous one, and in addition the running time of each job is also known when that job is available. Thus the only on-line feature is the lack of knowledge of jobs arriving in the future.

Sometimes algorithms that know the running time of a job as soon as it arrives are called *clairvoyant*, in contrast to *non-clairvoyant* algorithms that correspond to the previous paradigm of unknown running times.

Interval scheduling. All the previous paradigms assume that a job may be delayed. Contrary to that, the paradigm of interval scheduling assumes that each job has to be executed in a precisely given time interval; if this is impossible it may be rejected. This scenario is very different from the previous three. For example, it is meaningless to measure the length of the schedule, as it is essentially fixed; instead we measure the weight (or the number) of accepted jobs. This paradigm is studied for example in the papers [83, 62, 32]. We will not cover it in this chapter.

2.3 Objective functions

The most common objective function is the *makespan*, which is the length of the schedule, or equivalently the time when the last job is completed. In one of the variations we allow jobs to be rejected at a certain penalty, in which case we minimize the sum of the makespan and the penalties of rejected jobs. These objective functions formalize the viewpoint of the owner of the machines. If the makespan is small, the utilization of his machines is high; this captures the situation when the benefits of the owner are proportional to the work done. Penalties are intended to capture the situation when this is not true: if a job has a long running time and small benefit (i.e., there is a small penalty for not scheduling it), it is better to reject it.

If we turn our attention to the viewpoint of a user, the time it takes to finish individual jobs may be more important; this is especially true in interactive environments. Thus, if many short jobs are postponed after some long job, it is unacceptable to the user of the system even if the makespan is optimal. For that reason other objective functions are studied, namely the *total completion time*, the *total flow time* (also called response time), and the *total waiting time*. The completion time of a job is the time when this particular job is completed; thus the makespan equals the maximal completion time. The flow time of a job is the time the job is in the system, i.e., the completion time minus the time when it becomes first available. The waiting time is the flow time minus the running time of a job. The objective functions are the sums of these values over all jobs. (Equivalently, we can take the average values, as they always differ by a factor of n.) The competitive ratio w.r.t. the total completion time is at most the competitive ratio w.r.t. the total flow time, as the flow times are smaller by an additive term which is equal for both on-line and optimal schedules; similarly the competitive ratio w.r.t. the total flow time is at most the competitive ratio w.r.t. the total waiting time. Thus the best algorithms are those competitive w.r.t. the total waiting time. The weighted variant of these measures is also studied; in this case each job has its weight and we take the weighted sums. This captures the situation when some jobs are more important than others. Maximal waiting and flow time are also reasonable objective functions, but they were not considered in the context of on-line scheduling algorithms so far.

Another possibility is to consider general L_p norms of the vector of loads of the machines, which is studied for load-balancing, where in particular the L_2 norm has a natural interpretation [5, 4].

If we allow preemptions (see Section 2.6), we usually want to minimize the number of times we preempt a job. However, this is always a secondary criterion. This means that we are interested to quantify how many preemptions we need to use to obtain optimal or almost optimal competitive ratio w.r.t. some other objective function.

2.4 Randomization

In most cases we are interested in both deterministic and randomized algorithms. If we allow randomization, we consider the expected objective value, where the expectation is taken over the random choices of the algorithm. A randomized algorithm is σ-competitive if for each instance this expectation is within a factor of σ of the optimal objective value. This corresponds to the so-called oblivious adversary [12], which has to commit to an input instance beforehand, without any knowledge of the random bits or actions of the algorithm.

2.5 Release times and deadlines, precedence constraints and conflicting jobs

Each job may have an individual *release time*, which is the earliest time when it may be scheduled, and in the on-line setting also the time when it becomes known. The dual notion of individual *deadlines* is common in off-line scheduling, but not in the on-line case. The reason is that individual deadlines only define feasible solutions, which is not compatible with the goal of the on-line algorithm which is to do reasonably well on some global measure. This intuition can even be stated formally. In the case of preemptive scheduling of jobs arriving over time with known running times there exists an on-line algorithm which schedules all jobs before their common deadline if this is feasible; as soon as we allow two different deadlines no such on-line algorithm exists [48].

An often considered variant assumes that there are some *precedence constraints* between the jobs. They are generally given by a directed acyclic graph on the jobs; each directed edge indicates that one job has to be scheduled before another one. In the on-line framework a job is known only after all its predecessors in the dependency graph are processed by the on-line algorithm. A related model considers the case of *conflicting jobs* when the jobs may conflict with each other, but the order in which such pairs have to be scheduled is not given; in such a case the conflict graph is undirected.

2.6 Preemptions and restarts

In many problems it is assumed that a running job may be *preempted*, i.e., its processing may be stopped and resumed later on the same or on different machine(s). In the on-line setting, there is yet another possibility which is meaningless for off-line algorithms. Namely, the running of a job can be stopped and *restarted* later on the same or on different machine(s); thus in order to finish, it has to be assigned to the same machine(s) for its whole running time without an interruption.

2.7 Parallel jobs

Parallel jobs are those jobs that have to be scheduled on some number of machines at the same time. They are characterized by two parameters, the running time t and the number of requested machines p. We consider two variants. In the first the jobs are non-malleable, which means that they have to be scheduled on the requested number of machines. On the other hand, malleable jobs may be scheduled on fewer machines, at the cost of increasing the processing time. Most of the time we consider ideally malleable jobs, where the increase of the running time is proportional to the decrease of the number of machines, i.e., scheduling on $q \leq p$ machines takes time tp/q.

2.8 Different speeds of machines

The speeds of the machines can be different. The three most common models of this situation were both studied in the on-line setting.

In the variant of *uniformly related machines* the ith machine has speed $s_i > 0$. If a job with running time t is scheduled on ith machine, its processing takes time t/s_i. (Here we possibly deviate from our convention that processing a job takes always at least its running time; we can always repair this by scaling the speeds so that $s_i \leq 1$.)

The next variant is that of *unrelated machines* where the vector of speeds is possibly different for each job. However, even if the running time is unknown, we assume that the speeds are known for each job (i.e., for each job we know the relative speeds of the machines).

The last variant is the *restricted assignment*. Here the machines have identical speed, but each job can be executed only on some subset of processors. This can be thought of as a special case of unrelated machines, where the speeds are always 1 or infinitely small. This variant is not comparable to uniformly related machines.

2.9 Shop scheduling

In shop scheduling the job has several tasks (operations) that have to be processed on different machines. The running time of each task is a separate parameter. The different tasks of the same job cannot be scheduled at the same time. According to additional restrictions, we distinguish *open shop*, when the different tasks of a job may be scheduled in any order (in addition, if preemption is allowed, different tasks may be even interleaved), *flow shop*, if the order of the tasks is fixed and it is the same for all jobs, and *job shop*, if the order of the tasks is fixed and possibly different for each job.

2.10 Variants not covered in this chapter

A very different model of on-line scheduling in real-time systems, where each job has its deadline and value and the goal is to maximize the value of jobs finished before their deadline, is considered in several papers [28, 11, 52, 65].

Other papers study models with additional obstacles. On-line scheduling in the presence of processor faults is studied in [51, 53]. On-line scheduling in presence of delays in communication between processors is studied in [26, 27].

3 General results and history

The first proof of competitiveness of an on-line algorithm for a scheduling problem, and perhaps for any problem, was given by Graham already in 1966 [43]. He studied a simple deterministic greedy algorithm, now commonly called List Scheduling. The studied model is the basic one, where we have m identical machines and a sequence of sequential jobs characterized by their running times. The objective is to minimize the makespan. The algorithm was designed for the case of precedence constraints, but it can be easily modified to handle also release times [46]. Preemption is not used.

Algorithm List Scheduling

(i) The jobs are ordered in an arbitrary list (sequence).

(ii) Whenever some machine is idle, we schedule on it the first job on the list which is available (i.e., it is not scheduled yet, the current time is greater than its release time, and all its predecessors in the precedence graph are finished).

This algorithm works in all three on-line paradigms we study. It is already formulated in the model with release times, and since it does not use any information about running times, it carries over to the paradigm of unknown running times. For the paradigm of scheduling jobs one by one we simply present the jobs in the order of the input list, and when a job is presented, we schedule it on a machine with the smallest load so far. This works only for the case with no precedence constraints, but we do not study precedence constraints in this paradigm (cf. Section 4).

Theorem 1 [43, 46]. *The competitive ratio of List Scheduling is* $2 - \frac{1}{m}$.

Proof. First we show that the competitive ratio of List Scheduling is not better than $2-\frac{1}{m}$. Consider the sequence of $m(m-1)$ jobs with running time 1 followed by one job with running time m. There are no precedence constraints or release times (and hence the lower bound is true in any of the three on-line paradigms). List Scheduling schedules this sequence in time $2m-1$, while the optimal schedule has makespan m.

Next we show that List Scheduling is $(2 - \frac{1}{m})$-competitive if there are no precedence constraints and no release times. Consider the job that finishes as the last one, suppose it was started at time τ and its running time is T. At all times before τ, all the machines are busy, as otherwise the last job would be scheduled earlier. Hence the optimal makespan T_{opt} is at least $T_{\text{opt}} \geq \tau + \frac{T}{m}$, as the optimal schedule has to schedule all the jobs. We certainly have $T_{\text{opt}} \geq T$, as the optimal schedule takes time T to process even this single job. Combining these two inequalities we get that the makespan of the on-line solution, which is $\tau + T$, is bounded by $\tau + T = \tau + \frac{T}{m} + (1 - \frac{1}{m})T \leq (2 - \frac{1}{m})T_{\text{opt}}$.

For the general case with precedence constraints and release times we need a different bound on the time when some machine is idle. We define a sequence of jobs $J_1,\ldots,J_k$ inductively as follows. Let J_1 be the job that finishes last. If no predecessor of J_i in the dependency graph finishes after the release time of J_i, we stop. Otherwise J_{i+1} is defined as the latest-finishing predecessor of J_i. It follows that whenever some machine is idle, either one of the jobs $J_1,\ldots,J_k$ is running or it is before the release time of J_k (otherwise one of the jobs J_i would be available and hence scheduled). The optimal schedule has to schedule these jobs sequentially from J_k to J_1, and it can start only after J_k is released. This proves that the total time when some machine is idle is bounded by the optimal makespan, and the rest of the argument is the same as without precedence constraints. □

While we now interpret Graham's result as a proof of competitiveness of List Scheduling, it should be stressed that his analysis was deeper. He gives examples where we can increase the makespan by making the problem easier ("timing anomalies" stands for this paradox), namely by either increasing the number of machines, or decreasing the running time of some job, by relaxing the precedence constraints, or, finally, by reordering the list. He proves that in all of these cases the makespan can change by almost a factor of 2, giving the tight bounds in all cases. The case of reordering the list amounts to the competitiveness analysis given in Theorem 1, as the optimal schedule can be obtained by some particular ordering of the list.

In the follow-up paper [44] Graham shows that the factor of 2 decreases if we modify the algorithm so that some number of long jobs is scheduled first using an optimal schedule, and the rest is scheduled by List Scheduling. Clearly, this algorithm is no longer on-line in any of the paradigms we study.

Other two early papers that contain results about on-line scheduling algorithms are [66, 23]. The first one gives an optimal algorithm for minimizing the makespan of a preemptive schedule on identical machines where jobs arrive over time, and mentions that the algorithm is on-line. The second paper is to our best knowledge the first one that states explicitly a lower bound on the performance ratio of any on-line algorithm for some scheduling problem, namely the bound of $\Omega(\sqrt{n})$ for non-preemptive scheduling jobs with unknown running times on uniformly related machines; the paper even mentions the possible usefulness of restarts, which later indeed proved to be quite useful in this case [75].

Around 1990 new results were discovered concerning many variants of on-line scheduling, both old and new. Most of the results use the makespan as the objective function, consequently our understanding of this measure is most complete. Recently other objective functions are drawing more attention, perhaps also because in many practical applications they are more important.

4 Scheduling jobs one by one

This paradigm corresponds most closely to the standard model of request sequences in competitive analysis. It can be formulated in the language of on-line load balancing as the case where the jobs are permanent and the load is their only parameter corresponding to our running time (cf. [5]).

In this paradigm we do not allow release times and precedence constraints, as these restrictions appear to be unnatural with scheduling jobs one by one. In most of the variants it is also sufficient to assign each job to some machine(s) for some length of time, but it is not necessary to specify the actual time slot(s), in other words it is not necessary or useful to introduce idle time on any machine.

We first give the results considering minimizing the makespan, only in Sections 4.7 and 4.8 we briefly mention results for other objective functions, namely minimizing the L_p norm and the total completion time.

4.1 The basic model

We have m machines and a sequence of jobs characterized by their running times. The jobs are presented one by one, and we have to schedule each job before we see the next one. Performance is measured by the makespan. Each job is assigned to a single machine. There are no additional constraints, preemption is not allowed, all the machines have the same speed, and the objective function is the makespan. In this section we are interested in deterministic algorithms.

By Theorem 1 it follows that the competitive ratio of List Scheduling is $2-\frac{1}{m}$. This is provably the best possible for $m = 2$ and $m = 3$ [31], but for larger m it is possible to develop better algorithms.

From the proof of Theorem 1 it is clear what is the main issue in designing algorithms better than List Scheduling. If all machines have equal loads and a job with long running time is presented, we create a schedule which is almost twice as long as the optimal one. This is a problem if the scheduled jobs are sufficiently small, and the optimal schedule can distribute them evenly on $m-1$ machines in parallel with the last long job on the remaining machine. Thus, to achieve better results, we have to create some imbalance and keep some machines lightly loaded, to be used by long jobs.

Let us suppose that we want to achieve competitive ratio σ. When a job is presented, we can schedule it on any machine such that after this step the competitive ratio is at most σ. Suppose that we choose always the most loaded of these machines, to create as large imbalance as possible. This seems to be a natural idea to prevent the previous problems, however, it turns out that it does not work, either. If this algorithm is presented with a long list of jobs with the same running time, it distributes them almost evenly on a constant fraction of the machines, with only one job scheduled on each of the remaining machines. Now we can continue with a sequence of long jobs, first making the load distributed evenly on all machines, and then forcing the schedule to be too long. Thus this method cannot give a better competitive ratio than List Scheduling.

To design a good algorithm, we need to avoid both of these extremes. Current results use two different approaches. One is to schedule each job on one of the two currently least loaded machines [37, 17]. This gives better results than List Scheduling for any $m \geq 4$, and achieves the currently best upper bounds for small m. However, for large m, the competitive ratio still approaches 2. This approach leaves at most one lightly loaded machine, hence after two long jobs we get a long schedule and the competitive ratio is at least $2 - \frac{2}{m}$. To keep the competitive ratio bounded away from 2 even for large m it is necessary to keep some constant fraction of machines lightly loaded. Such an algorithm was first developed in [8], later better algorithms based on this idea were designed in [55, 1] to give the currently best upper bounds for large m. The analysis of all these algorithms is relatively complicated.

The current state of our knowledge is summarized in Table 1. For comparison we include also the competitive ratio of List Scheduling. (See Section 4.2 for a discussion of results for randomized algorithms.) The observation that List Scheduling is optimal for $m = 2, 3$ is due to [31]. The other lower bounds for small m are due to [17]. The lower bound for large m is due to [1], improving upon [9]. Very recently R. Chandrasekaran claimed a lower bound of $\sqrt{3} \approx 1.7321$ for $m = 4$, which would significantly decrease the gap in this case.

4.2 Randomized algorithms

Much less is known about randomized algorithms for the basic model studied in Section 4.1. Only for the case of $m = 2$ we know an optimal randomized algorithm. A 4/3-competitive randomized algorithm for two machines was presented in [8]. First we prove that this is best possible.

	deterministic			randomized	
m	lower bound	upper bound	LS	lower bound	upper bound
2	1.5000	1.5000	1.5000	1.3333	1.3334
3	1.6666	1.6667	1.6667	1.4210	1.5567
4	1.7310	1.7333	1.7500	1.4628	1.6589
5	1.7462	1.7708	1.8000	1.4873	1.7338
6	1.7730	1.8000	1.8333	1.5035	1.7829
7	1.7910	1.8229	1.8571	1.5149	1.8169
∞	1.8520	1.9230	2.0000	1.5819	–

Table 1. Current bounds for algorithms scheduling jobs one by one with no constraints.

Theorem 2 [8]. *No randomized algorithm for 2-machine scheduling can be better than* 4/3*-competitive.*

Proof. Consider the sequence of three jobs with running times 1, 1, and 2. Suppose we have an algorithm which is better than 4/3-competitive. We schedule the first two jobs, order the machines according to their load and consider the expected load of the more loaded machine and of the less loaded one, where the expectation is taken over the random choices of the algorithm. After scheduling the first two jobs, the expected load of the more loaded machine is less than 4/3, as it is equal to the expected makespan and the optimal makespan is 1. Hence the expected load of the less loaded machine is more than 2/3, and even if the last job is always scheduled on the least loaded machine, the expected makespan after scheduling of all three jobs is more than 8/3, contradicting the assumption that the algorithm is better than 4/3-competitive. □

In the proof we can replace the first two jobs by an arbitrary sequence of jobs with total running time 2. Hence the proof actually shows that in any 4/3-competitive algorithm, the expected load of the more loaded machine has to be at least twice as much as the expected load of the other machine at all times. This has to be tight whenever we can partition the jobs into two sets with exactly the same sum of running times. The most natural way to design an algorithm with this in mind is to keep the desired ratio of expected loads at all times, and this in fact works.

Algorithm Random

(i) If possible, schedule the job randomly so that afterwards the expected makespan equals 2/3 of the total running time.

(ii) Otherwise schedule the job always on the less loaded machine.

To implement this algorithm it is necessary to keep track of all possible schedules and their probabilities. The naive way of doing this uses 2^{n-1} configurations after n jobs, but it is possible to implement the algorithm with only n configurations [8].

Theorem 3 [8]. *The algorithm Random is 4/3-competitive for two machines.*

Proof. After scheduling some sequence of jobs, let a be the expected makespan, b be the total running time of all jobs scheduled so far, and let T be the longest running time among all jobs scheduled so far. We prove by induction that at any time the following is true:

$$a \geq \frac{2}{3}b, \text{ and} \tag{1}$$

$$\text{If } a > \tfrac{2}{3}b \text{ then } T \geq \frac{3}{4}a. \tag{2}$$

From this condition it follows that the algorithm is 4/3-competitive, as the optimal makespan is at least $\max(T, b/2)$. Both conditions are trivially true before any job is scheduled.

Let t be the running time of the job that has to be scheduled next, and let $a(t)$ be the expected makespan after scheduling this job deterministically on the less loaded machine. Consider what happens if we change the probability that the job is scheduled on the more loaded machine continuously from 0 to 1. The expected makespan increases continuously from $a(t)$ to $a + t$. By the induction assumption (1), $a + t \geq \frac{2}{3}b + t \geq \frac{2}{3}(b + t)$. Hence if $a(t) \leq \frac{2}{3}(b + t)$, we can schedule the job so that afterwards the expected makespan equals to $\frac{2}{3}(b + t)$, as required in the step (i) of the algorithm, and both (1) and (2) are satisfied.

Thus we only need to consider the case when $a(t) > \frac{2}{3}(b + t)$. In this case the next job is always scheduled on the less loaded machine, and (1) is satisfied since the expected makespan is $a(t)$. It remains to prove the condition (2). We distinguish three cases according to the value of t.

First suppose that $t \leq 2T - b$. This means that $T \geq t + b - T$, i.e., T is larger than the total running time of all the other jobs including the next one. Hence by scheduling the new job on the less loaded machine we do not change the makespan and (2) remains satisfied.

Next suppose that $t \geq b$. This means that the next job has longer running time than all the previous jobs together, and hence the machine on which it is scheduled always becomes the more loaded one. Hence $a(t) = t + b - a$. Since $b - a \leq b/3$ by the induction assumption (1), it follows that $a(t) \leq t + b/3 \leq \frac{4}{3}t$, and (2) remains true, as t is now the longest running time.

Now consider the remaining case when $\max(0, 2T - b) \leq t \leq b$. We prove that in this case $a(t) \leq \frac{2}{3}(b + t)$, hence by the previous considerations the next job is scheduled by the step (i) of the algorithm and the inductive conditions are satisfied. We prove that $a(t) \leq \frac{2}{3}(b+t)$ at the endpoints of the interval allowed for t and that $a(t)$ is a convex function of t; the inequality then follows for every t in the interval, since the right hand side is linear in t. First consider the endpoints of the interval. If $t = b$, we have seen in the previous paragraph that $a(t) \leq t + b/3$, which equals $\frac{2}{3}(b+t)$. If $t = 2T - b \geq 0$, we have $a(t) = a \leq \frac{4}{3}T = \frac{2}{3}(b+t)$, using the induction assumption (2). If $t = 0 > 2T - b$, we know that $T < \frac{1}{2}b \leq \frac{3}{4}a$ by the assumption (1); therefore it has to be the case that $a = \frac{2}{3}b = \frac{2}{3}(b+t)$ if (2) is satisfied. It remains to prove that $a(t)$ is convex. The derivative of $a(t)$ at point t is equal to the probability that after scheduling the job with running time t on the less loaded machine it becomes the more loaded one. It is easy to see that this probability is non-decreasing in t, hence $a(t)$ is convex. (To be more precise, we should notice that the derivative may be undefined at finitely many points. However, this does not change the conclusion.) □

Very recently new randomized algorithms for small m were developed in [71, 70]. It is provably better than any deterministic algorithm for $m = 3, 4, 5$ and better than the currently best deterministic algorithm for $m = 6, 7$. It always assigns the new job on one of the two least loaded machines, similarly to the deterministic algorithms for small m from [37, 17]. Consequently, its competitive ratio approaches two as m grows. The analysis of this algorithm is again difficult, even involving extensive computations to obtain the best results.

The idea of the lower bound for two machines can be extended to arbitrary number of machines [16, 72, 74]. It turns out that for m machines, the expected loads should be in geometric sequence with the ratio $m : (m-1)$, if the machines are always ordered so that their loads are non-decreasing. (For example, for $m = 3$ the ratio of loads is $4 : 6 : 9$.) This leads to a lower bound of $1/(1-(1-1/m)^m)$, which approaches $e/(e-1) \approx 1.5819$ for large m and increases with increasing m.

For any $m > 2$ it is an open question whether there exists an algorithm matching this lower bound. (Seiden [71] demonstrated that his algorithm does not match this bound.) The insight from the proof of the lower bound leads to a natural invariant that should be preserved by any algorithm matching it. Namely, such an algorithm should preserve the ratio of expected loads described above. An algorithm based on this invariant would be a natural generalization of the optimal algorithm for two machines from [8]; it would also follow the suggestion from [18] (see Section 4.3). This faces several problems.

First of all, it is not clear at all that we would be able to handle long jobs similarly as for $m = 2$. A job is long if its running time is more than the $1/(m-1)$ fraction of the sum of running times of all previous jobs (intuitively this means that its running time determines the optimal makespan). For $m = 2$ this means that the running time of a long job is more than the total running time of all previous jobs, therefore we know that no matter on which machine it is scheduled, this machine will become the most loaded one; we used this fact significantly in the proof of Theorem 3. For three machines this is no longer true, and hence the structure of the problem is much more difficult.

Second, it is not clear whether we would be able to preserve the invariant ratio of expected loads even if all jobs are small. In [74] it is demonstrated that even for $m = 3$ it is impossible to preserve this invariant inductively, meaning that there exists a probability distribution on the configurations such that the expected loads have the desired ratio, but after the next job this ratio cannot be maintained; moreover this configuration is reachable so that the ratio is kept invariant at all times, except for the first few jobs. We want to keep the ratio invariant, so this means that for a design of a matching algorithm we should use a stronger inductive invariant; at present we do not know if this is possible.

Third, for $m = 2$ the lower bound proof implies that the ratio of loads has to be at least $2 : 1$ at all times. For $m = 3$ this is only true if the previous jobs can be exactly balanced on two machines, and for $m > 3$ we get even more such balancing conditions. This means that we have some more freedom in the design of the algorithm, and hence it is harder to improve the lower bounds.

To summarize, we have the optimal algorithm for $m = 2$, and an improvement of the deterministic algorithms for small m. However, for $m > 7$ we have no randomized algorithm with a better competitive ratio than known deterministic algorithms; this means that we do not know how to make use of randomization for large m. See Table 1.

4.3 Preemptive scheduling [18]

In this model preemption is allowed. Each job may be assigned to one or more machines and time slots (the time slots have to be disjoint, of course), and this assignment has to be determined completely as soon as the job is presented. It should be noted that in this model the off-line case is easily solved, and the optimal makespan is the maximum of the maximal running time and the sum of the running times divided by m (i.e., the average load of a machine).

It is easy to see that the lower bounds from Section 4.2 hold in this model, too, as they only use the arguments about expected load. This again leads to a lower bound of $1/(1 - (1 - 1/m)^m)$, which approaches $e/(e - 1) \approx 1.5819$ for large m, valid even for randomized algorithms. The proof shows that expected loads in the optimal algorithm have to be in geometric sequence with the ratio $m : (m - 1)$, if the machines are always ordered so that their loads are non-decreasing and there are no long jobs; in this case this has to be true at all times, as using preemption we can always balance the machines exactly.

Interestingly, for this model there exists a deterministic algorithm matching this lower bound. It essentially tries to preserve the invariant above, with some special considerations for large jobs.

Thus, in this model both deterministic and randomized cases are completely solved, giving the same bounds as the randomized lower bounds in Table 1. Moreover, we know that randomization does not help. This agrees with intuition. In the basic model randomization can serve us to spread the load of a job among more machines, but we still have the problem that the individual configurations cannot look exactly as we would like. With preemption, we can spread the load as we wish, while still keeping just one configuration with the ideal spread of the total load, and this makes it more powerful than randomization.

4.4 Scheduling with rejections

In this version jobs may be rejected at a certain penalty. Each job is characterized by the running time and the penalty. A job can either be rejected, in which case its penalty is paid, or scheduled on one of the machines, in which case its running time contributes to the completion time of that machine (as usual). The objective is to minimize the makespan of the schedule for accepted jobs plus the sum of the penalties of all rejected jobs. Again, there are no additional constraints and all the machines have the same speed.

The main goal of an on-line algorithm is to choose the correct balance between the penalties of the rejected jobs and the increase in the makespan for the accepted jobs. At the beginning it might have to reject some jobs, if the penalty for their rejection is small compared to their running time. However, at some point it would have been better to schedule some of the previously rejected jobs since the increase in the makespan due to scheduling those jobs in parallel is less than the total penalty incurred. Thus this on-line problem can be seen as a non-trivial generalization of the well-known ski rental problem.

We first look at deterministic algorithms in the case when preemption is not allowed [10]. At first it would seem that a good algorithm has to do well both in deciding which jobs to accept, and on which machines to schedule the accepted jobs. However, it turns out that after the right decision is made about rejections, it is sufficient to schedule the accepted jobs using List Scheduling. This is certainly surprising, as we know that without rejections List Scheduling is not optimal, and hence it is natural to expect that any algorithm for scheduling with rejections would benefit from using a better algorithm for scheduling the accepted jobs.

We can solve this problem optimally for $m = 2$ and for unbounded m, the competitive ratios are ϕ and $1+\phi$. However, the best competitive ratio for fixed $m \geq 3$ is not known. It certainly tends to $1 + \phi$, which is the optimum for unbounded m, but the rate of convergence is not clear: while the upper bound is $1 + \phi - 1/m$ (i.e., the same rate of convergence as for List Scheduling), the lower bound is only $1 + \phi - 1/O(\log m)$.

The optimal algorithm for two machines is extremely simple: if a job with running time t and penalty p is presented, we reject it if $t \geq \phi p$; otherwise we schedule it using List Scheduling. The optimal algorithm for arbitrary m uses two rules for rejecting jobs: (i) a job is rejected whenever $t \geq mp$, and (ii) a job is rejected if $t \geq \phi(P + p)$, where P is the total penalty of all jobs rejected so far by the rule (ii); accepted jobs are again scheduled by List Scheduling.

The lower bounds for small m from [10] work also for preemptive deterministic algorithms, but for large m yield only a lower bound of 2. An improved algorithm for deterministic preemptive scheduling was designed in [69]. It achieves competitive ratio 2.3875 for all m. The scheme for rejecting jobs is similar as in the previous case, but the optimal algorithm for preemptive scheduling is used instead of List Scheduling. An interesting question is whether a better than 2-competitive algorithm can be found for $m = 3$: we now know several different 2-competitive algorithms even without preemption, but the lower bound does not match this barrier.

Randomized algorithms for this problem were designed in [70, 69]. The general idea is to use modifications of the deterministic algorithms where the thresh-

m	deterministic lower bounds	deterministic upper bounds		randomized upper bounds	
		non-preemptive	preemptive	non-preemptive	preemptive
2	$\phi \approx 1.6180$	ϕ	ϕ	1.5000	1.5000
3	1.8392	2.0000	2.0000	1.8358	1.7774
4	1.9276	2.1514	2.0995	2.0544	2.0227
5	1.9660	2.2434	2.1581	2.1521	2.0941
∞	$1 + \phi \approx 2.6180$	$1 + \phi$	2.3875	–	–

Table 2. Current bounds for algorithms scheduling jobs one by one with possible rejection.

olds for rejection are parameterized, and certain random choice of these parameters is made. In the non-preemptive case the competitive ratios are 1.5, 1.8358, and 2.0545 for $m = 2, 3$, and 4. With preemption better upper bounds can be achieved. No algorithms better than the deterministic ones are known for large m. The lower bounds for randomized scheduling without rejection (Table 1) clearly apply here (set the penalties infinitely large), and no better lower bounds are known.

The results are summarized in Table 2. The deterministic lower bounds apply both for algorithms with and without preemption, with the exception of arbitrary m where the lower bound is only 2 with preemption.

4.5 Different speeds

For related machines, a simple doubling strategy leads to a constant competitive ratio [2]. We guess an estimate on the makespan, and schedule each job on the slowest machine such that the current makespan does not exceed the estimate; if this fail we double the estimate and continue. The competitive ratio can be improved by using more sophisticated techniques instead of doubling, but its precise value is not known, see [5] for more references.

For restricted assignment the optimal competitive ratio is $\Theta(\log m)$ both for deterministic and randomized algorithms [6]. For unrelated machines with no restriction it is also possible to obtain an $O(\log m)$-competitive deterministic algorithm [2, 60]. By the previous lower bound this is optimal, too.

It is interesting that both for related and unrelated machines the optimal algorithms are asymptotically better than List Scheduling. Here List Scheduling is modified so that the next job is always scheduled on that machine on which it will finish earliest (for the case of identical speed this is clearly equivalent to the more usual formulation that the next job is scheduled on the machine with the smallest load). For unrelated machines the competitive ratio of List Scheduling is exactly n [2]. For related machines the competitive ratio of List Scheduling is asymptotically $\Theta(\log m)$ [22, 2] (the lower and upper bounds, respectively).

The exact competitive ratio for $m = 2$ is ϕ and for $3 \le m \le 6$ it is equal to $1 + \sqrt{(m-1)/2}$ [22]; moreover for $m = 2, 3$ it can be checked easily that there is no better deterministic algorithm.

For two machines we are able to analyze the situation further [30]. Suppose that the speeds of the two machines are 1 and $s \ge 1$. It is easy to see that List Scheduling is the best deterministic on-line algorithm for any choice of s. For $s \le \phi$ the competitive ratio is $1 + s/(s+1)$, increasing from $3/2$ to ϕ. For $s \ge \phi$ the competitive ratio is $1 + 1/s$, decreasing from ϕ to 1; this is the same as for the algorithm which puts all jobs on the faster machine. It turns out that this is also the best possible randomized algorithm for $s \ge 2$. On the other hand, for any $s < 2$ randomized algorithms are better than deterministic ones. If we consider deterministic preemptive scheduling, the competitive ratio is better than for non-preemptive randomized scheduling for any $s > 1$, moreover, it is also always better than for the identical machines ($s = 1$), in contrast without preemption the worst competitive ratio (both deterministic and randomized) is achieved for some $s > 1$.

4.6 Shop scheduling [21]

On-line shop scheduling was so far considered mainly for two machines. This variant of scheduling is somewhat different from all the ones we considered before, since here it may be necessary to introduce idle times on the machines. Hence we need to specify also the time slots for each job, not only the machine(s) on which it runs as before. As a consequence, this variant no longer corresponds to load balancing.

For flow shop and job shop scheduling it turns out that no deterministic algorithm is better than 2-competitive. To design 2-competitive algorithms is trivial: just reserve for each job the needed time on both machines. For both problems, there exist matching lower bounds of 2. However, we do not know whether it is possible to extend these lower bounds to randomized algorithms. Also the situation where preemption is allowed remains unclear.

For the open shop problem the situation is more interesting. If preemption is allowed, the optimal competitive ratio is 4/3. As far as randomization is concerned, the situation is similar as in the basic model with preemption: the 4/3-competitive algorithm is deterministic, while the lower bound holds also for randomized algorithms. Without preemption we have a 1.875-competitive deterministic algorithm and a lower bound of $\phi \approx 1.6180$ for deterministic algorithms. Nothing is known about the power of randomization in this case.

Only a few observations are known about open shop scheduling for $m \ge 3$. Joel Wein observed that for preemptive open shop scheduling there exists a 2-competitive algorithm for arbitrary m. Gerhard Woeginger and the author observed that the randomized lower bound from the basic model which approaches $e/(e-1) \approx 1.5819$ (see Table 1) can be modified to work for open shop, too.

4.7 Minimizing the L_p norm [3]

Here we minimize the L_p norm of the load vector, instead of the makespan, which is equivalent to the L_∞ norm. Of special interest is the Euclidean L_2 norm, the square root of the sum of squares of loads, which has a natural interpretation in load balancing [5, 4]. For L_2 norm, List Scheduling is $\sqrt{4/3}$ competitive, and this is optimal. The performance of List Scheduling is not monotone in the number of machines. It is equal to $\sqrt{4/3}$ only for m divisible by 3, otherwise it is strictly better.

More surprisingly, there exists an algorithm which is for sufficiently large m better than $\sqrt{4/3} - \delta$ for some $\delta > 0$, which means that also the optimal competitive ratio is not monotone in m. For a general p, the same approach leads also to an algorithm better than List Scheduling for large m.

4.8 Minimizing the total completion time [36]

In this variant it is necessary to use idle times, as we have to finish the jobs with short running times first to minimize the total completion time. Even on a single machine it is hard to design a good algorithm and the competitive ratio depends on the number of jobs logarithmically. More precisely, there exists a deterministic $(\log n)^{1+\varepsilon}$-competitive algorithm on a single machine without preemptions, but no $\log n$-competitive algorithm exists even if preemption is allowed.

4.9 Open problems

Randomized algorithms We understand very little about the power of randomization in this on-line paradigm. We know that randomization does not help in most of the variations with preemption. It is open whether randomization helps for shop scheduling and in the model with rejections for large m.

In the basic model, we only know the optimal randomized algorithm for $m = 2$, but for large m we even know no better randomized algorithms than deterministic ones. We conjecture that randomized algorithms are provably better than deterministic ones for every m, and also for m tending to infinity. The following two problems seem to be most interesting.

Open Problem 4.1 *Design an optimal randomized algorithm for 3-machine scheduling (in the basic model).*

Open Problem 4.2 *Design a randomized algorithm for arbitrary number of machines which is provably better than any such deterministic algorithm.*

Asymptotic behavior of the competitive ratio In the basic model we do not know optimal deterministic algorithms for any fixed $m > 3$. A major open problem is this.

Open Problem 4.3 *Determine the optimal competitive ratio for deterministic scheduling algorithms in the basic model working for arbitrary m.*

Considering this problem, we usually assume that the competitive ratio increases with increasing m, and hence the hardest case is for m large. However, the following problem is open.

Open Problem 4.4 *Prove that the optimal competitive ratio for m is less than or equal to the optimal competitive ratio for $m+1$.*

Not only that, we even cannot prove that the competitive ratio increases if the number of machines for example doubles (this seems to be a more reasonable goal, as we could avoid some anomalies that occur when the increase of the number of machines is small, cf. [43]). The lack of our knowledge is demonstrated by the fact that we even cannot exclude that the maximal competitive ratio is actually attained for some $m < \infty$. The problems about behavior of the competitive ratio as a function of m are equally open for randomized scheduling.

To compare, for scheduling with rejections we know the limiting value for large m, and we also know that if we increase the number of machines exponentially, the competitive ratio actually increases. On the other hand, if we minimize the L_2 norm instead of the makespan, the maximal competitive ratio is achieved for $m = 3$, and the limit for large m is strictly smaller.

5 Unknown running times

In this on-line paradigm the running time of a job is unknown until the job finishes. This is motivated by the situation of a scheduling algorithm which gets the jobs from different users and has no way of saying how long each job will take. We first focus on minimizing the makespan and later in Section 5.4 we discuss other objective functions.

We are interested in the variants with jobs released over time, either at their release times or according to the precedence constraints, but also in the variant of *batch-style scheduling* where all the jobs are given at time 0. The next general reduction theorem explains why the batch-style algorithms are so important.

Theorem 4 [75]. *Suppose that we have a batch-style σ-competitive algorithm (w.r.t. the makespan). Then there exists a 2σ-competitive algorithm which allows release times.*

Proof. Consider an on-line algorithm that works in phases as follows. In each phase all jobs available at the beginning of the phase are scheduled using the batch-style algorithm. The next phase starts immediately after all these jobs are processed, or, if no jobs are available at that time, at the time the next job is released. This describes a legal algorithm, as at the beginning of each phase no job is running and hence we can use the batch-style algorithm.

Now consider a schedule generated by this algorithm. Let T_3 be the time spent in the last phase, T_2 the time spent in the last but one phase, and T_1 be the time of all the previous phases. We know that $T_2 \leq \sigma T_{\text{opt}}$, as the jobs scheduled during T_2 must be scheduled by the optimal schedule, too, and the batch-style algorithm is σ-competitive. Similarly, $T_1 + T_3 \leq \sigma T_{\mathrm{opt}}$, as the jobs scheduled during T_3 can be scheduled in the optimal schedule only after the time T_1 (if they are released earlier, the on-line algorithm would schedule them in one of the earlier phases), and the batch-style algorithm takes at most σ times longer than the optimal one. The theorem now follows. □

The above reduction is completely satisfactory if we are interested only in the asymptotic behavior of the competitive ratio. However, if the competitive ratio is a constant, we may be interested in a tighter result. In [34] it is proved that for a certain class of algorithms the competitive ratio is increased only by 1, instead of the factor of 2 in the previous theorem; this class of algorithms includes all algorithms that use a greedy approach similar to List Scheduling.

The intuition beyond these reductions is that if the release times are fixed, the optimal algorithm cannot do much before the last release time. In fact, if the on-line algorithm would know which job is the last one, it could wait until its release, then use the batch-style algorithm once, and achieve the competitive ratio of $\sigma + 1$ easily.

In the basic model where the only characteristic of a job is the running time, there is not much we can do if we do not know it. Theorem 1 shows that List Scheduling is $2 - \frac{1}{m}$ competitive also if release times are allowed (hence we do not lose anything in the competitive ratio, unlike in the reductions above), the same bound is true even with precedence constraints. This competitive ratio is tight for deterministic algorithms and almost tight for randomized algorithms, even restricted to the batch-style model.

Theorem 5 [75]. *For batch-style scheduling with unknown running times, no deterministic algorithm is better than* $(2 - \frac{1}{m})$*-competitive and no randomized algorithm is better than* $(2 - O(\frac{1}{\sqrt{m}}))$*-competitive.*

Proof. In the deterministic case we use the same instance as in Theorem 1. The algorithm is given $m(m-1)+1$ jobs. All of them have running time 1, except the one scheduled last, which has running time m. We can assign the running times in this way, since the algorithm is deterministic and we can simulate it on the instance where all the jobs have running time 1. After we see which job is scheduled last, we change its running time to m. As the algorithm does not see the running times, it has to behave identically. The last job is scheduled at time at least $m-1$, hence the on-line schedule has makespan at least $2m-1$, while the optimal makespan is m.

For the randomized case we consider the instance with $\sqrt{m}$ jobs of running time m and $m(m-\sqrt{m})$ jobs of running time 1, permuted randomly. As the algorithm does not see the running times, whenever it schedules a job, it in fact chooses one of the remaining ones at random. A standard computation shows that the probability that the last $cm^{3/2}$ jobs contain no long jobs is at most α^c for some constant α. Hence the expected time when the last long job is scheduled is at least $m - O(\sqrt{m})$ and the expected makespan of the on-line algorithm is at least $2m - O(\sqrt{m})$, while the optimal makespan is m. □

5.1 Different speeds

Here we consider both variants, uniformly related machines and unrelated machines. In the case of related machines the speed of each machine is the same for all jobs and given in advance. For unrelated machines the speeds are different for each job. However, we assume that the speeds are known for each job, only the running time is not known (i.e., for each job we know the relative speeds of machines).

If no restarts are allowed, a simple example shows that the best competitive ratio is $\Omega(\sqrt{m})$ [23], even for uniformly related machines. Consider the case of m jobs to be scheduled on m machines, one with speed $\sqrt{m}$ and the rest with speed 1. Whenever a job is scheduled on one of the slow machines, we assign it long running time, and scheduling it on the fast machine is $\sqrt{m}$ times faster; if all jobs are scheduled on the fast machine we assign them the same running time, and lose the same factor. A matching, $O(\sqrt{m})$-competitive, algorithm is known even for unrelated machines [23].

Next we consider the case when restarts are allowed, studied in [75]. In this case we can use a similar general reduction as Theorem 4 to convert an arbitrary off-line algorithm into an on-line algorithm. (We can use either the optimal algorithm, or, if we require a polynomial time algorithm, we can use the approximation schemes known for the considered problems to obtain the same asymptotic bounds.) Since we do not know the running time, we guess that all jobs have some chosen running time, then run the appropriate schedule. If any job is not finished in the guessed time, we stop it, double the estimate, and repeat the procedure for all such jobs. This method, together with additional improvements, yields for uniformly related machines an algorithm with competitive ratio $O(\min(\log m, \log R))$, where R is the ratio between the largest and smallest speed. A matching lower bound shows that this is optimal. Also in the restricted assignment case there exists $O(\log m)$-competitive algorithm, but it is not known whether this is tight; in fact the best lower bound is only $2 - \frac{1}{m}$. For unrelated machines similar methods yield an $O(\log n)$-competitive algorithm, where n is the number of jobs. It would be interesting to know if there exists an on-line algorithm with a competitive ratio independent of n.

5.2 Parallel jobs

In this variant each job is characterized by its running time and the number of machines (processors) it requests. We consider two variants, batch-style algorithms and algorithms for instances with precedence constraints. While the running times are unknown, the number of machines a job requests is known as soon as it becomes available. All machines have the same speed and no preemptions or restarts are allowed.

Consider the simplest greedy approach for batch-style algorithms: whenever there are sufficiently many machines idle, we schedule some job on as many machines as it requests. This leads to $(2 - \frac{1}{m})$-competitive algorithm, regardless of the rule by which we choose the job to be scheduled (note that here we have a meaningful choice, as we know how many machines each job requests) [35]. This is optimal by Theorem 5, as the basic model corresponds to the special case when each job requests only one machine. Moreover, this algorithm works even for non-malleable jobs.

If we allow precedence constraints, no reasonable on-line algorithm exists for non-malleable parallel jobs. Consider the following situation. At the beginning there are available m jobs requesting one machine; one of them has running time 1 and all other 0. There is one parallel job requesting all machines and running time 0; this job is dependent on one of the jobs with running time 0. The on-line algorithm cannot distinguish the jobs available at the beginning, so that it may happen that the parallel job can be scheduled only after the job with running time 1 is finished. If we iterate this properly, we obtain a lower bound of m on the competitive ratio; a trivial algorithm which at each time schedules only one job achieves this [33]. This argument works also for randomized algorithms, and gives a lower bound of $m/2$ in this case [72].

Hence we turn our attention to ideally malleable jobs. It turns out that the optimal competitive ratio for deterministic algorithms is $1 + \phi \approx 2.6180$, and it is achieved by the following simple algorithm [33].

Algorithm Parallel

(i) If an available job requests p machines and p machines are idle, schedule this job on p machines.

(ii) If less than m/ϕ machines are busy and some job is available, schedule it on all available machines.

Note that this algorithm uses the fact that jobs are malleable only for large jobs. Accordingly, if there is an upper bound on the number of machines a job can use, we can get better algorithms and also algorithms for non-malleable jobs. The tight tradeoffs are given in [33]. It is also interesting that this result improves on the best previously known off-line algorithm, which only achieves an approximation ratio 3 [82].

5.3 Parallel jobs on specific networks

Here we consider a similar model as in the last section with an additional restriction. We require that each parallel job is scheduled on some subset of machines with a specific structure, not an arbitrary subset as before. This is motivated by the situation in which parallel jobs are designed for specific multiprocessor systems and may use the specific properties of the network connecting the individual processors. We consider three topologies of this network. If the network is a hypercube, each parallel job can only be scheduled on a subhypercube of the network (in particular the number of processors a job requests must be a power of two). If the network is a linear array, each job must be scheduled on a contiguous segment of this line. If the network is a two-dimensional mesh, a job must be scheduled on a rectangle of given dimensions. The previous case with no restriction on the set of machines may be viewed as the case of PRAM or a

Network	Batch-style		With precedence constraints	
	Deterministic	Randomized	Deterministic	Randomized
Hypercube	$2 - \frac{1}{m}$	$\leq 2 - \frac{1}{m}$	$O(\frac{\log m}{\log \log m})$	$O(\frac{\log m}{\log \log m})$
Linear array	≤ 2.5	≤ 2.5	$\Theta(\frac{\log m}{\log \log m})$	$\Theta(\frac{\log m}{\log \log m})$
Two-dimensional mesh	$O(\sqrt{\log \log m})$	$O(1)$	$O((\frac{\log m}{\log \log m})^2)$	$O((\frac{\log m}{\log \log m})^2)$

Table 3. Summary of results for scheduling parallel jobs with unknown running times on specific networks.

complete graph, where every two machines are directly connected, and therefore there is no preference among the subsets.

Table 3 summarizes the results in this model. The results for deterministic batch-style algorithms are from [35], the results for deterministic algorithm with precedence constraints from [33], and the results on randomized algorithms are from [72, 73].

For batch-style algorithms we have seen in the previous section that with no restriction given by the network the simple greedy approach works. This is no longer true for specific networks; the problem is that a few jobs using one machine can make the whole system unusable for larger jobs. If we allow preemptions or restarts, we can solve this easily by rearranging the jobs into some compact area, but if it is impossible to stop a job, the situation is more difficult. It is always essential to sort the jobs according to their sizes. If we then schedule the jobs greedily from the largest ones (i.e., those requesting most machines), we get the optimal batch-style algorithm for hypercubes. For linear array this leads to 3-competitive algorithm, the 2.5-competitive algorithm needs more careful placement of the jobs.

For the two-dimensional mesh the situation is most interesting. It is no longer possible to start from the largest jobs, as for example 10×10 and 5×20 meshes are not comparable. Instead, we divide the jobs into $(\log m)^2$ classes so that the jobs in each class require meshes of similar sizes, and deal with each of them separately. The optimal deterministic algorithm always schedules all of them at once in equal partitions of the mesh, and repeats this procedure several times. Interestingly, any greedy approach that tries from the beginning to use the whole mesh fails and leads to an algorithm whose competitive ratio is the square of the optimal one; to achieve the optimal results we have to start by using only a small portion of the mesh.

In the case of two-dimensional meshes we know that randomization decreases the competitive ratio from a non-constant one to a constant. The basic idea is to sample the running times in each class of jobs, and then to schedule each class in an area of the mesh proportional to the estimate of the total work (running time times the number of machines requested) in that class (unlike the deterministic case where we use the same area for each class). To make this work, it is essential to bound the probability that some estimate is wrong by a constant. Since the number of classes is non-constant, this requires a trick: we use the fact that from the classes of jobs requiring less machines we can sample more jobs in the same time and thus we get more accurate estimates.

As in the previous section, all the batch-style algorithms work for non-malleable jobs, but no better algorithms exist even for malleable jobs. Again, with precedence constraints, we always need to use malleable jobs to obtain non-trivial upper bounds.

Even then in the presence of precedence constraints we cannot use the same ideas as for the batch-style algorithms. Even if we process at the beginning all available large jobs, we cannot exclude that later on more of them become available. Thus the best algorithms we can design simply set aside groups of machines for each size of the jobs. For example, for linear array we divide the line of machines into $\log m / \log \log m$ segments, divide the jobs according to the sizes so that in each class the number of requested machines differs by at most a factor of $\log m / \log \log m$, and schedule each class in its segment greedily, using malleability if necessary. Whenever some job is available and not running, one of the segments is fully used, and no job is slowed down by a larger factor than $\log m / \log \log m$ due to malleability; this together gives the upper bound. It is interesting that this simple approach is optimal and even randomization does not help to improve it; the proof of this result is tedious.

Many results in Table 3 are tight, but a few gaps remain. For batch-style algorithms we do not know the exact competitive ratio for the case of linear array. This is interesting, since the case of linear array is an on-line version of the strip packing, where we have to pack given rectangles into a strip of fixed width and as small height as possible. For a long time the best algorithm for this off-line problem gave approximation ratio 2.5 [76], which was matched by the on-line algorithm mentioned in the table. Later, the off-line solution was improved to approximation ratio 2 [77], and it would be interesting to see if this can be achieved by an on-line algorithm, too. The strip packing problem was also studied in the setting equivalent to our paradigm of scheduling jobs one by one as a variant of two-dimensional bin packing, see [38].

For scheduling with precedence constraints on two-dimensional meshes, both deterministic and randomized, there is a gap between the lower bound which follows from the lower bound for linear arrays and is $\Omega(\log m / \log \log m)$ and the upper bound which is the square of the lower bound. For hypercubes there is no non-trivial lower bound, the claim of a tight bound for this case in [33] is incorrect.

5.4 Other objective functions

In this section we consider the competitive ratio w.r.t. the total waiting time and completion time on identical machines. To minimize these objectives off-line, we have to schedule first the jobs with small running times. If there are no preemptions, we clearly cannot do this at all for unknown running times even in batch-style scheduling: consider a sequence of $n - 1$ jobs with running time

0 and one job with running time t on a single machine. If we do not know the running time, it may happen that we schedule as the first job the long one (even in the randomized case with probability $1/n$), and the resulting total waiting time is $(n-1)t$, while the optimal schedule has waiting time 0. Thus, we have to use preemption, and even then it is surprising that we can design competitive algorithms at all. We also assume that there is a minimal running time for each job (the above example also shows that this is necessary).

First we consider batch style algorithms for sequential jobs. In turns out that the optimal competitive ratio is obtained by the simple Round Robin algorithm. It cycles through all unfinished jobs and assigns to each of them to one of the machines and time slot of length τ fixed beforehand. This algorithm has competitive ratio 2 [63] even w.r.t. the total waiting time. (More precisely, if τ approaches zero, the competitive ratio approaches 2 from above.) This is optimal even for randomized algorithms, and even if we consider the total completion time instead of waiting time [63].

It is somewhat unsatisfactory that in the previous algorithm the number of preemptions is $\Theta(t)$ for a job with running time t. It turns out that for sequential jobs we can increase τ during the Round Robin algorithm so that the values in the successive cycles form a geometric sequence. Then the competitive ratio is the same, approaching 2 from above for small starting τ and a small step of the geometric sequence, and the number of preemptions decreases to $\Theta(\log t)$ [63]. This is optimal for deterministic algorithms, as any algorithm with $o(\log t)$ preemptions has a competitive ratio of at least $\Omega(n)$ [63].

Interestingly, the results for batch-style scheduling can be generalized to parallel jobs. Here we consider only total completion time (note that waiting time is not well defined for malleable parallel jobs). For ideally malleable jobs there exists a 2-competitive deterministic algorithm [25], matching the performance for sequential jobs (consequently, randomization cannot help in this case).

A wide range of types of non-ideally malleable jobs together with various restrictions on the number of preemptions is studied in [29]. It is even possible to obtain algorithm for non-ideally malleable jobs under the restriction that the speedup function is non-decreasing and sublinear, which means that allocating an extra processor cannot decrease the actual processing time and cannot decrease the work done for this job; we can even allow that this parallelism profile changes over time for each job. Here the simple algorithm which assigns the same number of processors to each unfinished job has competitive ratio at most $2+\sqrt{3} \approx 3.74$; the number of preemptions is n for each job, but it can be decreased to $\log n$ at the cost of a constant factor increase in the competitive ratio [29]. (Note that here the jobs are allowed to change parallelism profile; in other models this would be treated e.g. as a sequence of distinct different jobs and the number of jobs n could increase significantly.) Another type of parallel jobs is studied in [24]. Here each job is represented as a directed graph of sequential (sub)jobs, and the competitive ratio achieved is 4.

For algorithms with release times, we know that there are no good on-line algorithms w.r.t. total flow time even for sequential jobs. The competitive ratio is at least $\Omega(n^{1/3})$ for deterministic algorithms and $\Omega(\log n)$ for randomized algorithms [63].

5.5 Open problems

This on-line paradigm seems to be understood relatively well, including such issues like randomization. Perhaps the most interesting problem concerns the general reduction in Theorem 4.

Open Problem 5.1 *Find a variant of scheduling for which the optimal competitive ratio (w.r.t. the makespan) for algorithms with release times is twice the optimal competitive ratio for batch-style algorithms.*

In particular, this possibility is open for parallel jobs with no specific network, on hypercubes, and on linear arrays, where the best competitive ratios in the presence of release times given by Theorem 4 are 4, 4, and 5, respectively. It would be interesting to improve any of these bounds. Note that if we allow malleable jobs in the case with no specific network, the algorithm Parallel gives the upper bound of $1+\phi$, which is strictly below the upper bound of 4 obtained by Theorem 4 for this case.

6 Jobs arriving over time

In this paradigm the only feature unknown to the on-line algorithm is the existence of the jobs whose release time did not pass yet. In the results surveyed in this section we will see that from many viewpoints such algorithms can do almost as well as off-line algorithms.

We first consider the objective of minimizing the makespan, and then turn to other objective functions and scheduling of conflicting jobs.

6.1 Minimizing the makespan

Here a batch-style algorithm has full information, and hence it can schedule the jobs optimally. Thus from Theorem 4 we get the following result.

Theorem 6 [75]. *For any variant of on-line scheduling of jobs arriving over time (with all characteristics known), there exists a 2-competitive algorithm w.r.t. the makespan.*

As most of the scheduling variants are NP-hard, algorithms obtained by the previous theorem may not be computationally feasible. However, instead of an optimal algorithm we can use an off-line ρ-approximation algorithm, in which case we obtain a 2ρ-competitive on-line algorithm.

For the basic scheduling problems we can achieve even better results than using this reduction. The optimal, i.e., 1-competitive, on-line algorithm for preemptive scheduling on identical machines is given in [42, 48]. The idea of the algorithm is simple: whenever a new job arrives, we reschedule the unfinished parts of previous jobs and all unscheduled jobs so that they are finished as early as possible. For uniformly related machines with different speeds an optimal on-line algorithm exists if and only if the speeds satisfy $s_{i-1}/s_i \leq s_i/s_{i+1}$, where s_i is the speed of ith fastest machine [80, 81]. All these algorithms use $\Theta(mn)$ preemptions, and this is actually necessary [80, 81].

For uniformly related machines with arbitrary speeds there exist optimal algorithms that are nearly on-line [66, 58]. This means that at each time we know when the next job will be released, in addition to the running times of already released jobs. Any nearly on-line optimal algorithm can be easily transformed into a $(1+\varepsilon)$-competitive on-line algorithm for arbitrary $\varepsilon > 0$: The on-line algorithm chooses a small $\delta > 0$, changes the instance by introducing new jobs with release dates at each multiple of δ and running time 0 (or sufficiently small) and by delaying each release time of an original job by δ. Now it uses the nearly on-line scheduler on the new instance; it always knows the next release time which is at most δ away. The result is an on-line algorithm which produces a makespan of $T_{\text{opt}} + \delta$, i.e., with only a small additive constant; to achieve a small relative error, choose δ proportional to the running time of the first released job with non-zero running time. Note that this on-line algorithm is 1-competitive if we allow an additive term in the definition of competitiveness, but it is not optimal and not 1-competitive if we allow no additive term. Thus, using the results of [80, 81], for uniformly related machines with certain speed ratios we have a curious situation where $(1+\varepsilon)$-competitive algorithms do exist for any $\varepsilon > 0$, but no optimal on-line algorithm exists.

If we consider scheduling without preemptions, we no longer can get a 1-competitive algorithm, even for identical machines. The best upper bound was obtained for the simple algorithm which always schedules the available job with the longest running time; this algorithm is 1.5-competitive [19, 81]. A lower bound of 1.3473 shows that this is close to the best possible [19, 81].

For open shop scheduling on two machines the greedy algorithm achieves the competitive ratio of 3/2; this is optimal for scheduling without preemptions [20]. With preemption, a 5/4 competitive algorithm exists and this is optimal [20]. Note that the greedy algorithm can also be used in the paradigm with unknown running times; in that case it is optimal even if preemptions are allowed [20].

6.2 Minimizing the total weighted completion time

For minimizing the total completion time, and even total weighted completion time, it is possible to give a similar general theorem as Theorem 6 [47, 45, 14]. More surprisingly, it is possible to design schedules that are close to optimal simultaneously for the total completion time and the makespan [14]. For technical reasons we assume that all running times are at least 1. The algorithm for this general reduction is the following.

Algorithm Greedy-Interval

for $i := 0, 1, \ldots$ **do**

At time $\tau = 2^i$ consider all the jobs released by the time τ and not scheduled yet.

(i) Find a schedule with the optimal makespan for these jobs. If it is shorter than τ, use it starting at time τ (i.e., at time 2τ all the jobs will be finished).

(ii) Otherwise find a schedule with makespan at most τ which schedules the jobs with the largest possible total weight and use it starting at time τ.

Theorem 7 [47, 45, 14]. *For any variant of on-line scheduling of jobs arriving over time, Greedy-Interval is 4-competitive w.r.t. the total weighted completion time and simultaneously 3-competitive w.r.t. the makespan.*

Proof. First we prove that Greedy-Interval is competitive w.r.t. the total weighted completion time. Fix an optimal schedule. The rule (ii) of the algorithm guarantees that the total weight of the jobs that Greedy-Interval completes by the time 2^{l+1} is at least the total weight of jobs finished by time 2^l in the optimal schedule. Hence the weight of jobs Greedy-Interval finishes by an arbitrary time τ is at least the weight the optimal schedule finishes by time $\tau/4$. The bound now follows as the total completion time can be equivalently expressed as the sum of the weight of unfinished jobs over all times.

Next we prove that Greedy Interval is competitive w.r.t. the makespan. Suppose that the optimal makespan T_{opt} satisfies $2^i \leq T_{\text{opt}} < 2^{i+1}$. All jobs are released by 2^{i+1}, so in the next iteration Greedy-Interval is able to schedule all jobs and finds the optimal schedule. Hence its makespan is at most $T_{\text{opt}} + 2^{i+1} \leq 3T_{\text{opt}}$. □

For preemptive scheduling on a single machine w.r.t. the total completion time it is easy to construct an optimal schedule on-line by always running the job with shortest remaining processing time. The same rule yields a 2-competitive algorithm on identical machines [64].

In the case of single-machine non-preemptive scheduling it is also possible to get better bounds than in Theorem 7. For minimizing the total completion time 2-competitive algorithms were given [64, 49, 78], moreover this is optimal [49, 78]. (An open question is whether it helps if we allow restarts, a lower bound for deterministic algorithms is 1.112 [81].) For minimizing the total weighted completion time a slight modification of Greedy-Interval yields a competitive ratio of 3 [45]; later this was improved to $1 + \sqrt{2} \approx 2.414$ by α-point scheduling discussed below [40].

As in Theorem 6, it is possible to use an approximation algorithm instead of the infeasible optimal one and obtain accordingly larger competitive ratios. Here the goal of the approximation algorithm is not to schedule all jobs as fast as possible, but it needs to solve a dual problem, namely within a given time, to schedule jobs with as large weight as possible; we then use this algorithm also in the step (i) of Greedy-Interval instead of the optimal makespan schedule. This gives competitive ratio 4ρ w.r.t. both the makespan and the total completion time, if ρ is the approximation ratio of the algorithm we use. A number of results that follows from using such approximation algorithms is described in [47, 45, 14]; for example for minimizing the total weighted completion time on identical machines there exists an $(1+\varepsilon)$-approximation polynomial time algorithm, and hence we obtain $(4+\varepsilon)$-competitive polynomial time algorithm [45].

A general randomization technique can be used to improve upon the deterministic algorithm Greedy-Interval. If we use $\tau = \beta 2^i$ in the algorithm for β chosen uniformly between 1/2 and 1, the competitive ratios will be 2.89 w.r.t. the total completion time and 2.45 w.r.t. the makespan [14]. (Note that since the randomized competitive ratio is actually an expectation, we cannot guarantee that the schedule is actually simultaneously within the given factor of both objective functions. However, this proves e.g. that there always exists a schedule which is within a factor of 2.89 of the optimal total completion time and within a factor of 3 of the optimal makespan.)

One method to obtain deterministic 2-competitive algorithms w.r.t. the total completion time is to take the optimal preemptive schedule (which is easy to compute even on-line) and schedule the jobs in the order of their completion in this auxiliary schedule [64]. This idea led to a generalization which turned to be very useful for off-line approximation algorithms and also for randomized on-line scheduling.

Call an α-point of a job the first time when α fraction of this job is finished. Now schedule the jobs in the order of α-points for some α [47]. (Thus the method of [64] is simply scheduling in the order of 1-points.) After using α-point scheduling for off-line algorithms in [47], it was observed that choosing α randomly, under a suitable distribution and starting from a suitable preemptive schedule that can be computed on-line, leads to new randomized on-line algorithms [15, 40]. These methods generally lead not only to c-competitive algorithms for non-preemptive scheduling, they in fact guarantee that the produced non-preemptive schedule is within the factor c of the optimal preemptive schedule.

In the case of a single machine α-point scheduling leads to a randomized algorithm with competitive ratio w.r.t. the total completion time $e/(e-1) \approx 1.5819$ [15]; this is also optimal [79, 81]. For the total weighted completion time α-point scheduling gives a randomized 2-competitive algorithm for a single machine [40]. Recently, this has been improved to 1.6853-competitive algorithm, using a further modification that the α is chosen randomly not once for the whole schedule but independently for each job [41]. Similar methods can be used also for other problems. If preemption is allowed, a competitive ratio of 4/3 w.r.t. the total weighted completion time for a single machine can be achieved [67]. On parallel identical machines without preemptions a randomized algorithm 2-competitive w.r.t. the total weighted completion time was given in [68].

6.3 Minimizing the total flow time

Minimizing the total flow time is much harder than to minimize the total completion time also with known running times.

Without preemption we have strong lower bound even for a single machine. Clearly, no deterministic algorithm can be better then $n-1$ competitive: consider an instance where one job with running time arrives at time 0 and $n-1$ jobs with running time 0 arrive just after this job was scheduled. Even if the algorithm is randomized, no algorithm is better than $\Omega(\sqrt{n})$ competitive [79, 81], and if a deterministic algorithm is allowed to restart jobs, the lower bound is $\Omega(\sqrt[4]{n})$ [81]. Note that also the off-line problem is hard, it is NP-hard to achieve an approximation ratio $n^{\frac{1}{2}-\varepsilon}$ for any $\varepsilon > 0$ [57].

With preemptions the optimal competitive ratio still depends on the number of jobs; it is $\Theta(\log(n/m))$ [61]. Only in the case when the ratio between the maximum and the minimum running time is bounded by P, we can obtain a bound independent of n, namely $\Theta(\log P)$; this is again tight [61].

6.4 Conflicting jobs

The last variant we consider in this section is very different than all the ones considered before. Here some jobs may conflict with each other, in which case we cannot schedule them at the same time. These conflicts are given by a conflict graph, which means that at any time we are allowed to schedule only an independent set in this graph; we assume that on a given node of the conflict graph there may be more jobs which then have to be scheduled one by one (this can be modeled by a clique of these jobs, but this generalization is important if we consider restricted graphs). We assume that we have infinitely many machines, which allows us to focus on the issue of conflicts, and also corresponds to some practical motivation, cf. [50]. We assume that the jobs have integral release times and all the running times are 1; however, we can clearly relax this to arbitrary times if we allow preemptions, as job with running time t is equivalent to t jobs of time 1 on the same node in the conflict graph.

For the makespan, the optimal competitive ratio is 2, even for randomized algorithms, and it is achieved by the following simple algorithm. At any time we find the coloring of the available jobs by the smallest number of colors and schedule one of these colors. This bound follows directly from Theorem 4, even for arbitrary known running times, and it was rediscovered in [63], who also proved the matching lower bound.

If we consider maximal flow time instead of the makespan, some partial results were obtained by [50]. They show that for conflict graphs that are either interval graphs or bipartite graph, there is an on-line algorithm with the maximal response time bounded by $O(v^3A^2)$, where A is the optimal objective value and v is the number of vertices in the conflict graph. It would be interesting to obtain a competitive ratio which is a function of only v, i.e., a performance guarantee linear in A. The same paper gives a lower bound of $\Omega(v)$ for the competitive ratio of deterministic algorithms which applies even to the interval and bipartite graphs.

6.5 Open problems

Theorems 6 and 7 give very good general algorithms for the makespan and the total weighted completion time. Moreover, for the total completion time we have tight results at least on a single machine.

To our best knowledge, no lower bounds for total weighted completion time are known that are better than the bounds for the unweighted total completion time, and also no lower bounds for total completion time on identical machines are better than the bounds on a single machine.

Open Problem 6.1 *Prove for some scheduling problem that the competitive ratio on identical machines w.r.t. total weighted completion time is strictly larger than the competitive ratio on a single machine w.r.t. total completion time.*

Another open problem is to investigate the other objective functions, total flow time and total waiting time even in some restricted cases (as we have seen that to minimize these objectives is hard in general).

7 Relaxed notions of competitiveness

For several variants of scheduling we have seen quite strong negative results. However, it turns out that if we allow the on-line algorithm to use some additional information or slightly more resources than the off-line algorithm, we can sometimes overcome these problems and obtain reasonable algorithms.

7.1 Algorithms with more computational power

If we allow the on-line algorithm to use machines with speed $1 + \varepsilon$, there exists a $(1 + 1/\varepsilon)$-competitive algorithm for minimizing total flow time with preemptions and unknown running times on one machine [52]; in contrast without the additional power we have seen that the competitive ratio has to depend on the number of jobs. For scheduling of jobs arriving over time with known running times several results of this kind are obtained by [65]; they either use $O(\log n)$ machines instead of one in the non-preemptive one-machine version or increase the speed of the machines by a factor of two in the preemptive m-machine version, and in both cases they obtain the optimal sum of flow times. In the second case they also show that increasing the speed by $1+\varepsilon$ for some small ε is not sufficient. The positive results should again be contrasted with the negative results if we do not allow additional resources.

7.2 Algorithms with additional knowledge

Another way to improve the performance of the on-line algorithms is to give them some additional information. If we use List Scheduling on a sequence of jobs with non-increasing running times arriving one by one, the competitive ratio is $4/3 - 1/(3m)$, an improvement from $2 - 1/m$ [44]. For deterministic scheduling of jobs arriving one by one on two machines several such possibilities are considered in [56]. They show that the competitive ratio decreases from 3/2 to 4/3 in any of the following three scenarios. First, we know the total running time of all jobs. Second, we have a buffer where we can store one job (if we allow buffer for more jobs, we do not gain either). Third, we are allowed to produce two solutions and choose the better one afterwards (equivalently, this means that we get one bit of a hint in advance). If the optimum is known, the problem is also called bin-stretching (because we know that the jobs fit into some number of bins of some height, and we ask how much we need to "stretch" the bins to fit the jobs on-line), and is studied in [7]. For two machines once again 4/3 is the correct and tight answer and for more machines a 1.625-competitive algorithm is presented.

8 Conclusions

We have seen a variety of on-line scheduling problems. Many of them are understood satisfactorily, but there are also many interesting open problems. Studied scheduling problems differ not only in the setting and numerical results, but also in the techniques used. In this way on-line scheduling illustrates many general aspects of competitive analysis.

Acknowledgements

I am grateful to Bo Chen, Andreas Schulz, David Shmoys, Martin Skutella, Leen Stougie, Arjen Vestjens, Joel Wein, Gerhard Woeginger, and other colleagues for many useful comments, pointers to the literature, and manuscripts. Without them this chapter could not possibly cover as many results as it does. This work was partially supported by grant A1019602 of AV ČR.

References

1. S. Albers. Better bounds for on-line scheduling. In *Proc. of the 29th Ann. ACM Symp. on Theory of Computing*, 130–139. ACM, 1997.
2. J. Aspnes, Y. Azar, A. Fiat, S. Plotkin, and O. Waarts. On-line load balancing with applications to machine scheduling and virtual circuit routing. *J. ACM*, 44(3):486–504, 1997.
3. A. Avidor, Y. Azar, and J. Sgall. Ancient and new algorithms for load balancing in the L_p norm. In *Proc. of the 9th Ann. ACM-SIAM Symp. on Discrete Algorithms*, 426–435. ACM-SIAM, 1998.
4. B. Awerbuch, Y. Azar, E. F. Grove, M.-Y. Kao, P. Krishnan, and J. S. Vitter. Load balancing in the l_p norm. In *Proc. of the 36th Ann. IEEE Symp. on Foundations of Computer Sci.*, 383–391. IEEE, 1995.
5. Y. Azar. On-line load balancing. Chapter 8 in *On-Line Algorithms: The State of the Art*, eds. A. Fiat and G. Woeginger, Lecture Notes in Comput. Sci. Springer-Verlag, 178–195, 1998.

6. Y. Azar, J. Naor, and R. Rom. The competitiveness of on-line assignments. *J. of Algorithms*, 18:221–237, 1995.
7. Y. Azar and O. Regev. On-line bin-stretching. Manuscript, 1997.
8. Y. Bartal, A. Fiat, H. Karloff, and R. Vohra. New algorithms for an ancient scheduling problem. *J. Comput. Syst. Sci.*, 51(3):359–366, 1995.
9. Y. Bartal, H. Karloff, and Y. Rabani. A new lower bound for m-machine scheduling. *Inf. Process. Lett.*, 50:113–116, 1994.
10. Y. Bartal, S. Leonardi, A. Marchetti-Spaccamela, J. Sgall, and L. Stougie. Multiprocessor scheduling with rejection. In *Proc. of the 7th Ann. ACM-SIAM Symp. on Discrete Algorithms*, 95–103. ACM-SIAM, 1996. To appear in SIAM J. Disc. Math.
11. S. Baruah, G. Koren, B. Mishra, A. Raghunatan, L. Roiser, and D. Sasha. On-line scheduling in the presence of overload. In *Proc. of the 32nd Ann. IEEE Symp. on Foundations of Computer Sci.*, 100–110. IEEE, 1991.
12. S. Ben-David, A. Borodin, R. M. Karp, G. Tardos, and A. Wigderson. On the power of randomization in on-line algorithms. *Algorithmica*, 11:2–14, 1994.
13. A. Borodin and R. El-Yaniv. *On-line Computation and Competitive Analysis.* Cambridge University Press, 1998.
14. S. Chakrabarti, C. A. Phillips, A. S. Schulz, D. B. Shmoys, C. Stein, and J. Wein. Improved scheduling algorithms for minsum criteria. In *Proc. of the 23th International Colloquium on Automata, Languages, and Programming, Lecture Notes in Comput. Sci. 1099*, 646–657. Springer-Verlag, 1996.
15. C. Chekuri, R. Motwani, B. Natarajan, and C. Stein. Approximation techniques for average completion time scheduling. In *Proc. of the 8th Ann. ACM-SIAM Symp. on Discrete Algorithms*, 609–618. ACM-SIAM, 1997.
16. B. Chen, A. van Vliet, and G. J. Woeginger. A lower bound for randomized on-line scheduling algorithms. *Inf. Process. Lett.*, 51:219–222, 1994.
17. B. Chen, A. van Vliet, and G. J. Woeginger. New lower and upper bounds for on-line scheduling. *Oper. Res. Lett.*, 16:221–230, 1994.
18. B. Chen, A. van Vliet, and G. J. Woeginger. An optimal algorithm for preemptive on-line scheduling. *Oper. Res. Lett.*, 18:127–131, 1995.
19. B. Chen and A. P. A. Vestjens. Scheduling on identical machines: How good is lpt in an on-line setting? *Oper. Res. Lett.*, 21:165–169, 1998.
20. B. Chen, A. P. A. Vestjens, and G. J. Woeginger. On-line scheduling of two-machine open shops where jobs arrive over time. *J. of Combinatorial Optimization*, 1:355–365, 1997.
21. B. Chen and G. J. Woeginger. A study of on-line scheduling two-stage shops. In D.-Z. Du and P. M. Pardalos, editors, *Minimax and Applications*, 97–107. Kluwer Academic Publishers, 1995.
22. Y. Cho and S. Sahni. Bounds for list schedules on uniform processors. *SIAM J. Comput.*, 9(1):91–103, 1980.
23. E. Davis and J. M. Jaffe. Algorithms for scheduling tasks on unrelated processors. *J. ACM*, 28(4):721–736, 1981.
24. X. Deng and P. Dymond. On multiprocessor system scheduling. In *Proc. of the 7th Ann. ACM Symp. on Parallel Algorithms and Architectures*, 82–88. ACM, 1996.
25. X. Deng, N. Gu, T. Brecht, and K. Lu. Preemptive scheduling of parallel jobs on multiprocessors. In *Proc. of the 7th Ann. ACM-SIAM Symp. on Discrete Algorithms*, 159–167. ACM-SIAM, 1996.

26. X. Deng and E. Koutsoupias. Competitive implementation of parallel programs. In *Proc. of the 4th Ann. ACM-SIAM Symp. on Discrete Algorithms*, 455–461. ACM-SIAM, 1993.
27. X. Deng, E. Koutsoupias, and P. MacKenzie. Competitive implementation of parallel programs. To appear in Algorithmica, 1998.
28. M. Dertouzos and A. Mok. Multiprocessor on-line scheduling with release dates. *IEEE Transactions on Software Engineering*, 15:1497–1506, 1989.
29. J. Edmonds, D. D. Chinn, T. Brecht, and X. Deng. Non-clairvoyant multiprocessor scheduling of jobs with changing execution characteristics. In *Proc. of the 29th Ann. ACM Symp. on Theory of Computing*, 120–129. ACM, 1997.
30. L. Epstein, J. Noga, S. S. Seiden, J. Sgall, and G. J. Woeginger. Randomized online scheduling for two related machines. Work in progress, 1997.
31. U. Faigle, W. Kern, and G. Turan. On the performance of on-line algorithms for partition problems. *Acta Cybernetica*, 9:107–119, 1989.
32. U. Faigle and W. M. Nawijn. Note on scheduling intervals on-line. *Discrete Applied Mathematics*, 58:13–17, 1995.
33. A. Feldmann, M.-Y. Kao, J. Sgall, and S.-H. Teng. Optimal on-line scheduling of parallel jobs with dependencies. In *Proc. of the 25th Ann. ACM Symp. on Theory of Computing*, 642–651. ACM, 1993. To appear in a special issue of *J. of Combinatorial Optimization* on scheduling.
34. A. Feldmann, B. Maggs, J. Sgall, D. D. Sleator, and A. Tomkins. Competitive analysis of call admission algorithms that allow delay. Technical Report CMU-CS-95-102, Carnegie-Mellon University, Pittsburgh, PA, U.S.A., 1995.
35. A. Feldmann, J. Sgall, and S.-H. Teng. Dynamic scheduling on parallel machines. *Theoretical Comput. Sci.*, 130(1):49–72, 1994.
36. A. Fiat and G. J. Woeginger. On-line scheduling on a single machine: Minimizing the total completion time. Technical Report Woe-04, TU Graz, Austria, 1997.
37. G. Galambos and G. J. Woeginger. An on-line scheduling heuristic with better worst case ratio than Graham's list scheduling. *SIAM J. Comput.*, 22(2):349–355, 1993.
38. G. Galambos and G. J. Woeginger. On-line bin packing – a restricted survey. *ZOR – Mathematical Methods of Operations Research*, 42:25–45, 1995.
39. M. R. Garey and D. S. Johnson. *Computers and Intractability: a Guide to the Theory of NP–completeness*. Freeman, 1979.
40. M. X. Goemans. Improved approximation algorithms for scheduling with release dates. In *Proc. of the 8th Ann. ACM-SIAM Symp. on Discrete Algorithms*, 591–598. ACM-SIAM, 1997.
41. M. X. Goemans, M. Queyranne, A. S. Schulz, M. Skutella, and Y. Wang. Manuscript, 1997.
42. T. F. Gonzales and D. B. Johnson. A new algorithm for preemptive scheduling of trees. *J. ACM*, 27:287–312, 1980.
43. R. L. Graham. Bounds for certain multiprocessor anomalies. *Bell System Technical J.*, 45:1563–1581, Nov. 1966.
44. R. L. Graham. Bounds on multiprocessor timing anomalies. *SIAM J. Appl. Math.*, 17(2):416–429, 1969.
45. L. A. Hall, A. S. Schulz, D. B. Shmoys, and J. Wein. Scheduling to minimize average completion time: Off-line and on-line approximation algorithms. *Mathematics of Operations Research*, 22:513–544, 1997.

46. L. A. Hall and D. B. Shmoys. Approximation schemes for constrained scheduling problems. In *Proc. of the 30th Ann. IEEE Symp. on Foundations of Computer Sci.*, 134–139. IEEE, 1989.
47. L. A. Hall, D. B. Shmoys, and J. Wein. Scheduling to minimize average completion time: Off-line and on-line algorithms. In *Proc. of the 7th Ann. ACM-SIAM Symp. on Discrete Algorithms*, 142–151. ACM-SIAM, 1996.
48. K. S. Hong and J. Y.-T. Leung. On-line scheduling of real-time tasks. *IEEE Transactions on Computers*, 41(10):1326–1331, 1992.
49. J. A. Hoogeveen and A. P. A. Vestjens. Optimal on-line algorithms for single-machine scheduling. In *Proc. of the 5th Workshop on Algorithms and Data Structures, Lecture Notes in Comput. Sci. 1084*, 404–414. Springer-Verlag, 1996.
50. S. Irani and V. Leung. Scheduling with conflicts, and applications to traffic signal control. In *Proc. of the 7th Ann. ACM-SIAM Symp. on Discrete Algorithms*, 85–94. ACM-SIAM, 1996.
51. B. Kalyanasundaram and K. R. Pruhs. Fault-tolerant scheduling. In *Proc. of the 26th Ann. ACM Symp. on Theory of Computing*, 115–124. ACM, 1994.
52. B. Kalyanasundaram and K. R. Pruhs. Speed is as powerful as clairvoyance. In *Proc. of the 36th Ann. IEEE Symp. on Foundations of Computer Sci.*, 214–221. IEEE, 1995.
53. B. Kalyanasundaram and K. R. Pruhs. Fault-tolerant real-time scheduling. In *Proc. of the 5th Ann. European Symp. on Algorithms, Lecture Notes in Comput. Sci. 1284*, 296–307. Springer-Verlag, 1997.
54. D. Karger, C. Stein, and J. Wein. Scheduling algorithms. To appear in *Handbook of Algorithms and Theory of Computation*, M. J. Atallah, editor. CRC Press, 1997.
55. D. R. Karger, S. J. Phillips, and E. Torng. A better algorithm for an ancient scheduling problem. *J. of Algorithms*, 20:400–430, 1996.
56. H. Kellerer, V. Kotov, M. G. Speranza, and Z. Tuza. Semi on-line algorithms for the partition problem. To appear in Oper. Res. Lett., 1998.
57. H. Kellerer, T. Tautenhahn, and G. J. Woeginger. Approximability and nonaproximability results for minimizing total flow time on a single machine. In *Proc. of the 28th Ann. ACM Symp. on Theory of Computing*, 418–426. ACM, 1996. To appear in SIAM J. Comput.
58. J. Labetoulle, E. L. Lawler, J. K. Lenstra, and A. H. G. Rinnooy Kan. Preemptive scheduling of uniform machines subject to release dates. In W. R. Pulleyblank, editor, *Progress in Combinatorial Optimization*, 245–261. Academic Press, 1984.
59. E. L. Lawler, J. K. Lenstra, A. H. G. Rinnooy Kan, and D. B. Shmoys. Sequencing and scheduling: Algorithms and complexity. In S. C. Graves, A. H. G. Rinnooy Kan, and P. Zipkin, editors, *Handbooks in Operations Research and Management Science, Vol. 4: Logistics of Production and Inventory*, 445–552. North-Holland, 1993.
60. S. Leonardi and A. Marchetti-Spaccamela. On-line resource management with application to routing and scheduling. In *Proc. of the 22th International Colloquium on Automata, Languages, and Programming, Lecture Notes in Comput. Sci. 944*, 303–314. Springer-Verlag, 1995. To appear in Algorithmica.
61. S. Leonardi and D. Raz. Approximating total flow time with preemption. In *Proc. of the 29th Ann. ACM Symp. on Theory of Computing*, 110–119. ACM, 1997.
62. R. J. Lipton and A. Tomkins. On-line interval scheduling. In *Proc. of the 5th Ann. ACM-SIAM Symp. on Discrete Algorithms*, 302–305. ACM-SIAM, 1994.
63. R. Motwani, S. Phillips, and E. Torng. Non-clairvoyant scheduling. *Theoretical Comput. Sci.*, 130:17–47, 1994.

64. C. Philips, C. Stein, and J. Wein. Minimizing average completion time in the presence of release dates. In *Proc. of the 4th Workshop on Algorithms and Data Structures, Lecture Notes in Comput. Sci. 955*, 86–97. Springer-Verlag, 1995. To appear in Math. Programming.
65. C. A. Phillips, C. Stein, E. Torng, and J. Wein. Optimal time-critical scheduling via resource augmentation. In *Proc. of the 29th Ann. ACM Symp. on Theory of Computing*, 140–149. ACM, 1997.
66. S. Sahni and Y. Cho. Nearly on line scheduling of a uniform processor system with release times. *SIAM J. Comput.*, 8(2):275–285, 1979.
67. A. S. Schulz and M. Skutella. Scheduling-LPs bear probabilities: Randomized approximations for min-sum criteria. Technical Report 533/1996, Department of Mathematics, Technical University of Berlin, Berlin, Germany, 1996 (revised 1997).
68. A. S. Schulz and M. Skutella. Scheduling-LPs bear probabilities. In *Proc. of the 5th Ann. European Symp. on Algorithms, Lecture Notes in Comput. Sci. 1284*, 416–429. Springer-Verlag, 1997.
69. S. S. Seiden. More multiprocessor scheduling with rejection. Technical Report Woe-16, TU Graz, Austria, 1997.
70. S. S. Seiden. *Randomization in On-line Computation.* PhD thesis, University of California, Irvine, CA, U.S.A., 1997.
71. S. S. Seiden. A randomized algorithm for that ancient scheduling problem. In *Proc. of the 5th Workshop on Algorithms and Data Structures, Lecture Notes in Comput. Sci. 1272*, 210–223. Springer-Verlag, 1997.
72. J. Sgall. *On-Line Scheduling on Parallel Machines.* PhD thesis, Technical Report CMU-CS-94-144, Carnegie-Mellon University, Pittsburgh, PA, U.S.A., 1994.
73. J. Sgall. Randomized on-line scheduling of parallel jobs. *J. of Algorithms*, 21:149–175, 1996.
74. J. Sgall. A lower bound for randomized on-line multiprocessor scheduling. *Inf. Process. Lett.*, 63(1):51–55, 1997.
75. D. B. Shmoys, J. Wein, and D. P. Williamson. Scheduling parallel machines on-line. *SIAM J. Comput.*, 24:1313–1331, 1995.
76. D. D. Sleator. A 2.5 times optimal algorithm for packing in two dimensions. *Inf. Process. Lett.*, 10:37–40, 1980.
77. A. Steinberg. A strip-packing algorithm with absolute performance bound 2. *SIAM J. Comput.*, 26(2):401–409, 1997.
78. L. Stougie. Personal communication, 1995.
79. L. Stougie and A. P. A. Vestjens. Randomized on-line scheduling: How low can't you go? Manuscript, 1997.
80. A. P. A. Vestjens. Scheduling uniform machines on-line requires nondecreasing speed ratios. Technical Report Memorandum COSOR 94-35, Eindhoven University of Technology, 1994. To appear in Math. Programming.
81. A. P. A. Vestjens. *On-line Machine Scheduling.* PhD thesis, Eindhoven University of Technology, The Netherlands, 1997.
82. Q. Wang and K. H. Cheng. A heuristic of scheduling parallel tasks and its analysis. *SIAM J. Comput.*, 21(2):281–294, 1992.
83. G. J. Woeginger. On-line scheduling of jobs with fixed start and end times. *Theoretical Comput. Sci.*, 130:5–16, 1994.

10

On-line Searching and Navigation

PIOTR BERMAN

1 Introduction

In this chapter we review problems where an algorithm controls movements of an agent (e.g. a robot) in an unknown environment, and receives the sensory inputs of the agent, which may be endowed with vision, touch, sense of location etc. In a navigation problem, a target in a specified position has to be reached; in a searching problem, the target is recognizable, but its position is not specified. Other problems that fit in this category include exploring (mapping) and localization (reverse of mapping: the map is known but the position is not).

The research on navigation and searching investigates the computational problems that can be encountered while operating in an unknown environment. The earliest result on searching is due to Ariadne and Theseus [2]. Later a number of papers considered the navigation and searching in the context of robotics. A futuristic example can be exploration of a surface of a distant planet or a moon by robots [34]. Without a doubt, robots could be used with many times smaller cost and risk; however, even within the solar system the distances are so great that on-line control will not be possible. Therefore the robots should operate in a highly autonomous manner. Other, more mundane motivations include operating automatic warehouses or automatic missions on the ocean bottom. Moreover, some practical on-line problems can be expressed as searching in an unknown graph.

As is the case with other on-line problems, the most common measure of the quality of an algorithm (or, of the intrinsic difficulty of a problem) is the competitive ratio, i.e., the ratio of the lengths of the path traversed by the algorithm and the shortest path that can achieve the specified goal. In the case of a navigation problem, the goal is to reach a target at a specified position from the starting point. Such a problem has four aspects:

- the target (a point, an infinite line, the center of the rectangle that contains obstacles, level in a layered graph etc.);
- the character of the environment in which the motion is planned, it can be a graph, or a geometric scene characterized by
  - the placement of the obstacles (can be unrestricted, within a rectangle that has the starting point as one of its corners etc.),
  - the character of the obstacles (oriented rectangles, arbitrary rectangles, arbitrary convex figures etc.);
- and, lastly, the senses of the agent (e.g. vision, touch, sense of position).

The complexity parameter n usually is the distance between the starting point and the target; sometimes it is the number of nodes that define the scene. In graphs, the complexity parameter usually measures some characteristic of the graph that may look a-priori ad hoc, but which actually measures the difficulty of the problem. In geometric scenes, we may have to assure that there is a relationship between the distance and the competitive ratio, e.g. by assuming that every obstacle contains a unit circle.

The *searching* problems are similar to navigation problems, except that the position of the target is not specified. Instead, it can be discovered by the algorithm using the senses (e.g. "I will know it when I see it").

Another type of problem that involves collecting the data through sensory inputs is *exploring*, i.e. creating the map of the environment. Usually, an exploring problem assumes the sense of position. A reverse problem is that of *localization*; here the map is given, but the initial location is not; instead of sense of location the agent has the exact control of motion, i.e. the relative position with respect to the initial one, the goal is to establish the agent's position on the map.

2 Navigation in geometric scenes

The earliest result on navigation is due to Ariadne and Theseus [2]. Later a number of papers considered the navigation and searching in the context of robotics (like [34]). The earliest result that estimated the *efficiency* of the proposed solution is credited to Lumelsky and Stepanov [29] [30] who considered navigation to a given point on a plane in the presence of arbitrary polygonal obstacles. Their strategy was quite simple, namely the robot either walks directly toward the target, or, if an obstacle is encountered, searches the contour of the obstacle for the point nearest to the target. This strategy guarantees that the path traversed is no larger than 1.5 times the sum of perimeters of obstacles that intersect the straight line between the starting point and the target, with the exclusion of those parts of the perimeters that are further from the target than the starting point. While in terms of competitive ratio this approach is as bad as one can get, it is still the best in the most general types of scenes: with arbitrary polygonal obstacles and with arbitrary convex obstacles (in the latter case, there exists a better *randomized* algorithm, [8]).

In 1989, Papadimitriou and Yannakakis [32] introduced competitive analysis to navigation problems. They considered a variety of problems where the obstacles are oriented rectangles (with sides parallel to the axes of the coordinate system). They showed that point-to-point navigation in the presence of square obstacles must have competitive ratio of at least 1.5, and showed an algorithm with the ratio $1.5 + o(1)$. They introduce the *wall* problem, where the target is a straight line. They have shown that a deterministic algorithm cannot have ratio better than $\sqrt{n}$. Since the wall problem was a subject of several other papers, it is worthwhile to sketch their arguments.

We form first a scene where $1 \times \sqrt{n}$ rectangles form a pattern of bricks in a wall. We let the deterministic algorithm wander in this scene until it reaches the target line, and then we remove all the obstacles not touched by the robot. This leaves $b \geq n$ bricks, and the algorithm traversed at least $b\sqrt{n}$. Now it suffices to find a path of length $3b$. We consider all paths of the following form: walk for a distance $k\sqrt{n} \leq b$ parallel to the target line and then follow the direct line toward the target, going around all the encountered obstacles. If on one of these paths we encounter at most $b/2\sqrt{n}$ obstacles, its length is at most $b + n + (b/2\sqrt{n})2\sqrt{n} \leq 3b$. If no such path exists, we have $2b/\sqrt{n}$ paths, each encountering a separate set of more than $b/2\sqrt{n}$ obstacles, which implies that there are more than $b^2/n > b$ obstacles, a contradiction.

One can generalize this approach for the robot with the sense of vision. The easiest way is to make a somewhat more complicated pattern of rectangles than the simple brick wall, so that each brick is surrounded by "fog" which is easy to traverse but completely opaque (see [5] for the example of such a pattern).

This lower bound was matched two years later by an algorithm of Blum, Raghavan and Schieber [11]. One should note that this lower bound does not apply in the following situations: the length of bricks or their aspect ratio is $o(\sqrt{n})$, or the algorithm is randomized (in the latter case, we do not know which bricks can be removed from the scene). For the case each obstacle is either shorter than f or it has its aspect ratio below f, Mei and Igarashi [31] have shown an upper bound of $1 + 0.6f$.

The already quoted [11] describes a randomized algorithm with competitive ratio of $\exp(O(\sqrt{\log n \log\log n}))$ for a version of the wall problem that has the above deterministic lower bound (the robot has vision and the obstacles are located in such a way that at every time the robot has seen all the obstacles that are no closer to the target line than its present location). In particular, the problem was shown to be equivalent to an n-server problem for a metric space consisting of $n + 1$ points evenly spaced on a line. Very recently, Bartal *et al* [9] have obtained an algorithm for the latter problem with competitive ratio $O(\log^3 n)$. Moreover, Karloff *et al.* [27] have obtained an $\Omega(\log\log\ n)$ lower bound for this server problem, thus providing the only lower bound for randomized competitive ratio for the wall problem.

If the obstacles are not aligned, the vision sense of the robot can be rendered useless and with it the randomized algorithm mentioned above. Still, randomization has been shown to be helpful, if in a much smaller degree. In 1996, a group of six authors [7] showed a randomized algorithm with expected competitive ratio $O(n^{4/9})$. Concurrently, Berman and Karpinski investigated the version of the wall problem with arbitrary convex obstacles [8] and obtained a randomized algorithm with expected competitive ratio $O(n^{3/4})$. The wall problem has two widely open cases. In the case of oriented rectangular obstacles, for randomized algorithms there exists a gap between the above upper bound and a lower bound which is lower than $\log n$. In the case of arbitrary convex obstacles, no better lower bounds were proven, and consequently the gap is even wider. At the moment, the competitive ratios are so high that the advice to designers of a planetary rover should be: find a way to create good maps. Nevertheless, it is conceivable that a practical randomized algorithm exists.

Both [11] and [7] consider a three dimensional version of the wall problem as well, where the target is a plane, and obstacles are cylinders with bases parallel to the target plane and without holes. In [11] it is shown that the competitive ratio for deterministic algorithms for this problem is $\Theta(n^{2/3})$, while [7] claims that a randomized algorithm can be faster by a factor of n^{ϵ}.

An interesting variation of the wall problem was considered by Blum and Chalasani [10]. They investigated a scenario when the trip from the starting point to the target line is repeated k times (the obstacles were oriented rectangles). This way the algorithm has a chance to investigate the scene more thoroughly to find a better route to the target. In [10] the authors show that both upper and lower bounds for this problem are proportional to $\sqrt{kn}$. One should expect that the technique used in the algorithm that provided the $O(\sqrt{kn})$ upper bound can be combined with the technique of [7] to get randomized acceleration by an n^{ϵ} factor.

The paper of Blum, Raghavan and Schieber [11] introduced also several other navigation problems. One was the *room* problem, where all obstacles are within a $n \times n$ square, the starting point is located on the perimeter and the target in the middle. Two versions of the problem were considered. In the first, where the obstacles are arbitrary rectangles, Blum *et al.* showed a deterministic lower bound $\Omega(\sqrt{n})$ (adapting the technique of [32]), and a randomized algorithm with $O(\sqrt{n})$ ratio. In the second, where the obstacles are oriented rectangles, their results were superseded by Bar-Eli *et al.* [5], who obtained a tight bound of $\Theta(\log n)$.

The last problem tackled by Blum, Raghavan and Schieber concerned point to point navigation in the presence of arbitrary polygonal (i.e. concave) obstacles. Here they defined the complexity parameter n to be the number of nodes that the obstacles have. Both upper and lower bounds were shown to be proportional to n. The lower bound here is actually very simple: the cavities of the single obstacle form $n/4$ parallel corridors, and all corridors make a turn. Only one corridor leads to the chamber containing the target. Clearly, even a randomized algorithm with vision must inspect on average half of the corridors before finding the correct one.

3 Searching problems

The last result could be interpreted as dismal news to anyone encountering concave obstacles. However, the usual experience (like driving through American suburbs) is not quite *that* bad. For that reason a variety of scenes was investigated, where the obstacles, while concave, are much easier to handle.

In particular, the following scenario would form a *street* problem: the robot can move inside a simple polygon, and is equipped with vision; the task is to find the target point that is visibly marked (i.e. its location is not disclosed a priori, hence in our nomenclature this is a searching problem). Klein [23] defined a street to be a polygon whose perimeter consists of two connected curves — the sides of the street), so that from each point on one side of the street a point on the other side is visible. Klein showed upper and lower bounds on the competitive ratio to be $1+1.5\pi$ and $\sqrt{2}$ respectively. Later, Kleinberg [24] improved the upper bound to 2.61 ($\sqrt{4+\sqrt{8}}$). Very recently, Semrau derived a bound of $\pi/2$ (=1.57). Datta and Icking [12] considered a *generalized* street where from every point on one side of the street one can see a point on a horizontal line that connects the two sides. This would yield a strictly larger class of polygons, except that they make a restriction that all sides of the polygon are oriented. The algorithm in their paper has proven competitive ratio close to 9; since they also prove that 9 is a *lower* bound on the ratio this result is nearly tight! Another generalization of a street is HV-street introduced by Datta *et al.* [13]. There from every point on one side of the street one can see a point on a horizontal *or vertical* line that connects the two sides of the street (which are also oriented). They obtained competitive ratio of 14.5 for searching in HV-streets; this result is best possible. Other varieties of generalized streets were also considered (e.g. in [13]), but also with the restriction to oriented polygons. López-Ortiz and Schuierer [28] give a 80-competitive strategy in arbitrarily oriented, nonrectilinear generlaized streets.

Baeza-Yates *et al.* [4] considered searching on the plane, without obstacles, for a line that is recognized by touch. By providing (or not) partial information about the unknown line, one obtains several variants of the problem. This information can give the slope of the line, a pair of possible slopes (e.g. horizontal or vertical), or none. Given the lack of obstacles and visual inputs, the problem is that of designing an optimum searching path. Baeza-Yates *et al.* provided optimal or nearly optimal solutions to several versions (including the most general, where no information is given), and discussed the earlier results, several of which were published in *Naval Research Logistics Quarterly*, which hints at a possible area of applications.

In the process of investigating the optimal search paths on the plane, the authors of [4] introduced the problem of searching for a point in m concurrent rays, later called the *cow path* problem. Because this problem is widely applied in various contexts, it is worthwhile to discuss it in detail. Using the terms of [2], the *cow-path* problem can be described as follows. A beautiful young cow, Ariadne, faces the entrance of a simple labyrinth which branches into w disjoint corridors. Somewhere in the dark waits handsome Minotaur. What strategy will bring Ariadne fastest to the connubial bliss? Baeza-Yates *et al.* [4] stated the problem as "searching for a point in w concurrent rays" and showed an optimal deterministic solution. In this algorithm, one enters the "corridors" in a round-robin manner and follows for some length, only to retreat if the target is not found. Each such foray is by factor $w/(w-1)$ longer than the previous. The resulting competitive ratio is $1+2w(\frac{w}{w-1})^{w-1}$. Kao *et al.* [26] refined this strategy by replacing the factor $w/(w-1)$ with an optimal factor $r(w)$ (solution of an equation) and randomizing the strategy by making the mantissa of the length of a foray uniformly random. This makes the *expected* competitive ratio nearly two times better.

The last problem shows how the concept of controlling the motion of an agent and receiving sensory input can be extended from subsets of a Euclidean space to graphs. However, unlike in Euclidean space, there does not exist a way to describe a location in an unknown graph that would be of any use to the algorithm. Therefore a typical problem in the graph context is that of searching, where the algorithm moves its agent in the graph until it reaches the target.

A deceptively simple generalization of the cow path problem is searching through a layered graph (as the number of the target can be given, this is in our nomenclature a navigation problem). As in the previous problem, the robot has the following senses: of traversed distance, of distinguishing the edges leaving a node and of encountering the target. The graph where the robot moves is now more general: the nodes are divided into layers $\{s\} = L_0, L_1, L_2, \ldots$ (where s is the starting point) each of size at most w (the width of the graph), while edges (of variable length) straddle adjacent pairs of layers. The target is one of the layers. The robot has also a new sense akin to vision: upon reaching layer L_i, it can map the subgraph induced by layers from L_0 to L_i. It closely resembles the situation described above for the version of the wall problem with a polylog competitive randomized solution.

The problem was introduced in [32] who provided a 9-competitive solution for $w = 2$ and an erroneous solution for larger widths. Fiat et al. [17] investigated this problem quite thoroughly. Their results include deterministic bounds on the competitive ratio of $\Omega(2^w)$ and $O(9^w)$ and a randomized lower bound of $w/2$. For the graphs that have the simple form from the cow path problem, they showed a tight bound of $\Theta(\log w)$. Ramesh [33] improves the deterministic upper bound of [17] to $O(w^3 2^w)$, thus matching the lower bound up to a polynomial factor. He also gives an $O(w^{13})$ randomized upper bound and an $\Omega(w^2/\log^{1+\varepsilon} w)$ randomized lower bound.

4 Exploration problems

A natural problem to be solved in an unknown environment is to create its map. Moreover, for a searching problem, sometimes the best strategy is to make a complete map of the environment: it is possible that the target is not noticed until the entire environment is inspected. This leads to exploration problems which were investigated both in subsets of the Euclidean plane and in graphs.

Deng and Papadimitriou [15] investigated the following model. The environment is a strongly connected directed graph G, and traversing an edge has a unit cost. When positioned on a node, the robot can sense the outgoing edges and identify each edge visited before. Therefore a tour of the robot accomplishes the task of mapping/exploration of G if and only if it traverses all of its edges. The authors showed that the *deficiency* d of G, defined as the minimal number of edges that have to be added to make G Eulerian, is a crucial parameter of the difficulty of the problem. For the case when $d = 0$ and 1 they show that the competitive ratios of 2 and 4 are achievable and necessary. For larger values of d, they show a lower bound $d/(2 \log d)$ and an algorithm with competitive ratio $3(d+1)^{2d+1}$. They also present a very simple randomized algorithm: "while there is an unexplored edge, follow the nearest one, breaking the ties randomly". They conjecture that this algorithm has a very good competitive ratio. Albers and Henzinger [1] improve the competitive ratio from $3(d+1)^{2d+1}$ to $(d+1)^6 d^{2 \log d}$, thus getting the competitive ratio down to subexponential.

Bender and Slonim [6] studied also the problem of exploring strongly connected directed graphs, but with the following assumptions. First, each node has outdegree d and the entrances to the edges leaving a node are labeled $1, \ldots, d$. Second, there is no regular way to distinguish whether a node was already visited. Third, the algorithm controls and receives input from two robots; crucially, the robots can sense each other when present on the same node. It can be trivially observed that a graph cannot be mapped in this model with a single robot. On the other hand the two-robot assembly can explore any graph in time $O(d^2 n^5)$. Bender and Slonim present also results on the exploration by a single robot with a constant number of pebbles (possible, but it takes exponential time), and on graphs with *high conductance*, where a randomized algorithm can be much more efficient.

Deng, Kameda and Papadimitriou [14] investigated later the exploration of geometric scenes with a robot endowed with a sense of location and vision. The environment is a polygon (possibly with holes) and a tour forms a solution if every point of the polygon is visible from some point on the tour. In computational geometry, this is called a *watchman tour*. The authors assume that the algorithm starts this tour at a specified point and, for technical reasons, that it also has to finish at a (possibly different) specified point. This extra requirement does not change the competitive ratios significantly. For simple rectilinear polygons their paper provided the lower and upper bounds of $(1 + \sqrt{2}/2$ and $2\sqrt{2}$. If the exit requirement is dropped, the upper bound improves to 2. Later Kleinberg [24] provided a randomized algorithm for this problem with expected competitive ratio 5/4. For arbitrary simple polygons, the authors sketch an argument that a greedy algorithm has the competitive ratio of 2016, and that for any fixed number of holes in the polygon, the competitive ratio is constant as well. A better strategy with competitive ratio below 133 was found by Hoffman *et al.* [19]; recently, this strategy has been further improved by the same set of authors [20] and the competitive ratio went down to $(18\sqrt{2} + 1) \leq 26.5$.

Kalyanasundaram and Pruhs [22] consider a problem of graph exploring with the following model. The graph is undirected and the edges have different lengths. Edges are like straight corridors with unmarked doors at either side. The robot has a sense of vision so when it visits a node, it can identify the edges already visited and learn the lengths of all adjacent edges. Therefore a tour forms a solution iff it visits all the nodes, in other words if it is a *traveling salesman tour*. An algorithm presented in this paper, called ShortCut, achieves the competitive ratio of 16.

5 Localization problems

This author once made a wrong turn on a highway near Boston and found himself on the corner of Maple Street and Broadway. Quick consultation with the map revealed that *every* town in the area has such a corner. Therefore he had to solve a *localization* problem: given the map of the environment, establish the location by moving and observing the local features. The problem arises naturally in robotics and guidance systems. Guibas *et al.* [18] addressed the related problem in computational geometry: given a map in the form of a simple polygon $\mathcal{P}$ (possibly, with holes), and the shape $\mathcal{V}$ of the part of $\mathcal{P}$ visible from some unknown point, find the list of possible locations of the unknown point. For simple polygons, they provide a data structure to answer such queries in optimal time; for general polygons (i.e. with holes), they provide a solution that is slower, but which does not require expensive preprocessing.

Later Kleinberg [25] considered the complexity of localization from the point of view of the necessary motions of the robot. Here the benchmark for the competitive ratio is the shortest path that allows one to pinpoint the initial location given the observations of the kind discussed by Guibas *et al.* [18] that can be performed on the path. Kleinberg discussed two kinds of geometric scenes where the robots move on narrow paths between obstacles (consequently, the "free space" forms a graph). In the first kind of environment, this graph forms a tree, and the competitive ratio obtained is $O(n^{2/3})$. In the second kind, the obstacles are oriented rectangles, and the competitive ratio is $O(n(\log n/\log\log n)^{-1/2})$. For both cases Kleinberg proved a lower bound of $\Omega(\sqrt{n})$.

It appears that the results of Kleinberg are of the "dismal" nature: even though the types of scenes he has considered appear rather simple, the competitive ratio is necessarily very high. A potentially more optimistic result was obtained by Dudek *et al.* [16]. The environment they consider is an arbitrary simple polygon $\mathcal{P}$. They measure complexity of an instance with the size of H, which is the locus of the points with the same shape $\mathcal{V}$ of the visible region as the initial placement of the robot. They show a strategy with competitive ratio $|H| - 1$ and prove that this is optimal.

6 Beyond competitive analysis

While competitive analysis has a very clear mathematical definition and practical motivation, other models are investigated as well. First, there exists the *excess distance ratio* model, where the benchmark is the sum of the perimeters of the obstacles in the scene, rather than the off-line optimum. In particular, this is the original model of Lumelsky and Stepanov. Angluin *et al.* [3] study the implications of depriving the robot of the ability to stop and measure its position in the middle of a monotone fragment of motion. In some cases the same ratio is achieved as before, e.g. in the problem of reaching a given point that is *assuredly* accessible, in some other cases the original task becomes impossible, for example the problem of finding consciously the obstacle that contains the target point.

In general, given that we have great flexibility in specifying the task of navigation: reaching a target, mapping, searching, localization, or a hybrid reaching/learning task of [10], the competitive analysis almost always is sufficient. Moreover, we could manipulate the notion of the cost. For example, Lumelsky and Stepanov, as well as Angluin *et al.* [3] measure the competitive ratio of the cost of reaching the target, where the cost includes a credit for mapping new obstacles.

Acknowledgements

I am grateful to Joseph Mitchell for many helpful comments on a preliminary version of this chapter.

References

1. S. Albers and M.R. Henzinger, *Exploring unknown environments*, Proc. 29th STOC, 416–425 (1997).
2. Ariadne and Theseus, *Path planning for a hero moving in an unknown labyrinth*, Proc. Panhellenic Symp. on Attic Antics and Antiquities (1300 B.C.).
3. D. Angluin, J. Westbrook and W. Zhu, *Robot Navigation with range queries*, Proc. 28th STOC, 469–478 (1996).
4. R. A. Baeza-Yates, J. C. Coulbertson and G. J. E. Rawlings, *Searching in the plane*, Information and Computation, 106, 234–252 (1993).
5. E. Bar-Eli, P. Berman, A. Fiat and P. Yan, *On-line navigation in a room*, Journal of Algorithms 17, 319–341 (1994).
6. M. A. Bender and D. K. Slonim, *The power of team exploration: two robots can learn unlabeled directed graphs*, Proc. 35th FOCS, 75–85 (1994).
7. P. Berman, A. Blum, A. Fiat, H. Karloff, A. Rosen and M. Saks, *Randomized robot navigation algorithms*, Proc. 7th SODA, 75–84 (1996).
8. P. Berman and M. Karpinski, *Randomized Navigation to a Wall through Convex Obstacles*, Bonn University Tech. Rep. 85118-CS (1994).
9. Y. Bartal, A. Blum, C. Burch and A. Tomkins, *A polylog(n)-competitive algorithm for metrical task systems*, Proc. 29th STOC, 711–719 (1997).
10. A. Blum and P. Chalasani, *An on-line algorithm for improving performance in navigation*, Proc. 34th FOCS, 2–11 (1993).
11. A. Blum, P. Raghavan, and B. Schieber, *Navigation in unfamiliar terrain*, Proc. 23rd STOC, 494–504 (1991).
12. A. Datta and C. Icking, *Competitive searching in a generalized street*, Proc. 10th ACM Symp. on Computational Geometry, 175–182 (1994).

13. A. Datta, C. Hipke and S. Schuierer, *Competitive searching in polygons–beyond generalized streets*, Proc. 6th ISAAC, 32–41 (1995).
14. X. Deng, T. Kameda and C. Papadimitriou, *How to learn an unknown environment*, Proc. 32nd FOCS, 298–303 (1991).
15. X. Deng, C. H. Papadimitriou, *Exploring an unknown graph*, Proc. 31st FOCS, 355–361 (1990).
16. G. Dudek, K. Romanik and S. Whitesides, *Localizing a robot with minimum travel*, Proc. 6th SODA, 437–446 (1995).
17. A. Fiat, D. Foster, H. Karloff, Y. Rabani , Y. Ravid and S. Vishwanathan, *Competitive algorithms for layered graph traversal*, Proc. 32nd FOCS, 288–297 (1991).
18. L. J. Guibas, R. Motwani and P. Raghavan, *The robot localization problem in two dimensions*, SIAM J. of Computing, 26, 1120–1138 (1997).
19. F. Hoffmann, C. Icking, R. Klein and K. Kriegel, *A competitive strategy for learning a polygon*, Proc. 8th SODA, 166–174 (1997).
20. F. Hoffmann, C. Icking, R. Klein and K. Kriegel, *The polygon exploration problem: a new strategy and a new analysis technique*, to appear in Proc. Workshop on the Algorithmic Foundations of Robotics, (1998)
21. C. Icking and R. Klein, *Searching for the kernel of a polygon - a competitive strategy*, Proc. 11th ACM Symp. on Computational Geometry, 258–266 (1995).
22. B. Kalyanasundaram and K. R. Pruhs, *Constructing competitive tours from local information*, Proc. 20th ICALP, 102–113 (1993).
23. R. Klein, *Walking an unknown street with bounded detour*, Computational Geometry: Theory and Applications, 1, 325–351 (1992).
24. J. M. Kleinberg, *On-line search in a simple polygon*, Proc. 5th SODA, 8–15 (1994).
25. J. M. Kleinberg, *The localization problem for mobile robots*, Proc. 35th FOCS, 521–531 (1994).
26. M. Kao, J. H. Reif and S. R. Tate, *Searching in an unknown environment: an optimal randomized algorithm for the cow path problem*, Proc. 4th SODA, 441–447 (1993).
27. H. J. Karloff, Y. Rabani and Y. Ravid, *Lower bounds for randomized server algorithms*, Proc. 23rd STOC, 278–288 (1991).
28. A. López-Ortiz and S. Schuierer, *Generalized streets revisited* Proc. 4th ESA, 546–558 (1996).
29. V. J. Lumelsky, A. A. Stepanov, *Dynamic path planning for a mobile automaton with limited information on the environment*, IEEE Transactions on Automatic Control, AC-31, 1058–1063, 1986.
30. V. J. Lumelsky and A. A. Stepanov, *Path-planning strategies for a point mobile automaton moving amidst unknown obstacles of arbitrary shape*, Algorithmica, 2, 403–430. (1987).
31. A. Mei and Y. Igarashi, *Efficient strategies for robot navigation in unknown environment*, Information Processing Letters, 52, 51–56 (1994).
32. C. H. Papadimitriou and M. Yannakakis, *Shortest paths without a map*, Theoretical Computer Science, 84, 127–150 (1991).
33. H. Ramesh, *On traversing layered graphs on-line*, Journal of Algorithms, 18, 480–512 (1995).
34. C. N. Shen and G. Nagy, *Autonomous navigation to provide long distance surface traverses for Mars rover sample return mission*, Proc. IEEE Symp. on Intelligent Control, 362–367 (1989).

11

On-line Network Routing

STEFANO LEONARDI

1 Introduction

The problems associated with routing communications on a network come in a large variety of flavors. Many of these problems would be uninteresting if link bandwidth was unlimited and switching speeds were imperceivable.

One often distinguishes between *circuit routing* and *packet routing*. The main difference between the two is that packet routing allows one to store transmissions (packets) in transit and forward them later whereas circuit routing as it is conceived does not. In fact, one can today perform streaming video (a virtual circuit) on IP (a packet routing protocol) using ATM virtual circuits that are themselves transmitted as packets, so confusion is inherently unavoidable. To simplify our discussion, in this chapter we deal with idealized models for circuit routing and packet routing.

A technological advance with significant consequences regarding routing is the use of optical fiber and optical switching elements. We will distinguish between electrical routing (the "standard model") and optical routing, the two models are quite different. Optical routing itself has several variant models, depending on the technological assumptions used. These differing assumptions range from available hardware to futuristic assumptions about what could possibly be done, all these assumptions are motivated by wavelength division multiplexing [28].

One obviously has to define the goal of a routing algorithm, *i.e.*, the objective function by which its performance will be measured. One possible objective is to reduce the *load* in the network, e.g., route calls so that the maximal load on any link is minimized. Load can be taken to mean different things in different contexts. In electrical routing networks, load could be taken to mean the utilized bandwidth on a link compared with the maximal link capacity, *i.e.*, the percent utilized. Reducing the load in this context means reducing the highest percent utilized, over all links. In the context of optical routing, load could be taken to mean the number of different wavelengths in use in the network. Reducing the load means routing the same calls while using less wavelengths.

We will present competitive on-line algorithms whose goal is to reduce the load defined as percent utilization of a link, or number of wavelengths required in an optical network. *I.e.*, algorithms whose load requirements will never be more than c times the load requirements of an optimal solution, where c is the appropriate competitive ratio.

A potential problem with the load measure is that nothing prevents the load requirements to exceed the (real) available resources. As long as no load requirement ever exceeds 100% the available resources, we are fine. Once the load requirement exceeds 100% of the available resources (link bandwidth, or wavelength capacity of optical fiber) this measure becomes problematic.

Obviously, whenever the resources required to perform some task exceed the available resources, something has to give. One possible interpretation of a routing algorithm that uses load greater than actually available is that the bandwidth allocated to every relevant call goes down, a slowdown is introduced and rather than give a call the bandwidth requested, the quality of service goes down.

In some cases this might make sense, but certainly not when performance guarantees are required (*e.g.*, streaming video). An alternative to reducing the quality of service is to reject (some) communication requests yet give others (those accepted) the quality of service (*e.g.*, bandwidth) they require.

The *throughput* objective function seeks to maximize the benefit derived from accepting communication requests, while remaining within the constraints of the available resources. Benefit can be various functions of the communication requests, such as a dollar value attached to every such request, a function of the geographical distance, etc. The primary benefit function considered in the algorithms described herein is the total number of communication requests serviced. *I.e.*, a benefit of one for every call accepted.

Generally, one can distinguish between *cost problems* where the goal of the algorithm is to reduce the cost of dealing with some sequence of events σ, and *benefit problems* where the goal of the algorithm is to increase the benefit associated with dealing with some sequence of events σ. The competitive ratio of an on-line algorithm ON for a cost problem is defined to be

$$\sup_{\sigma} \frac{\mathrm{Cost}_{\mathrm{ON}}(\sigma)}{\mathrm{Cost}_{\mathrm{OPT}}(\sigma)},$$

while for a benefit problem is defined to be

$$\sup_{\sigma} \frac{\mathrm{Benefit}_{\mathrm{OPT}}(\sigma)}{\mathrm{Benefit}_{\mathrm{ON}}(\sigma)},$$

where $\mathrm{ON}(\sigma)$ and $\mathrm{OPT}(\sigma)$ denote the on-line and the optimal solution for a sequence σ. When considering randomized algorithms, the definition of competitive ratio is in terms of the expected cost or benefit of the algorithm.

The paging problem is a cost problem, as are the load variants of the various routing problems. The throughput variants are benefit problems. Note that we adapt the convention that the competitive ratio is always greater or equal to one.

A typical throughput problem we would like to deal with is as follows:

Given a network, $G = (V, E)$, with link capacities $u : E \mapsto R$, the routing algorithm receives requests of the form: create a virtual circuit from $v \in V$ to $w \in V$ for a duration of d time units and of bandwidth b. The algorithm can do one of the following:

1. Assign the virtual circuit request (call) a route $v, x_1, x_2, \ldots, x_k, w$ in the network such that $(v, x_1) \in E$, $(x_i, x_{i+1}) \in E$, $(x_k, w) \in E$, and such that the utilized capacity of every one of these edges, including the new call, does not exceed the edge capacity.
2. Reject the call.

The algorithm that deals with the problem above should try to ensure that accepting the call and routing it via the route chosen will not create serious problems in the future in that it will make many other calls impossible to route. Because of the nature of the competitive ratio, what we really care about is that the set of calls to be accepted should be close (in benefit) to the set of calls accepted by the all powerful adversary.

Accepting and rejecting calls, if allowed in the model, is also known as *Call control* or *Admission control*. In some special networks *e.g.*, a single link from A to B, there are no real "routing" decisions to be made. All calls between A and B make use of the (single) link between them. Here, call control is the only consideration. However, we will call all these problems jointly routing problems, whether route selection is an issue or not, and whether call control is allowed or not.

Throughput competitive routing algorithms on electrical networks can be considered as belonging to one of two categories:

1. Low bandwidth requirement: all calls require no more than a small fraction of the minimal link capacity. Assuming the low bandwidth requirement allows us to present a competitive algorithm for any network topology. This algorithm is deterministic.
2. High bandwidth requirement: otherwise. Deterministic algorithms cannot work well on high bandwidth calls in any network topology. Randomized algorithms give good competitive ratios for some topologies, but not all.

There are a great many other issues and models that have been studied in the context of competitive routing. We will at least try to mention such work.

Despite the vast number of models and results, we can isolate certain combinatorial problems that underly many of the problems. The load problems on electrical networks and throughput competitive algorithms for low bandwidth calls are intimately related to the problem of multicommodity flow [34]. The high bandwidth throughput competitive algorithms, the load problems on optical networks, and the throughput problems on optical networks are all strongly connected to the problem of finding edge disjoint paths in a graph [21]. Optical routing, almost by definition, is related to the problem of coloring paths in graphs.

1.1 Structure of the chapter

Algorithms for electrical networks, and in particular for the virtual circuit routing problem, are covered in Section 2. In this section we will present competitive algorithms for the load balancing version and for the throughput version with small bandwidth requirement, and for the throughput version with high bandwidth requirement on specific topologies. Optical routing is presented in Section 3.

2 On-line virtual circuit routing

Given a weighted network $G = (V, E)$ of $|V| = n$ vertices and $|E| = m$ edges, where $u(e)$, $e \in E$, represents the capacity of link e, we can define many on-line throughput problems on this graph.

Perhaps the most general version of the various routing problems can be described as consisting of a sequence of 5-tuples, $\sigma = \{(s_j, t_j, r_j, d_j, b_j)\}$ where $s_j, t_j \in V$, $r_j \in R^+$, and $t_j, b_j \in Z^+$. Every 5-tuple $(s_j, t_j, r_j, d_j, b_j)$ represents a request for the establishment of a virtual circuit between s_j and t_j with bandwidth requirment r_j for a duration d_j. The b_j entry gives the benifit associated with this request. I.e., if this request is satisfied then b_j profit is made, otherwise no profit is made from this request.

In the *load balancing* version all the pairs s_j, t_j of the sequence must be connected with a virtual circuit. The only decision to be made is how to route the circuit, *i.e.*, which of the possibly many paths is to be chosen. The load of an edge is defined as the ratio between the sum of the transmission rates of the virtual circuits that include that edge and the capacity of the edge. The goal is to minimize the maximum load of an edge. Thus, for load balancing, there is no significance to the benifit element associated with a call, as all calls have to be routed.

The throughput version of this problem is also known as the *call control* problem, and was first studied from the point of view of competitive analysis by Garay and Gopal [22] and by Garay, Gopal, Kutten, Mansour and Yung [23]. Some justification for this model was the emerging ATM protocol, where a guaranteed quality of service is enforced by reserving the required bandwidth on a virtual circuit connecting the communicating parties on which the whole information stream is sent.

For throughput problems, two different decisions have to be made for every call:

1. *Admission control*, i.e., the choice of whether to accept the call or not, and

2. *Route selection*, i.e., deciding upon the route used to establish the virtual circuit of an accepted call.

In any case, one must obey the constraint that the sum of the transmission rates of the virtual circuits that cross a given edge do not exceed the capacity of that edge. The goal is to maximize the overall benefit obtained from the sequence of calls. We remark that the admission control issue is important even if route selection is trivial, *e.g.*, if the network only allows one path between every s_j, t_j pair.

The throughput version of the problem turns out to be considerably easier if the transmission rate of any call is always small when compared to the minimum capacity of an edge. In this case deterministic on-line algorithms that work for any network topology have been designed and proved to achieve a good competitive ratio [7].

The algorithm for the load balancing version and for the throughput version with small bandwidth requirement are both applications of a common idea. This idea, first used in the context of competitive algorithms for routing by Aspnes et al. in [1], is to associate every edge with a cost that grows exponential with the fraction of the capacity of that edge already assigned. For this reason we present these two algorithms and other direct applications of the use of the exponential cost function jointly in Section 2.1. (The exponential function also finds applications to on-line load balancing on parallel machines; cf. Chapter 8 of this book).

As mentioned before, the nature of the problem changes completely if the transmission rate requested by calls is a significant part of the minimum capacity of an edge. An appealing aspect of this case is that in its basic form (when link capacities are 1, call requests bandwidth 1, and durations are infinite), it is an on-line version of the extensively studied combinatorial problem of connecting a maximum number pairs with edge-disjoint paths. In this case deterministic algorithms can achieve only a very poor competitive ratio even for simple network topologies like lines or trees [7].

In Section 2.2 we describe randomized competitive algorithms for several network topologies and high bandwidth call requirements. The idea of using an exponential function actually plays a central role also in this context. Some of these special-topology algorithms use variations of the algorithm for the throughput version with small bandwidth requirement as a subroutine. However, there is no randomized on-line algorithm that is efficient for all the network topologies, since there are specific topologies where also randomized on-line algorithms achieve a very poor competitive ratio [15].

Many variations of the basic virtual circuit routing problem have been studied. Other network problems have also been studied using a similar approach. We will try to survey part of this work in Section 2.3.

2.1 The applications of the exponential function

An on-line virtual circuit algorithm based on the use of an exponential cost function was first presented by Aspnes, Azar, Fiat, Plotkin and Waarts [1] for the load balancing version (the AAFPW algorithm) and later used by Awerbuch, Azar and Plotkin [7] for the throughput version with small bandwidth requirement (the AAP algorithm).

The idea is to assign with every edge of the network a cost that is exponential in the fraction of the capacity of that edge assigned to on-going circuits. We describe an algorithm based on this idea and show how it applies to both load balancing and throughput.

For simplicity, we restrict the description to the case where calls are permanent, or equivalently have infinite duration. Moreover, without loss of generality we assume that the benefit b_j of call j, is in the range $[1, B]$.

Let $\mathcal{A}$ denote the set of calls accepted by the on-line algorithm and let $\mathcal{A}^*$ denote the set of calls accepted by the optimal off-line algorithm but rejected by the on-line algorithm. (Recall that in the load balancing version no call is ever rejected.) If the jth call is accepted, let $P(j)$ and $P^*(j)$ denote the path on which the virtual circuit for call j is established by the on-line and optimal algorithms, respectively.

Define $l_e(j) = \frac{\sum_{j \in \mathcal{A}, i<j: e \in P(j)} r_j}{u(e)}$ and $l_e^*(j) = \frac{\sum_{j \in \mathcal{A}^*, i<j: e \in P^*(j)} r_j}{u(e)}$, i.e., $l_e(j)$ gives the on-line load on edge e when the j-th call of the sequence is presented, and $l_e^*(j)$ represents the analogous value for the optimal algorithm.

Let $\Lambda = \max_{e \in E} l_e(f)$ and $\Lambda^* = \max_{e \in E} l_e^*(f)$ be the on-line and the optimal maximum load on a link of the network at the end of the sequence. The maximum load is 1 for the throughput version since the capacity of an edge cannot be exceeded. For the load balancing version the objective function is to minimze the maximum load.

Let μ be a parameter that we will specify later for the load balancing and the throughput algorithms. The algorithm for call j is the following:

1. Let $c_e(j) = \mu^{\frac{l_e(j)}{\Lambda}}$ be the cost of edge e when call j is presented.
2. Find path $P^{ON}(j)$ from s_j to t_j with minimum cost

$$C^{ON}(j) = \sum_{e \in P^{ON}(j)} \frac{r(j)}{u(e)} c_e(j).$$

3. If $C^{ON}(j) \leq 2mb_j$ then accept call j on path $P^{ON}(j)$, else reject call j.

Load Balancing. We describe the proof of Aspnes, Azar, Fiat, Plotkin and Waarts [1] of the competitive ratio of the algorithm for the load balancing problem. The algorithm achieves for load balancing on any network topology a maximum load that is at most an $O(\log n)$ factor greater than the optimum maximum load. Assume for the moment that the algorithm knows the optimal load Λ^*. We show that if we assign $\mu = 2m$, $\Lambda = 2\Lambda^* \log \mu$, then the algorithm does not reject any call and the load on every edge is at most Λ.

Consider the time at which call (s_j, t_j) is presented. The cost of any path $P(j)$ connecting s_j to t_j is at most $Z(j) = \sum_{e \in E} c_e(j)$, the sum of the exponential costs of all the edges of the network when call j is presented. We prove in the following that $Z(j) \leq 2m$, subject to accepting every call $i < j$ on a minimum cost path $P^{ON}(i)$. This immediately implies, $\forall e \in E$:

$$\mu^{\frac{l_e(j)}{\Lambda}} \leq Z(j) \leq 2m,$$

and then $l_e(j) \leq \Lambda$.

We use the following potential function to prove the claim :

$$\Phi(j) = \sum_{e \in E} c_e(j)(1 - \frac{l_e^*(j)}{2\Lambda^*}).$$

We have $Z(j) \leq 2\Phi(j) \leq 2(\Phi(j) - \Phi(0)) + 2m$. $Z(j)$ is then bounded by $2m$ if Φ does not increase when a call $i < j$ is accepted. The proof is as follows:

$$\begin{aligned}
\Phi(i+1) - \Phi(i) \leq & \sum_{e \in P^{ON}(i)} (\mu^{\frac{l_e(i+1)}{\Lambda}} - \mu^{\frac{l_e(i)}{\Lambda}}) \\
& - \frac{1}{2\Lambda^*} \sum_{e \in P^*(i)} (\mu^{\frac{l_e(i+1)}{\Lambda}} l_e^*(i+1) - \mu^{\frac{l_e(i)}{\Lambda}} l_e^*(i)) \\
= & \sum_{e \in P^{ON}(i)} (\mu^{\frac{l_e(i)+\frac{r_i}{u(e)}}{\Lambda}} - \mu^{\frac{l_e(i)}{\Lambda}}) \\
& - \frac{1}{2\Lambda^*} \sum_{e \in P^*(i)} (\mu^{\frac{l_e(i)+\frac{r_i}{u(e)}}{\Lambda}} (l_e^*(i) + \frac{r_i}{u(e)}) - \mu^{\frac{l_e(i)}{\Lambda}} l_e^*(i)) \\
\leq & \sum_{e \in P^{ON}(i)} (\mu^{\frac{r_i}{u(e)\Lambda}} - 1)\mu^{\frac{l_e(i)}{\Lambda}} - \frac{1}{2\Lambda^*} \sum_{e \in P^*(i)} \frac{r_i}{u(e)} \mu^{\frac{l_e(i)}{\Lambda}} \\
\leq & \sum_{e \in P^{ON}(i)} (2^{\frac{r_i \log \mu}{2u(e)\Lambda^* \log \mu}} - 1)\mu^{\frac{l_e(i)}{\Lambda}} - \frac{1}{2\Lambda^*} \sum_{e \in P^*(i)} \frac{r_i}{u(e)} \mu^{\frac{l_e(i)}{\Lambda}} \\
\leq & \frac{1}{2\Lambda^*} \sum_{e \in P^{ON}(i)} \frac{r_i}{u(e)} \mu^{\frac{l_e(i)}{\Lambda}} - \frac{1}{2\Lambda^*} \sum_{e \in P^*(i)} \frac{r_i}{u(e)} \mu^{\frac{l_e(i)}{\Lambda}} \\
\leq & \, 0.
\end{aligned}$$

The last two equations follow since $r_i \leq u(e)$, $2^x - 1 \leq x$ if $0 \leq x \leq 1$, and $P^{ON}(i)$ is the minimum cost path.

To complete the description of the result we are left to remove the assumption that the algorithm knows the optimal load Λ^*. This is done [1] by using a doubling technique based on the observation that if the algorithm exceeds load Λ then we do not have a correct estimation of Λ^*. We initially run a copy of the algorithm with Λ^* equal to the ratio between the transmission rate of the first call and the maximum capacity of an edge. Every time the algorithm exceeds the maximum load, we double the estimation of Λ^* and start a new copy of the algorithm. This results in a multiplicative factor of 4 in the competitive ratio of the algorithm.

The $O(\log n)$ competitive ratio is the best achievable up to a constant factor. A deterministic lower bound has been shown in [1] for a directed network. Bartal and Leonardi [16] show an $\Omega(\log n)$ randomized lower bound for an undirected network.

The algorithm can be extended to calls with limited duration. If D is the ratio between the maximum and the minimum duration, an $O(\log nD)$ competitive deterministic algorithm is presented in [13]. In this algorithm a distinct cost is associated with a link for every unit of time, to model the assignment of a link to different circuits for different units of time. A matching $\Omega(\log nD)$ lower bound for this algorithm is *not* known.

The problem changes dramatically if the duration of calls is not known in advance. In this case an $\Omega(n^{1/4})$ randomized lower bound is derived by the lower bound for load balancing on parallel machines with jobs of unknown duration proved by Azar, Broder and Karlin [12]. This lower bound is obtained on a sequence with exponentially growing holding times and therefore does not exclude the existence of an $O(\log D)$ competitive algorithm. An $\Omega(\min(n^{1/4}, D^{1/3}))$ randomized lower bound was later presented by Ma and Plotkin [38]. An $O(\sqrt{n})$ upper bound is given by [13]. Therefore, there is still a gap between the $O(\sqrt{n})$ upper bouund and the $\Omega(n^{1/4})$ lower bound for the problem.

Throughput. We give in the following the proof of Awerbuch, Azar and Plotkin for the competitive ratio of the AAP algorithm. Recall that the goal is to maximize the total benifit of calls accepted, subject to the capicity constraints on the edges. However, for the AAP algorithm to work we also require the following additional restriction on the ratio between the maximal transmission rate of a call and the minimal bandwidth of a link of the network:

$$\frac{1}{P} \leq \frac{r_j}{u(e)} \leq \frac{1}{\log \mu}. \tag{1}$$

We assume in the algorithm $\Lambda = 1$ and $\mu = 4mPB$.

We first show that when a new call $j \in \mathcal{A}$ is accepted by the algorithm, the capacity of the edges is not exceeded.

Lemma 1. *The solution given by the algorithm is feasible.*

Proof. Assume by contradiction that the capacity of edge $e \in P^{ON}(j)$ is violated for the first time when call $j \in \mathcal{A}$ is accepted. Since the solution was feasible when call j was presented, and $\frac{r_j}{u(e)} \leq \frac{1}{\log \mu}$, we have that $\forall e \in E$, $l_e(j) > 1 - \frac{1}{\log \mu}$. Then, for $C^{ON}(j)$ it holds that

$$C^{ON}(j) > \frac{r(j)}{u(e)} \mu^{1-\frac{1}{\log \mu}} \geq \frac{1}{P} \frac{\mu}{2} = 2mB,$$

a contradiction, since call j has been accepted. □

The next lemma gives an upper bound over the benefit obtained by the optimal algorithm on $\mathcal{A}^*$, as a function of the sum of the exponential costs of the edges in the network at the end of the sequence. Let $c_e(f)$ denote the cost of edge e at the end of the sequence.

Lemma 2. $\sum_{j\in\mathcal{A}^*} b_j \leq \frac{1}{2m}\sum_{e\in E} c_e(f)$.

Proof. All calls $j \in \mathcal{A}^*$ have been rejected by the on-line algorithm. We can then write:

$$\sum_{j\in\mathcal{A}^*} b_j < \frac{1}{2m}\sum_{j\in\mathcal{A}^*}\sum_{e\in P^*(j)} \frac{r_j}{u(e)} c_e(j)$$
$$\leq \frac{1}{2m}\sum_{e\in E} c_e(f) \sum_{j\in\mathcal{A}^*:e\in P^*(j)} \frac{r_j}{u(e)}$$
$$\leq \frac{1}{2m}\sum_{e\in E} c_e(f),$$

where the last equation follows since the edge capacity cannot be exceeded by any throughput algorithm.

□

The next lemma gives a lower bound on the benefit obtained by the on-line algorithm.

Lemma 3. $\frac{1}{2m}\sum_{e\in E} c_e(f) \leq (1+\log\mu)\sum_{j\in\mathcal{A}} b_j$.

Proof. It is easy to verify that it is sufficient to prove that for any call $j \in \mathcal{A}$, $\sum_{e\in P^{ON}(j)}(c_e(j+1) - c_e(j)) \leq 2mb_j \log\mu$. This follows from:

$$\sum_{e\in P^{ON}(j)} (c_e(j+1) - c_e(j)) = \sum_{e\in P^{ON}(j)} \left(\mu^{l_e(j)+\frac{r_j}{u(e)}} - \mu^{l_e(j)}\right)$$
$$= \sum_{e\in P^{ON}(j)} \mu^{l_e(j)} \left(2^{\log\mu \frac{r_j}{u(e)}} - 1\right)$$
$$\leq \log\mu \sum_{e\in P^{ON}(j)} \frac{r_j}{u(e)} \mu^{l_e(j)}$$
$$\leq 2mb_j \log\mu,$$

where the third inequality follows since $2^x - 1 \leq x$ if $0 \leq x \leq 1$, and the fourth inequality stems from the fact that call j has been accepted by the on-line algorithm.

□

We then conclude:

Theorem 4. *The algorithm for the throughput version of on-line virtual circuit routing is $O(\log\mu)$ competitive.*

Proof. Let OPT and ON denote the optimal and the on-line benefit, respectively, over the sequence. The benefit of the optimal algorithm OPT is no more than $\sum_{j \in \mathcal{A} \cup \mathcal{A}^*} b_j$, while the on-line benefit is $ON = \sum_{j \in \mathcal{A}} b_j$. From Lemmata 2 and 3 we derive $OPT \leq (2 + \log \mu) \sum_{j \in \mathcal{A}} b_j \leq (2 + \log \mu) ON$, thereby proving the theorem. □

The competitive ratio of an algorithm for this problem can actually be improved to $O(\log L)$ [7], where L is the diameter of the network. This algorithm is also easily extended to the case of calls of limited duration, at the expense of an additive $O(\log D)$ factor in the competitive ratio [7], where D is the ratio between the maximum and the minimum duration.

Awerbuch, Azar and Plotkin [7] also prove lower bounds for the throughput version of the problem. A matching randomized lower bound $\Omega(nPBD)$ on a line network shows that the proposed algorithm is optimal up to a constant factor with respect to the various parameters of the problem, when condition (1) is satisfied.

On the contrary, if the transmission rate of a call can be up to a fraction $\frac{1}{k}$ of the bandwidth of a link, the authors show an $\Omega(n^{1/k} + B^{1/k} + D^{1/k})$ lower bound on the competitive ratio of deterministic algorithms for a line network of n vertices [7].

Throughput competitive algorithms have not been proposed for the case in which the range of variation of the parameters of the problem is not known in advance.

Probabilistic Analysis and Experimental Results. Kamath, Palmon and Plotkin [30] study the virtual circuit routing problem in an intermediate model between worst case competitive analysis and probabilistic assumptions on the network traffic. The requests arrive following a Poisson distribution while call durations are exponentially distributed. However the matrix of the transmission rates requested between any pair of vertices of the network is unknown and chosen by the adversary.

Let r be the maximum ratio between the transmission rate of a call and the capacity of an edge, and let $\epsilon = \sqrt{r \log n}$. For the load balancing version, the authors propose algorithms that achieve a worst case ratio, over all the traffic matrices, between the expected algorithm's congestion and the expected optimal congestion bounded by $1 + O(\epsilon)$. For the throughput version, the authors prove a worst case rejection ratio of $R^* + \epsilon$, where R^* is the optimal expected rejection ratio. Observe that the performance of the algorithm improves if r is smaller, a behaviour that is not observed in the AAP algorithm.

Algorithms based on these ideas have been implemented and tested on commercial networks [24]. A less conservative approach has been followed for these implementations. For instance, a smaller value of μ is used for the exponential cost. The results obtained from the authors suggest that these strategies can behave better than greedy or reservation-based strategies in many practical scenarios.

Multicast Routing. On-line routing problems have been studied beyond point-to-point connection. Awerbuch and Azar [3] considered multicast routing. Several multicast groups are active in a network. A tree connects all the nodes registered in a single group to a source node that transmits information to all the members of the group. At every step, one of the nodes of the network asks to join one of the group. The goal is to update the tree while minimizing the maximum load on a link of the network, i.e. the maximum ratio between the sum of the bandwidth request of the active multicasts including an edge, and the capacity of that edge. The authors present an $O(\log n \log d)$ competitive algorithm, where d is the maximum number of nodes of the network that join a single group.

The throughput version of the multicast routing problem has first been addressed by Awerbuch and Singh [11]. In the throughput version the goal is to maximize the number of accepted subscriptions to a set of multicast groups that are active in the network. To accept a new subscription it is required to select a path that connects the request node of the network to a spanning tree rooted at the source of the multicast group. This problem is qualitative different from the simple point-to-point connection where the algorithm knows the benefit obtained from a single connection when the corresponding circuit is assigned. In the multicast case, the investment of deploying a multicast tree in a given area of the network may be compensated only in the future, when a number of subscriptions from nodes of that area will be issued. It is fairly easy to see that a deterministic algorithm cannot obtain good performances even though the bandwidth required by every connection is small when compared with link capacity.

The throughput version of the multicast routing problem obviously contains the virtual circuit routing problem. It also contains the following problem: A set of items are presented on-line. Every item belongs to one of different sets. The algorithm achieves a benefit only once it commits to one of the set. The benefit of the algorithm is the number of items in the selected set presented after the commitment. Awerbuch, Azar, Fiat and Leighton [5] introduced this problem, denoted as "picking the winner", and proposed randomized algorithms with polylogarithmic competitive ratio for its solution.

The algorithm proposed by Awerbuch and Singh [11] for the multicast routing problem is restricted to the case of connection requests requiring a bandwidth that is at most a logarithmic fraction of the capacity of a link, and it is restricted to input sequences where the subscriptions to different multicast groups are not interleaved. The algorithm uses ideas from the algorithms for the throughput version of virtual circuit routing and for the "picking the winner" problems, to achieve a competitive ratio polylogarithmic in the number of multicast groups and in the number of vertices of the network.

Since subscriptions to multicasts are not interleaved, the algorithm is involved at every time in the construction of a single multicast tree. The multicast routing problem in which subscriptions to different multicast groups are interleaved is later studied by Goel, Henzinger and Plotkin [27], that show an algorithm that can simultaneously build a tree for all the multicast groups active in the network, while maintaining a polylogarithmic competitive ratio for the problem. However, there is still a logarithmic gap between the $O(\log \mathcal{M}(\log n \log\log \mathcal{M}) \log n)$ upper bound and the $\Omega(\log n \log \mathcal{M})$ lower bound known for the problem, where $\mathcal{M}$ is the number of multicast groups active in the network. Multicast routing is still an open problem if connection requests can require a big fraction of the capacity of a link.

Further Applications of the Exponential Function. Algorithms based on the idea of associating to every distinct resource of the network a cost that is exponential in the fraction of the resource currently assigned, are designed for various network problems other than virtual circuit routing.

The algorithms described in this section for the on-line virtual circuit routing problem are centralized algorithm. Awerbuch and Azar [2] propose admission control and route selection distributed algorithms, and evaluate the quality of the solution and the number of steps required to compute it. They show how a distributed algorithm can achieve a logarithmic competitive ratio for the load balancing version of the on-line virtual circuit routing problem. They also study the flow control problem, where every pair is presented together a fixed path that connects the two endpoints. The goal is to decide how much traffic to route on every path connecting a pair, with the goal of maximizing the overall flow in the network.

An on-line variant of packet routing is considered by Awerbuch, Azar and Fiat [4]. Every packet injected in the network is specified by a source node, a destination node and a time of release. There are buffers of limited capacity at the nodes, while we assume that a packet takes one unit of time to cross a link. The goal is to minimize the delay of packets to reach their destination, while keeping limited the load on the links and the number of packets in the buffers. The authors present an algorithm that is $O(1)$ competitive with respect to the average and the maximum delay, while the capacity of the links and of the buffers is increased by at most an $O(\log nT)$ factor, where T is the maximum delay of the off-line schedule.

Leonardi and Marchetti-Spaccamela [35] show how a general algorithm based on the use of the exponential function solves an on-line version of positive linear programming, a model suitable for a large number of routing and scheduling problems in their load balancing and throughput version. A constraint is associated with every distinct resource, for instance the use of a link for one unit of time, and every of the requests can be satisfied if one among a set of combinations of the available resources is assigned to the request. The proposed algorithm is $O(\log m)$ competitive, where m is the number of distinct resources in the system.

2.2 The edge-disjoint path problem

We have described in the previous section efficient algorithms for the virtual circuit routing problem for any network topology for its load balancing version, and the throughput version restricted to the case of small bandwidth call request (see condition (1) of Section 2). In this section we describe throughput competitive algorithms when calls can request bandwidth up to the whole capacity of a link. In this case, a polynomial lower bound on the competitive ratio of deterministic on-line algorithms is given in [7], and holds even for a restricted topology, specifically for a line network. For some specific topologies such as expander graphs [32] deterministic competitive algorithms with logarithmic competitive ratio are possible.

A great deal of effort has been devoted to the study of *randomized* on-line algorithms for network topologies such as line, trees and meshes.

The throughput version of the on-line virtual circuit routing problem contains as a special case the on-line version of the edge-disjoint path problem. A general technique described below allows one to reduce the problem on a network with edges of uniform capacities to the on-line edge-disjoint path problem. Randomized logarithmically competitive on-line algorithms have been devised for a line network [23], trees [9, 10, 36], meshes [10, 33], and for "densely embedded, nearly Eulerian" planar graphs [33]. However the question of the existence of an efficient randomized algorithm for any network topology has been shown to be impossible, since a polynomial lower bound can be proved for a specific topology [15]. We will describe these results in the rest of this section.

From Virtual Circuit Routing to Edge-disjoint Path. Awerbuch, Bartal, Fiat and Rosèn [9] show how the design of randomized algorithms for the problem on a network with edges of *uniform capacities* can be reduced to the design of randomized algorithms for the edge-disjoint path problem. The parameters involved in the problem are the benefit, the duration and the bandwidth. The authors show how to deal with every one of these parameters at the expenses of a factor $O(\log \Delta)$ in the competitive ratio, where Δ is the ratio between the maximum and the minimum value of the parameter.

This result is obtained using a "classify and randomly select" technique. Assume that the individual benefit of a call is ranging between 1 and B, and that a c competitive algorithm is available if calls have equal benefits. We partition the interval $[1, B]$ into intervals $[2^i, 2^{i+1}]$, $i = 0, \ldots, \lceil \log B \rceil - 1$. A call is said of class i if its benefit is in the ith interval. The algorithm chooses an integer uniformly at random in the set $\{0, \ldots, \lceil \log B \rceil - 1\}$. A c competitive algorithm for calls of equal benefit is then applied to calls of class i when presented, while calls of other classes are rejected.

An $O(c \log B)$ competitive ratio for the whole algorithm can be shown as follows. Denote with ON_i and with OPT_i the benefit of the on-line and of the optimal algorithm when restricted to deal only with calls of the ith class. If the algorithm selects the ith class, then its benefit is at least $\frac{1}{2c} OPT_i$, since the benefits of calls of class i differ by at most a factor of 2. We then have for the overall algorithm:

$$E(ON) \geq \sum_{i=0}^{\lceil \log B \rceil - 1} E(ON_i)$$

$$\geq \frac{1}{2c}\frac{1}{\lceil \log B \rceil} \sum_{i=0}^{\lceil \log B \rceil - 1} OPT_i$$

$$\geq \frac{1}{2c\lceil \log B \rceil} OPT$$

A similar technique also allows to deal with varying duration and bandwidth requests. Lipton and Tomkins [37] devise a technique suitable for the case of parameters whose range of variation is not known at the beginning of the execution of the algorithm. This technique obtains an $O(\log^{1+\epsilon} \Delta)$ competitive ratio, with $\epsilon > 0$ arbitrarily small, where Δ is the range of variation of the parameter observed during the whole sequence.

At the present status of the art, no randomized algorithm for any of the network topologies studied until now, is able to deal with varying benefit, duration, and bandwidth requirement at the expenses of an additive logarithmic factor in the competitive ratio.

Edge-disjoint Path on Trees. The problem for trees has first been addressed by Awerbuch, Bartal, Fiat and Rosèn [9]. They propose an $O(\log n)$ competitive algorithm, also based on the classify a randomly select technique.

All the vertices of the tree are partitioned into $O(\log n)$ different classes, on the basis of a recursive application of a balanced tree separator.

A balanced tree separator [41] is a vertex whose removal disconnects the tree into pieces of at most $\frac{2}{3}n$ vertices. A separator for the initial tree $T = T_0$ is called a level 0 separator. Removing the separator from the tree leaves us a forest of trees of level 1, each of which contains some separator of level 1. In general, we recurse at most logarithmically many times until we are left with constant sized trees.

Calls are then partitioned in $O(\log n)$ classes. A call is assigned to class i if the vertex of lowest level in the path connecting the two endpoints has level i. One of the classes is chosen at random at the beginning, and the algorithm accepts only calls of that class using a greedy algorithm strategy. The greedy strategy accepts a call if it does not intersect any previously accepted call. Two calls are said to be consistent if they share no edges. When restricted to calls of a class i, the greedy algorithm is 2 competitive since a call of class i includes at most two edges that are adjacent to a vertex of level i, and therefore could prevent at most 2 other consistent calls of level i from being accepted.

Following the type of analysis presented in the previous paragraph, the algorithm can be easily proved to be $O(\log n)$ competitive.

Borodin and El-Yaniv [18] mention that for the specific case of trees an $O(\log n \log P)$ competitive algorithm exists, where calls can have arbitrary bandwidth requests and links have arbitrary capacity, and where P is the ratio between the maximum and the minimum fraction of the link capacity required by a call. A similar result it is not known for general networks.

An algorithm with $O(\log D)$ competitive ratio, where D is the diameter of the tree, was given by Awerbuch, Gawlick, Rabani and Leighton [10]. A matching $\Omega(\log D)$ randomized lower bound is given for trees of diameter D [9, 10].

These algorithms have an asymptotically good competitive ratio, but suffer the major drawback that they achieve a good solution only with small probability, while the main contribution to the average is given by a very high benefit achieved with low probability. An $O(\log D)$-competitive algorithm that achieves a benefit close to the expectation with good probability is proposed by Leonardi, Marchetti-Spaccamela, Presciutti and Rosèn [36].

In a first step, an on-line deterministic filter is applied to the input sequence. Every call that is not filtered out is called a *candidate call*. The authors prove the existence of an on-line deterministic filter that produces a set of candidate calls $\mathcal{C}$ whose number is at least 1/4 of the optimal solution. Moreover, every candidate call is shown to intersects at most $O(\log D)$ candidate calls that occur earlier in the sequence. (Such a filter can also be based on the AAP algorithm when the on-line algorithm is assumed to have logarithmic more capacity than the optimal solution.) The set $\mathcal{C}$ forms a new input sequence to an on-line randomized selection procedure that accept every candidate call with probability $p = O(\frac{1}{\log D})$, if not intersecting any previously considered candidate, otherwise reject.

For an appropriate value of p, the algorithm is shown to have $O(\log D)$ competitive ratio and achieves any constant fraction of the expected benefit with at least constant probability. For a different tuning of p, the algorithm has a competitive ratio of $O(\log^{1+\epsilon} D)$, for arbitrary $\epsilon > 0$, but achieves any constant fraction of the expected benefit with probability that tends asymptotically to 1 as the size of the optimal solution and D grow.

Edge-disjoint Path on Meshes. An $O(\log n)$ competitive algorithm for meshes is proposed by Kleinberg and Tardos.

Awerbuch et al. [10] gave a previous $O(\log n \log\log n)$ algorithm and an $\Omega(\log n)$ randomized lower bound.

The algorithm proposed by Kleinberg and Tardos, KT in the following, partitions the $n \times n$ mesh into $\frac{n^2}{\log^2 n}$ squares of size $\log n \times \log n$. Calls that have both endpoints in the same square are *short calls*, calls that have the two endpoints in different squares are *long calls*. KT first tosses a coin and decides to accept only short calls, or only *long calls*.

The basic idea of KT is to deal with long calls using an auxiliary network, called "simulate network", with logarithmic capacity on the edges, and running a version of the AAP algorithm on this network.

The simulate network is constructed by associating a vertex with every submesh, and connecting with edges vertices associated with adjacent submeshes. The logarithmic capacity on the edges is to model that at most a logarithmic number of edge-disjoint paths can be routed through adjacent submeshes.

A part of the mesh selected at random is then dedicated to a crossbar structure used to route long calls.

Short calls are routed using a greedy algorithm that is easily proved to be $O(\log n)$ competitive.

Long calls are rejected if one of the endpoints are within the crossbar structure because all the paths leading to this endpoint can be easily blocked from other calls routed in that region. A long call is also rejected if a previous long call with an endpoint in the same submesh has been accepted. This is possible while still keeping a logarithmic competitive ratio since at most a logarithmic number of long calls with an endpoint in a given submesh can be accepted in a solution. If not, the call in the original mesh is transformed into a call in the simulate network between the two vertices associated with the two submeshes that contain the two endpoints of the call.

The sequence of calls in the simulate network so formed is the input of the $O(\log n)$ competitive AAP algorithm. The algorithm returns a path in the simulate network for those calls that are accepted. The path in the simulate network indicates a sequence of submeshes to cross to connect the two endpoints. This path is then transformed into a path in the original mesh that is edge-disjoint with all the previously accepted calls.

Kleinberg and Tardos extend the ideas in this algorithm to obtain an $O(\log n)$ competitive algorithm also for "densely embedded, nearly Eulerian" planar graphs [33].

Leonardi, Marchetti-Spaccamela, Presciutti e Rosèn [36] give a variant of the Kleinberg-Tardos algorithm that obtains any constant fraction of the expected solution with probability that tends to 1 as the size of the optimal solution grows.

A Lower Bound on Randomized Algorithms for Arbitrary Networks. The obvious question was if there exists an efficient, polylogarithmic, randomized competitive algorithm for any network topology. Bartal, Fiat and Leonardi [15] proved an $\Omega(n^\epsilon)$ lower bound on on-line edge-disjoint path for a specific network.

An $\Omega(n^\epsilon)$ lower bound is first devised for an on-line version of the independent set problem. A graph G from which the vertices of the sequence are drawn is known since the beginning to the algorithm. Every time a new vertex is presented, the algorithm can accept the vertex only if it is not adjacent to any previously accepted vertex.

The lower bound for the on-line independent set problem is then transformed into a lower bound for the on-line edge-disjoint path problem. This is done by embedding the graph and the input sequence for the independent set lower bound into a set of calls on a planar network of degree 3 shaped as a "brick wall". Every vertex of the graph is mapped into a call in the network. The embedding satisfies the property that the vertices of any independent set of the graph are mapped into a set of calls that can be connected with mutually edge-disjoint paths, while two adjacent vertices of the graph are mapped into two calls that can only be connected with edge-intersecting paths. An $\Omega(n^\epsilon)$, with $\epsilon = 2/3(1 - \log_4 3)$ is devised in this way for the problem.

The best upper bound known for this problem on a general network is $O(\sqrt{m})$ [17].

2.3 Other models

In the previous sections we have described on-line virtual circuit routing algorithms where connection requests are accepted or rejected when presented, and the benefit obtained is either arbitrary or equal for all the communication requests.

In this section we describe the results obtained for different models. I.e., calls can be preempted and/or rerouted at some later time after their acceptance, or the benefit accrued from a call can be proportional to the amount of resources assigned to this call.

The competitiveness of on-line call control algorithms was first studied by Garay and Gopal [22] when call preemption is allowed and the benefit of a call is its holding time. Calls that are accepted can be later preempted, while calls that are rejected cannot be resumed in the future. Call preemption may compensate in part for mistakes made by the on-line algorithm. However, the authors prove that if the holding time of a call is unknown at the time the call request is presented, no competitive algorithm can exist for this model. On the other hand, if the benefit obtained from a call until its preemption is not lost, an on-line algorithm is able to perform as well as the optimal solution.

In many practical scenarios this is not the case, if a call is interrupted, the corresponding gain is lost, or the connection is started again from the beginning. Garay, Gopal, Kutten, Mansour and Yung [23] consider this problem on a line network for different benefit functions. If the benefit obtained from calls that are terminated is the number of links of the route, the so called telephone model, then they present an algorithm with constant competitive ratio equal to $\frac{1}{2g+1}$ where g is the golden ratio. This is obtained by the following simple strategy: the algorithm accepts a new call if its benefit is at least g times the benefit lost for preempting calls. If the benefit of a call is the holding time of the call, then a competitive ratio of $1/4$ is achieved. If the benefit is constant for every call, independently from the length of the path connecting the endpoints, the competitive ratio is only logarithmic in the number of vertices of the network. They also prove that no on-line deterministic algorithm can achieve a better ratio.

Further work for preemptive algorithms has been done by Bar-Noy, Canetti, Kutten, Mansour and Schieber [14] assuming that the benefit obtained from calls that are terminated is the product bandwidth-holding time. They show how the best competitive ratio of an algorithm is parameterized by the maximum fraction δ of the bandwidth of a link that is requested by a single call. They first study the problem on a single link and present deterministic algorithms with almost optimal constant competitive ratio as soon as $\delta < 1$, while for $\delta = 1$ they prove that deterministic algorithms cannot achieve a bounded competitive ratio. Similar results are obtained on a line network if $\delta \leq 1/2$ and calls have constant duration. Canetti and Irani [20] prove a randomized logarithmic preemptive lower bound even for a single link, if calls have arbitrary durations and can potentially obtain the whole bandwidth ($\delta = 1$). Whether randomized preemptive algorithms for a line network with calls of constant benefit and infinite duration can achieve a competitive ratio better than a logarithmic factor is still an open problem.

Bartal, Fiat and Leonardi [15] prove an $\Omega(n^{\epsilon})$ preemptive randomized lower bound for arbitrary networks when the benefit of every call is constant, thus showing that it does not exist a polylogarithmic call control algorithm for any network even if preemption and randomization are allowed.

Finally, we mention the possibility to reroute virtual circuits. This has been proved useful when one seeks for minimizing the load in a network generated by calls with unknown holding times. Awerbuch, Azar, Plotkin and Waarts [8] prove that if an $O(\log n)$ number of rerouting is allowed, it is possible to design an $O(\log n)$ competitive algorithm for load balancing in arbitrary topology networks.

3 On-line routing in optical networks

On-line wavelength routing problems in all-optical networks have motivated several works in the area of on-line routing. All-optical communication networks exploit the optical technology for both switching and communication functions.

Information is transmitted as light rays from source to destination without any electronic conversion in between. The Wavelength Division Multiplexing technology (WDM) consists in partitioning the available bandwidth on every link in many channels, each at a different optical wavelength. This allows the parallel transmission on an optic fiber link of different data streams, with speed related to the assigned wavelength.

The main constraint the wavelength allocation must obey is that for a link, on a given wavelength, only one signal can be transmitted. Two data streams transmitted on a link must be assigned with different wavelengths. Conversion of data between different wavelengths is limited by the current technology. If we assume that wavelength conversion is not allowed, a data stream is then assigned to a single wavelength between the transmitter and the receiver.

In general a WDM network consists of routing photonic switches connected through point-to-point optic fiber links. In the following we will mainly consider *switched* optical networks. In these networks, routing switched are able to direct any pair of incoming data streams carried on the same wavelength to any pair of outgoing links where such wavelength is available.

In contrast, *switchless* optical networks have been considered. In this model, the routing pattern at every switch is fixed for every wavelength, meaning that every data stream entering the routing node on a given wavelength is directed to a predetermined fixed subset of the outgoing edges on which the wavelength is available. As a result, a wavelength cannot be used on all the links reached by the propagation of the light ray of an ongoing communication.

Unless specified otherwise, we will refer in the following to the switched model.

The Model. We model an optical network as a graph, whose vertices are switching routers with possible connected terminals, and links are optic fibers. Edges representing optic fibers should be directed in one single direction as a matter of fact that optic amplifiers are directed devices. This is reflected by modeling every optical link with two directed links in opposite direction rather than a single undirected link. However, many of the results presented for the undirected model, easily extend under certain restrictions (usually at the expenses of a constant factor in the competitive ratio) to the directed bidirectional model. Finally, the same set of wavelengths is considered potentially available on all the links of the network.

Every communication request in a sequence specifies a pair of vertices (s, t). An algorithm for wavelength routing in general deals with two kind of problems: *Wavelength selection* and *Routing assignment.* The wavelength selection problem consists in selecting a wavelength w for the communication. The routing assignment problem consists of choosing a path connecting node s to t such that every link of the path is not assigned to any other communication on wavelength w.

The routing problem on WDM switched networks has been often referred to as *path coloring.* Given a graph $G = (V, E)$ representing the network, we are given a sequence (s_i, t_i) of requests consisting of pairs of vertices of the graph G. The algorithm must assign to every pair (s_i, t_i) a color (wavelength), and a path connecting s_i to t_i such that two paths sharing an edge are not associated with pairs of same color.

As for the *virtual circuit routing* problem, a *load* version and a *throughput* version have been studied.

In the load version, all the communication requests must be accepted with the goal of minimizing the overall number of colors used.

In the throughput version, the number of available colors is considered fixed, while the goal is that of maximizing the number of communications that can be accepted given the network topology and the number of available wavelengths.

Although path coloring is at the basis of the study of optical routing, a number of additional features can be added to the problem. Communications can have limited durations, or, in the throughput version, varying benefits. We will not discuss further this second problem, that can be solved using an approach similar that used for virtual circuit routing [9].

3.1 The load version of path coloring

In the load version of the path coloring problem, all the communication requests must be accepted. A color and a path must be assigned to every call, with the goal of minimizing the number of used colors.

A natural question is the relation between the path coloring problem and the load balancing problem in networks of Section 2.1, where we seek to minimize the maximum number of paths crossing an edge. Given an instance for path coloring, e.g. a sequence of pairs of vertices in a network, the optimal solution for the corresponding load balancing problem is a clear lower bound on the optimal number of colors required for the path coloring problem. The size of the optimal load and of the optimal number of colors are actually equal for those topologies where a pair of vertices is connected by one single path (e.g. a line or a tree), while there are examples of networks where they can be as different as an $O(n)$ factor apart.

However, all the algorithms presented in the following for various network topologies are shown to be competitive against an adversary for which the optimal load is used as a lower bound on the optimal solution. For this reason, we will denote this measure by Λ^* throughout this section.

The on-line path coloring problem has been studied for the following network topologies: line, rings, trees, meshes. Efficient competitive algorithms have been devised for all these network topologies.

In contrast, an $\Omega(n^\epsilon)$ lower bound on the competitive ratio of randomized algorithms working for arbitrary network topologies has been shown by Bartal, Fiat and Leonardi [15]. This lower bound is achieved in a way similar to the lower bound for the on-line edge-disjoint path problem (see Section 2.2). A lower bound for the on-line graph coloring problem is first established and then it is turned into a lower bound for on-line path coloring in a network shaped as a brick wall.

Path Coloring on the Line. It turns out the the on-line path coloring problem on an undirected line network has been studied even before the notion of competitive analysis has been formally introduced. This problem is equal to the *on-line interval graph coloring* problem.

Every vertex of an interval graph is mapped to an interval on the line. Two vertices of the graph are adjacent if and only if the corresponding intervals are overlapping. Since an interval graph is a perfect graph, the optimal solution uses exactly Λ^* colors, and can be found in polynomial time for example with a simple divide and conquer technique.

The problem of coloring on-line an interval graph has been first studied by Kierstead and Trotter [31]. They present an optimal on-line deterministic algorithm whose competitive ratio is at most $3\Lambda^* - 2$ and prove that no deterministic algorithm can achieve a better ratio. These results immediately extend to path coloring on a line. More details on on-line interval graph coloring can be found in Chapter 13 of this book. If randomized algorithms can achieve a competitive ratio better than deterministic algorithm for path coloring on a line network is still unknown.

Path Coloring on Rings. In the wavelength routing problem on rings a communication between a pair of vertices can be routed with one of the two possible paths connecting the two vertices. A simple approach to the problem used by several authors (for instance [25] and [39]) consists of cutting the ring at any edge and then solving the problem on a line network. The number of paths that cross this edge in an optimal solution is at most equal to the optimal load. Then, the load on any other edge in the resulting problem on a line network is increased by at most Λ^* and then the optimal load is at most doubled. An algorithm for the line can now be applied yielding an algorithm for the ring whose competitive ratio is away from the competitive ratio of the algorithm for the line by at most a multiplicative factor of 2.

If every pair can be connected with only one of the two possible paths, then the problem reduces to the *circular arc graph coloring* problem. In a circular arc graph, every vertex can be mapped on on an arc of a ring, and two vertices are connected with an edge if and only if the two arcs are overlapping. For this problem, Šlusarek [40] has shown that an on-line algorithm can use a number of colors bounded by $3\Lambda^* - 2$, then yielding the same result for interval graph coloring.

Path Coloring on Trees. An $O(\log n)$-competitive algorithm for trees has been proposed by several authors [16, 19, 26]. The problem of on-line path coloring on a tree can be reduced to the problem of coloring on-line an $O(\Lambda^*)$-inductive graph. A graph is d-inductive if the vertices can be associated with numbers 1 through n in a way that every vertex is connected to at most d vertices with higher numbers. Given an instance of the path coloring problem on a tree, the so called *intersection graph* is built by associating a vertex with every path, and connecting two vertices with a link if the two corresponding paths are intersecting. The intersection graph can be shown to be $2(\Lambda^* - 1)$-inductive [16]. An algorithm by Irani [29] that shows how to color on-line a d inductive graph with $O(d \log n)$ colors can then be applied to yield an $O(\log n)$ competitive algorithm.

An almost matching deterministic $\Omega(\frac{\log n}{\log \log n})$ lower bound has also been proved by Bartal and Leonardi [16]. This lower bound has been obtained on a tree of depth $\log n$. This shows that an $O(\log \Delta)$ lower bound, where Δ is the diameter of tree, is not possible at least for deterministic algorithms. (In contrast, recall that an $O(\log \Delta)$ randomized competitive algorithm for edge disjoint paths on trees is possible). However, this lower bound does not exclude the existence of an on-line algorithm that uses $\Lambda^* + O(\log n)$ colors, since the optimal solution on the input sequence for the lower bound uses only 2 colors. An interesting open problem is to determine if randomized algorithm can beat deterministic algorithms for path coloring on trees.

Wavelength Assignment with Limited Duration. A natural variation of the problem consists of considering connection requests of limited duration. Pairs of vertices arrive and depart in an on-line fashion. A color must be assigned to the pair when the call arrives, the color can be reused after the call is finished. An algorithm with logarithmic competitive ratio has been devised for this problem for line, tree and ring topologies from Gerstel, Sasaki and Ramaswami [26].

The algorithm for line and tree topologies uses a balanced separator-based idea similar to that for the $O(\log n)$ algorithm for call control on trees [9]. In a preprocessing phase the graph is recursively separated through a balanced edge separator. The tree is broken in many pieces and a specific set of colors is reserved to those paths crossing the edge separator. A similar procedure is then applied recursively for a depth at most logarithmic until we end with trees formed by single edges.

Let Λ^* be the maximum number of pairs crossing a given edge that are active at the same time. One can prove that Λ^* colors for every level of separation are enough to color all the calls at their arrival. Λ^* is also an obvious lower bound on the optimal solution, then yielding the $O(\log n)$ competitive algorithm. The algorithm can also be extended to rings using again the trick of cutting the ring at any link.

Path Coloring on Meshes. The problem of path coloring on meshes has been approached by Bartal and Leonardi [16] using the idea of partitioning the meshes into submeshes of logarithmic size. This follows the solution proposed by Kleinberg and Tardos for the on-line edge-disjoint path problem on meshes [33]. However, in this case all the calls must be assigned with a color, while in the edge-disjoint path problem rejection is allowed.

Calls are divided between *short calls*, with both endpoints in the same submesh, and *long calls*, with endpoints in different submeshes. Disjoint set of colors are dedicated to long calls and short calls.

Short calls are routed through a shortest path connecting the two vertices. Short calls in different submeshes are then non-conflicting, and the colors can be reused in different submeshes. The colors are assigned to short calls in a greedy manner according to their relative length. This first part of the algorithm yields an $O(\log n)$ competitive ratio with respect to an optimal algorithm for short calls.

The part of the algorithm that deals with long calls transforms the problem into a *path coloring problem with more bandwidth* in a *simulated network* where every fiber optic connection between two vertices is formed by a link with logarithmic capacity. This results in the fact that a logarithmic number of paths with same color can include a link.

A vertex of the simulated network is associated with every submesh, while two vertices are connected by an edge with logarithmic capacity if the two corresponding submeshes are adjacent. This is to model that at most a logarithmic number of paths with same color can be routed through two adjacent submeshes.

An $O(\log n)$-competitive algorithm for the path coloring problem with logarithmic bandwidth is then obtained. This algorithm uses an exponential function as in the algorithm for on-line load balancing (see Section 2.1. The proposed algorithm is an $O(\log n)$ competitive algorithm on general networks if a logarithmic bandwidth is allowed on any color.

The instance in the original network is transformed into an instance in the simulated network by replacing every call in the original mesh with a call between the two vertices in the simulated network that represent the two submeshes containing the two vertices of the call. Every call in the simulated network is then given to the algorithm for path coloring with more bandwidth, that assigns a color and a route in the simulated network. This route is then converted on-line into a route in the original mesh preserving the property that paths with same color are edge-disjoint.

The part of the algorithm dealing with long calls has also a logarithmic competitive ratio then yielding together with the algorithm for short calls an $O(\log n)$ competitive algorithm for the problem.

The $\Omega(\log n)$ lower bound on the competitive ratio of randomized on-line algorithms for the on-line load balancing problem for virtual circuit routing [16] also implies that the algorithm for path coloring on meshes is optimal up to a constant factor.

3.2 The throughput version of path coloring

The throughput version of path coloring consists of maximizing the number of communication requests that can be accepted for a given set of wavelengths and a given network topology. Let us assume that for a given network topology, a competitive algorithm is available for the on-line edge-disjoint path problem described in Section 2.2. This can also be considered an algorithm for the throughput version of the path coloring problem when one single wavelength is available. A simple technique proposed by Awerbuch, Azar, Fiat, Leonardi and Rosèn [6] allows one to transform an algorithm for a single wavelength into an algorithm for the multi wavelength case at the expenses of an additive term of 1 in the competitive ratio.

The algorithm uses a first fit approach. Every one of the Λ colors are assigned with a number from 1 to Λ. A different copy of the algorithm for on-line virtual circuit routing is then executed for every color. Every time a new call is presented, it is given as an input to the algorithm for the first color. If this algorithm accepts the call, then it is assigned with the first color and the route chosen by the algorithm. If the algorithm for the first color rejects the call, this is given as input to the algorithm for the second color, and so on until the call is eventually accepted by an algorithm for a color, or rejected by all the colors. In this last case the call is rejected by the overall algorithm.

Let c be the competitive ratio of the base algorithm (deterministic or randomized) for virtual circuit routing. The algorithm described above for the throughput version of on-line path coloring achieves a competitive ratio of at most $c+1$. This allows to state the existence of an $O(\log n)$ competitive algorithm for those topologies for which there exists a competitive $O(\log n)$ competitive algorithm for on-line virtual circuit routing.

If we look to algorithms for general networks, an $\Omega(n^{\epsilon})$ lower bound immediately follows from the analogous lower bound for the on-line virtual circuit routing problem [15].

The throughput version has also been studied for other models of optical networks, namely switchless optical networks [6]. An $O(\log n)$ randomized competitive algorithm for trees has been obtained by devising an $O(\log n)$-competitive algorithm for the single wavelength case and then extending it to deal with an arbitrary number of wavelengths.

4 Conclusions

In this chapter we have described competitive on-line algorithms for on-line network routing problems. We have concentrated on routing in electrical and optical networks, presented algorithms for load minimization and throughput maximization problems, and mentioned some of the most popular open problems in the area.

We have often referred to idealized models of communication networks. It is still open the issue of using many of these ideas to design more efficient algorithms in practical existing networks. A first relevant issue in this direction is to devise efficient distributed implementations of competitive on-line network routing algorithms.

Acknowledgements

This work was partly supported by EU ESPRIT Long term Research Project ALCOM-IT under contract n. 20244, and by Italian Ministry of Scientific Research Project 40% "Algoritmi, Modelli di Calcolo e Strutture Informative". Part of this work has been done while the author was visiting the Max-Planck Institute für Informatik, Im Stadtwald, 66123 Saarbrücken, Germany.

References

1. J. Aspnes, Y. Azar, A. Fiat, S. Plotkin, and O. Waarts. On-line load balancing with applications to machine scheduling and virtual circuit routing. In *Proceedings of the 25th ACM Symposium on the Theory of Computing*, pages 623–631, 1993.
2. B. Awerbuch and Y. Azar. Local optimization of global objectives: competitive distributed deadlock resolution and resource allocation. In *Proceedings of the 35th Annual IEEE Symposium on Foundations of Computer Science*, pages 240–249, 1994.
3. B. Awerbuch and Y. Azar. Competitive multicast routing. *Wireless Networks*, 1:107–114, 1995.
4. B. Awerbuch, Y. Azar, and A. Fiat. Packet routing via min-cost circuit routing. In *Proceedings of the 4th Israeli Symposium on Theory of Computing and Systems*, pages 37–42, 1996.
5. B. Awerbuch, Y. Azar, A. Fiat, and T. Leighton. Making commitments in the face of uncertainty: How to pick a winner almost every time. In *Proceedings of the 28th Annual ACM Symposium on Theory of Computing*, pages 519–530, 1996.
6. B. Awerbuch, Y. Azar, A. Fiat, S. Leonardi, and A. Rosén. On-line competitive algorithms for call admission in optical networks. In *Proceedings of the 4th Annual European Symposium on Algorithms, Lecture Notes in Compter Science 1136*, pages 431–444. Springer-Verlag, 1996.
7. B. Awerbuch, Y. Azar, and S. Plotkin. Throughput-competitive online routing. In *Proceedings of the 34th IEEE Symposium on Foundations of Computer Science*, pages 32–40, 1993.
8. B. Awerbuch, Y. Azar, S. Plotkin, and O. Waarts. Competitive routing of virtual circuits with unknown duration. In *Proceedings of the 5th ACM-SIAM Symposium on Discrete Algorithms*, pages 321–327, 1994.
9. B. Awerbuch, Y. Bartal, A. Fiat, and A. Rosén. Competitive non-preemptive call control. In *Proceedings of the 5th ACM-SIAM Symposium on Discrete Algorithms*, pages 312–320, 1994.
10. B. Awerbuch, R. Gawlick, T. Leighton, and Y. Rabani. On-line admission control and circuit routing for high performance computing and communication. In *Proceedings of the 35th Annual IEEE Symposium on Foundations of Computer Science*, pages 412–423, 1994.
11. B. Awerbuch and T. Singh. On-line algorithms for selective multicast and maximal dense trees. In *Proceedings of the 29th Annual ACM Symposium on Theory of Computing*, pages 354–362, 1997.

12. Y. Azar, A. Broder, and A. Karlin. On-line load balancing. In *Proceedings of the 33rd Annual IEEE Symposium on Foundations of Computer Science*, pages 218–225, 1992.
13. Y. Azar, B. Kalyanasundaram, S. Plotkin, K. Pruhs, and O. Waarts. Online load balancing of temporary tasks. In *Proceedings of the 3rd Workshop on Algorithms and Data Structures*, LNCS, pages 119–130. Springer-Verlag, 1993.
14. A. Bar-Noy, R. Canetti, S. Kutten, Y. Mansour, and B. Schieber. Bandwidth allocation with preemption. In *Proceedings of the 27th ACM Symposium on Theory of Computing*, pages 616–625, 1995.
15. Y. Bartal, A. Fiat, and S. Leonardi. Lower bounds for on-line graph problems with application to on-line circuit and optical routing. In *Proceedings of the 28th ACM Symposium on Theory of Computing*, pages 531–540, 1996.
16. Y. Bartal and S. Leonardi. On-line routing in all-optical networks. In *Proceedings of the 24th International Colloqium on Automata, Languages and Programming*, LNCS 1256, pages 516–526. Springer-Verlag, 1997.
17. A. Blum, A. Fiat, H. Karloff, and Y. Rabani, 1993. Personal comunication.
18. A. Borodin and R. El-Yaniv. Call admission and circuit-routing. In *Online Computation and Competitive Analysis.* Cambridge University Press, 1997.
19. A. Borodin, J. Kleinberg, and M. Sudan, 1996. Personal communication.
20. R. Canetti and S. Irani. Bounding the power of preemption in randomized scheduling. In *Proceedings of the 27th ACM Symposium on Theory of Computing*, pages 616–615, 1995.
21. A. Frank. Packing paths, cuts, and circuits – a survey. In B. Korte, L. Lovász, H.J. Proemel, and A. Schrijver, editors, *Paths, Flows, and VLSI-Layout*, pages 49–100. Springer-Verlag, 1990.
22. J.A. Garay and I.S. Gopal. Call preemption in communications networks. In *Proceedings of INFOCOM '92*, pages 1043–1050, 1992.
23. J.A. Garay, I.S. Gopal, S. Kutten, Y. Mansour, and M. Yung. Efficient on-line call control algorithms. In *Proceedings of the 2nd Israeli Symposium on Theory of Computing and Systems*, pages 285–293, 1993.
24. R. Gawlick, A. Kamath, S. Plotkin, and K. Ramakrishnan. Routing and admission control of virtual circuits in general topology networks. Technical Report BL011212-940819-19TM, AT&T Bell Laboratories, 1994.
25. O. Gerstel and S. Kutten. Dynamic wavelength allocation in WDM ring networks. In *Proceedings of ICC '97*, 1997.
26. O. Gerstel, G.H. Sasaki, and R. Ramaswami. Dynamic channel assignment for WDM optical networks with little or no wavelength conversion. In *Proceedings of the 34th Allerton Conference on Communication, Control, and Computin*, 1996.
27. A. Goel, M. R. Henzinger, and S. Plotkin. Online throughput-competitive algorithms for multicast routing and admission control. In *Proceedings of the 9th ACM-SIAM Symposium on Discrete algorithms*, pages 97–106, 1998.
28. P.E. Green. *Fiber-Optic Communication Networks.* Prentice Hall, 1992.
29. S. Irani. Coloring inductive graphs on-line. *Algorithmica*, 11:53–72, 1994.
30. A. Kamath, O. Palmon, and S. Plotkin. Routing and admission control in general topology networks with poisson arrivals. In *Proceedings of the 7th ACM-SIAM Symposium on Discrete Algorithms*, pages 269–278, 1996.
31. H. A. Kierstead and W. T. Trotter. An extremal problem in recursive combinatorics. *Congressus Numerantium*, 33:143–153, 1981.

32. J. Kleinberg and R. Rubenfield. Short paths in expander graphs. In *Proceedings of the 37th Annual IEEE Symposium on Foundations of Computer Science*, pages 86–95, 1996.
33. J. Kleinberg and E. Tardos. Disjoint paths in densely embedded graphs. In *Proceedings of the 36th Annual IEEE Symposium on Foundations of Computer Science*, pages 52–61, 1995.
34. T. Leighton, F. Makedon, S. Plotkin, C. Stein, E. Tardos, and S. Tragoudas. Fast approximation algorithms for multicommodity flow problems. In *Proceedings of the 23rd ACM Symposium on Theory of Computing*, pages 101–111, 1991.
35. S. Leonardi and A. Marchetti-Spaccamela. On-line resource management with applications to routing and scheduling. In *Proceedings of the 23rd International Colloqium on Automata, Languages and Programming*, LNCS 955, pages 303–314. Springer-Verlag, 1995.
36. S. Leonardi, A. Marchetti-Spaccamela, A. Presciutti, and A. Rosèn. On-line randomized call-control revisited. In *Proc. of the 9th Annual ACM-SIAM Symposium on Discrete Algorithms*, pages 323–332, 1998.
37. R. J. Lipton and A. Tomkins. Online interval scheduling. In *Proceedings of the 5th ACM-SIAM Symposium on Discrete Algorithms*, pages 302–311, 1994.
38. Y. Ma and S. Plotkin. Improved lower bounds for load balancing of tasks with unknown duration. *Information Processing Letter*, 62:31–34, 1997.
39. M. Mihail, C. Kaklamanis, and S. Rao. Efficient access to optical bandwidth. In *Proceedings of the 36th Annual IEEE Symposium on Foudations of Computer Science*, pages 548–557, 1995.
40. M. Ślusarek. Optimal on-line coloring of circular arc graphs. *RAIRO Journal on Informatique Theoretique et Applications*, 29(5):423–429, 1995.
41. Jan van Leeuwen ed. *Handbook of theoretical computer science, Vol A, Algorithms and Complexity.* The MIT Press, 1990.

12

On-line Network Optimization Problems

BALA KALYANASUNDARAM
KIRK PRUHS

1 Introduction

This chapter surveys results on on-line versions of the standard network optimization problems, including the minimum spanning tree problem, the minimum Steiner tree problem, the weighted and unweighted matching problems, and the traveling salesman problem. The goal in these problems is to maintain, with minimal changes, a low cost subgraph of some type in a dynamically changing network. In the early 1920's Otakar Borøuvka was asked by the Electric Power Company of Western Moravia (EPCWM) to assist in EPCWM's electrification of southern Moravia by solving from a mathematical standpoint the question of how to construct the most economical electric power network [9]. In 1926 Borůvka initiated the study of network optimization problems, by publishing an efficient algorithm for constructing a minimum spanning tree of a fixed network [9]. Certainly since the 1920's the underlying collection of sites that require electrification in southern Moravia has changed frequently as new sites require service, and perhaps occasionally some sites drop service. It would not be reasonable for EPCWM to continually maintain the absolute minimum spanning tree since the addition of a site might radically change the lines in the minimum spanning tree, and there would be significant cost to removing old lines. Hence, the real problem faced by EPCWM would be how to maintain, with minimal changes, a small weight spanning tree of a dynamic network.

In this chapter we survey algorithmic results for problems related to maintaining, with minimal changes, a low cost subgraph H of some type in a dynamically changing network G. For example, H may be required to be a maximum matching in an on-line matching problem. These problems are on-line in nature because the algorithm is unaware of future changes to G. The particular types of subgraphs we consider are unweighted matchings, weighted matchings, tours, spanning trees, k-connected spanning graphs, Steiner trees, and generalized Steiner trees. Generally these problems are formalized in the following manner. Let N be some fixed global network which may or may not be known a priori, and G_0 some initial subgraph of N. G_{i+1} is formed from either adding a vertex to, or possibly deleting a vertex from, G_i. The goal of the on-line algorithm is to minimally modify the subgraph H_i of G_i to yield the subgraph H_{i+1} of G_{i+1}.

The first case generally considered for these problems is when G and H are constrained to grow monotonically, i.e. each vertex in G_i is a vertex in G_{i+1} and $H_i \subseteq H_{i+1}$. The competitive ratio of an algorithm A in this model is then the maximum over all sequences I of changes to G, of the ratio of the cost of the final H subgraph constructed by the on-line algorithm divided by the cost of the optimal H subgraph.

If vertices may depart from G_i, or edges may be deleted from H_i, then there are two parameters one would like to optimize, the quality of H_i and the cost of restructuring H_i. Notice how both of these costs are captured in the cost of the final H in the monotone model. If one wishes to use competitive analysis, it seems necessary to fix one of these two parameters. Generally you can not hope to be competitive on both parameters at once since this would require that the on-line algorithm be competitive, in terms of restructuring cost, against an adversary that refuses to restructure its H subgraph, while at the same time, being competitive, in terms of solution quality, against an adversary that always maintains the optimal H subgraph.

In what we call the *fixed quality model*, we assume that we fix a priori a quality parameter β, and that there is a cost function $\alpha(H_i, H_{i+1})$ that gives the cost of changing from H_i to H_{i+1}. Most commonly, if N is unweighted, the cost might be the number of edges in the symmetric difference of H_i and H_{i+1}, and if N is weighted, the cost might be the aggregate weight of the edges in the symmetric difference of H_i and H_{i+1}. The problem is then to minimize the cost of handling the sequence I of changes, while maintaining an H_i that has cost at most β times the cost of optimal solution for G_i. The competitive ratio in this case is then the cost the on-line algorithm incurs for handling I divided by the optimal cost of maintaining a subgraph within a multiplicative factor of β of optimum. One example where it might be important to keep the quality of H high is the spanning tree example where there is a yearly maintenance cost proportional to the total line length.

Alternatively, in what we call the *fixed cost model*, we fix a cost α that we are willing to spend (in the amortized sense) each time G changes. If N is unweighted, this cost might be the number of edges in the symmetric difference between H_i and H_{i+1}. Here the problem would be to minimize the cost of the final H subgraph subject to the constraint that only $\alpha\, k$ can be spent, where k is the number of changes made to G.

When considering randomized algorithms we consider only oblivious adversaries, i.e. the adversary must specify the sequence I of changes a priori. Generally, adaptive adversaries, which can specify the ith change after seeing the on-line algorithm's response to the $(i-1)st$ change, rob the the randomized algorithm of the power of randomization.

This chapter is organized as follows. In section 2 we consider both weighted and unweighted matching problems. In section 3 we consider subgraphs involving connectivity, such as generalized Steiner trees and spanning trees. In section 4, we consider traveling salesman tours and paths. In each section we relate the main results that have been obtained to date. Where it is possible to do so succinctly, we give a description of the proposed on-line algorithms, along with the bounds obtained. We assume the reader is familiar with basic graph theoretic concepts, and with basic concepts concerning on-line algorithms. We generally end the discussion of each paper with the most appealing questions left open in that paper. In the conclusion we briefly discuss directions for future research.

2 Matching

The on-line transportation (weighted matching) problem is formulated as follows. The initial setting consists of a collection $\{s_1, \ldots, s_m\}$ of server sites which are vertices in a underlying network N. Each server site s_j has a positive integer capacity c_j. The on-line algorithm sees over time a sequence $\{r_1, \ldots r_n\}$ of requests for service, with each request being a point in N. In response to the request r_i, A must select a site $s_{\sigma(i)}$ to service r_i. The cost/distance incurred to service r_i from s_j would be the weight $d(s_j, r_i)$ on the edge (s_j, r_i) in N. One can assume that each nonedge in N has weight $+\infty$. Each site s_j can service at most c_j requests. The goal is minimize the total cost incurred in servicing the requests. Here the acceptable H subgraphs are maximum matchings in G.

It sometimes helps to have some concrete applications of a problem. One could imagine the server sites as fire stations, the servers as fire trucks, the requests as fires, the cost as the distance/time required to get from a fire station to a fire, and the problem is to minimize the average response time to a fire. Or one could imagine the server sites as schools, the servers as seats in the schools, the requests as the arrival of a student in a school district, the cost as the distance from the student's home to the school, and the problem is to minimize the average distance a student has to travel to school.

On-line matching is quite similar to the k-server problem. The difference is that the servers in the k-server problems are mobile, while in on-line matching they have fixed sites (cf. Chapter 4 on server problems for more information). Also many on-line load balancing problems can be phrased as on-line matching problems (cf. Chapter 8 on on-line load balancing for more information).

2.1 Unweighted matching

In unweighted matching problems it is assumed that all edges have weight 1 or $+\infty$. Under this assumption it is easier to think about the $+\infty$ edges as not being present, and the problem is constructing a maximum cardinality matching. In this setting the competitive ratio is the number of request vertices matched by the on-line algorithm divided by the size of the maximum matching. Notice that the competitive ratio is at most 1, and to be consistent with other papers we follow this notation instead of considering the inverse of this ratio. Karp, Vazirani, and Vazirani [19] consider the standard matching problem, with each server capacity c_j is equal to 1, in the monotonic model. Since every maximal matching is at least half of the size of a maximum matching, every deterministic algorithm that matches a request, when it is possible to do so, is $\frac{1}{2}$-competitive. It is easy to see that no deterministic algorithm can be have a competitive ratio better than $\frac{1}{2}$. [19] gives a randomized algorithm RANKING whose competitive ratio approaches $1 - 1/e$ for large n. Initially, RANKING numbers the sites randomly so that each of the $n!$ orderings are equally likely. Then in response to a request, RANKING selects the highest ranked unused site to service that request. [19] also show that RANKING is optimally competitive, up to lower order terms, for randomized algorithms. Kao and Tate [18] give some partial results for the case that points arrive in groups of t at a time.

Kalyanasundaram and Pruhs [17] continue the study in the monotonic model for the case where each capacity c_j is equal to some fixed integer k. [17] give a deterministic algorithm BALANCE whose competitive ratio is $1 - \frac{1}{(1+\frac{1}{k})^k}$, which approaches $1-1/e$ for large n and large k. Among all server sites that can service a request r_j, BALANCE selects an arbitrary site that has a maximal number of unused servers at that site. The competitive ratio of BALANCE is shown to be optimal (including low order terms) among deterministic algorithms.

Open Question: What is the optimal randomized algorithm and the optimal randomized competitive ratio for the case where each capacity c_j is equal to some fixed integer k. One obvious candidate algorithm, BALANCE-RANKING, would initially rank the server sites randomly, and then in response to a request r_j select the highest ranked server site among the set of sites with the maximal number of unused servers among the sites that can service r_j.

Motivated by problems in mobile computing, Grove, Kao, Krishnan, and Vitter [10] consider the unweighted matching problem in essentially the fixed quality model, with the quality parameter $\beta = 1$. That is they wish to always maintain a maximum matching. Strictly speaking, [10] assume that a request r_j must be serviced when it arrives, if it is possible to do so without terminating service to another request, but there is no requirement that r_j receive service at some later point of time if it should become possible to provide service. However, this seems to be equivalent to the fixed quality model in the sense that bounds for one model hold for the other. The setting for mobile computation is many roving mobile computers (the requests) that each must stay in contact with a mobile support station, or MSS (the server sites). Each MSS has a limit on the number of mobile computers it can support, and a maximum range of service (so not every MSS can support every mobile computer). The event of a mobile computer leaving the coverage area of its current MSS can be simulated by a request departure, followed by a request arrival. Motivated by this mobile computing application, [10] make the natural assumption that the cost of switching the server site handling a request is approximately the same as the cost of handling a request. So generally speaking, servicing a request requires a sequences of switches along an augmenting path, with the cost being the length of the augmenting path.

All of the results presented in [10] assume that the degree of each request vertex is only two. They argue that this should be the worst case because more paths intuitively help the on-line algorithm more than the adversary. When each $c_j = 1$ and no vertices leave G, the greedy algorithm (that assigns the request to an arbitrary server site that will minimize the number of switches) is $\Theta(\log n)$ competitive and this is optimal among deterministic algorithms. However, in the case that vertices can leave G, the competitive ratio of greedy is $\Omega(n)$. Define RAND to be the algorithm that switches along an augmenting path with probability inversely proportional to the length of the path. So RAND is essentially a randomized harmonic form of the greedy algorithm. In the case that each $c_j = 1$ and vertices are allowed to leave G, the competitive ratio of RAND is $O(\sqrt{n})$ and is $\Omega(n^{1/3})$. [10] show that the competitive ratio of every deterministic algorithm is $\Omega(\sqrt{n})$, and implicitly show that the competitive ratio of every randomized algorithm is $\Omega(\log n)$ for the general problem in the fixed quality model with $\beta = 1$. Optimal algorithms for some specific metric spaces, and lower bounds for some other natural algorithms can also be found in [10].

Open Question: The obvious open question is to find optimally competitive deterministic and randomized algorithms for the general problem in the fixed cost model. Perhaps allowing β to be slightly larger than 1 might make the on-line algorithm's task easier. Note that the problem becomes almost trivial for $\beta = 2$.

2.2 Weighted matching

Kalyanasundaram and Pruhs [14] and Khuller, Mitchell, and Vazirani [20] considered the on-line assignment problem, i.e. the on-line transportation problem where each capacity $c_j = 1$ (so $n \leq m$). If the costs are arbitrary then [14, 20] show that no algorithm (deterministic or randomized) can have a competitive ratio independent of the edge costs. So [14, 20] assume the costs are indeed distances, and satisfy the triangle inequality. [14, 20] give a deterministic on-line algorithm, called PERMUTATION in [14], with competitive ratio $2m - 1$. Let M_i be the minimum weight matching of the first i requests. In response to request r_i, PERMUTATION answers r_i with the unique unused server site in the same connected component as r_i in $M_{i-1} \cup M_i$. [14, 20] show that $2m - 1$ is the lower bound on the competitive ratio of any deterministic algorithm in a "star" metric space with the servers initially on the leaves. A star metric space has one center point of distance 1 from the leaves and each leaf is distance 2 from every other leaf.

The natural greedy algorithm that matches a request with the closest unused server has a competitive ratio of $2^m - 1$ [14]. Assuming that the points are uniformly scattered on the unit disk in the Euclidean plane, Tsai, Tang and

Chen [22] showed that the expected cost for greedy is at most $O(\sqrt{m})$ times the cost of the minimum matching. Other variants of on-line assignment have also been considered. For the on-line bottleneck matching problem, where you want to minimize the maximum weight of an edge in the matching, PERMUTATION is still $2m-1$ competitive and Idury and Schäffer [11] showed a lower bound on the competitive ratio of any deterministic algorithm that approaches $m/\ln 2 \approx 1.44\, m$ for large m. For the problem of maximizing the weight of the matching, [14] showed that the greedy algorithm, which matches each server to the farthest available server, is 3-competitive and that this is optimal for deterministic algorithms.

Kalyanasundaram and Pruhs [16] consider the on-line transportation problem under the assumption that the capacity of each site c_j is equal to some integer k. The competitive ratio for greedy is still 2^m-1, but the competitive ratio for PERMUTATION rises to $\Theta(km)=\Theta(n)$. Notice that greedy can be asymptotically better than PERMUTATION if $2^m=o(n)$.

Open Question: What is the optimal competitive ratio in terms of m, the number of sites, for the on-line transportation problem? It seems that there should be a $2m-1$ competitive algorithm.

In [16] the weak adversary model is assumed. In this model, the on-line algorithm A that has c_i servers at s_i is compared to an off-line line adversary that has only $a_i < c_i$ servers at s_i. The weak competitive ratio of an on-line algorithm A is the supremum over all instances I, with $n \le \sum_{i=1}^m a_i$ requests, of the cost incurred when the assignment is made by A on I divided by the minimum possible cost of any assignment for the instance I given only a_i servers at s_i. The weak adversary model is motivated by the desire to exclude adversarial inputs, and the desire to determine the additional resources required by the on-line algorithm to counteract the adverse effect of nonclairvoyance. Further motivation can be found in [16]. In [16] only the case $a_i = c_i/2$ is considered. [16] show that the weak-competitive ratio of the greedy algorithm is $\Theta(\min(m, \log C))$, where $C=\sum_{i=1}^m c_i$ is the sum of the capacities. If the server capacity of each site is constant, then the weak competitive ratio is logarithmic in m, a significant improvement over the exponential traditional competitive ratio for greedy. Furthermore, [16] give an on-line algorithm, BALANCE, that is a simple modification of the greedy algorithm, and that has a constant weak competitive ratio. Define the pseudo-distance $pd(s_i, r_j)$ to handle a request r_j by s_i to be $d(s_i, r_j)$ if less than half the servers at s_i have been used, and $\gamma \cdot d(s_i, r_j)$ otherwise. Here γ is some sufficiently large constant. BALANCE then services each request r_j with the available server site s_i that minimizes $pd(s_i, r_j)$.

Open Question: What is the weak competitive ratio for a general a_i? We know of no nonconstant lower bound on the weak competitive ratio even in the case that each $a_i = c_i - 1$.

Open Question: The optimal competitive ratio for the on-line transportation problem depends critically on the underlying metric space N. It would be interesting to determine the optimal competitive algorithms for common metric spaces such as the line and the Euclidean plane. The on-line matching problem on a line can also be formulated in the following way. A ski rental shop must match skis to skiers as the skiers walk through the door, with each skier ideally getting skis of height equal to their height. The problem is to minimize the average discrepancy between skier's heights and the heights of their assigned skis.

The underlying combinatorics for the on-line assignment problem seems to be captured by an on-line search game, that we call hide and seek. To begin, the hider and the seeker are at some origin point. The hider, unbeknownst to the seeker, then moves to some new point x in the metric space N. The seeker then traverses N until it arrives at x. The seeker's goal is minimize the ratio of the distance that it travels divided by the distance between the origin and x. This corresponds to a lazy adversary that only moves one server. There is an $O(\sqrt{n})$ competitive algorithm for hide and seek in the plane [13] and a constant competitive algorithm for the line. We believe these are also the optimal competitive ratios for on-line assignment in these metric spaces. One possible candidate algorithm for the line is the generalized work function algorithm. The generalized work function algorithm picks a collection S_i of servers, to handle the collection R_i of the first i requests, that minimizes $\gamma\, MM(S_i, R_i) + MM(S_i, S_{i-1} \cup \{r_i\})$, where $MM(X, Y)$ denotes the minimum matching between X and Y, and $\gamma > 1$. This algorithm is a linear combination of PERMUTATION (the first term) and greedy (the second term). As circumstantial evidence that the generalized work function algorithm is a reasonable candidate for the line, we note that it is essentially optimal against a lazy adversary.

Ultimately, it would be nice to have a "universal" metric space dependent algorithm for on-line weighted matching. A universal algorithm would have a competitive ratio that is at most a constant times optimal for every metric space. Since there seems to be no natural description for the optimal on-line algorithm for on-line assignment with 3 server sites on a line [14], it seems unlikely that there is a simple algorithm that would assure absolute optimality on every metric space. Note that it is unlikely that the generalized work function algorithm will be universal since it uses nearest neighbor search to search configurations that have the same optimal cost. Nearest neighbor is known to be only $\Theta(\log n)$ competitive for on-line TSP [21]. Although, a proof that the generalized work function algorithm is within $O(\log n)$ of being universal would be a good first step toward finding a universal algorithm.

Open Question: What is the optimal randomized competitive algorithm for the on-line assignment problem (and on-line transportation problem)? Suggestive evidence that an $O(\log n)$-competitive randomized algorithm might exist can be found from the fact that the optimal competitive ratio for the star metric space (the worst case for the deterministic algorithm) is $2H_k - 1$ against an oblivious adversary.

3 Generalized Steiner trees

Consider the design of a reliable communication/transportation network of low cost. Usually, reliability means that we would like to maintain connections between given pairs of sites even if some number of sites/links are down. Krarup (see [23]) introduced the generalized Steiner problem GSP to study many variants of such design problems.

Given an undirected graph G with positive weights on the edges, the goal in GSP is to construct a minimum weight subgraph of G that satisfies a given connectivity requirement. More precisely, in the node-GSP (edge-GSP) problem the connectivity requirement is specified by assigning a nonnegative integer $r(u, v)$ for every pair of distinct nodes u and v in G, and the goal is to construct a minimum weight subgraph of G with at least $r(u, v)$ node-disjoint (edge-disjoint) paths between u and v.

In the on-line version of GSP, the on-line algorithm sees a sequence of requests of the form $(u_i, v_i, r(u_i, v_i))$ where u_i, v_i are vertices in the underlying network N, which is assumed to be known a priori. The acceptable H subgraphs are those where there are at least $r(u_j, v_j)$ disjoint paths between u_j and v_j for all $j \leq i$. Note that in general the subgraph H need not be a single connected component. With the exception of the paper of Imase and Waxman [12], all the papers assume the monotonic model. We use n to denote the number of nodes in the underlying network N, s to denote the number of requests, and k to denote the number of distinct vertices that appeared in at least one of the s requests.

Westbrook and Yan [23] showed that there is no randomized algorithm that achieves a competitive ratio better than $(k + 2)/4$ for on-line node-GSP, and that there is no randomized algorithm that achieves a competitive ratio better than $(s + 1)/4$ for on-line edge-GSP, even if $r(u, v) = 2$. On the other hand, the greedy algorithm, which adds the minimum cost collection of edges that will maintain the connectivity requirement, is trivially $O(s)$ competitive. Westbrook and Yan showed that the greedy algorithm is $O(k)$-competitive for node-GSP if each $r(u, v) = 2$. For edge-GSP, the greedy algorithm is $O(k)$-competitive if the connectivity requirements form a nonincreasing sequence.

Open Question: Can the competitive ratios be improved if the underlying graph is a metric space, i.e. it is complete graph and the edge weights satisfy the triangle inequality?

In the case that each $r(u, v) = 1$, Westbrook and Yan [24] showed that the greedy algorithm has a competitive ratio of $O(\sqrt{k} \log k)$. In contrast, the competitive ratio of every deterministic on-line algorithm is $\Omega(\log k)$ [12, 24]. Awerbuch, Azar and Bartal [4] proved that the greedy algorithm is $O(\log^2 k)$ competitive. Subsequently [7] gave a $O(\log k)$ competitive on-line algorithm.

Open Question: What is the competitive ratio of the greedy algorithm?

Imase and Waxman [12] considered the dynamic Steiner tree problem, which is a restricted case of GSP, where nodes in G are revealed on-line and the goal is to maintain a Steiner tree that contains all of the revealed nodes. They considered both the monotonic model and the fixed cost model. They showed a lower bound of $\Omega(\log k)$ on the deterministic competitive factor for the monotonic case. They showed that the greedy algorithm is $O(\log k)$-competitive.

Chandra and Viswanathan [8] extended the results of Imase and Waxman by considering the problem of constructing a d-connected subgraph of the revealed nodes in the monotone model. They showed that the greedy algorithm is $8d^2 \log k$ competitive. They gave a lower bound of $(d/4)\log(k/d - 1)$ on the competitive ratio of every deterministic algorithm. They showed that any on-line algorithm A, with competitive factor c, can be modified to yield another on-line algorithm B with competitive factor at most $2c$, and B maintains an H subgraph with maximum degree at most $3d$. Notice that a minimum degree bound of d is required to maintain a d-connected graph. Unfortunately, the diameter of this subgraph is not small. They showed that the degree bound of $2d$ is necessary for any deterministic on-line algorithm to maintain a d-connected subgraph. Analogous to reducing the degree bound, they also showed how to augment the selection of edges by the greedy algorithm so that the diameter is at most 3 times that of the off-line optimal diameter. It appears that the diameter is reduced in this construction at the cost of increasing the degrees of the nodes.

Open Question: Is there an on-line algorithm which is close to optimal in size, diameter, and degree?

Alon and Azar [1] considered the on-line Steiner tree problem where the underlying graph is the Euclidean plane. Based on a construction by Bentley and Saxe [6], they proved a lower bound of $\Omega(\log k/ \log\log k)$ on the competitive ratio of any randomized on-line algorithm.

The one result that does not assume the monotone model is by Imase and Waxman [12], who considered the fixed cost model. If k points have been added they allow $O(k^{3/2})$ edge deletions/additions to the H subgraphs, or in other words, $O(k^{1/2})$ amortized changes per revealed vertex. They showed that this many changes is sufficient to maintain a Steiner tree that has cost at most a constant times optimum.

Open Question: Is it possible to reduce the number of edge deletions to constant per change to G?

Finally we mention some results that are closely related to GSP. Bartal, Fiat and Rabani [5] used competitive analysis to measure the efficiency of algorithms for the file allocation problem which is closely related to GSP. Awerbuch, Azar and Bartal [4] generalized GSP to the *Network Connectivity Leasing Problem*, where the on-line algorithm can either buy or lease an edge while constructing the graph. Obviously the cost of buying is more than that of leasing. They gave $O(\log^2 n)$-competitive randomized algorithm for this problem.

4 Traveling salesman problems

A traveling salesman path (TS path) is a walk that visits each vertex in G, and a tour is a TS path that returns to the point of origin. The on-line versions of TSP in the literature involve a salesman physically moving through G attempting to visit all the vertices of G. It might help to imagine the edges as streets and the vertices as intersections. Since these problems fall more within the paradigm of search problems (see Chapter 10 on search problems) they do not fit neatly within in the models proposed in the introduction. Ausiello, Feuerstein, Leonardi, Stougie, and Talamo [2, 3] consider a variant of the TSP problem with release times. Let t_i be the time that the ith point p_i is added to G. It is required that the salesman must visit p_i after time t_i. One can imagine the salesman as occasionally receiving additional assignments via a cellular phone. When a new point arrives the salesman can, at no cost, recompute a new path to visit the remaining points. The competitive ratio is then just the length of the path traveled by the on-line traveling salesman divided by the optimal path that satisfies the release time constraints.

[2, 3] give the following results. The graph G is always assumed to be a metric space. In [2] they first considered the TS path problem, i.e. the salesman is not constrained to return to the origin. For an arbitrary metric space, the greedy algorithm has a competitive ratio of $5/2$. If a new point p arrives, greedy recomputes, as its future planned route, the shortest route T that visits the remaining unvisited points. Note that greedy requires superpolynomial time per query unless $P = NP$. By using the minimum spanning tree approximation for T one can obtain a 3-competitive algorithm that requires only polynomial time computation. The competitive ratio for every deterministic algorithm is at least 2, even if the underlying metric space N is the real line. If N is the real line, the algorithm that always attempts to visit the extreme point that is nearest the origin is $7/3$ competitive.

In the tour problem, where the salesman is required to return to the origin, [3] shows that life is slightly easier for the on-line algorithm. For an arbitrary metric space, a modification to the greedy algorithm is 2-competitive. If the salesman is at the origin, it computes a the optimal tour T that visits the remaining unvisited points in G and then returns to the origin, and then the salesman begins traversing T. If a new point x arrives in G that is farther from the origin than the current position of the salesman, then the salesman returns on the shortest path to the origin, ignoring all point arrivals on the way. New points p that are closer to the origin than the salesman's position when p arrives are ignored until the salesman returns to the origin. There is an algorithm CHR that is 3-competitive and requires only polynomial computation time. If a new point arrives, CHR uses Christofides approximation algorithm to find a path T through the remaining unvisited points and returns to the origin. The competitive ratio for any deterministic algorithm on a general metric space is at least 2, and on a line is at least $(9 + \sqrt{17})/8 \approx 1.64$. There is a rather complicated algorithm PQR that is shown to have a competitive ratio of $7/4$ on a line.

Motivated by the problem of a salesman attempting to visit all the towns in some rural state for which he does not have a map, Kalyanasundaram and Pruhs [15] define the following version of on-line TSP. When the salesman visits a vertex v, it learns of each vertex w adjacent to v in N, as well as the length $|vw|$ of the edge vw. The salesman's goal is to visit all of the vertices in N, with his/her path being as short as possible. Note that the salesman controls to some extent how N is revealed. [15] gives a deterministic algorithm ShortCut that is at most 16-competitive when N is planar. Note that N need not even be a metric space, all that is required is that the unweighted version of N can be drawn in the plane so that no edges cross.

ShortCut is a modification of depth first search that occasionally shifts the search from one portion of G to another via a known path in G. At any particular time, let $d(v, w)$ denote the length of the shortest path known between vertices v and w using only the known part of N. An edge xy is called a boundary edge if x is explored but y is not. A boundary edge xy blocks a boundary edge vw if $|xy| < |vw|$ and $d(v, x) + |xy| < (1+\delta)|vw|$. A boundary edge vw is a shortcut if no other boundary edge blocks vw. In ShortCut the searcher begins as if it were performing a standard depth first search on G. Assume that the searcher is at a vertex v and is considering whether to traverse a boundary edge vw. The edge vw will be traversed at this time if and only if vw is a shortcut. Assume that the searcher just traversed a boundary edge xy, causing y to become visited. It may then be the case that some other boundary edge vw, whose traversal was delayed at some previous point in time, now becomes a shortcut. In this case, a so-called jump edge is added from y to w. If at some time, ShortCut directs the searcher to traverse this jump edge, then the searcher will actually traverse the shortest path that it is aware of from y to w.

Open Question: We do not know of any example showing that ShortCut is not constant competitive for an arbitrary N. Note that in [15] the planarity of N is only required in the proof that ShortCut is constant competitive. The obvious question is to determine if ShortCut, or some other on-line algorithm, is constant competitive for an arbitrary N.

5 Conclusion

There has been a reasonable amount of work on on-line network optimization in the monotone model. Usually one first analyzes the greedy algorithm that lets H_{i+1} be the cheapest extension of H_i. Some reasonable fraction of the time the greedy algorithm turns out to be close to optimal. Even if it isn't, the analysis usually provides good intuition for developing a better algorithm. Often this better algorithm is relatively a minor modification to greedy.

There has been much less work in the fixed quality and fixed cost models. This is probably the area where there is the greatest need for further investigation. The obvious greedy algorithm in the fixed quality model is to have H_{i+1} be the cheapest modification of H_i that satisfies the quality constant. Analogously, the obvious greedy algorithm in the fixed cost model is to have H_{i+1} be the solution closest to optimal that can be obtained from H_i within the cost bound. As a first step, one should probably try to analyze these greedy algorithms. The scant evidence available to date [10, 12] suggests that perhaps the greedy algorithms in the fixed cost and fixed quality models will be less likely to be close to optimally competitive than in the monotone model.

Acknowledgements

We thank Eddie Grove and Stefano Leonardi for helpful suggestions.

References

1. N. Alon and Y. Azar. On-line Steiner trees in the Euclidean plane. *Discrete and Computational Geometry*, 10:113–121, 1993. Also in *Proc. 8th ACM Symposium on Computational Geometry*, 1992, pp. 337-343.
2. G. Ausiello, E. Feuerstein, S. Leonardi, L. Stougie, and M. Talamo. Serving requests with online routing. In *Proceedings of Scandinavian Workshop on Algorithm Theory*, volume 824 of *Lecture Notes in Computer Science*, 1994.
3. G. Ausiello, E. Feuerstein, S. Leonardi, L. Stougie, and M. Talamo. Competitive algorithms for the on-line traveling salesman problem. In *Proceedings of International Workshop on Algorithms and Data Structures*, volume 955 of *Lecture Notes in Computer Science*, 1995.
4. B. Awerbuch, Y. Azar, and Y. Bartal. On-line generalized Steiner problem. In *Proc. of 7th ACM-SIAM Symposium on Discrete Algorithms*, pages 68–74, 1996.
5. Y. Bartal, A. Fiat, and Y. Rabani. Competitive algorithms for distributed data management. In *Proc. of the 24th Symposium on Theory of Computation*, pages 39–48, 1992.
6. J. Bentley and J. Saxe. An analysis of two heuristics for the Euclidean traveling salesman problem. In *Proceedings of the Allerton Conference on Communication, Control, and Computing*, pages 41–49, 1980.
7. P. Berman and C. Coulston. Online algorithms for Steiner tree problems. In *Proceedings of the 29th Annual ACM Symposium on Theory of Computing*, 1997.
8. B. Chandra and S. Viswanathan. Constructing reliable communication networks of small weight. *Journal of Algorithms*, 18(1):159–175, 1995.
9. R. Graham and P. Hell. On the history of the minimum spanning tree problem. *Annals of the History of Computing*, 7:43–57, 1985.
10. E. Grove, M. Kao, P. Krishnan, and J. Vitter. Online perfect matching and mobile computing. In *Proceedings of International Workshop on Algorithms and Data Structures*, 1995.
11. R. Idury and A. Schaffer. A better lower bound for on-line bottleneck matching. manuscript, 1992.
12. M. Imase and B. M. Waxman. Dynamic Steiner tree problems. *SIAM J. Discrete Math.*, 4:369–384, 1991.
13. B. Kalyanasundaram and K. Pruhs. A competitive analysis of algorithms for searching unknown scenes. *Computational Geometry: Theory and Applications*, 3:139–155, 1993. Preliminary version appeared in STACS '92.
14. B. Kalyanasundaram and K. Pruhs. Online weighted matching. *Journal of Algorithms*, 14:478–488, 1993. Preliminary version appeared in SODA '91.
15. B. Kalyanasundaram and K. Pruhs. Constructing competitive tours from local information. *Theoretical Computer Science*, 130:125–138, 1994. Preliminary version appeared in ICALP '93.
16. B. Kalyanasundaram and K. Pruhs. The online transportation problem. In *Proceedings of European Symposium on Algorithms*, volume 979 of *Lecture Notes in Computer Science*, pages 484–493, 1995.

17. B. Kalyanasundaram and K. Pruhs. An optimal deterministic algorithm for online b-matching. manuscript, April 1996.
18. M.-Y. Kao and S. R. Tate. Online matching with blocked input. *Information Processing Letters*, 38:113–116, 1991.
19. R. M. Karp, U. V. Vazirani, and V. V. Vazirani. An optimal algorithm for on-line bipartite matching. In *Proceedings of the 22nd Annual ACM Symposium on Theory of Computing*, pages 352–358, Baltimore, Maryland, May 1990.
20. S. Khuller, S. G. Mitchell, and V. V. Vazirani. On-line algorithms for weighted bipartite matching and stable marriages. In *Proc. 18th Int. Colloquium on Automata, Languages and Programming*, pages 728–738. Lecture Notes in Computer Science 510, Springer-Verlag, Berlin, 1991.
21. D. Rosenkrantz, R. Stearns, and P. Lewis. An analysis of several heuristics for the traveling salesman problem. *SIAM Journal of Computing*, 6:563–581, 1977.
22. Y. Tsai, C. Tang, and Y. Chen. Average performance of a greedy algorithm for the on-line minimum matching problem on Euclidean space. *Information Processing Letters*, 51:275–282, 1994.
23. J. Westbrook and D. Yan. Linear bounds for on-line Steiner problems. *Information Processing Letters*, 55:59–63, 1995.
24. Jeffery Westbrook and Dicky C.K. Yan. Lazy and greedy: On-line algorithms for Steiner problems. In *Proc. 1993 Workshop on Algorithms and Data Structures*, 1993. To appear in *Mathematical Systems Theory*.

13

Coloring Graphs On-line

Hal A. Kierstead

1 Introduction

This chapter presents a survey of three types of results concerning on-line graph coloring: The first type deals with the problem of on-line coloring k-chromatic graphs on n vertices, for fixed k and large n. The second type concerns fixed classes of graphs whose on-line chromatic number can be bounded in terms of their clique number. Examples of such classes include interval graphs and the class of graphs that do not induce a particular radius two tree. The last type deals with classes of graphs for which First-Fit performs reasonably well in comparison to the best on-line algorithms. Examples of such classes include interval graphs, the class of graphs that do not induce the path on five vertices, and d-degenerate graphs.

An *on-line graph (digraph)* is a structure $G^{\prec} = (V, E, \prec)$, where $G = (V, E)$ is a graph (digraph) and $\prec$ is a linear order on V. (Here V will always be finite.) The ordering $\prec$ is called an *input sequence*. Let $G_n^{\prec}$ denote the on-line graph induced by the $\prec$-first n elements $V_n = \{v_1 \prec \cdots \prec v_n\}$ of V. An algorithm A that properly colors the vertices of the on-line graph $G^{\prec}$ is said to be an *on-line coloring algorithm* if the color of the n-th vertex v_n is determined solely by the isomorphism type of $G_n^{\prec}$. Intuitively, the algorithm A colors the vertices of G one vertex at a time in the externally determined order $v_1 \prec \cdots \prec v_n$, and at the time a color is irrevocably assigned to v_n, the algorithm can only see G_n. For example, the on-line coloring algorithm First-Fit (FF) colors the vertices of $G^{\prec}$ with an initial sequence of the colors $1, 2, \ldots$ by assigning the vertex v the least color that has not already been assigned to any vertex adjacent to v. The number of colors that an algorithm A uses to color $G^{\prec}$ is denoted by $\chi_A(G^{\prec})$. For a graph G the maximum of $\chi_A(G^{\prec})$ over all input sequences $\prec$ is denoted by $\chi_A(G)$. If Γ is a class of graphs, the maximum of $\chi_A(G)$ over all G in Γ is denoted by $\chi_A(\Gamma)$. The on-line chromatic number of Γ, denoted by $\chi_{ol}(\Gamma)$, is the minimum over all on-line algorithms A of $\chi_A(\Gamma)$.

For a graph $G = (V, E)$, the chromatic number, clique number, and independence number of G are denoted by $\chi(G)$, $\omega(G)$, and $\alpha(G)$. Let u and v be vertices in G. If u and v are adjacent, we may write $u \sim v$. Let $N(v) = \{w \in V : v \sim w\}$ and $d(v) = |N(v)|$. If $G^{\prec} = (V, E, \prec)$ is an on-line graph, then $N^{\prec}(v) = \{w \in V : v \sim w \text{ and } w \prec v\}$ and $d^{\prec}(v) = |N^{\prec}(v)|$. If G is isomorphic to H we may write $G \approx H$. The set $\{1, 2, \ldots, n\}$ is denoted by $[n]$. For a sequence $\sigma = (\sigma_1, \ldots, \sigma_n)$ a subsequence of the form $\sigma = (\sigma_1, \ldots, \sigma_i)$ is called an *initial* sequence of σ and a subsequence of the form $(\sigma_i, \ldots, \sigma_n)$ is called a *final* sequence of σ. Let $|\sigma|$ be the length of σ.

Our goal is to find on-line coloring algorithms that perform well on various classes of graphs. To see what this might mean, we begin by considering some simple examples. In later sections we explore in more detail the issues raised by these examples. We will include some illustrative proofs.

Example 1. (Gyárfás and Lehel [9]).
For every positive integer k there exists a tree T_k on 2^{k-1} vertices such that for every on-line coloring algorithm A, $\chi_A(T_k) \geq k$.

Proof. We begin by defining the tree T_k. Let $D = \{\sigma : \sigma \text{ is a strictly decreasing sequence of positive integers}\}$. For $\sigma \in D$, let $V_\sigma = \{\tau \in D : \sigma \text{ is an initial segment of } \tau\}$. Let T_σ be the tree on the vertex set V_σ such that τ is adjacent to τ' iff $|\tau| + 1 = |\tau'|$ and $\tau' \in V_\tau$ or vice versa. We shall call σ the root of T_σ and abbreviate $T_{(t_1,\ldots,t_i)}$ by $T_{t_1,\ldots,t_i}$. In particular $T_k = T_{(k)}$. Note that if τ is a final segment of σ, then there exists an isomorphism from T_τ to T_σ that maps $\tau^\wedge\rho$ to $\sigma^\wedge\rho$. So $T_k - (k) = T_{k,1} + \ldots + T_{k,k-1}$. Putting (k) back, we see that (1) $T_k \approx T_{k-1} + T_{k-1} + e$, where e is an edge joining the roots of the two copies of T_{k-1}, and the root of $T_{k-1} + T_{k-1} + e$ can be either one of the endpoints of e. In particular, $|V_k| = 2^{k-1}$.

Let $S_{k,i} = T_{k,1} + \ldots + T_{k,i}$. The key property of T_k that we exploit is that (2) for any $i < k$, there exists an embedding of $S_{k,i}$ into T_k that maps (k, i) to (k) and is extendible to an automorphism of T_k. It follows that an on-line algorithm that has only seen a subgraph isomorphic to $S_{k,i}$ cannot distinguish between (k, i) and (k). Property (2) is easily proved by induction on $k - i$. The base step $k - i = 1$ follows immediately from (1). The induction step follows from the induction hypothesis applied to the pair $\{k - 1, i\}$ and from the base step.

Let P_k be a partial ordering on $S_{k,k-1}$ such that $x\, P_k\, y$ iff $x \in T_{k,i}$, $y \in T_{k,j}$, and $i < j$. We claim that for every positive integer k and on-line coloring algorithm A, there exists a total ordering $\prec_k$ of $S_{k,k-1}$ such that $\prec_k$ extends P_k and A assigns each of the vertices $(k, 1), \ldots, (k, k-1)$ a distinct color when $S_{k,k-1}$ is presented in the order $\prec_k$. It then follows that if (k) is presented last, then A uses a k-th color to color (k). Arguing inductively, assume that we have shown this for the case $k = m$ and consider the case $k = m+1$. Since $S_{k,m-1} \approx S_{m,m-1}$, there exists an ordering $\prec_m$ such that when A is applied to $S_{k,m-1}$ in the order

$\prec_m$, A uses distinct colors on the set $Q = \{(k,1), \ldots, (k, m-1)\}$. Let B be an on-line algorithm that colors $T_{k,m}$ in the same way that A colors $T_{k,m} \approx T_m$ after first coloring $S_{k,m-1}$ in the order $\prec_m$. By the induction hypothesis applied to B, instead of A, there exists an ordering $\prec^*$ of $T_{k,m}$ extending the preimage of P_m, such that when A is applied to $S_{k,m}$ in the order $\prec_m + \prec^*$, A uses distinct colors on the set $R = \{(k,m,1), \ldots, (k,m,m-1)\}$. If A uses the same colors on both Q and R, then (k,m) gets a new color and we are done. Otherwise some root (k,m,i) gets a new color α. Because $\prec^*$ extends the preimage of P_m, at the time (k,m,i) is colored by the algorithm B has only seen a subgraph of $T_{k,m} \approx T_m$ which is isomorphic to $S_{m,i}$. So by (2) we can reorder $V_{k,m}$ so that (k,m) looks like $(k,m,m-1)$ and is colored with α by A. □

This example shows that we cannot bound the on-line chromatic number of a graph solely in terms of its chromatic number, even in the case of trees. In Section 2 we will obtain non-trivial bounds on the on-line chromatic number of graphs on n vertices in terms of their chromatic number and n. Because of the following example, our emphasis will be on fixed k and large n.

Example 2. (Szegedy [34]).
For every on-line algorithm A and positive integer k, there exists an on-line graph $G^{\prec}$ on n vertices such that $\chi(G^{\prec}) \le k$, $n \le k2^k$, and $\chi_A(G^{\prec}) \ge 2^k - 1$.

Proof. We construct $G^{\prec}$ in stages; $G^{\prec}_s$ is constructed at the s-th stage, which consists of three steps. First we introduce a new vertex v_s together with all edges from v_s to previous vertices. Next we determine the color $A(v_s) = c$, that A assigns v_s. We may assume that $c \in \{c_1, \ldots, c_{2^k-1}\}$. Finally we assign a color $f(v_s) \in \{r_1, \ldots, r_k\}$ to v_s. Let C_j be the set of vertices that A has colored c_j, R_i be the set of vertices that we have colored r_i, and $X_{ij} = R_i \cap C_j$. We shall try to maintain the following induction hypothesis:

(1) f is a proper k-coloring and
(2) $|X_{ij}| \le 1$, for all $i \in [k]$ and $j \in [2^k - 1]$.

By (1) $G^{\prec}$ is k-colorable and by (2) $G^{\prec}$ has less than $k2^k$ vertices. Thus it suffices to show that we can maintain (1) and (2) until A uses $2^k - 1$ colors.

Let $S \subset [k]$. We say that S is *represented* if there exists j such that $i \in S$ iff $X_{ij} \ne \emptyset$. If every non-empty subset of $[k]$ is represented then A has already used $2^k - 1$ colors and we are done. Otherwise, suppose S is a non-empty subset of $[k]$ which is not represented. Let v_s be adjacent to v iff $v \in R_i$ and $i \notin S$. Suppose A colors v_s with c_j. Then $X_{ij} = \emptyset$ for all $i \notin S$. Thus, since S is neither empty nor represented, there exists $i \in S$ such that $X_{ij} = \emptyset$. Let $f(v_s) = i$. □

After the first two negative examples one might wonder whether there are any interesting on-line coloring algorithms. Our next example is a simple on-line coloring algorithm with a nontrivial performance bound.

Example 3. (Kierstead [13]).
For every positive integer n, there exists an on-line algorithm B such that

$\chi_B(G) \leq 2n^{1/2}$, for any graph G on n vertices that contains neither C_3 nor C_5.

Proof. Consider the input sequence $v_1 \prec v_2 \prec \cdots \prec v_n$ of an on-line graph $G^\prec$ that contains neither C_3 nor C_5. Initialize by setting $W_i = \emptyset$ for all $i \in [2n^{1/2}] - [n^{1/2}]$. At the s-th stage the algorithm processes the vertex v_s as follows.

1. If there exists $i \in [2n^{1/2}]$ such that v_s is not adjacent to any vertex colored i, then let j be the least such i and color v_s with j.
2. Otherwise, if there exists $i > n^{1/2}$ such that $v_s \in N(W_i)$, then let j be the least such i and color v_s with j.
3. Otherwise, let j be the least integer $i > n^{1/2}$ such that $W_i = \emptyset$. Set $W_j = \{v \in N^\prec(v_s) : \text{the color of } v \text{ is at most } n^{1/2}\}$ and color v_s with color j. (Note that $|W_j| \geq n^{1/2}$, since Case 1 does not hold. Also, for all $i < j$, $W_i \cap W_j = \emptyset$, since Case 2 does not hold.)

Suppose for a contradiction that two adjacent vertices x and y, with $x \prec y$, have the same color j. Clearly y is not colored by Step 1. Thus $j > n^{1/2}$, and thus x is not colored by Step 1. Since only the first vertex colored j can be colored by Step 3, y must be colored by step 2. If x is colored by Step 3, then $W_j \subset N(x)$ and $y \in N(W_j)$, and so x and y have a common neighbor in W_j. But then G contains C_3, a contradiction. If x is colored by Step 2, then both x and y are in $N(W_j)$ and so either they have a common neighbor in W_j, and we are done as before, or they have distinct neighbors in W_j, each of which is adjacent to the first vertex colored j. In this case G contains C_5, a contradiction. So B produces a proper coloring. At most $n^{1/2}$ colors are used in Step 1. Since the W_j are disjoint and have size at least $n^{1/2}$, at most $n^{1/2}$ colors are used for Steps 2 and 3 combined. Thus $\chi_B(G^\prec) \leq 2n^{1/2}$. □

In Section 3 we study special classes of graphs that have the property that the on-line chromatic number of any graph in the class can be bounded in terms of its chromatic number, in fact even in terms of its clique size. The following was probably the first such result.

Example 4. (Kierstead and Trotter [25]).
There exists an on-line coloring algorithm A such that for any interval graph G, $\chi_A(G) \leq 3\omega(G) - 2$; moreover for any on-line coloring algorithm A and any positive integer k, there exists an interval graph G such that $\omega(G) = k$ and $\chi_A(G) \geq 3k - 2$.

Proof. We shall only prove the first statement. First we prove by induction that for all k there exists an on-line algorithm A_k such that if $G^\prec$ is an on-line interval graph with $\omega(G) = k$ then $\chi_{A_k}(G^\prec) \leq 3\omega(G) - 2$. The base step $k = 1$ is trivial, so consider the induction step $k > 1$. Consider the input sequence $v_1 \prec \ldots \prec v_n$ of G. The algorithm A_k will maintain an on-line partition of V. When a new vertex v_s is presented, A_k puts v_s into a set of B, if $\omega(B \cup \{v_s\}) < k$; otherwise

A_k puts v_s into C. If v_s is put into B, it is colored by A_{k-1} applied to B using the set of colors $[3k-5]$; otherwise v_s is colored by First-Fit applied to C using colors greater than $3k-5$. It suffices to show that First-Fit uses at most 3 colors on C. To prove this, we will show that the maximum degree of C is at most 2. For each vertex x of G, let I_x be the interval that corresponds to x in some interval representation of G. If $x \in C$, then x is in a k-clique K such that $K - \{x\} \subset B$. Let p_x be a point in the intersection of all intervals corresponding to vertices in K. Note that $p_x \notin I_y$ for any other vertex $y \in C$, since otherwise $K \cup \{y\}$ would be a $(k+1)$-clique in G. Suppose for a contradiction that x is adjacent to three vertices in C. Without loss of generality we can assume that for two of them, say y and z, $p_x < p_y < p_z$. Then $I_y \subset I_x \cup I_z$, since I_x intersects I_z and I_y is contained in the interval from p_x to p_z. But then $p_y \in I_x \cup I_z$, which is a contradiction.

The algorithm A guesses that $\omega(G) \leq k$ and uses A_k to color G until a vertex v_s is presented that forms a $(k+1)$-clique. At this time the algorithm switches to A_{k+1}. This does not cost any colors because A_{k+1} would have also used A_k to color the first $s-1$ vertices anyway. □

In Section 4, we shall study classes of graphs for which First-Fit performs well. The next two examples show that the class of trees has this property, but the class of 2-colorable graphs certainly does not.

Example 5. (Gyárfás and Lehel [9]).
For any tree T, $\chi_{\text{ol}}(T) = \chi_{\text{FF}}(T)$.

Proof. Notice that the maximum degree of the tree T_k constructed in Example 1 is $k-1$. Thus $\chi_{\text{FF}}(T_k) \leq k$, and so by Example 1, $\chi_{\text{FF}}(T) = k$. We shall show by induction on k, that for any tree T, if First-Fit colors a vertex of v of T with color k, then T contains a copy of T_k with root v. It follows that First-Fit is an optimal on-line coloring algorithm for trees. The base step is trivial, so consider the induction step. If First-Fit colors v with $k+1$, then for all positive integers $i \leq k$, v is adjacent to a vertex v_i that First-Fit has colored i. By the induction hypothesis, v_i is a root of a copy U_i of T_i in $T - v$. Since T is acyclic, distinct U_i are in distinct components of $T - v$. It follows that $\{v\} \cup \bigcup_{i \leq k} U_i$ is a copy of T_{k+1}. □

Example 6. For every positive integer n there exists a 2-colorable graph G on n vertices such that $\chi_{\text{FF}}(G) = n/2$.

Proof. Let $B_n = K_{n,n} - M$, where $M = \{a_i b_i : i \in [n]\}$, is a perfect matching in $K_{n,n}$. Let $\prec$ be the input sequence $(a_1, b_1, a_2, b_2, \ldots, a_n, b_n)$. Then First-Fit colors each a_i and b_i with the color i. □

2 Performance bounds for general graphs

In this section we consider the problem of finding good on-line coloring algorithms for the class of all graphs. Let $\phi(k, n)$ be the least integer $t\,(\leq n)$

for which there exists an on-line algorithm A such that $\chi_A(G) \le t$, for any k-colorable graph G on n vertices. We have already seen in Example 2 that $\phi(k, k2^k) \ge 2^k - 1$. Here we shall be interested in the case where k is fixed and n is much larger than $k2^k$. In the definition of ϕ, the algorithm A is allowed to depend on n. In other words, the algorithm knows the number of vertices of G ahead of time as in Example 3. This makes the statement of some algorithms simpler, but does not change the order of ϕ as the following doubling technique shows.

Lemma 1. *Let Γ be a class of graphs and g be an integer valued function on the positive integers such that $g(x) \le g(x+1) \le g(x)+1$, for all x. If for every n, there exists an on-line coloring algorithm A_n such that for every graph $G \in \Gamma$ on n vertices, $\chi_{A_n}(G) \le g(n)$, then there exists a fixed on-line coloring algorithm A such that for every $G \in \Gamma$ on n vertices, $\chi_A(G) \le 4g(n)$.*

Proof. Choose a sequence of integers $c_0 = 1, c_1, c_2, \ldots$ such that $2g(c_i) = g(c_{i+1})$. Color the first c_0 vertices using A_{c_0}, then color the next c_1 vertices, using A_{c_1} and a new palette, then color the next c_2 vertices, using A_{c_2} and a new palette, etc. This algorithm will color every graph $G \in \Gamma$ on $\sum_{0 \le h \le i} c_h$ vertices, with at most $2g(c_i)$ colors. To see this, argue by induction on i. The base step $i = 0$ is trivial, so consider the induction step $i = j+1$. We use at most $2g(c_j) = g(c_i)$ colors on the first $\sum_{0 \le h \le j} c_h$ vertices by the induction hypothesis, and at most $g(c_i)$ colors on the last c_i vertices. So we use at most $2g(c_i)$ colors in all.

Now suppose that $G \in \Gamma$ is a graph on n vertices with $\sum_{0 \le h < i} c_h \le n < \sum_{0 \le h \le i} c_h$. After coloring $\sum_{0 \le h < i} c_h$ vertices we guess that there are going to be $\sum_{0 \le h \le i} c_h$ vertices, which we will be able to color with the allotted number of colors, because by the claim we have accumulated a surplus of $2g(c_{i-1}) = g(c_i)$ colors. Thus we use at most $4g(c_{i-1}) \le 4g(n)$ colors. □

We begin our study of ϕ with the case $k = 2$. In Example 1 we saw that $\phi(2, n) \le \lg n$. The next theorem shows that $\phi(2, n) = \Theta(\log n)$, a quite satisfactory answer.

Theorem 2. (Lovász, Saks, and Trotter [27]).
There exists an on-line algorithm A such that for every on-line 2-colorable graph G on n vertices, $\chi_A(G) \le 2 \lg n$.

Proof. Consider the input sequence $v_1 \prec v_2 \prec \cdots \prec v_n$ of an on-line 2-colorable graph $G^\prec$. When v_i is presented there is a unique partition (I_1, I_2) of the connected component of $G_i^\prec$ to which v_i belongs, into independent sets such that $v_i \in I_1$. The algorithm A assigns v_i the least color not already assigned to some vertex of I_2.

It suffices to show that if A uses at least t colors on any connected component of $G_i^\prec$, then that connected component contains at least $2^{\lfloor t/2 \rfloor}$ vertices. We argue by induction on i and note that the base step is trivial. For the induction step, observe that if A assigns v_i color $k+2$, then A must already have assigned color k to some vertex $v_p \in I_2$ and color $k+1$ to some other vertex in I_2. Thus A must have assigned color k to some vertex $v_q \in I_1$. Since A assigned v_p and v_q the same color, v_p and v_q must be in separate components of $G_r^{\prec}$, where $r = \max\{p, q\}$. Thus by the induction hypothesis, each of these connected components must have at least $2^{\lfloor k/2 \rfloor}$ vertices and so the component of v_i in $G_i^{\prec}$ has at least $2^{\lfloor (k+2)/2 \rfloor}$ vertices. □

The situation is not nearly as clear for $k > 2$. Vishwanathan generalized the lower bound in the case $k = 2$, showing that $\phi(k,n) = \Omega\left(\log^{k-1} n\right)$.

Theorem 3. (Vishwanathan [36]).
For all integers k and n, $\phi(k,n) \geq (\lg n/(4k))^{k-1}$.

Proof. In order to simplify calculations, we will prove the weaker result that for every k, there exists $\varepsilon_k > 0$ such that for all n, $\phi(k,n) \geq \varepsilon_k (\lg n)^{k-1}$. The key idea of the proof is to show that there exists a function $f(k,n)$ satisfying the initial conditions $f(2,n) \geq \varepsilon_2 \log(n)$, $f(k,k) = k$, and recurrence relation $f(k+1, 3n) = f(k+1, n) + \frac{1}{2} f(k,n)$, such that for every on-line algorithm A, there exists an on-line k-colorable graph $G^{\prec}$ on n vertices and a proper k-coloring c of G such that A uses at least $f(k,n)$ colors on some color class of c. It then follows that $\phi(k,n) \geq f(k,n) \geq \varepsilon_k (\log n)^{k-1}$, for the some constant $\varepsilon_k > 0$.

We argue by double induction on k and then n. Fix an on-line algorithm A. Using Example 1, the base steps follow easily. We shall construct a $(k+1)$-colorable on-line graph $G^{\prec}$ on $3n$ vertices and a proper $(k+1)$-coloring c^* of G such that A uses at least $f(k,n) + \frac{1}{2} f(k,n)$ colors on some color class of c^*. By the secondary induction hypothesis there exists a $(k+1)$-colorable on-line graph $X^{\prec}$ on n vertices and a proper $(k+1)$-coloring c of X such that A uses at least $f(k+1, n)$ colors on some color class I of c. Let A' be the on-line algorithm that colors an on-line graph $H^{\prec}$ in the same way that A would color H after first coloring a disjoint copy of $X^{\prec}$. Then again using the secondary induction hypothesis, there exists an on-line $(k+1)$-colorable graph $Y^{\prec}$ on n vertices so that $Y^{\prec}$ is disjoint from $X^{\prec}$ and there exists a proper $(k+1)$-coloring c' of Y such that A' uses at least $f(k+1, n)$ colors on some color class I' of c'. Then when A is presented with $X^{\prec}$ followed by $Y^{\prec}$, A uses a set C of at least $f(k+1, n)$ colors on I and another set D of at least $f(k+1, n)$ colors on I'. If $|C \cup D| \geq f(k+1, n) + \frac{1}{2} f(k,n)$, then we are done since $I \cup I'$ is a color class of the $(k+1)$-coloring $c^* = c \cup c'$. Otherwise, $|C \cap D| \geq f(k+1, n) - \frac{1}{2} f(k,n)$. Let A'' be the on-line algorithm that colors an on-line graph $H^{\prec}$ in the same way that A would after first coloring a disjoint copy of $X^{\prec}$ followed by a disjoint copy of $Y^{\prec}$, if every vertex in H were adjacent to every vertex in I. By the primary induction hypothesis there exists a k-colorable on-line graph $Z^{\prec}$ on n vertices and a proper k-coloring c'' of Z such that A'' uses at least $f(k,n)$ colors on some color class I'' of c''. Let $G^{\prec}$ be $X^{\prec}$ followed by $Y^{\prec}$ followed by $Z^{\prec}$ together with all possible edges from Z to I. Then none of the colors A uses on I'' are used on I, and so A uses at least $f(k+1, n) + \frac{1}{2} f(k,n)$ colors on $I' \cup I''$. Since G has a proper $(k+1)$-coloring c^* with a color class containing $I' \cup I''$, we are done. □

Until recently, the best upper bound on $\phi(k,n)$ was given by the following theorem, where $\lg^{(k)}$ is the lg function iterated k times.

Theorem 4. (Lovász, Saks, and Trotter [27]).
There exists an on-line algorithm A such that for every k-colorable on-line graph $G^{\prec}$ on n vertices, $\chi_A(G^{\prec}) = O\left(n \lg^{(2k-3)} n / \lg^{(2k-4)} n\right)$. □

Their proof made use the following combinatorial lemma, whose proof follows easily from Inclusion-Exclusion or Lovász [26].

Lemma 5. *Let n be a positive integer and let δ be a positive real less than one. If F is a family of subsets of $[n]$ such that for all distinct $D, E \in F$, $|E| \geq \delta n$ and $|D \cap E| < \delta^2 n/2$, then $|F| < 2/\delta$.* □

Very recently the author used the same lemma to obtain a $O\left(n^{1-1/k!}\right)$ upper bound on $\phi(k,n)$.

Theorem 6. (Kierstead [13]).
For every positive integer k, there exists an on-line algorithm A_k and an integer N such that, for every on-line k-colorable graph $G^{\prec}$ on $n \geq N$ vertices, $\chi_{A_k}(G^{\prec}) \leq n^{1-1/k!}$.

Proof. To avoid messy calculations, we shall prove a somewhat weaker statement, but the full strength of the theorem can be obtained from the proof we give by being a little more careful with the details and initial conditions. We shall show that for all positive integers k there exist positive constants C and ε such that for all positive integers n there exists an on-line coloring algorithm $A_{k,n}$ such that for all k-colorable graphs G on n vertices $\chi_{A_{k,n}}(G) \leq Cn^{1-\varepsilon}$. We argue by induction on k. The base step is trivial; for the induction step assume that we have proved that there exist positive constants C and ε such that for all $i \leq k$, there exists an on-line algorithm $A_{i,n}$ such that $\chi_{A_{i,n}}(G) \leq Cn^{1-\varepsilon}$, for all i-colorable graphs G on n vertices.

Fix n. We shall describe $A = A_{k+1,n}$ in terms of two parameters α and δ, where $0 < \alpha,\ \delta < 1$. For $i \leq k$, let $A_i = A_{i,n'}$, where $n' = n^{\alpha}$. Set $\delta_i = 2^{-i}\delta$, $s_0 = n$, and $s_i = \delta_{i-1}s_{i-1}$, for $i \leq k$. Later we shall apply Lemma 5 with $n = s_i$ and $\delta = \delta_i$. Let $G^{\prec} = (V, E, \prec)$ be an on-line $(k+1)$-colorable graph on n vertices, and let Z be a subset of V.

First we describe a dynamic data structure in terms of the life cycle of the strange mythical species of *witnesses*. *Male* witnesses are *witness vertices* in $V - Z$. *Female* witnesses are certain *witness sets* contained in Z. A *witness tree* records the female genealogy of witnesses starting from the original witness set Z (Eve). From time to time witness sets will *spawn* large *litters* of *daughters*. Each of the daughters in a litter is a subset of her *mother*. Each daughter D in the litter has a distinct (!) *father* $F(D)$, who is a witness vertex that is adjacent in G to every vertex in D. Once a witness set is spawned, it will never gain or lose elements. However Eve is special in that Eve was not spawned and will gain, but not lose, elements. The witness sets form a tree with Eve at the 0-th level, the daughters of Eve at the 1-st level, their daughters at the 2-nd level, and so on, through the k-th generation. A witness set at the i-th level is called an i-witness set. For all $i > 0$, an i-witness set has size s_i. At some times some of the witness sets may *die*. Once they die, they will never *live* again. If they never die, they are *immortal*; otherwise they are *mortal*. If all the daughters in a single litter die, then the mother also dies (of grief).

Next we describe the on-line coloring algorithm A, using the above data structure. For any i-witness set W, with $i < k$, let $N^*(W) = \{v \in V - Z : |N(v) \cap W| \geq s_{i+1}\}$. If W is a k-witness, then $N^*(W) = N(W)$. The algorithm will maintain a partition $\{S_W : W \text{ is a witness set}\}$ of $V - Z$ such that each $S_W \subset N^*(W)$. Each S_W will be partitioned by $P_W = \{X_W(j) : j \in [t_W]\}$. The last part $X_W(t_W)$ of this partition is called the *active* part. When new elements enter S_W they will be put in the active part. Let $X = \bigcup\{P_W : W \text{ is a witness set}\}$. Call $X_W(j) \in X$ *small* if it has size less than n^α. Otherwise it is *large*. The algorithm will partition V into Z, at most $n^{1-\alpha}$ large subsets of size n^α, and a bounded number of small subsets. Each of these subsets will be colored from disjoint palettes of colors. The palette for Z will have δn colors and each of the other palettes will have $Cn^{\alpha(1-1/\varepsilon)}$ colors.

Consider the input sequence $v_1 \prec v_2 \prec \cdots \prec v_n$ of $G^\prec$. At the s-th stage the algorithm processes the vertex v_s as follows.

1. If $d^\prec(v_s) < \delta n$, then put v_s in Z. Color v_s by First-Fit applied to Z, using a palette of size δn.

2.1. Otherwise v_s is a witness vertex. Find a *live* i-witness set W, with i as large as possible subject to the condition that $v_s \in N^*(W)$. Such a witness set exists by the fact that $|N(v_s) \cap Z| \geq \delta n$ and so $v_s \in N^*(Z)$, provided we can prove (Lemma 7) that Eve is immortal.

2.2. Put v_s in the active part $X_W(t)$, $t = t_W$, of P_W. Color v_s by A_i applied to $X_W(t)$, using a palette of size $Cn^{\alpha(1-\varepsilon)}$. (By step 2.3, $|X_W(t)| \leq n^\alpha$.)

2.3. If after the addition of v_s, $|X_W(t)| = n^\alpha$, then set $t_W = t+1$ and set $X_W(t_W) = \emptyset$. Then $X_W(t)$ is large.

2.4. If $n^{\alpha(1-\varepsilon)}$ colors have been used on $X_W(t)$, then we have a proof that $\chi(X_W(t)) \geq i+1$. Set $t_W = t+1$ and set $X_W(t+1) = \emptyset$. (We may have just done this.) In this case, if $i = k$, then W dies. (This may cause some female ancestors of W to die of grief.) Otherwise $i < k$ and W spawns a new litter $\{D_v : v \in X_W(t)\}$, where each daughter D_v is a s_{i+1}-subset of $N(v) \cap W$. The father of D_v is v, for each $v \in X_W(t)$.

This completes the description of the algorithm A. To show that the algorithm is well defined, we need the following Lemma.

Lemma 7. *Eve is immortal.*

Proof. Suppose that Z is mortal. Let c be a proper $(k+1)$-coloring of G. First consider any mortal $(i-1)$-witness set M, with $i \in [k]$. Since M is mortal, M has a litter L such that every daughter $D \in L$ is mortal. When L is spawned, $\chi(\{F(D) : D \in L\}) \geq i$. Thus $|\{c(F(D)) : D \in L\}| \geq i$. It follows that, setting

$W_0 = Z$, we can find a collection $\{W_i : i \in [k]\}$ such that W_i is a mortal daughter of W_{i-1} and $|\{c(F(W_i)) : i \in [k]\}| = k$. Every father in the set $\{F(W_i) : i \in [k]\}$ is an ancestor of W_k and so is adjacent to every vertex in W_k. Thus c must color every vertex in W_k with the same color. It follows that c restricted to $N(W_k)$ is a proper k-coloring. Since W_k is mortal A_k must use at least $n^{\alpha(1-\varepsilon)}$ colors on the k-colorable graph induced by $N(W_k)$, which is a contradiction. □

Clearly A produces a proper coloring of G. It remains to bound the number of colors that A uses. The key step is the next lemma that bounds the number of litters a witness set can spawn.

Lemma 8. *Every i-witness set M has less than $2/\delta_i$ litters.*

Proof. We may assume that M is alive since after M dies, M will have no more litters. Then each litter of M contains a live $(i+1)$-witness set. Suppose W and U are two live daughters of M from distinct litters. Then there exist distinct j and j' such that $F(W) \in X_M(j)$ and $F(U) \in X_M(j')$. Say $j < j'$. At the stage that $F(U)$ is processed, all the vertices in W have already been processed. Thus $|W \cap U| < s_{i+2} = \frac{\delta_i^2}{2} s_i$, since otherwise $F(U)$ would be put in S_W instead of S_M. Thus by Lemma 5, M has less than $2/\delta_i$ litters. □

Let $Q = \{X_W(j) \in X : X_W(j) \text{ is small}\}$. We claim that $|Q| \leq 2^{k^2}(n^\alpha/\delta)^k$. For any i-witness set W, $|\{X_W(j) \in P_W : X_W(j) \text{ is small}\}|$ is at most one more than the number of litters of W. Note that a k-witness set spawns $l_k = 0$ litters and, by Lemma 8, an i-witness set spawns less than $l_i = 2/\delta_i$ litters. Let w_i be the number of i-witness sets. Then $w_0 = 1$ and $w_{i+1} \leq w_i l_i n^\alpha$, since each of the w_i i-witness sets spawns less than l_i litters of at most n^α daughters. It follows that $w_i \leq 2^{i(i+1)/2}(n^\alpha/\delta)^i$. So $|Q| \leq \sum_{0 \leq i \leq k} w_i l_i \leq 2w_k \leq 2^{k^2}(n^\alpha/\delta)^k$.

The algorithm A partitions V into at most $2^{k^2}(n^\alpha/\delta)^k$ small pieces and at most $n^{1-\alpha}$ large pieces. Each piece is colored with at most $Cn^{\alpha(1-\varepsilon)}$ colors, except that Z is colored with at most δn colors. Thus, setting $\delta = n^{-\alpha\varepsilon}$ and $\alpha = 1/(k+1+k\varepsilon)$, the number of colors used by the algorithm is at most

$$\begin{aligned} \delta n + Cn^{\alpha(1-\varepsilon)}\left(2^k n^\alpha/\delta\right)^k + Cn^{1-\alpha}n^{\alpha(1-\varepsilon)} &= \delta n + C2^{k^2}n^{\alpha(k+1+k\varepsilon-\varepsilon)} + Cn^{1-\alpha\varepsilon} \\ &= n^{1-\alpha\varepsilon} + 2^{k^2}Cn^{1-\alpha\varepsilon} + Cn^{1-\alpha\varepsilon} \\ &= \left(2C + 2^{k^2}\right)n^{1-\alpha\varepsilon}. \end{aligned}$$ □

We could improve the performance of the last algorithm if we could further limit the number of witness sets. One way to do this is to improve the bounds on the size of the litters. The size of the litters of an i-witness set W is determined by the least t_i such that whenever A_i uses more than a given number c_i of colors on the active part X of P_W, X contains a subgraph on t_i vertices whose chromatic number is at least $i+1$. Clearly $t_0 = 1$, $t_1 = 2$, and by Example 3 we can take $t_2 = 5$, when $c_0 = 0$, $c_1 = 1$, and $c_2 = 2n^{1/2}$. These observations lead to the more elegant bounds in the statement of Theorem 6 and, for small values of k, the following noticeable improvements in the algorithm.

Theorem 9. (Kierstead [13]).
There exists an on-line algorithm A_3 such that for every on-line 3-colorable graph $G^{\prec}$ on n vertices, $\chi_{A_3}(G^{\prec}) < 20n^{2/3}\log^{1/3} n$.

Theorem 10. (Kierstead [13]).
There exists an on-line algorithm A_4 such that for every on-line 4-colorable graph $G^{\prec}$ on n vertices, $\chi_{A_4}(G^{\prec}) < 120n^{5/6}\log^{1/6} n$.

These algorithms are not only on-line, but also run in polynomial time. From this point of view they are quite good since the best off-line algorithms for polynomial time coloring of 3-colorable graphs use $n^{3/14}$ colors ([2]). The author and Kolossa obtained much tighter bounds in the case of perfect graphs. Let $\pi(k, n)$ be the least integer t for which there exists an on-line algorithm A such that $\chi_A(G) \leq t$, for any k-colorable perfect graph G on n vertices.

Theorem 11. (Kierstead and Kolossa [19]).

$$\Omega\left(\log^{k-1} n\right) = \pi(k, n) = O\left(n^{10k/\log\log n}\right)$$

□

Example 3 illustrates the basic idea behind the proof of Theorem 11, but this idea must be iterated $\log^{(3)} n$ times and extended to graphs that do not induce any odd cycles. The actual proof is much more difficult and requires many on-line coloring techniques. The lower bound on π is derived from Vishwanathan's construction (Theorem 3). This suggests that the known lower bound on $\phi(k, n)$ is far from tight. But maybe this lower bound is close to the truth for $\pi(k, n)$. In the case of chordal graphs, Irani has an upper bound of the form $O(k \log n)$ (see Theorem 34). The following problems remain open and very interesting.

Problem 12. Find tighter bounds on $\phi(k, n)$ for fixed k and large n, especially for $k \in \{3, 4, 5\}$. Does $\phi(3, n) = O\left(n^{1/2}\right)$?

Problem 13. Find tighter bounds on $\pi(k, n)$ for fixed k and large n. Does there exist a function $p(k)$ such that $\pi(k, n) < \log^{p(k)} n$?

Vishwanathan studied randomized on-line algorithms. His lower bounds (Theorem 3) were actually proved for these more powerful algorithms.

Theorem 14. (Vishwanathan [36]).
There exists a randomized on-line algorithm A such that for every k-colorable on-line graph $G^{\prec}$ on n vertices, the expected value of $\chi_A\left(G^{\prec}\right) = O\left(k2^k n^{(k-2)/(k-1)}(\lg n)^{1/(k-1)}\right)$. Moreover, for any randomized on-line algorithm B, there exists a k-colorful on-line graph $G^{\prec}$ on n vertices such that the expected value of $\chi_B\left(G^{\prec}\right) = \Omega(1/(k-1)(\lg n/(12(k+1)) + 1)^{k-1})$. □

3 On-line χ-bounded classes

In this section we consider classes Γ of graphs, for which there exists an on-line algorithm A such that for all $G \in \Gamma$, $\chi_A(G)$ can be bounded by a function of $\omega(G)$ ($\leq \chi(G)$), regardless of the number of vertices in G. More precisely, we say that Γ is *on-line χ-bounded* iff there exists an on-line algorithm A and a function $g(k)$, such that $\chi_A(G^{\prec}) \leq g(\omega(G))$, for any on-line presentation $G^{\prec}$ of any $G \in \Gamma$. Similarly, Γ is *χ-bounded* if there exists a function $f(k)$ such that $\chi(G) \leq f(\omega(G))$, for all $G \in \Gamma$. Most of the time we will not be concerned with the size of the function g; the important point is that it does not depend on the number of vertices of G.

The results of this section have their roots in the author's previous work in recursive combinatorics and a beautiful graph theoretical conjecture formulated independently by Gyárfás and Sumner. The problems the author considered in recursive combinatorics can be very roughly described as follows. Given a countably infinite graph G, design an algorithm to color each vertex v of G using only certain types of local information (in particular, only finitely much information) about v. Depending on the amount of information allowed, in increasing order, the graphs may be recursive, highly recursive, or decidable. Usually, results about coloring recursive graphs, such as Bean [1], Kierstead [14], and Kierstead and Trotter [25] translate immediately to on-line results, while results on highly recursive or decidable graphs such as Kierstead [15] [16], Manaster and Rosenstein [28], and Schmerl [31] do not. The starting point for the work of this section is Theorem 15, which we will state after introducing some more terminology.

An *on-line ordered* set is an on-line digraph $D^{\prec}$ such that D is an ordered set, i.e., D is transitive, antisymmetric, and antireflexive. The *comparability* graph of an ordered set $D = (V, A)$ is the undirected graph $G = (V, E)$, where $E = \{xy : (x, y) \in A \text{ or } (y, x) \in A\}$. Similarly the *cocomparability* graph of D is the undirected graph $G^c = (V, E^c)$, where $E^c = \{xy : (x, y) \notin A \text{ and } (y, x) \notin A \text{ and } x \neq y\}$. A *chain (antichain)* in D is an independent set in the cocomparability graph G^c (comparability graph G). The *height (width)* of an ordered set is the number of vertices in the maximum chain (antichain). Notice that the height of D is the clique size of G and the width of D is the clique size of G^c. It is well known that G (and hence G^c) is perfect.

Theorem 15. (Kierstead [14]).
There exists an on-line algorithm A that will partition the vertices of any on-line ordered set of width w into at most $(5^w - 1)/4$ chains. □

Theorem 15 does not assert the existence of an on-line coloring algorithm that will color every on-line cocomparability graph $G^{\prec}$ with $(5^w - 1)/4$ colors. The problem is that the algorithm of Theorem 15 receives as input the digraph D of an ordered set; in general this provides more information than the cocomparability graph. This was shown rigorously by Penrice [29]. The best lower bounds for Theorem 15 were obtained by Szemerédi.

Theorem 16. (Szemerédi [35]).
For every integer w and on-line chain partitioning algorithm A, there exists an on-line ordered set $D^{\prec}$ with width w such that A partitions D into at least $\binom{w+1}{2}$ chains. □

The author [14] had previously derived a super linear lower bound and shown that at least five chains were necessary in the case $w = 2$. Recently Felsner has improved the upper bound in the case $w = 2$. (This also gives a slight improvement in the general upper bound.)

Theorem 17. (Felsner [5]).
There exists an on-line algorithm A that will partition the vertices of any on-line ordered set of width 2 into at most 5 chains. □

Over the last fifteen years Theorem 17 is the only progress on the following natural problem.

Problem 18. Let $p(w)$ be the least integer such that there exists an on-line algorithm A such that A will partition the vertices of any on-line ordered set of width w into at most $p(w)$ chains. Improve the bounds $\binom{w+1}{2} \leq p(w) \leq (5^w - 1)/4$. Is $p(w)$ polynomial? □

The analogous problem for antichains is much simpler.

Theorem 19. (Schmerl [30]).
There exists an on-line algorithm that will partition any on-line ordered set of height h into at most $\binom{h+1}{2}$ antichains; moreover for every positive integer h and on-line algorithm A, there exists an on-line ordered set $D^{\prec}$ such that A cannot partition $D^{\prec}$ into fewer than $\binom{h+1}{2}$ antichains.

Proof. We only prove the upper bound. Consider the input sequence $v_1, \ldots, v_n$ of an on-line ordered set D with height h. At stage s we process the vertex v_s by putting v_s into the antichain $A_{a,b}$ where $a = a(s)$ is the number of vertices in the longest chain (at stage s) with maximum element v_s and $b = b(s)$ is the number of vertices in the longest chain (at stage s) with minimum element v_s. Then $2 \leq a + b \leq h + 1$. It follows that there are at most $\binom{h+1}{2}$ choices for the sets $A_{a,b}$. To see that the sets $A_{a,b}$ really are antichains, consider two comparable vertices v_s and v_t with $s < t$. If $v_s < v_t$ in D, then $a(s) < a(t)$. Otherwise $v_s > v_t$ in D and $b(s) < b(t)$. □

Note that the class of comparability graphs is not on-line χ-bounded since it contains the class of trees which is not on-line χ-bounded by Example 1. The following conjecture of Schmerl from 1978 motivated a lot of work, including Example 4.

Conjecture 20. (*Schmerl*).
The class of cocomparability graphs is on-line χ-bounded. □

A solution of Schmerl's Conjecture required some purely graph theoretical results. For a graph H, let Forb(H) be the class of all graphs that do not contain an induced copy of H. Quite independently of any work on on-line algorithms, and independently of each other, Gyárfás and Sumner made the following conjecture.

Conjecture 21. (Gyárfás [7], Sumner [33]).
For any tree T, Forb(T) is χ-bounded. □

The conjecture is essentially the strongest possible. It would be false if T were replaced by a graph H that contained a cycle (say of length t), since any graph with girth greater than t is in Forb(H) and Erdös and Hajnal [4] have shown that there are graphs with arbitrarily large girth and arbitrarily large chromatic number. Moreover if the conjecture is true for trees it is easy to show that it is also true for forests. Gyárfás [8] gave an easy proof to show that it is true for paths. The author and Penrice built on earlier work of Gyárfas, Szemerédi, and Tuza [11], to prove the following off-line theorem. The general conjecture is still open.

Theorem 22. (Kierstead and Penrice [20]).
For any tree T with radius at most two, Forb(T) is χ-bounded.

Gyárfás and Lehel [10] made a fundamental and unexpected breakthrough in on-line coloring when they proved the following theorem.

Theorem 23. (Gyárfás and Lehel [10]).
Forb (P_5) is on-line χ-bounded, but Forb (P_6) is not on-line χ-bounded.

Theorem 23 was the first hint of a connection between Conjectures 20 and 21. Then Gyárfás made the following observation. Let S be the radius two tree obtained by subdividing each edge of a star on four vertices. It is well known that no cocomparability graph induces S and so the class of cocomparability graphs is contained in Forb(S). Thus to prove Conjecture 20, it suffices to show that Forb(S) is χ-bounded. With this challenge, the author, Penrice and Trotter proved the following very general theorem.

Theorem 24. (Kierstead, Penrice, and Trotter [22]).
For any tree T, Forb(T) is on-line χ-bounded iff T has radius at most two.

The proof of Theorem 24 is much too long to present here. However the proof for the special case of Forb(S), which provided most of the motivation for attacking the general theorem, is considerably simpler and illustrates many of the key techniques needed to prove the general theorem.

Corollary 25. *Forb(S) is on-line χ-bounded.*

Proof. We shall first prove that Forb(S) is off-line χ-bounded and then use the proof as a basis for constructing an on-line algorithm. Let $R(\omega, \alpha)$ be the Ramsey function such that for any graph G on $R(\omega, \alpha)$ vertices either $\omega(G) \geq \omega$ or $\alpha(G) \geq \alpha$. Let f be the function on the positive integers defined inductively by:

$$f(1) = 1 \text{ and } f(\omega) = f(\omega-1)+\omega+\omega f(\omega-1)\left(\omega^2\left(R\left(\omega, R(\omega+1,3)\right)+1\right)^2+1\right)$$

We shall prove by induction on $\omega(G)$, that $\chi(G) \leq f(\omega(G))$, for every graph $G = (V, E)$ in Forb(S). The base step, $\omega = 1$, is trivial, so consider a graph $G \in$ Forb(S) with $1 < \omega(G) = \omega$.

Let $\{Q_1, \ldots, Q_t\}$ be a maximal collection of ω-cliques in G such that for all distinct $i, j \in [t], N(Q_i) \cap N(Q_j) = \emptyset$. For each $i \in [t]$, let $N_i = N(Q_i) - \bigcup_{j<i} N(Q_j)$. Let $Q = \bigcup_{j \in [t]} Q_j$, $N = \bigcup_{j \in [t]} N_j$, and $X = V - Q - N$. The ω-cliques Q_i are called *templates* and the sequence $Q_1, N_1, \ldots, Q_t, N_t$ is called a *template sequence.* Since Q is a union of disjoint ω-cliques, we can color Q with ω colors. Since $\omega(X) < \omega$, by the induction hypothesis, we can color X with $f(\omega - 1)$ new colors. These $\omega + f(\omega - 1)$ colors will not be used on any of the vertices of $V - Q - X$, so it suffices to show that $\chi(N) \leq \omega f(\omega - 1)\left(\omega 2\left(R\left(\omega, R(\omega+1,3)\right)+1\right)^2+1\right)$.

Since each $N_i \subset \bigcup_{q \in Q} N(q)$, it can be colored with $\omega f(w-1)$ colors. However, if we try to color each of the N_i with the same set of $\omega f(w - 1)$ colors, two adjacent vertices $v_i \in N_i$ and $v_j \in N_j$ may be assigned the same color. To avoid this problem each vertex of N will be assigned a two coordinate color. The first coordinate, called the *local* color and assigned as above, will insure that two adjacent vertices in the same N_i are assigned different colors. The second coordinate, called the *global* color will take care of the problem of adjacency between vertices in different N_i. Since t is unbounded, we can not simply use disjoint sets of colors for the different N_i. We shall need the following lemma, which is the only place in the proof of the theorem that we use the hypothesis that $G \in$ Forb(S).

Lemma 26. *Let* $d(s) = 2\left(R\left(\omega, R(\omega+1,3)\right)+1\right)^{s-1}$. *Then for every vertex v in G, v is connected to vertices in at most $d(s)$ templates by paths with at most s edges.*

Proof. We argue by induction on s. The base step $s = 1$ follows from the fact $G \in$ Forb(S): Suppose $v \sim q_i$, where $q_i \in Q_i$, for $i \in \{j_1 < j_2 < j_3\}$. Since each Q_i is a maximum clique there exist vertices $y_i \in Q_i$ such that not $v \sim y_i$, for $i \in \{j_1 < j_2 < j_3\}$. But then $\{v, q_i, y_i : i \in \{j_1 < j_2 < j_3\}\}$ induces S in G, which is a contradiction.

For the induction step, suppose that a vertex v is connected to vertices in $d+1$ distinct templates by paths with at most s edges, where $d = d(s)$. Chose a minimal set of vertices $F \subset N(v)$ such that each of these $d+1$ templates contains a vertex which is either connected to some vertex in F by an induced path with exactly $s-1$ edges or is connected to v by a path with at most $s-1$ edges. By the induction hypothesis $(|F|+1)\,d(s-1) \geq d+1$, and so $|F| \geq R(\omega, R(\omega+1,3))$. Using Ramsey's Theorem, and the fact that every vertex in F is adjacent to v, there exists an independent subset $F_0 = \{v_1, \ldots, v_p\} \subset F$ with cardinality $p = R(\omega+1,3)$. Using the minimality of F, for each $v \in F_0$ there exists a template T_i such that v_i is the only vertex in $F \cup \{v\}$ to which any vertex of T_i is connected by an induced path (say R_i) with exactly $s-1$ edges. Say $v \sim v_i \sim y_i$ in R_i. Then neither $v \sim y_j$ nor $v_i \sim y_j$, if $i \neq j$. Using Ramsey's Theorem again, there exist $j_1 < j_2 < j_3$ such that $\{y_i : i \in \{j_1, j_2, j_3\}\}$ is an independent set. But then $\{v, v_1, y_i : i \in \{j_1, j_2, j_3\}\}$ induces S in G, which is a contradiction. □

In order to assign a global color to the vertices in N, we construct an auxiliary graph B. The vertex set of B is the set of templates $\{Q_1, \ldots, Q_t\}$. Two templates Q_x and Q_y are adjacent if there is a path from one to the other with at most three edges. By Lemma 26, the maximum degree of B is at most $\omega d(3)$, since each of the ω vertices in Q_i can be connected to at most $d(3)$ templates by paths with at most three edges. Thus B can be colored with $\omega d(3)+1$ colors. The global color of a vertex $v \in N_i$ is the color of Q_i in B. To see that this gives a proper coloring, consider two adjacent vertices x and y in N. If x and y have the same local color, there exist distinct indices i and j such that $x \in N_i$ and $y \in N_j$. But then Q_i is adjacent to Q_j in B, so x and y are assigned different global colors. This completes the proof of the off-line case.

We still must show that there exists a function g and an on-line algorithm A such that $\chi(G) \leq g(\omega(G))$, for every graph $G \in \mathrm{Forb}(S)$. It suffices to show by induction on ω that there exist on-line algorithms A_ω, for $\omega = 1, 2, \ldots$ and a function $h(\omega)$ such that $\chi_A(G) \leq h(\omega)$, for every graph $G \in \mathrm{Forb}(S)$ such that $\omega(G) = \omega$: First guess that $\omega = 1$ and use A_1. If a 2-clique is found, guess that $\omega(G) = 2$ and start using A_2 with a new set of colors, etc. Then $g(\omega) = \sum_{j \leq \omega} h(\omega)$. The base step is trivial, so consider the induction step.

The major problem in developing an on-line algorithm from the proof of the off-line case is that we cannot possibly construct the template sequence on-line. We can only add a template to the sequence after we have seen all the vertices in the clique, but by this time we may have missed some of its neighbors. The key idea is to consider the neighbors of neighbors of vertices in the templates. A minor problem will be that we can not maintain the auxiliary graphs on-line. Our on-line algorithm $A = A_\omega$ will maintain a list of templates $Q_1, \ldots, Q_{t(s)}$, where $Q_i = \{x_{i,1} \prec \cdots \prec x_{i,\omega}\}$. A template Q_i will enter the end of the list at the time $x_{i,\omega}$ is presented. Once a template has entered the list it will not change position or leave the list. When a new vertex v_s is presented, A will assign v_s to exactly one of the sets N, L, D, or H. The sets N and D are more finely partitioned as: $N = \bigcup\{N_{i,j} : i \in \{1, \ldots, t(s)\}, j \in \{1, \ldots, \omega\}\}$ and $D = \bigcup\{D_i : i \in \{1, \ldots, t(s)\}\}$. Then each of the sets of vertices N, L, D, H, will be colored with disjoint sets of colors.

(N) If v_s is adjacent to some vertex in some template in the current template list, let i be the least such index such that $v_s \sim x_{i,k}$, for some k, and let j be the least such k. Put v_s in $N_{i,j}$.

(L) Otherwise, if v_s is in an ω-clique Q in $G - \left(N \cup \bigcup_{1\le i\le t(s)} Q_i\right)$, then add $Q = Q_{t(s)+1}$ to the template list and put v_s in L.

(D) Otherwise, if v_s is connected to some vertex in some template in the template list by a path with two edges, let i be the *largest* index such that, for some j, v_s is connected to some $x_{i,j}$ by a path with two edges. Put v_s in D_i.

(X) Otherwise put v_s in X.

Clearly $\omega(X) < \omega$. Thus the on-line algorithm A can use the on-line algorithm $A_{\omega-1}$ to color the vertices of H with one set of $h(\omega - 1)$ colors. Also L is an independent set, so we can use one special color to color L. Thus it remains to color the vertices of N and D. As in the off-line proof, each vertex in N, and also D, will be assigned a two coordinate color. The local color insures that two adjacent vertices in $N_i = \bigcup_{1\le j\le\omega} N_{i,j}$ or D_i are assigned different colors, while the global color insures that two adjacent vertices with the same local color are assigned different colors.

We consider the local coordinate. For N the local coordinate is assigned as in the off-line case, using the on-line algorithm $A_{\omega-1}$. In order to use $A_{\omega-1}$ to assign a local coordinate to the vertices of D_i, we must first show that $\omega(D_i) < \omega$. Suppose $Q = \{q_1 \prec \ldots \prec q_\omega\}$ is an ω-clique in D_i and consider the situation at the time q_ω was presented. Since q_ω was added to D_i instead of N, q_ω is not adjacent to any vertex in any template in the template list. Since q_ω was not added to L, some $q \in Q$ must be adjacent to some vertex in a template Q_j. Thus q_ω is connected to Q_j by a path through q with two edges. Since q is not in N, Q_j must have been added to the template list after q was presented and thus $i < j$. But then q_ω would have been assigned to D_j. We conclude that $\omega(D_i) < \omega$, for all i, and assign the local coordinate to each vertex in D using $A_{\omega-1}$.

It remains to determine the global color. First consider the vertices of N. If we could color the vertices of the auxiliary graph B on-line, we would be done. However this is not possible since the auxiliary graph is not presented on-line. Two templates Q_x and Q_y may start out being non-adjacent, but when a new vertex of $G^{\prec}$ is presented they may suddenly become adjacent. On the other hand, the degree of a template in B can only increase $\omega d(3, \omega)$ times. The on-line algorithm A will maintain a two coordinate, $(\omega d(3) + 1)^2$-coloring of B such that (i) the first coordinate of a template is the current degree of the template in B, (ii) the second coordinate insures that two templates which are adjacent in B and have the same degree in B are assigned different colors, and (iii) the second coordinate of a color assigned to a template will only change when the degree of the template changes. The global color of a vertex in N_i will be the color assigned to Q_i in B by A at the time the vertex is presented. To assign the global color to a vertex in D, we define another auxiliary graph B' on the templates, where two templates Q_x and Q_y are adjacent iff there is a path from a vertex in Q_x to a vertex in Q_y with at most five edges. Since each template has ω vertices, the maximum degree of B' is bounded by $\omega d(5, \omega)$. Thus as above we need only $(\omega d(5, \omega) + 1)^2$ colors for the global color of vertices in D. Thus $h(\omega)$

is defined recursively by $h(1) = 1$ and

$$h(\omega) = \omega h(\omega - 1)\left(\omega d(3, \omega) + 1\right)^2 + 1 + h(\omega - 1)\left(\omega d(5, \omega) + 1\right)^2 + h(\omega - 1) \;.\square$$

4 First-Fit χ-bounded classes

In this section we consider classes of graphs for which First-Fit performs reasonably well in comparison to the best on-line algorithm for the class. We begin by continuing our study of classes of graphs defined by forbidding certain induced subgraphs. Later we consider various classes of d-degenerate graphs such as trees, interval graphs, and chordal graphs. A First-Fit coloring of a graph $G = (V, E)$ with the colors $[t]$ produces a structure called a *wall*. A wall W in G is a partition $\{L_1, \ldots, L_t\}$ of V into independent sets such that for all vertices $v \in L_j$, there exists a vertex $u \in L_i$ such that v is adjacent to u, whenever $i < j$. The independent sets L_j are called *levels* of the wall. The *height* of W is t. It is easy to see that G contains a wall of height t iff $t \leq \chi_{\mathrm{FF}}(G)$.

A class of graphs Γ is First-Fit χ-bounded if there exists a function f such that for all graphs $G \in \Gamma$, $\chi_{\mathrm{FF}}(G) \leq f(\omega(G))$. It follows immediately from Ramsey's theorem that if S is a star, then Forb(S) is First-Fit χ-bounded. Also it is well known that any graph in Forb(P_4) is perfect, and moreover First-Fit produces an optimal coloring, where P_k is the path on k vertices. Gyárfás and Lehel [9] showed that if T is a tree that is not in Forb$(K_1 \cup K_1 \cup K_2)$, then Forb(T) is not First-Fit χ-bounded. Since P_5 is the only tree in Forb$(K_1 \cup K_1 \cup K_2)$ that is neither a star nor P_4, they asked whether Forb(P_5) was χ-bounded. This was answered affirmatively by the author, Penrice and Trotter.

Theorem 27. (Kierstead, Penrice and Trotter [23]).
The class Forb(P_5) is First-Fit χ-bounded, and thus, for any tree T, Forb(T) is First-Fit χ-bounded iff T does not contain $K_1 \cup K_1 \cup K_2$ as an induced subgraph.

Proof. We must show that there exists a function f such that for all $G \in$ Forb(P_5), $\chi(G) \leq f(\omega(G))$. We argue by induction on $\omega(G)$. The base step $\omega(G) = 1$ is trivial, so consider the induction step $\omega(G) = k$. Let $R(a_1, a_2, a_3)$ be the Ramsey function such that for any 3-coloring of the edges of the complete graph on $R(a_1, a_2, a_3)$ vertices, there exists $i \in [3]$ and a complete subgraph on a_i vertices whose edges are all colored i. The following Lemma provides the main technical tool for defining $f(k)$.

Lemma 28. *For any integer t, if a graph $G \in$ Forb(P_5) with clique size k contains a wall W of height $1 + R(f(k-1)+1, t, 3)$, then for every vertex x in the top level of W, there exists an induced subgraph H contained in $G - x$ that has a wall W' of height t such that the top level of W' has exactly one vertex y, which is the only vertex of H adjacent to x.*

By the lemma, for sufficiently large t, we can let $f(k) = 1 + R(f(k-1)+1, t, 3)$, since if there were a graph $G \in$ Forb(P_5) with clique size k that contained a wall

W of height $1+R(f(k-1)+1,t,3)$, then by iterating the lemma four times we could find vertices $x_1 \sim y_1 = x_2 \sim y_2 = x_3 \sim y_3 = x_4 \sim y_4 = x_5$ that induce P_5. This would be a contradiction.

Proof of Lemma 28. Fix an integer t, a graph $G \in \text{Forb}\,(P_5)$, and a vertex x such that $\omega(G) = k$ and G contains a wall $W = \{L_1, \ldots, L_{t'}\}$ of height $t' = 1+R(f(k-1)+1,t,3)$ with $x \in L_{t'}$. Then for every $s \in [t'-1]$, $N(x) \cap L_s \neq \emptyset$. Define a function g on the 2-subsets of [t-1] by

$$\begin{aligned}
&\text{if } \forall v \in L_j \cap N(x) \exists u \in L_i \cap N(x) v \sim u, \text{ then } g(i<j) = 1\\
&\text{else if } \forall v \in L_j \cap N^c(x) \exists u \in L_i \cap N^c(x) v \sim u, \text{ then } g(i<j) = 2\\
&\text{else } g(i<j) = 3\ .
\end{aligned}$$

Then by the choice of t', there exists a subset $S \subset [t'-1]$ such that all pairs in S have the same color α and either (1) $\alpha = 1$ and $|S| = f(k-1)+1$ or (2) $\alpha = 2$ and $|S| = t$ or (3) $\alpha = 3$ and $|S| = 3$. Note that if (1) holds, then $\{L_s \cap N(x) : s \in S\}$ is a wall, and if (2) holds, then $\{L_s \cap N^c(x) : s \in S\}$ is a wall.

Suppose (1) holds. Let F be the subgraph of G induced by $N(x)$. Since $\{L_s \cap N(x) : s \in S\}$ is a wall in F of height $f(k-1)+1$, the induction hypothesis implies that $\omega(F) \geq k$. Since x is adjacent to every vertex in F, $\omega(F) \geq k+1$, which is a contradiction. So (1) is impossible.

Suppose (3) holds. Say $S = \{q < r < s\}$. Since $g(q<r) \neq 1$ and $g(r<s) \neq 1$, there exist $v \in L_r \cap N(x)$ and $w \in L_s \cap N(x)$ such that v is not adjacent to any vertex in $L_q \cap N(x)$ and w is not adjacent to any vertex in $L_r \cap N(x)$. Thus $L_q \cap N^c(x) \cap N(v)$ and $L_r \cap N^c(x) \cap N(w)$ are nonempty. Let $u' \in L_q \cap N^c(x) \cap N(v)$. Since $1 \neq g(q<r) \neq 2$, there exists $v' \in L_r \cap N^c(x)$ such that v' is not adjacent to any vertex in $L_q \cap N^c(x)$. Thus v' is not adjacent to u' and there exists $u \in L_q \cap N(x) \cap N(v')$. Since v is not adjacent to any vertex in $L_q \cap N(x)$, v is not adjacent to u. Since L_q and L_r are independent sets, u is not adjacent to u' and v is not adjacent to v'. Thus (v', u, x, v, v') induces P_5, which is a contradiction. So (3) is impossible.

Thus (2) must hold. Let s be the largest element in S and let $y \in L_s \cap N(x)$. Let H be the subgraph of G induced by $\{y\} \cup \bigcup_{r \in S-\{s\}} \{L_r \cap N^c(x)\}$. Then y is the only vertex of H adjacent to x and $\{y\} \cup \{L_r \cap N^c(x) : r \in S - \{s\}\}$ is a wall in H of height t with top level $\{y\}$.

This completes the proofs of Lemma 28 and Theorem 27. □

Theorem 27 completely answers the question of whether Forb(T) is First-Fit χ-bounded for any tree T. Next consider the problem in which a tree and some other graphs are forbidden. The following theorem of this type is a key result in the proof of Theorem 24. Its proof requires a special Ramsey theoretic result due to Galvin, Rival, and Sands [6].

Theorem 29. (Kierstead and Penrice [21]).
For every tree T and complete bipartite graph $K_{t,t}$, Forb$\,(T, K_{t,t})$ is First-Fit χ-bounded.

Let B_t be the graph introduced in Example 6. Let D_k be the tree obtained by adding an edge between the roots of two disjoint copies of S_k, where S_k is the star on k vertices whose root is its only nonleaf.

Theorem 30. (Kierstead, Penrice and Trotter [23]).
The classes Forb $(P_{5,1}, B_t)$ *and Forb* (D_k, B_t) *are First-Fit* χ*-bounded.*

Consider a graph $G \in \text{Forb}(P_{5,1})$ on which First-Fit uses a huge number of colors. Theorem 30 explains why First-Fit uses at least a large numbers of colors on G. The reason is that G contains an induced subgraph that is either a large clique or a large induced B_t. Since First-Fit is known to require a large number of colors on either of these subgraphs, First-Fit requires a large number of colors on G. It would be very nice to prove a theorem of the following form: If $\chi_{\text{FF}}(G) > g(k)$, then G induces a graph in Q_k, where Q_k is a finite set of graphs such that $\chi_{\text{FF}}(H) \geq k$, for all $H \in Q_k$. The author, Penrice and Trotter [23] generalized the construction of B_t as follows. Call a graph $H = (V, E)$ *t-bad* if there exist sets $A_1, \ldots, A_t$ such that:

(1) $V = A_1 \cup \ldots \cup A_t$;
(2) $A_j = \{a_{1,j}, \ldots, a_{j,j}\}$ is an independent set of vertices for $j \in [t]$;
(3) $A_j \cap A_{j+1} = \emptyset$, for $j \in [t]$;
(4) $a_{i,k} \not\sim a_{i,j}$, whenever $1 \leq i \leq k < j \leq t$; and
(5) $a_{i,j} \sim a_{k,j+1}$, whenever $1 \leq i < k \leq j+1 \leq t$.

If H is t-bad then $\chi_{\text{FF}}(H) \geq t$. Note that it is not required that $A_k \cap A_j = \emptyset$, if $|j-k| \geq 2$. If $a_{i,j} = a_{i,j+2}$, whenever $1 \leq i \leq j \leq t-2$, then H is just B_t with one vertex removed. It is easy to see that it is possible to present the vertices so that $a_{i,j}$ precedes $a_{r,s}$ if $i < r$ and that when the vertices are presented in such an order, First-Fit uses t colors.

Problem 31. Let $Q_k = \{T_k\} \cup \{H : H \text{ is } k\text{-bad}\}$. Does there exist a function $g(k)$ such that if $\chi_{\text{FF}}(G) > g(k)$, then G contains an induced subgraph H such that $H \in Q_k$? □

We have already seen in Examples 5 and 6 that First-Fit performs optimally on trees. For interval orders, First-Fit is not optimal. Example 4 showed that there is an on-line algorithm that will color any interval graph G with $3\omega(G) - 2$ colors, while Chrobak and Slusarek showed the following.

Theorem 32. (Chrobak and Slusarek [3]).
There exists a constant C such that for every positive integer k, there exists an interval graph G with $\omega(G) = k$ such that First-Fit uses at least $4.4k - C$ colors on G. □

However the author answered a question of Woodall [37] by proving the following theorem that shows that First-Fit is close to optimal for interval graphs.

Theorem 33. (Kierstead [17]).
For any interval graph G, $\chi_{\text{FF}}(G) \leq 40\omega(G)$.

Proof. Identify the vertices of G with the intervals of an interval representation of G. Let $L \subset V$. An interval I has *density* $d = d(I/L)$ in L if every point in I is in at least d intervals in L. If K is a k-clique in G, then some interval in the representation of K has density at least $d/2$. To see this, alternately remove from K the intervals with the least left endpoint and the greatest right endpoint. When this process terminates the last interval will be contained in each pair of removed intervals and thus will have density at least $d/2$. Next we define the notion of *centrality.* Consider two adjacent intervals I and J in a set of intervals L. Let $N = N(J) \cap L$. Then N is a set of intervals and so defines an interval order P. Let $\lambda(I/J, L)$ be the length of the longest chain of intervals less than or equal to I in P. Note that if $I' \in N$ satisfies $\lambda(I/J, L) = \lambda(I'/J, L)$, then I is adjacent to I'. Similarly, let $\rho(I/J, L)$ be the length of the longest chain of intervals greater than or equal to I in P. Again, if $\rho(I/J, L) = \rho(I'/J, L)$, then I is adjacent to I'. The *centrality* of I in J with respect to L is $c(I/J, L) = \min\{\lambda(I/J, L), \rho(I/J, L)\}$. Note that $c(I/J, L) > 0$ iff $I \subset J$.

Let W be the wall of height h associated with a First-Fit coloring of G. We shall actually prove that W contains an interval with density $h/40$. Begin by setting I_1 equal to any interval in the top level of W and $d_1 = 1$. Now suppose that we have constructed a sequence $S = (I_1, \ldots, I_k)$ of intervals and a sequence $D = (d_1, \ldots, d_k)$ of integers. (I_1 and d_1 may have changed.) Let X_1 be the top $40d_1$ levels of W, X_2 be the next $40d_2$ levels of W, etc. Let T_t be the top $20d_t$ levels of X_t and B_t be the bottom $20d_t$ levels of X_t, for all $t \in [k]$. Also, let $B_{-1} = \phi$. We shall write X_t^*, T_t^*, and B_t^* for $\bigcup X_t$, $\bigcup T_t$, and $\bigcup B_t$, respectively. So for example X_t is a set of levels, while X_t^* is the set of intervals contained in the levels in X_t. Suppose further that the pair (S, D) is *acceptable*, i.e.,

(1) $c\left(I_{t+1}/I_t, B_t^* \cup T_{t+1}^*\right) \geq 2$, for all $t \in [k-1]$;
(2) $d\left(I_t/B_{t-1}^* \cup T_t^*\right) \geq d_t$, for all $t \in [k]$; and
(3) $d_{t+1} \geq 3^{-c+2}$, where $c = c\left(I_{t+1}/I_t, B_t^* \cup T_{t+1}^*\right)$, for all $t \in [k-1]$.

Then $d\left(I_k/X\right) \geq d = \sum D = \sum_{t \in [k]} d_t$, where X is the union of $\{X_1, \ldots, X_k\}$. It suffices to show that if $d < h/40$, then we can find a new acceptable pair (S', D') with $d' > d$, where $d' = \sum D'$. Note that in this case B_k is well defined and we have not yet used any of the intervals in B_k. We make further progress by taking advantage of these unused intervals.

Consider $N = N\left(I_k\right) \cap B_k^*$. As above N can be partitioned into cliques $K_0, K_0', \ldots, K_c, K_c'$ so that for all indices i, $\lambda\left(I/I_k, B_t^*\right) = i = c\left(I/I_k, B_t^*\right)$, for all intervals $I \in K_i$ and $\rho\left(I/I_k, B_t^*\right) = i = c\left(I/I_k, B_t^*\right)$, for all intervals $I \in K_i'$. Since $|N| \geq 20d_k = 2(1+2+2+5)$, either

(i) $|K_i| \geq 2\left\lceil 3^{-i+2} d_k \right\rceil$ or $|K_i'| \geq 2\left\lceil 3^{-i+2} d_k \right\rceil$, for some i such that $3 \leq i \leq \log_3 d_k$,
(ii) $|K_2| \geq 2d_k$, $|K_2'| \geq 2d_k$, $|K_1| \geq 2d_k$, or $|K_1'| \geq 2d_k$,
(iii) $|K_0| \geq 5d_k$ or $|K_{0'}| \geq 5d_k$.

(We can improve the argument at this point by showing that $|N|$ is actually considerably larger than $20d_k$.) In each case, we assume without loss of generality that the condition is met by K_i.

Case (i). The clique K_i contains an interval I_{k+1} such that $d\left(I_{k+1}/B_k^*\right) \geq 3^{-i+2}d_k$ and $c\left(I/I_k, B_t^*\right) = i \geq 2$. Thus setting $S' = (I_1, \ldots, I_k, I_{k+1})$ and $D' = \left(d_1, \ldots, d_k, \lceil 3^{-i+2}d_k \rceil\right)$ yields an acceptable pair (S', D') with $d' > d$.

Case (ii). Let $|K_i| \geq 2d_k$, where $i \in \{1,2\}$. Then K_i contains an interval I such that $d\left(I/B_k^*\right) \geq d_k$ and $c\left(I/I_k, B_k^*\right) \geq 1$. Thus $I \subset I_k$. It follows that $d\left(I/B_{k-1}^* \cup T_k^* \cup B_k^*\right) \geq 2d_k$ and $c\left(I/I_{k-1}, B_{k-1}^* \cup T_k^* \cup B_k^*\right) \geq c(I_k/I_{k-1}, B_{k-1}^* \cup T_k^* \cup B_k^*) \geq 2$. Note that $T_k \cup B_k$ consists of the next $21^* \left(2d_k\right)$ levels after B_{k-1}. Thus setting $S' = (I_1, \ldots, I_{k-1}, I)$ and $D' = (d_1, \ldots, d_{k-1}, 2d_k)$ yields an acceptable pair (S', D') with $d' > d$.

Case (iii). The left endpoint of d_k is contained in $5d_k$ intervals from $K_0 \subset B_k^*$ and d_k additional intervals from $B_{k-1}^* \cup T_k^*$ that witness that $d(I_k/B_{k-1}^* \cup T_k^*) \geq d_t$. Thus for one of these intervals I, $d(I/B_{k-1}^* \cup T_k^* \cup B_k^*) \geq 3d_k$. Let J be the interval in $B_{k-1}^* \cup T_k^*$ with $c(J/I_{k-1}, B_{k-1}^* \cup T_k^*) = c-1$, where $c = c(I_k/I_{k-1}, B_{k-1}^* \cup T_k^*)$. If $J \subset I$, let $I' = J$; otherwise let $I' = I$. In either case $c(I'/I_{k-1}, B_{k-1}^* \cup T_k^* \cup B_k^*) \geq c-1$ and $d(I'/B_{k-1}^* \cup T_k^* \cup B_k^*) \geq 3d_k$. First suppose that $c > 2$. Note that $T_k \cup B_k$ is contained in the first 21^*3d_k levels below B_{k-1}. Thus setting $S' = (I_1, \ldots I_{k-1}, I')$ and $D' = (d_1, \ldots, d_{k-1}, 3d_k)$ yields an acceptable pair (S', D') with $d' > d$. Otherwise $c = 2$. Then $d_k \geq d_{k-1}$ and $I' \subset I_{k-1}$. Thus $d(I'/B_{k-2}^* \cup T_{k-1}^* \cup B_{k-1} \cup T_k^* \cup B_k^*) \geq 2d_{k-1} + 2d_k$ and $c(I'/I_{k-2}, B_{k-2}^* \cup T_{k-1}^* \cup B_{k-1}^* \cup T_k^* \cup B_k^*) \geq c(I_{k-1}/I_{k-2}, B_{k-2}^* \cup T_{k-1}^* \cup B_{k-1}^* \cup T_k^* \cup B_k^*)$. Thus setting $S' = (I_1, \ldots, I_{k-2}, I')$ and $D' = (d_1, \ldots, d_{k-2}, 2d_{k-1} + 2d_k)$ yields an acceptable pair (S', D') with $d' > d$. □

The proof of Theorem 4.7 can be strengthened to obtain a better constant. The author and Qin [24] showed that at most $26\omega(G)$ colors are used by First-Fit to color interval graphs. I strongly believe that the real constant is less than ten.

A graph $G = (V, E)$ is *d-degenerate* (sometimes called d-inductive) if there exists an ordering $v_1 \prec v_2 \prec \cdots \prec v_n$ of the vertices of G so that $|\{w \in V : w \sim v \text{ and } w \prec v\}| \leq d$, for all $v \in V$. For example planar graphs are 5-inductive and trees are 1-inductive. If G is a d-degenerate graph, then clearly $\chi(G) \leq d+1$. The graph G is chordal if every cycle in G of length at least four contains a chord, i.e, an edge between two nonconsecutive vertices. There are two well known characterizations of chordal graphs. The first is that G is chordal iff there exists an ordering $v_1 \prec v_2 \prec \cdots \prec v_n$ of the vertices of G so that $\{w \in V : w \sim v \text{ and } w \prec v\}$ is a clique, for all $v \in V$. Thus if G is chordal, then G is $(\omega(G) - 1)$-degenerate. It follows that chordal graphs are perfect. The other characterization is that G is chordal iff there exists a mapping f of the vertices of G to subtrees of a tree T such that two vertices u and v are adjacent iff $E(f(u)) \cap E(f(v)) \neq \emptyset$. Thus interval graphs are chordal since intervals can be represented as subpaths of a path. Irani showed that First-Fit performs close to optimally on the class of d-degenerate graphs.

Theorem 34. (Irani [12]).
There exists a constant C such that First-Fit uses at most $Cd\log n$ colors to

color any d-degenerate graphs on n vertices. Moreover there exists a constant C' such that for every on-line algorithm A and all integers d and n with $n > d^3$, there exists a d-degenerate graph G on n vertices such that $\chi_A(G) \geq C'd\log n$.

We end this section with two more applied problems. First we consider the Broadcast Problem, which is a generalization of Path Coloring on Trees.

Broadcast Problem.
INSTANCE: Tree T and a set $S = \{T_1, \ldots, T_n\}$ of subtrees of T.
PROBLEM: Color the elements of S with as few colors as possible subject to the condition that two subtrees that share an edge must receive different colors.

By the discussion above, the Broadcast Problem is equivalent to coloring chordal graphs. Thus by Theorem 34 First-Fit uses at most $\chi \log n$ colors, where χ is the optimal number of colors. If the underlying tree of the Broadcast problem is a path, then the problem reduces to interval graph coloring and First-Fit uses at most 42χ colors. Irani left open the problem of whether there exists a chordal graph G on n vertices for which First-Fit uses $\Omega(\omega(G)\log n)$ colors.

The second application is the following storage problem that was shown to be NP-complete by Stockmeyer [32]. Here we shall only be interested in the off-line version; however we will use on-line interval graph coloring to obtain a polynomial time approximation for this off-line version.

Dynamic Storage Allocation (DSA).
INSTANCE: Set A of items to be stored, each $a \in A$ having a positive integer size $s(a)$, a non-negative integer arrival time $r(a)$, and a positive integer departure time $d(a)$, and a positive storage size D.
PROBLEM: Is there a feasible allocation of storage for A, i.e., a function $\sigma : A \to \{1, 2, \ldots, D\}$ such that for every $a \in A$ the allocated storage interval $I(a) = [\sigma(a), \sigma(a)+1, \ldots, \sigma(a)+s(a)-1]$ is contained in $[1, D]$ and such that, for all $a, a' \in A$, if $I(a) \cap I(a')$ is nonempty then either $d(a) \leq r(a')$ or $d(a') \leq r(a)$?

Notice that if all the items have the same size, then DSA is just interval graph coloring. The following approximation algorithm for DSA has been well known since the late sixties. First put each item into the smallest possible box whose size is a power of two. Then order the boxes by decreasing size and use First-Fit to assign the boxes to storage locations. Clearly this algorithm produces a proper assignment of storage locations. Let ω^* be the maximum, over all times t, of the sum of the sizes of the boxes that must be stored at time t. Then ω^* is a lower bound on the amount of storage that must be used. It is easy to show that, because of the uniformity in box size and the use of First-Fit, this algorithm uses at most $f(\omega^*)$ storage locations, where $f(k)$ is the maximum number of colors used by First-Fit to color an interval graph with clique number k. Thus Theorem 34 has the following corollary.

Corollary 35. (Kierstead [17]).
There is a polynomial time approximation algorithm for Dynamic Storage Allocation with a constant performance ratio of 80. □

It later turned out that a slight modification of the algorithm in Example 4 had this same property. So we have the following corollary to Example 4.

Corollary 36. (Kierstead [18]).
There is a polynomial time approximation algorithm for Dynamic Storage Allocation with a constant performance ratio of six. □

This is an example of the following hypothetical situation. An off-line optimization problem (DSA) can be reduced to a simpler problem (interval graph coloring) if the data is preordered in a certain way (decreasing box size). However this preordering turns the simpler problem into an on-line problem. I know of no other examples of this situation, but if it turns out that there are other examples, this could be a very important application for on-line theory.

Acknowledgements

This work was partially supported by Office of Naval Research grant N00014-90-J-1206.

References

1. D. Bean. Effective coloration. *J. Symbolic Logic*, 41:469–480, 1976.
2. A. Blum and D. Karger. An $\tilde{O}(n^{3/14})$-coloring for 3-colorable graphs. Preprint.
3. M. Chrobak and M. Ślusarek. On some packing problems relating to Dynamical Storage Allocations. *RAIRO Informatique Theoretique*, 22:487–499, 1988.
4. P. Erdős and A. Hajnal. On chromatic numbers of graphs and set systems. *Acta Math. Sci. Hung.*, 17:61–99, 1966.
5. S. Felsner. On-line chain partitions of orders. To appear in *Theoret. Comput. Sci.*
6. F. Galvin, I. Rival, and B. Sands. A Ramsey-type theorem for traceable graphs. *J. Comb. Theory, Series B*, 33:7–16, 1982.
7. A. Gyárfás. On Ramsey covering numbers. In *Coll. Math. Soc. János Bolyai 10, Infinite and Finite Sets*, pages 801–816. North-Holland/American Elsevier, New York, 1975.
8. A. Gyárfás. Problems from the world surrounding perfect graphs. *Zastowania Matematyki Applicationes Mathemacticae*, 19:413–441, 1985.
9. A. Gyárfás and J. Lehel. On-line and first-fit colorings of graphs. *Journal of Graph Theory*, 12:217–227, 1988.
10. A. Gyárfás and J. Lehel. Effective on-line coloring of P_5-free graphs. *Combinatorica*, 11:181–184, 1991.
11. A. Gyárfás, E. Szemerédi, and Z. Tuza. Induced subtrees in graphs of large chromatic number. *Discrete Math*, 30:235–244, 1980.
12. S. Irani. Coloring inductive graphs on-line. *Algorithmica*, 11:53–72, 1994. Also in Proc. 31st IEEE Symposium on Foundations of Computer Science, 1990, 470-479.
13. H. A. Kierstead. On-line coloring k-colorable graphs. To appear in *Israel J. of Math.*
14. H. A. Kierstead. An effective version of Dilworth's theorem. *Trans. Amer. Math. Soc.*, 268:63–77, 1981.

15. H. A. Kierstead. Recursive colorings of highly recursive graphs. *Canadian J. Math.*, 33:1279–1290, 1981.
16. H. A. Kierstead. An effective version of Hall's theorem. *Proc. Amer. Math. Soc.*, 88:124–128, 1983.
17. H. A. Kierstead. The linearity of first-fit coloring of interval graphs. *SIAM Journal on Discrete Math*, 1:526–530, 1988.
18. H. A. Kierstead. A polynomial time approximation algorithm for dynamic storage allocation. *Discrete Math.*, 88:231–237, 1991.
19. H. A. Kierstead and K. Kolossa. On-line coloring of perfect graphs. To appear in *Combinatorica.*
20. H. A. Kierstead and S. G. Penrice. Recent results on a conjecture of Gyárfás. *Congressus Numerantium*, 79:182–186, 1990.
21. H. A. Kierstead and S. G. Penrice. Radius two trees specify χ-bounded classes. *J. Graph Theory*, 18:119–129, 1994.
22. H. A. Kierstead, S. G. Penrice, and W. T. Trotter. On-line graph coloring and recursive graph theory. *Siam Journal on Discrete Math.*, 7:72–89, 1994.
23. H. A. Kierstead, S. G. Penrice, and W. T. Trotter. First-fit and on-line coloring of graphs which do not induce P_5. *Siam Journal on Discrete Math.*, 8:485–498, 1995.
24. H. A. Kierstead and J. Qin. Coloring interval graphs with First-Fit. *Discrete Math.*, 144:47–57, 1995.
25. H. A. Kierstead and W. T. Trotter. An extremal problem in recursive combinatorics. *Congressus Numerantium*, 33:143–153, 1981.
26. L. Lovász. *Combinatorial Problems and Exercises.* Akadémiai Kiadó, Budapest, 1993. Problem 13.13.
27. L. Lovász, M. Saks, and W. T. Trotter. An on-line graph coloring algorithm with sublinear performance ratio. *Discrete Mathematics*, 75:319–325, 1989.
28. A. B. Manaster and J. G. Rosenstein. Effective matchmaking (recursion theoretic aspects of a theorem of Philip Hall). *Proc. Lond. Math. Soc.*, 25:615–654, 1972.
29. S. G. Penrice. On-line algorithms for ordered sets and comparability graphs. Preprint.
30. J. H. Schmerl. Private communication (1978).
31. J. H. Schmerl. Recursive colorings of graphs. *Canad. J. Math.*, 32:821–830, 1980.
32. I. J. Stockmeyer. Private communication (1976).
33. D. P. Sumner. Subtrees of a graph and chromatic number. In Gary Chartrand, editor, *The Theory and Applications of Graphs*, pages 557–576. John Wiley & Sons, New York, 1981.
34. M. Szegedy. Private communication (1986).
35. E. Szemerédi. Private communication (1981).
36. S. Vishwanathan. Randomized on-line graph coloring. *J. Algorithms*, 13:657–669, 1992. Also in Proc. 31st IEEE Symposium on Foundations of Computer Science, 1990, 464–469.
37. D. R. Woodall. Problem no. 4, combinatorics. In T. P. McDonough and V. C. Marvon, editors, *London Math. Soc. Lecture Note Series 13.* Cambridge University Press, 1974. Proc. British Combinatorial Conference 1973.

14

On-line Algorithms in Machine Learning

AVRIM BLUM

1 Introduction

The areas of On-Line Algorithms and Machine Learning are both concerned with problems of making decisions about the present based only on knowledge of the past. Although these areas differ in terms of their emphasis and the problems typically studied, there are a collection of results in Computational Learning Theory that fit nicely into the "on-line algorithms" framework. This chapter discusses some of the results, models, and open problems from Computational Learning Theory that seem particularly interesting from the point of view of on-line algorithms. This chapter is not meant to be comprehensive. Its goal is to give the reader a sense of some of the interesting ideas and problems in this area that have an "on-line algorithms" feel to them. The emphasis is on describing some of the simpler, more intuitive results, whose proofs can be given in their entirity. Pointers to the literature are given for more sophisticated versions of these algorithms.

We begin by describing the problem of "predicting from expert advice," which has been studied extensively in the theoretical machine learning literature. We present some of the algorithms that have been developed and that achieve quite tight bounds in terms of a competitive ratio type of measure. Next we broaden our discussion to consider several standard models of on-line learning from examples, and examine some of the key issues involved. We describe several interesting algorithms for on-line learning, including the Winnow algorithm and an algorithm for learning decision lists, and discuss issues such as attribute-efficient learning and the infinite attribute model, and learning target functions that change over time. Finally, we end with a list of important open problems in the area and a discussion of how ideas from Computational Learning Theory and On-Line Algorithms might be fruitfully combined.

To aid in the flow of the text, most of the references and discussions of history are placed in special "history" subsections within the article.

2 Predicting from expert advice

We begin with a simple, intuitive problem. A learning algorithm is given the task each day of predicting whether or not it will rain that day. In order to make this prediction, the algorithm is given as input the advice of n "experts". Each day, each expert predicts yes or no, and then the learning algorithm must use this information in order to make its own prediction (the algorithm is given no other input besides the yes/no bits produced by the experts). After making its prediction, the algorithm is then told whether or not, in fact, it rained that day. Suppose we make no assumptions about the quality or independence of the experts, so we cannot hope to achieve any absolute level of quality in our predictions. In that case, a natural goal instead is to perform nearly as well as the best expert so far: that is, to guarantee that at any time, our algorithm has not performed much worse than whichever expert has made the fewest mistakes to date. In the language of competitive analysis, this is the goal of being competitive with respect to the best single expert.

We will call the sequence of events in which the algorithm (1) receives the predictions of the experts, (2) makes its own prediction, and then (3) is told the correct answer, a *trial*. For most of this discussion we will assume that predictions belong to the set $\{0, 1\}$, though we will later consider more general sorts of predictions (e.g., many-valued and real-valued).

2.1 A simple algorithm

The problem described above is a basic version of the problem of "predicting from expert advice" (extensions, such as when predictions are probabilities, or when they are more general sorts of suggestions, are described in Section 2.3 below). We now describe a simple algorithm called the Weighted Majority algorithm. This algorithm maintains a list of weights $w_1, \ldots w_n$, one for each expert, and predicts based on a weighted majority vote of the expert opinions.

The Weighted Majority Algorithm (simple version)

1. Initialize the weights $w_1, \ldots, w_n$ of all the experts to 1.
2. Given a set of predictions $\{x_1, \ldots, x_n\}$ by the experts, output the prediction with the highest total weight. That is, output 1 if

$$\sum_{i:x_i=1} w_i \geq \sum_{i:x_i=0} w_i$$

and output 0 otherwise.

3. When the correct answer ℓ is received, penalize each mistaken expert by multiplying its weight by 1/2. That is, if $x_i \neq \ell$, then $w_i \leftarrow w_i/2$; if $x_i = \ell$ then w_i is not modified.
Goto 2.

Theorem 1. *The number of mistakes M made by the Weighted Majority algorithm described above is never more than $2.41(m + \lg n)$, where m is the number of mistakes made by the best expert so far.*

Proof. Let W denote the total weight of all the experts, so initially $W = n$. If the algorithm makes a mistake, this means that at least half of the total weight of experts predicted incorrectly, and therefore in Step 3, the total weight is reduced by at least a factor of 1/4. Thus, if the algorithm makes M mistakes, we have:

$$W \leq n(3/4)^M. \tag{1}$$

On the other hand, if the best expert has made m mistakes, then its weight is $1/2^m$ and so clearly:

$$W \geq 1/2^m. \tag{2}$$

Combining (1) and (2) yields $1/2^m \leq n(3/4)^M$ and therefore:

$$\begin{aligned} M &\leq \tfrac{1}{\lg(4/3)}(m + \lg n) \\ &\leq 2.41(m + \lg n) \square \end{aligned}$$

2.2 A better algorithm

We can achieve a better bound than that described above by modifying the algorithm in two ways. The first is by randomizing. Instead of predicting the outcome with the highest total weight, we instead view the weights as probabilities, and predict each outcome with probability proportional to its weight. The second change is to replace "multiply by 1/2" with "multiply by β" for a value β to be determined later.

Intuitively, the advantage of the randomized approach is that it dilutes the worst case. Previously, the worst case was that slightly more than half of the total weight predicted incorrectly, causing the algorithm to make a mistake and yet only reduce the total weight by 1/4. Now, there is roughly a 50/50 chance that the algorithm will predict correctly in this case, and more generally, the probability that the algorithm makes a mistake is tied to the amount that the weight is reduced.

A second advantage of the randomized approach is that it can be viewed as selecting an expert with probability proportional to its weight. Therefore, the algorithm can be naturally applied when predictions are "strategies" or other sorts of things that cannot easily be combined together. Moreover, if the "experts" are programs to be run or functions to be evaluated, then this view speeds up prediction since only one expert needs to be examined in order to produce the algorithm's prediction (although all experts must be examined in order to make an update of the weights). We now formally describe the algorithm and its analysis.

The Weighted Majority Algorithm (randomized version)

1. Initialize the weights $w_1, \ldots, w_n$ of all the experts to 1.
2. Given a set of predictions $\{x_1, \ldots, x_n\}$ by the experts, output x_i with probability w_i/W, where $W = \sum_i w_i$.
3. When the correct answer ℓ is received, penalize each mistaken expert by multiplying its weight by β.
   Goto 2.

Theorem 2. *On any sequence of trials, the expected number of mistakes M made by the Randomized Weighted Majority algorithm described above satisfies:*

$$M \leq \frac{m \ln(1/\beta) + \ln n}{1 - \beta}$$

where m is the number of mistakes made by the best expert so far.

For instance, for $\beta = 1/2$, we get an expected number of mistakes less than $1.39m + 2 \ln n$, and for $\beta = 3/4$ we get an expected number of mistakes less than $1.15m + 4 \ln n$. That is, by adjusting β, we can make the "competitive ratio" of the algorithm as close to 1 as desired, at the expense of an increase in the additive constant. In fact, by adjusting β dynamically using a typical "guess and double" approach, one can achieve the following:

Corollary 3. *On any sequence of trials, the expected number of mistakes M made by a modified version of the Randomized Weighted Majority algorithm described above satisfies:*

$$M \leq m + \ln n + O(\sqrt{m \ln n})$$

where m is the number of mistakes made by the best expert so far.

Proof of Theorem 2. Define F_i to be the fraction of the total weight on the *wrong* answers at the ith trial. Say we have seen t examples. Let M be our expected number of mistakes so far, so $M = \sum_{i=1}^{t} F_i$.

On the ith example, the total weight changes according to:

$$W \leftarrow W(1 - (1 - \beta)F_i)$$

since we multiply the weights of experts that made a mistake by β and there is an F_i fraction of the weight on these experts. Hence the final weight is:

$$W = n \prod_{i=1}^{t} (1 - (1 - \beta)F_i)$$

Let m be the number of mistakes of the best expert so far. Again, using the fact that the total weight must be at least as large as the weight on the best expert, we have:

$$n \prod_{i=1}^{t} (1 - (1-\beta)F_i) \geq \beta^m \tag{3}$$

Taking the natural log of both sides we get:

$$\begin{aligned}
\ln n + \sum_{i=1}^{t} \ln(1 - (1-\beta)F_i) &\geq m \ln \beta \\
- \ln n - \sum_{i=1}^{t} \ln(1 - (1-\beta)F_i) &\leq m \ln(1/\beta) \\
- \ln n + (1-\beta) \sum_{i=1}^{t} F_i &\leq m \ln(1/\beta) \\
M &\leq \frac{m \ln(1/\beta) + \ln n}{1-\beta}
\end{aligned}$$

Where we get the third line by noting that $-\ln(1-x) > x$, and the fourth by using $M = \sum_{i=1}^{t} F_i$. □

2.3 History and extensions

Within the Computational Learning Theory community, the problem of predicting from expert advice was first studied by Littlestone and Warmuth [28], DeSantis, Markowsky and Wegman [15], and Vovk [35]. The algorithms described above as well as Theorems 1 and 2 are from Littlestone and Warmuth [28], and Corollary 3, as well as a number of refinements, are from Cesa-Bianchi et al. [12]. Perhaps one of the key lessons of this work in comparison to work of a more statistical nature is that one can remove all statistical assumptions about the data and still achieve extremely tight bounds (see Freund [18]). This problem and many variations and extensions have been addressed in a number of different communities, under names such as the "sequential compound decision problem" [32] [4], "universal prediction" [16], "universal coding" [33], "universal portfolios" [13], and "prediction of individual sequences"; the notion of the competitiveness is also called the "min-max regret" of an algorithm. A web page uniting some of these communities and with a discussion of this general problem now exists at `http://www-stat.wharton.upenn.edu/Seq96`.

A large variety of extensions to the problem described above have been studied. For example, suppose that each expert provides a real number between 0 and 1 as its prediction (e.g., interpret a real number p as the expert's belief in the probability of rain) and suppose that the algorithm also may produce a real number between 0 and 1. In this case, one must also specify a loss function — what is the penalty for predicting p when the outcome is x? Some common loss functions appropriate to different settings are the absolute loss: $|p - x|$, the square loss: $(p - x)^2$, and the log loss: $-x \ln p - (1 - x) \ln(1 - p)$. Papers of Vovk [35, 36], Cesa-Bianchi et al. [12, 11], and Foster and Vohra [17] describe optimal algorithms both for these specific loss functions and for a wide variety of general loss functions.

A second extension of this framework is to broaden the class of algorithms against which the algorithm is competitive. For instance, Littlestone, Long, and Warmuth [27] show that modifications of the algorithms described above are constant-competitive with respect to the best linear combination of experts, when the squared loss measure is used. Merhav and Feder [29] show that one can be competitive with respect to the best off-line strategy that can be implemented by a finite state machine.

Another variation on this problem is to remove all semantics associated with specific predictions by the experts and to simply talk about losses. In this variation, the learning algorithm is required in each iteration to select an expert to "go with". For instance, suppose we are playing a 2-player matrix game as the row player. Each row can be viewed as an expert. To play the game we probabilistically select some expert (row) to use, and then, after the game is done, we find out our loss and that of each expert. If we are playing repeatedly against some adversary, we then would get another opportunity to probabilistically select an expert to use and so forth. Freund and Schapire show that extensions of the randomized Weighted Majority Algorithm discussed above can be made to fit nicely into this scenario [19] (see also the classic work of Blackwell [4]). Another scenario fitting this framework would be a case where each expert is a page-replacement algorithm, and an operating system needs to decide which algorithm to use. Periodically the operating system computes losses for the various algorithms that it could have used and based on this information decides which algorithm to use next.

Ordentlich and Cover [14] [30] describe strategies related to the randomized Weighted Majority algorithm for a problem of on-line portfolio selection. They give an on-line algorithm that is optimally competitive against the best "constant-rebalanced portfolio" (CRP). Their algorithm can be viewed as creating one expert for every CRP and then allocating its resources among them. This setting has the nice property that the market automatically adjusts the weights, so the algorithm itself just initially divides its funds equally among all infinitely-many CRPs and then lets it sit. A simple analysis of their algorithm with extensions to transaction costs is given in [10].

3 On-Line learning from examples

The previous section considered the problem of "learning from expert advice". We now broaden our focus to consider the more general scenario of on-line learning from examples. In this setting there is an example space $\mathcal{X}$, typically $\{0, 1\}^n$. Learning proceeds as a sequence of trials. In each trial, an example $x \in \mathcal{X}$ is presented to the learning algorithm. The algorithm then predicts either 1 or 0 (whether the example is positive or negative) and finally the algorithm is told the true label $\ell \in \{0, 1\}$. The algorithm is penalized for each mistake made; i.e., whenever its prediction differs from ℓ. Our goal is to make as few mistakes as possible. Typically, the presentation of examples will be assumed to be under the control of an adversary. This setting is also broadly called the Mistake Bound learning model.

The scenario described so far is not too different from the standard framework of On-Line Algorithms: we are given an on-line sequence of tasks and we want our penalty to be not too much larger than that of the best off-line algorithm. However, for the task of predicting labels, the "best we could do if there was no hidden information" would be to make zero mistakes, whereas no on-line algorithm could do better than make mistakes half the time if the labels were chosen randomly. Thus, some further restriction on the problem is necessary in order to make nontrivial statements about algorithms. Several natural restrictions are: (1) to restrict the labels to being determined by some "reasonable" function of the examples, (2) to restrict the off-line algorithms being compared against to some "reasonable" class of algorithms, and (3) to restrict the adversary to having some sort of randomness in its behavior. Each of these restrictions corresponds to a standard model studied in Computational Learning Theory, and we describe these in more detail below.

To describe these models, we need the notion of a *concept class*. A *concept class* $\mathcal{C}$ is simply a set of Boolean functions over the domain $\mathcal{X}$ (each Boolean function is sometimes called a *concept*), along with an associated representation of these functions. For instance, the class of *disjunctions* over $\{0, 1\}^n$ is the class of all functions that can be described as a disjunction over the variables $\{x_1, \ldots, x_n\}$. The class of *DNF formulas* contains all Boolean functions, each with a description length equal to the size of its minimum DNF formula representation. In the discussion below, we will use n to denote the description length of the examples, and $size(c)$ to denote the description length of some concept $c \in \mathcal{C}$.

We now describe three standard learning problems.

Learning a concept class $\mathcal{C}$ (in the Mistake Bound model): In this setting, we assume that the labels attached to examples are generated by some unknown target concept $c \in \mathcal{C}$. That is, there is some hidden concept c belonging to the class $\mathcal{C}$, and in each trial, the label ℓ given to example x is equal to $c(x)$. The goal of the learning algorithm is to make as few mistakes as possible, assuming that both the choice of target concept and the choice of examples are under the control of an adversary. Specifically, if an algorithm has the property that for any target concept $c \in \mathcal{C}$ it makes at most $poly(n, size(c))$ mistakes on any sequence of examples, and its running time per trial is $poly(n, size(c))$ as well, then we say that the algorithm *learns class $\mathcal{C}$ in the mistake bound model.* If, furthermore, the number of mistakes made is only $poly(size(c)) \cdot polylog(n)$ — that is, if the algorithm is robust to the presence of many additional irrelevant variables — then the algorithm is also said to be *attribute efficient.*

Algorithms have been developed for learning a variety of concept classes in the Mistake Bound model, such as disjunctions, k-DNF formulas, decision lists, and linear threshold functions. Below we will describe a very elegant and practical algorithm called the Winnow Algorithm, that learns disjunctions in the mistake bound model and makes only $O(r \log n)$ mistakes, where r is the number of variables that actually appear in the target disjunction. Thus, Winnow is attribute-efficient. This algorithm also has the property that it can be used to track a target concept that changes over time, and we will describe a sense in which the algorithm can be viewed as being $O(\log n)$ competitive for this task. We will also discuss a few general results on attribute-efficient learning and a model known as the infinite-attribute model.

Agnostic Learning / Being Competitive with the class $\mathcal{C}$: In this model, we make no assumptions about the existence of any relationship between the labels and the examples. Instead, we simply set our goal to be that of performing nearly as well as the best concept in $\mathcal{C}$. This is sometimes called the *agnostic* learning model and can be viewed as the problem of learning a concept class in the presence of adversarial noise. In this article, to use the terminology from On-Line Algorithms, we will call this the goal of being *competitive with respect to the best concept in $\mathcal{C}$*. Specifically, let us say that an algorithm is *α-competitive with respect to $\mathcal{C}$* if there exists a polynomial p such that for any sequence of examples and any concept $c \in \mathcal{C}$, the number of mistakes made by the algorithm is at most $\alpha m_c + p(n, size(c))$, where m_c is the number of mistakes made by concept c. The algorithm should have running time per trial polynomial in n and $size(c)$ where c is the best concept in $\mathcal{C}$ on the data seen so far.

If we consider the class $\mathcal{C}$ of single-variable concepts over $\{0,1\}^n$ (that is, $\mathcal{C}$ consists of n concepts $\{c_1, \ldots, c_n\}$ where $c_i(x) = x_i$), then this is really the same as the problem of "learning from expert advice" discussed in Section 2 (just think of the example as the list of predictions of the experts), and the algorithms of Section 2 show that for all $\epsilon > 0$, one can achieve $(1+\epsilon)$-competitiveness with respect to the best concept in this class.

It is worth noting that if we do not care about computational complexity (i.e., we remove the restriction that the algorithm run in polynomial time per trial) then we can achieve $(1+\epsilon)$-competitiveness for any concept class $\mathcal{C}$ over $\{0,1\}^n$. Specifically, we have the following theorem.

Theorem 4. *For any concept class $\mathcal{C}$ over $\{0,1\}^n$ and any $\epsilon > 0$ there is a non-polynomial time algorithm that on any sequence of examples, for all $c \in \mathcal{C}$, makes at most $(1+\epsilon)m_c + O(size(c))$ mistakes.*

Proof. We simply associate one "expert" with each concept $c \in \mathcal{C}$, and run the Randomized Weighted Majority algorithm described in Section 2 with the modification that the initial weight given to a concept c is $2^{-2size(c)}$. This assignment of initial weights means that initially, the total weight W is at most 1. Therefore, inequality (3) is replaced by the statement that for any concept $c \in \mathcal{C}$, after t trials we have:

$$\prod_{i=1}^{t}(1-(1-\beta)F_i) \geq \beta^{m_c} 2^{-2size(c)}$$

where m_c is the number of mistakes make by c. Solving this inequality as in the proof of Theorem 2 yields the guarantee that for any $c \in \mathcal{C}$, the total number of mistakes M made by the algorithm satisfies:

$$m \leq \frac{m_c \ln(1/\beta) + 2size(c)}{1-\beta}.$$

On the other hand, this algorithm clearly does not run in polynomial time for most interesting concept classes since it requires enumerating all of the possible concepts $c \in \mathcal{C}$.[1] □

A second fact worth noting is that in many cases there are NP-hardness results if we require the learning algorithm to use representations from the class $\mathcal{C}$. For instance, it is NP-hard, given a set S of labeled examples, to find the disjunction that minimizes the number of disagreements with this sample. However, this does not necessarily imply that it is NP-hard to achieve a competitive ratio approaching 1 for learning with respect to the class of disjunctions, since the hypothesis used by the learning algorithm need not be a disjunction.

As mentioned in the Open Problems section, it is unknown whether it is possible to achieve a good competitive ratio with respect to the class of disjunctions with a polynomial time algorithm.

$\mathcal{C}$ in the presence of random noise: This model lies somewhat in between the two models discussed so far. In this model, we assume that there is a target concept $c \in \mathcal{C}$ just like in the standard Mistake Bound model. However, after each example is presented to the learning algorithm, the adversary flips a coin and with probability $\eta < 1/2$, gives the algorithm the wrong label. That is, for each example x, the correct label $c(x)$ is seen with probability $1-\eta$, and the incorrect label $1-c(x)$ is seen with probability η, independently for each example. Usually, this model is only considered for the case in which the adversary itself is restricted to selecting examples according to some fixed (but unknown) distribution D over the instance space. We will not elaborate further on this model in this article, since the results here have less of an "on-line algorithms" feel to them, except to say that a very nice theory has been developed for learning in this setting, with some intriguing open problems, including one we list in Section 4.

One final point worth mentioning is that there are a collection of simple reductions between many standard concept classes. For instance, if one has an algorithm to learn the class of monotone disjunctions (functions such as $x_1 \vee x_5 \vee x_9$), then one can also learn non-monotone disjunctions (like $\overline{x}_1 \vee x_5$), conjunctions, k-CNF formulas for constant k, and k-DNF formulas for constant k, by just performing a transformation on the input space. Thus, if several classes are related in this way, we need only discuss the simplest one.

[1] In the PAC learning setting, there is a similar but simpler fact that one can learn any concept class in the presense of malicious noise by simply finding the concept in $\mathcal{C}$ that has the fewest disagreements on the sample.

3.1 Some simple algorithms

As an example of learning a class in the Mistake Bound model, consider the following simple algorithm for learning monotone disjunctions. We begin with the hypothesis $h = x_1 \vee x_2 \vee \ldots \vee x_n$. Each time a mistake is made on a negative example x, we simply remove from h all the variables set to 1 by x. Notice that we only remove variables that are guaranteed to not be in the target function, so we never make a mistake on a positive example. Since each mistake removes at least one variable from h, this algorithm makes at most n mistakes.

A more powerful concept class is the class of decision lists. A *decision list* is a function of the form: "if ℓ_1 then b_1, else if ℓ_2 then b_2, else if ℓ_3 then b_3, ..., else b_m," where each ℓ_i is a literal (either a variable or its negation) and each $b_i \in \{0, 1\}$. For instance, one possible decision list is the rule: "if $\overline{x}_1$ then positive, else if x_5 then negative, else positive." Decision lists are a natural representation language in many settings and have also been shown to have a collection of useful theoretical properties.

The following is an algorithm that learns decision lists, making at most $O(rn)$ mistakes if the target function has r relevant variables (and therefore has length $O(r)$). The hypotheses used by the algorithm will be a slight generalization of decision lists in which we allow several "if/then" rules to co-exist at the same level: if several "if" conditions on the same level are satisfied, we just arbitrarily choose one to follow.

1. Initialize h to the one-level list, whose level contains all $4n + 2$ possible "if/then" rules (this includes the two possible ending rules).
2. Given an example x, look at the first level in h that contains a rule whose "if" condition is satisfied by x. Use that rule for prediction (if there are several choices, choose one arbitrarily).
3. If the prediction is mistaken, move the rule that was used down to the next level.
4. Return to step 2.

This algorithm has the property that at least one "if/then" rule moves one level lower in h on every mistake. Moreover, notice that the very first rule in the target concept c will never be moved, and inductively, the ith rule of c will never move below the ith level of h. Therefore, each "if/then" rule will fall at most L levels, where L is the length of c, and thus the algorithm makes at most $O(nL) = O(nr)$ mistakes.

3.2 The Winnow algorithm

We now describe a more sophisticated algorithm for learning the class of (monotone) disjunctions than that presented in the previous section. This algorithm, called the Winnow Algorithm, is designed for learning with especially few mistakes when the number of relevant variables r is much less than the total number of variables n. In particular, if the data is consistent with a disjunction of r out of the n variables, then the algorithm will make at most $O(r \log n)$ mistakes. After describing this result, we then show how the Winnow algorithm can be used to achieve in essence an $O(\log n)$ competitive ratio for learning a disjunction that changes over time. We also discuss the behavior of Winnow in the agnostic setting. Variations on this algorithm can be used to learn Boolean threshold functions as well, but we will stick to the problem of learning disjunctions to keep the analysis simpler.

Like the Weighted Majority algorithm discussed earlier, the Winnow algorithm maintains a set of weights, one for each variable.

The Winnow Algorithm (a simple version)

1. Initialize the weights $w_1, \ldots, w_n$ of the variables to 1.
2. Given an example $x = \{x_1, \ldots, x_n\}$, output 1 if

$$w_1 x_1 + w_2 x_2 + \ldots + w_n x_n \geq n$$

   and output 0 otherwise.
3. If the algorithm makes a mistake:
   (a) If the algorithm predicts negative on a positive example, then for each x_i equal to 1, double the value of w_i.
   (b) If the algorithm predicts positive on a negative example, then for each x_i equal to 1, cut the value of w_i in half.
4. Goto 2.

Theorem 5. *The Winnow Algorithm learns the class of disjunctions in the Mistake Bound model, making at most* $2 + 3r(1 + \lg n)$ *mistakes when the target concept is a disjunction of* r *variables.*

Proof. Let us first bound the number of mistakes that will be made on positive examples. Any mistake made on a positive example must double at least one of the weights in the target function (the *relevant* weights), and a mistake made on a negative example will *not* halve any of these weights, by definition of a disjunction. Furthermore, each of these weights can be doubled at most $1 + \lg n$ times, since only weights that are less than n can ever be doubled. Therefore, Winnow makes at most $r(1 + \lg n)$ mistakes on positive examples.

Now we bound the number of mistakes made on negative examples. The total weight summed over all the variables is initially n. Each mistake made on a positive example increases the total weight by at most n (since before doubling, we must have had $w_1 x_1 + \ldots w_n x_n < n$). On the other hand, each mistake made on a negative example decreases the total weight by at least $n/2$ (since before halving, we must have had $w_1 x_1 + \ldots + w_n x_n \geq n$). The total weight never drops below zero. Therefore, the number of mistakes made on negative examples is at most twice the number of mistakes made on positive examples, plus 2. That is, $2 + 2r(1 + \lg n)$. Adding this to the bound on the number of mistakes on positive examples yields the theorem. □

How well does Winnow perform when the examples are not necessarily all consistent with some target disjunction? For a given disjunction c, let us define

m_c to be the number of mistakes made by concept c, and let A_c be the number of *attribute errors* in the data with respect to c, which we define as follows. For each example labeled positive but that satisfies no relevant variables of c, we add 1 to A_c; for each example labeled negative but that satisfies k relevant variables of c, we add k to A_c. So, if concept c is a disjunction of r variables, then $m_c \leq A_c \leq rm_c$. It is not hard to show that Winnow has the following behavior for agnostic learning of disjunctions.

Theorem 6. *For any sequence of examples and any disjunction c, the number of mistakes made by Winnow is $O(A_c + r \log n)$, where r is the number of relevant variables for c. Since $A_c \leq rm_c$, this means that Winnow is $O(r)$-competitive with respect to the best disjunction of r variables.*

In fact, by randomizing and tuning the Winnow algorithm to the specific value of r, one can achieve the following stronger statement.

Theorem 7. *Given r, one can tune a randomized Winnow algorithm so that on any sequence of examples and any disjunction c of r variables, the expected number of mistakes made by the algorithm is*

$$A_c + (2 + o(1))\sqrt{A_c r \ln(n/r)}$$

as $A_c/(r \ln \frac{n}{r}) \to \infty$.

These kinds of theorems can be viewed as results in a generalization of the "experts" scenario discussed in Section 2. Specifically, consider an algorithm with access to n "specialists". On every trial, each specialist may choose to make a prediction or it may choose to abstain (unlike the "experts" scenario in which each expert must make a prediction on every trial). That is, we can think of the specialists as only making a prediction when the situation fits their "specialty". Using a proof much like that used to prove Theorem 6, one can show that a version of the Winnow algorithm is constant-competitive with respect to the best *set* of specialists, where we charge a set one unit for every mistake made by a specialist in the set, and one unit whenever all specialists in the set abstain.

3.3 Learning drifting disjunctions

For the problem of learning a static target concept with no noise in the data, there is no real notion of "competitiveness". The algorithm just makes some fixed upper bounded number of mistakes. However, a natural variation on this scenario, which is also relevant to practice, is to imagine that the target function is not static and instead changes with time. For instance, for the case of learning a disjunction, we might imagine that from time to time, variables are added to or removed from the target function. In this case, a natural measure of "adversary cost" is the number of additions and deletions made to the target function, and the obvious goal is to make a number of mistakes that is not too much larger than the adversary's cost.

Specifically, consider the following game played against an adversary. There are n variables and a target concept that initially is the disjunction of zero of them (it says everything is negative). Then, each round of the game proceeds as follows.

Adversary's turn: The adversary may change the target concept by adding or removing some variables from the target disjunction. The adversary pays a cost of 1 for each variable added. (Since the number of variables removed over time is bounded by the number added over time, we may say that removing variables is free.) The adversary then presents an example to the learning algorithm.

Learner's turn: The learning algorithm makes a prediction on the example given, and then is told the correct answer (according to the current target concept). The algorithm is charged a cost of 1 if it made a mistake.

Consider the variation on the Winnow algorithm that never allows any weight to decrease below 1/2; that is, when a mistake is made on a negative example, only weights of value 1 or more are cut in half. Surprisingly, this Winnow variant guarantees that its cost is at most $O(\log n)$ times the adversary cost. So in a sense it is $O(\log n)$-competitive for this problem. Note that Theorem 5 can be viewed as a special case of this in which in its first move, the adversary adds r variables to the target function and then makes no changes from then on.

Theorem 8. *The Winnow variant described above, on any sequence of examples, makes at most $O(c_A \log n)$ mistakes, where c_A is the adversary's total cost so far.*

Proof. Consider the total weight on all the variables. The total weight is initially n. Each mistake on a positive example increases the total weight by at most n and each mistake on a negative example decreases the total weight by at least $n/4$ (because $\sum w_i x_i \geq n$ and at most $n/2$ of this sum can come from weights equal to $1/2$, so at least $n/2$ of the sum gets cut in half). Therefore, the number of mistakes on negative examples is bounded by $4(1 + M_p)$ where M_p is the number of mistakes made on positive examples. So, we only need to bound the number of mistakes on positives.

Let R denote the set of variables in the current target function (i.e., the currently relevant variables), and let $r = |R|$. Consider the potential function

$$\Phi = r \log(2n) - \sum_{i \in R} \lg w_i.$$

Consider now how our potential function Φ can change. Each time we make a mistake on a positive example, Φ decreases by at least 1. Each time we make a mistake on a negative example, Φ does not change. Each time the adversary adds a new relevant variable, Φ increases by at most $\log(2n) + 1$ ($\log(2n)$ for the increase in r and 1 for the possibility that the new weight w_i equals $1/2$ so $\lg w_i = -1$). Each time the adversary removes a relevant variable, Φ does not increase (and may decrease if the variable removed has weight less than $2n$). In summary, the only way that Φ can increase is by the adversary adding a new relevant variable, and each mistake on a positive example decreases Φ by at least 1; furthermore, Φ is initially zero and is always non-negative. Therefore, the number of mistakes we make on positive examples is bounded by $\log(2n) + 1$ times the adversary's cost, proving the theorem. □

3.4 Learning from string-valued attributes and the infinite attribute model

The discussion so far has focused on learning over the instance space $\mathcal{X} = \{0,1\}^n$. I.e., examples have Boolean-valued attributes. Another common setting is one in which the attributes are string-valued; that is, $\mathcal{X} = (\Sigma^*)^n$. For instance, one attribute might represent an object's color, another its texture, etc. If the number of choices for each attribute is small, we can just convert this to the Boolean case, for instance by letting "$x_1 =$ red" be a Boolean variable that is either true or false in any given example. However, if the number of choices for an attribute is large or is unknown apriori, this conversion may blow up the number of variables.

This issue motivates the "infinite attribute" learning model. In this model, there are infinitely many boolean variables $x_1, x_2, x_3, \ldots$, though any given example satisfies only finite number of them. An example is specified by listing the variables satisfied by it. For instance, a typical example might be $\{x_3, x_9, x_{32}\}$, meaning that these variables are true and the rest are false in the example. Let n be the size of (the number of variables satisfied by) the largest example seen so far. The goal of an algorithm in this setting is to make a number of mistakes polynomial in the size of the target function and n, but independent of the total number of variables (which is infinite). The running time per trial should be polynomial in the size of the target function and the description length of the longest example seen so far. It is not hard to see that this can model the situation of learning over $(\Sigma^*)^n$.

Some algorithms in the standard Boolean-attribute setting fail in the infinite attribute model. For instance, listing all variables and then crossing off the ones found to be irrelevant as in the simple disjunction-learning algorithm presented in Section 3.1 clearly does not work. The decision-list algorithm presented fails as well; in fact, there is no known polynomial-time algorithm for learning decision lists in this setting (see the Open Problems section).

On the other hand, algorithms such as Winnow can be adapted in a straightforward way to succeed in the infinite attribute model. More generally, the following theorem is known.

Theorem 9. *Let $\mathcal{C}$ be a projection and embedding-closed concept class*[2]. *If there is an attribute-efficient algorithm for learning $\mathcal{C}$ over $\{0,1\}^n$, then $\mathcal{C}$ can be learned in the Infinite-Attribute model.*

[2] This is just a "reasonableness condition" saying that one can take a concept in $\mathcal{C}$ defined on n_1 variables and embed it into a space with $n_2 > n_1$ variables and still stay within the class $\mathcal{C}$, and in the reverse direction, one can fix values of some of the variables and still have a legal concept. See [9] for details.

3.5 History

The Winnow algorithm was developed by Littlestone in his seminal paper [24], which also gives a variety of extensions and introduces the Mistake-Bound learning model. The Mistake Bound model is equivalent to the "extended equivalence query" model of Angluin [1], and is known to be strictly harder for polynomial-time algorithms than the PAC learning model of Valiant [34, 22] in which (among other differences) the adversary is required to select examples from a fixed distribution [6]. Agnostic learning is disussed in [23].

Littlestone [26] gives a variety of results on the behavior of Winnow in the presence of various kinds of noise. The improved bounds of Theorem 7 are from Auer and Warmuth [3]. The use of Winnow for learning changing concepts is folklore (and makes a good homework problem); Auer and Warmuth [3] provide a more sophisticated algorithm and analysis, achieving a stronger result than Theorem 8, in the style of Theorem 7. The Winnow algorithm has been shown to be quite successful in practical tasks as well, such as predicting links followed by users on the Web [2], and a calendar scheduling application [7].

The algorithm presented for learning decision lists is based on Rivest's algorithm for the PAC model [31], adapted to the Mistake Bound model by Littlestone [25] and Helmbold, Sloan and Warmuth [20]. The Infinite-Attribute model is defined in Blum [5] and Theorem 9 is from Blum, Hellerstein, and Littlestone [9].

4 Open problems

1. **Can the bounds of Corollary 3 be achieved and improved with a smooth algorithm?** The bound of Corollary 3 is achieved using a "guess and double" algorithm that periodically throws out all it has learned so far and restarts using a new value of β. It would seem more natural (and likely to work better in practice) to just smoothly adjust β as we go along, never restarting from scratch. Can an algorithm of this form be shown to achieve this bound, preferably with even better constants? (See [12] for the precise constants.)

2. **Can Decision Lists be learned Attribute-Efficiently?** Recall from Section 3.1 that a *decision list* is a function of the form: "if ℓ_1 then b_1, else if ℓ_2 then b_2, else if ℓ_3 then b_3, ..., else b_m," where each ℓ_i is a literal (either a variable or its negation) and each $b_i \in \{0,1\}$. We saw in Section 3.1 that decision lists with r relevant variables can be learned with at most $O(rn)$ mistakes in the mistake-bound model. An alternative approach using the Winnow algorithm makes $O(r^{2r} \log n)$ mistakes. *Can decision lists be learned attribute-efficiently? I.e., with mistake bound poly(r) $\cdot$ polylog(n)?*

3. **Can Parity functions be learned Attribute-Efficiently?** Let $\mathcal{C}_{parity}$ denote the class of functions over $\{0,1\}^n$ that compute the parity of some subset of variables. For instance, a typical function in $\mathcal{C}_{parity}$ would be $x_1 \oplus$

$x_5 \oplus x_{22}$. It is easy to learn C_{parity} in the mistake-bound model making at most n mistakes, by viewing each labeled example as a linear equality modulo 2 (each new example either is linearly dependent on the previous set and therefore its label can be deduced, or else it provides a new linearly independent vector). *Can C_{parity} be learned attribute-efficiently?*

4. **Can Decision Lists or Parity functions be learned in the Infinite Attribute model?** Can either the class of decision lists or the class of parity functions be learned in the Infinite Attribute model? For the case of decision lists, you may assume, if you wish, that none of the literals ℓ_i are negations of variables.

5. **Is there a converse to Theorem 9?**

6. **Can tolerance to random noise be boosted?** Suppose for some concept class C and some fixed constant noise rate $\eta > 0$ there exists a polynomial time algorithm A with the following property: for any target concept $c \in C$ and any distribution D on examples, A achieves an expected mistake rate less than $1/2 - 1/p(n)$ for some polynomial p after seeing polynomially many examples. Does this imply that there must exist a polynomial time algorithm B that succeeds in the same sense for *all* constant noise rates $\eta < 1/2$. (See Kearns [21] for related issues.)

7. **What Competitive Ratio can be achieved for learning with respect to the best Disjunction?** Is there a polynomial time algorithm that given any sequence of examples over $\{0,1\}^n$ makes a number of mistakes at most $cm_{disj} + p(n)$, where m_{disj} is the number of mistakes made by the best disjunction, for some constant c and polynomial p? How about $c = n^\alpha$ or $c = r^\alpha$ for some $\alpha < 1$, where r is the number of relevant variables in the best disjunction. (Making nm_{disj} mistakes is easy using any of the standard disjunction-learning algorithms, and we saw that the Winnow algorithm makes $O(rm_{disj})$ mistakes.)

8. **Can Disjunctions be Weak-Learned in the presence of adversarial noise?** For some polynomial $p(n)$ and some constant $c > 0$, does there exist an algorithm with the following guarantee: Given any sequence of t examples over $\{0,1\}^n$ such that at least a $(1-c)$ fraction of these examples are consistent with some disjunction over $\{0,1\}^n$, the algorithm makes at most $t[\frac{1}{2} - \frac{1}{p(n)}]$ mistakes (in expectation, if the algorithm is randomized). That is, given that there exists a disjunction that is "nearly correct" (say 99%) on the data, can the algorithm achieve a performance that is slightly ($1/poly$) better than guessing? The algorithm may require that $t \geq q(n)$ for some polynomial q.

9. **Can Linear Threshold Functions be Weak-Learned in the presence of adversarial noise?** Same question as above, except replace "disjunctions" with "linear threshold functions". An affirmative answer to this question would yield a quasi-polynomial ($n^{polylog(n)}$) time algorithm for learning

DNF formulas, and more generally for learning AC^0 functions, in the PAC learning model. This implication follows from standard complexity theory results that show that AC^0 can be approximated by low-degree polynomials.

5 Conclusions

This chapter has surveyed a collection of problems, models, and algorithms in Computational Learning Theory that look particularly interesting from the point of view of On-Line Algorithms. These include algorithms for combining the advice of experts, the model of on-line agnostic learning (or learning in the presence of worst-case noise) and the problem of learning a drifting target concept. It seems clear that a further crossover of ideas between Computational Learning Theory and On-Line Algorithms should be possible. Listed below are a few of the respective strengths and weaknesses of these areas where this crossover may prove to be especially fruitful.

The notion of state. The notion of an algorithm having a *state*, where there is a cost associated with changing state, is central to the area of On-Line Algorithms. This allows one to study problems in which the decisions made by an algorithm involve "doing something" rather than just predicting, and where the decisions made in the present (e.g., whether to rent or buy) affect the costs the algorithm will pay in the future. This issue has been virtually ignored in the Computational Learning Theory literature since that literature has tended to focus on prediction problems. In prediction problems the state of an algorithm is usually just its current hypothesis and there is no natural penalty for changing state. Nonetheless, as Computational Learning Theory moves to analyze more general sorts of learning problems, it seems inevitable that the notion of state will begin to play a larger role, and ideas from On-Line Algorithms will be crucial. Some work in this direction appears in [8].

Limiting the power of the adversary. In the On-Line Algorithms literature, it is usually assumed that the adversary has unlimited power to choose a worst-case sequence for the algorithm. In the machine learning setting, it is natural to assume there is some sort of regularity to the world (after all, if the world is completely random, there is nothing to learn). Thus, one often assumes that the world produces labels using a function from some limited concept class, or that examples are are drawn from some fixed distribution, or even that this fixed distribution is of some simple type. One can then parametrize one's results as a function of the adversary's power, producing especially good bounds when the adversary is relatively simple. This sort of approach may prove useful in On-Line Algorithms (in fact, it already has) for achieving less pessimistic sorts of bounds for many of the problems commonly studied.

Limiting the class of off-line algorithms being compared to. In the typical machine learning setup, if one does not restrict the adversary, then to achieve any non-trivial bound one must limit the class of off-line algorithms against which one is competing. This sort of approach may also be useful in On-Line Algorithms for achieving more reasonable bounds.

Acknowledgements

I would like to thank Yoav Freund for helpful discussions and pointers. This work was supported in part by NSF National Young Investigator grant CCR-9357793 and a Sloan Foundation Research Fellowship.

References

1. D. Angluin. Queries and concept learning. *Machine Learning*, 2(4):319–342, 1988.
2. R. Armstrong, D. Freitag, T. Joachims, and T. Mitchell. Webwatcher: A learning apprentice for the world wide web. In *1995 AAAI Spring Symposium on Information Gathering from Heterogeneous Distributed Environments*, March 1995.
3. P. Auer and M.K. Warmuth. Tracking the best disjunction. In *Proceedings of the 36th Annual Symposium on Foundations of Computer Science*, pages 312–321, 1995.
4. D. Blackwell. An analog of the minimax theorem for vector payoffs. *Pacific J. Math.*, 6:1–8, 1956.
5. A. Blum. Learning boolean functions in an infinite attribute space. *Machine Learning*, 9:373–386, 1992.
6. A. Blum. Separating distribution-free and mistake-bound learning models over the boolean domain. *SIAM J. Computing*, 23(5):990–1000, October 1994.
7. A. Blum. Empirical support for winnow and weighted-majority based algorithms: results on a calendar scheduling domain. In *Proceedings of the Twelfth International Conference on Machine Learning*, pages 64–72, July 1995.
8. A. Blum and C. Burch. On-line learning and the metrical task system problem. In *Proceedings of the 10th Annual Conference on Computational Learning Theory*, pages 45–53, 1997.
9. A. Blum, L. Hellerstein, and N. Littlestone. Learning in the presence of finitely or infinitely many irrelevant attributes. *J. Comp. Syst. Sci.*, 50(1):32–40, 1995.
10. A. Blum and A. Kalai. Universal portfolios with and without transaction costs. In *Proceedings of the 10th Annual Conference on Computational Learning Theory*, pages 309–313, 1997.
11. N. Cesa-Bianchi, Y. Freund, D. P. Helmbold, and M. Warmuth. On-line prediction and conversion strategies. In *Computational Learning Theory: Eurocolt '93*, volume New Series Number 53 of *The Institute of Mathematics and its Applications Conference Series*, pages 205–216, Oxford, 1994. Oxford University Press.
12. N. Cesa-Bianchi, Y. Freund, D.P. Helmbold, D. Haussler, R.E. Schapire, and M.K. Warmuth. How to use expert advice. In *Annual ACM Symposium on Theory of Computing*, pages 382–391, 1993.
13. T.M. Cover. Universal portfolios. *Mathematical Finance*, 1(1):1–29, January 1991.
14. T.M. Cover and E. Ordentlich. Universal portfolios with side information. *IEEE Transactions on Information Theory*, 42(2), March 1996.

15. A. DeSantis, G. Markowsky, and M. Wegman. Learning probabilistic prediction functions. In *Proceedings of the 29th IEEE Symposium on Foundations of Computer Science*, pages 110–119, Oct 1988.
16. M. Feder, N. Merhav, and M. Gutman. Universal prediction of individual sequences. *IEEE Transactions on Information Theory*, 38:1258–1270, 1992.
17. D.P. Foster and R.V. Vohra. A randomization rule for selecting forecasts. *Operations Research*, 41:704–709, 1993.
18. Y. Freund. Predicting a binary sequence almost as well as the optimal biased coin. In *Proceedings of the 9th Annual Conference on Computational Learning Theory*, pages 89–98, 1996.
19. Y. Freund and R. Schapire. Game theory, on-line prediction and boosting. In *Proceedings of the 9th Annual Conference on Computational Learning Theory*, pages 325–332, 1996.
20. D. Helmbold, R. Sloan, and M. K. Warmuth. Learning nested differences of intersection closed concept classes. *Machine Learning*, 5(2):165–196, 1990.
21. M. Kearns. Efficient noise-tolerant learning from statistical queries. In *Proceedings of the Twenty-Fifth Annual ACM Symposium on Theory of Computing*, pages 392–401, 1993.
22. M. Kearns, M. Li, L. Pitt, and L. Valiant. On the learnability of boolean formulae. In *Proceedings of the Nineteenth Annual ACM Symposium on the Theory of Computing*, pages 285–295, New York, New York, May 1987.
23. M. Kearns, R. Schapire, and L. Sellie. Toward efficient agnostic learning. *Machine Learning*, 17(2/3):115–142, 1994.
24. N. Littlestone. Learning quickly when irrelevant attributes abound: A new linear-threshold algorithm. *Machine Learning*, 2:285–318, 1988.
25. N. Littlestone. personal communication (a mistake-bound version of Rivest's decision-list algorithm), 1989.
26. N. Littlestone. Redundant noisy attributes, attribute errors, and linear-threshold learning using winnow. In *Proceedings of the Fourth Annual Workshop on Computational Learning Theory*, pages 147–156, Santa Cruz, California, 1991. Morgan Kaufmann.
27. N. Littlestone, P. M. Long, and M. K. Warmuth. On-line learning of linear functions. In *Proc. of the 23rd Symposium on Theory of Computing*, pages 465–475. ACM Press, New York, NY, 1991. See also UCSC-CRL-91-29.
28. N. Littlestone and M. K. Warmuth. The weighted majority algorithm. *Information and Computation*, 108(2):212–261, 1994.
29. N. Merhav and M. Feder. Universal sequential learning and decisions from individual data sequences. In *Proc. 5th Annual Workshop on Comput. Learning Theory*, pages 413–427. ACM Press, New York, NY, 1992.
30. E. Ordentlich and T.M. Cover. On-line portfolio selection. In *COLT 96*, pages 310–313, 1996. A journal version is to be submitted to *Mathematics of Operations Research*.
31. R.L. Rivest. Learning decision lists. *Machine Learning*, 2(3):229–246, 1987.
32. H. Robbins. Asymptotically subminimax solutions of compound statistical decision problems. In *Proc. 2nd Berkeley Symp. Math. Statist. Prob.*, pages 131–148, 1951.
33. J. Shtarkov. Universal sequential coding of single measures. *Problems of Information Transmission*, pages 175–185, 1987.
34. L.G. Valiant. A theory of the learnable. *Comm. ACM*, 27(11):1134–1142, November 1984.

35. V. Vovk. Aggregating strategies. In *Proceedings of the Third Annual Workshop on Computational Learning Theory*, pages 371–383. Morgan Kaufmann, 1990.
36. V. G. Vovk. A game of prediction with expert advice. In *Proceedings of the 8th Annual Conference on Computational Learning Theory*, pages 51–60. ACM Press, New York, NY, 1995.

15

Competitive Solutions for On-line Financial Problems

RAN EL-YANIV

1 Introduction

This chapter surveys work related to *on-line financial problems*. By that we refer to on-line problems (i.e. decision problems under uncertainty) related to the management of money and other assets. More precisely, we are concerned with on-line tasks whose natural modeling is in terms of *monetary* cost or profit functions. The family of such problems (and their related applications) is extremely broad, and the literature related is vast and originates from a number of disciplines including economics, finance, operations research and decision theory. In all these fields almost all of the related work is Bayesian. This means that all uncertain events are modeled as stochastic processes that are typically known to the decision maker[1] and the goal is to optimize the average case performance. Indeed, for financial problems this approach has been the dominating one over the last several decades and has led to the development of rich mathematical theory. In contrast, in this chapter we are concerned with non-Bayesian analyses of on-line financial problems while focusing on those using the *competitive ratio* optimality criterion or other similar worst-case measures. Not surprisingly, this reduces the number of relevant references from several hundreds (if not thousands) to only a dozen. Still however due to space limitations we shall focus here on only a subset of the work done in this field of study. Nevertheless we attempt at least to give pointers to most of the relevant problems and results.

Many of the financial problems studied in the literature and all of those discussed in this chapter are each a variant or an application of one of the

[1] Updates and new stochastic information is incorporated via Bayes' rule.

following elementary problems which are all described in terms of a one-player game against an *adversary* (sometimes called *nature*). The player is called the *on-line player*. Each one of these games takes place during some time horizon that may be continuous or discrete, divided into time *periods*. In all the games the on-line player's general objective is to minimize (resp. maximize) some monetary cost (resp. profit) function.

- *Search problems.* The on-line player is searching for the maximum (resp. minimum) price in a sequence of prices that unfolds sequentially. At the beginning of each time period the player can pay some *sampling cost* to obtain a *price quotation* p after which the player has to decide whether to *accept* p or continue sampling more prices. The game ends when the player accepts some price. His total return is the accepted price p minus the sum of all sampling costs incurred. Typical applications are: searches for jobs and employees, search for the best price of some product or asset, etc.
- *Replacement problems.* At each time period the player is engaged in one activity. Associated with each activity is a pair of numbers called the *changeover cost* and the *flow rate*. During the periods the player is engaged in some activity his budget is depleted at a rate corresponding to this activity's flow rate. From time to time new activities are offered as possible replacements for the current one. If the player decides to change over to a new activity he must pay the associated changeover cost. The question is when to switch from one activity to another so that the total cost, comprised of the sum of all flows and changeover costs, is minimized. Typical applications are: equipment replacement, jobs replacement, supplier replacement, mortgage refinancing, etc.
- *Portfolio selection.* In the basic setup of investment planning the on-line player has some initial wealth. At the start of each period the player must decide how to reallocate his current wealth among the available investment opportunities which may be commodities, securities and their derivatives.
- *Leasing problems.* The player needs to use some equipment during a number of time periods. This number is not known in advance but is made known on-line. Specifically, at the start of each period it becomes known whether or not the equipment will be needed in the current period and the player must choose whether to buy the equipment for a price b or to rent it for a price r with $r < b$. The game is over when the player purchases the equipment and/or when he no longer needs the equipment. Note that leasing problems can be presented as rudimentary forms of replacement problems. Due to lack of space the leasing problem is not discussed here. The reader is referred to the papers [21, 27] that study this problem using two different approaches.

1.1 General modeling issues

Each of the above problems attempts to model a particular economic situation in which an individual or an institution may be repeatedly involved within the modern market. Clearly, each of these problems abstracts away many factors while attempting to capture some of the essence embodied in the decision making task in question. With respect to each problem, various general or problem-specific features can be considered and incorporated in the model. In particular, the following are some features that may be considered: (i) finite vs. infinite time horizons; (ii) discrete vs. continuous time; (iii) discounting vs. non-discounting and (iv) consideration of various utility functions that represent the player's subjective rate of diminishing marginal worth. Clearly, the particular model chosen to formulate the problem determines the degree of its relevancy to the corresponding real-life problem while the addition of more features typically trades off its mathematical tractability.

1.2 Optimality criteria

This chapter is oriented towards results concerning *competitive* algorithms. That is, our primary concern is the analysis and design of on-line algorithms using the *competitive ratio* decision criterion for uncertainty (definitions follow in Sect. 2). A broader view encompasses (financial) decision making under uncertainty using other criteria such as the *minimax cost* (*maximin return*) and others. Indeed, occasionally we shall also discuss analyses of on-line algorithms via other optimality criteria including some criteria that relax the strict uncertainty assumptions typical of "pure" competitive analysis. Due to space limitations Bayesian solutions will not be surveyed at all but occasionally we point out key ideas or relevant features of Bayesian solutions that may bring further insight or motivation.

1.3 Descriptive vs. prescriptive theories

Descriptive decision making theories attempt to analyze and predict how individuals or institutions act. In contrast, *prescriptive* (or *normative*) theories attempt to describe how ideally rationalized decision makers *should* act. Although there may be an overlap between descriptive and prescriptive theories it is not in general expected that they will be identical. Of course, both kinds of theories have their own independent merits. For example, if one wishes to predict the behavior of a certain market based on the aggregate actions of the individual decision makers, then descriptive theories should be employed. On the other hand, prescriptive theories attempt to recommend to decision makers how to optimize their performance, sometimes even against their intuitive understanding. The "competitive" approach discussed in this chapter is, for the most part, a prescriptive approach.

1.4 Organization of the chapter

This chapter is organized as follows. Section 2 briefly presents definitions and notation that will be used throughout. Sections 3–5 survey results concerning the above basic financial problems in the order of their appearance above. For each problem (except for the leasing problem) we devote a special section that includes: a formal description of the basic problem and some of its variants, a discussion of some basic features of the problem and its notable applications, and a survey of selected known solutions. We conclude with Section 6, pointing out several directions for future research.

2 Preliminaries

2.1 On-line problems and competitive analysis

Consider a cost minimization problem $\mathcal{P}$ consisting of a set $\mathcal{I}$ of inputs and a cost function C. Associated with each input $I \in \mathcal{I}$ is a set of feasible outputs $F(I)$. For each input I and a feasible output $O \in F(I)$, the cost associated with I and O is $C(I, O) \in \mathbb{R}^+$. Let ALG be any algorithm for $\mathcal{P}$. Denote by $\text{ALG}[I]$ a feasible output produced by ALG given the input I. The cost incurred by ALG is denoted $\text{ALG}(I) = C(I, \text{ALG}[I])$. In the problems considered here each input I is a finite sequence, $I = i_1, i_2, \ldots, i_n$ and a corresponding feasible output is a finite sequence $O = o_1, o_2, \ldots, o_n$. An *on-line* algorithm for $\mathcal{P}$ must produce a feasible output in stages such that at the jth stage the algorithm is presented with the jth component of the input and must produce the jth component of (a feasible) output before the rest of the input is made known. Denote by OPT an optimal *off-line* algorithm for $\mathcal{P}$. That is, for each input I,

$$\text{OPT}(I) = \min_{O \in F(I)} C(I, O) \ .$$

An on-line algorithm ALG is *c-competitive* (or "attains a competitive ratio of c") if there exists a constant α such that for each input I,

$$\text{ALG}(I) - c \cdot \text{OPT}(I) \leq \alpha \ .$$

The smallest c such that ALG is c-competitive is called ALG's *competitive ratio.* Thus, a c-competitive algorithm is guaranteed to incur a cost no larger than c times the smallest possible cost (in hindsight) for each input sequence (up to the additive constant analogously. Specifically, with $\text{ALG}(I)$ (resp. $\text{OPT}(I)$) denoting the profit (or return) accrued by ALG (resp. OPT), ALG is c-competitive if there exists a constant α such that for all I,

$$c \cdot \text{ALG}(I) - \text{OPT}(I) \geq \alpha.$$

For *bounded* problems where the cost (resp. profit) function is bounded we require in the above definitions that the constant α is zero.

The extension of the competitive ratio definition to randomized on-line algorithms is straightforward under the assumption that the adversary that generate the input sequences is oblivious to the random choices made by the on-line player[2]. Specifically, the definition of the (expected) competitive ratio with respect to an *oblivious adversary* is the same as above with $E[\text{ALG}(I)]$ replacing $\text{ALG}(I)$ where $E[\cdot]$ is the expectation with respect to the random choices made by ALG.

[2] Such an adversary is called an *oblivious adversary.* Other kinds of adversaries are of *adaptive* type (see [6]). In this chapter we only consider oblivious adversaries.

2.2 Viewing on-line problems as two-player games

In analyses that use the competitive ratio measure it is often convenient to view an on-line problem as the following two-player game. The first player is the on-line player (running the on-line algorithm). The second player is called the *adversary* or the *off-line player*. The on-line player chooses an (on-line) algorithm ALG and makes it known to the adversary. Then, based on ALG, the adversary chooses an input so as to maximize the competitive ratio. The on-line player's objective is to minimize the competitive ratio (which means that this game is a zero-sum game). Thus, determining competitive-optimal algorithms is equivalent to determining an optimal strategy for the first player and the best possible competitive ratio is the value of this game, which in general is obtained using randomized on-line algorithms.

2.3 Competitive analysis

The use of competitive ratio for the evaluation of on-line algorithm is called *competitive analysis*. Competitive analysis was first used by computer scientists in the 70's in connection with approximation algorithms for NP-complete problems (see e.g. [24, 29, 30, 47]) and was more explicitly formulated in the 80's in the seminal work of Sleator and Tarjan [45] on *list accessing* and *paging* algorithms. Since then, competitive analysis has been extensively used to analyze and design on-line algorithms for many on-line optimization problems related to computer systems. Within the theoretical computer science community the competitive ratio has gained much recognition as a useful approach for the analysis of on-line problems. An axiomatization of the competitive ratio can be found in [19].

3 Search problems

3.1 The elementary search problem

The on-line player is searching for the maximum price in a sequence of prices that unfolds sequentially. At the beginning of each time period i the player can pay some *sampling cost* $c_i \geq 0$ to obtain a *price quotation* p_i after which the player has to decide whether to *accept* p_i or continue sampling more prices. The game ends when the player accepts some price p_j and the total return is then

$$p_j - \sum_{1 \leq i \leq j} c_i \ .$$

Search for the *minimum* price is defined analogously with respect to a cost function where the total cost of the player is $p_j + \sum_{1 \leq i \leq j} c_i$.

This elementary search problem has many extensions and variations. A search problem is termed *with recall* if all or some number k of the most recent price offers are retained and the player may choose any of the retained offers. We distinguish between search problems with or without recall, with or without discounting and discrete vs. continuous time. We assume finite time horizon but distinguish between a search of known or unknown duration. In the case of unknown duration we assume that the player is informed just before the last period that the game will end thus giving the player the opportunity to accept the last price if he has not accepted any price earlier.

3.2 Applications of search

Search is a most fundamental feature of economic markets. Let us mention some of its basic applications.

Job and employee search These two applications are of major importance to labor markets. In the *job search* application, the player is seeking employment. Each period the job seeker obtains one job offer (which corresponds to the above price quotation). The offer can be interpreted as the lifetime earning from the job. The sampling cost corresponds to the cost of generating an offer and includes all expenditures such as advertising and transportation and may include also the loss incurred from being currently unemployed. Thus each sampling cost may be modeled as a constant c.

Just as prospective employees are searching for jobs, employers are searching for employees to fill job vacancies. In the *employee search* application, the player is an employer searching for an employee with certain identifiable characteristics that should correspond to the employee's productivity (modeled via the above "price quotation").[3] This modeling has an underlying assumption that there are such characteristics that are correlated to the employee's productivity and that these characteristics can be tested by the employee.

Search for the lowest price of goods Here the player considers the purchase of some goods that are sold at different stores in different prices. The player can elicit quotations from the various sellers by paying the sampling fee that accounts for traveling or phone costs, the time wasted for sampling and may include the cost of managing without the goods since the start of the search. Like the job (and employee) search applications, this application is of fundamental value to economics because optimal searching rules determine the demand function that sellers face which in turn determines (some of) the nature of the markets themselves.

One-way trading In this application the player is a trader who is considering the exchange of some initial wealth w_0, given in some currency (say, dollars), to some other asset or currency (say, yen). Each period starts when a new price quotation is made available and the trader should decide whether to accept

[3] This application is closely related to the well known *secretary problem* [23, 1], the difference being that in secretary problems the objective is typically to accept one or more "secretaries" of best ordinal value among an ordered set of all secretaries.

it or wait for a better price. In this application the sampling cost is typically negligible as prices are widely available in banks, newspapers and quotation services. Nevertheless, the trader is required to pay some transaction fees that are typically some fixed percent of the return. Note that in this application the trader may partition the initial wealth and exchange w_0 sequentially in parts. One-way trading algorithms can be applied in various economic situations. For instance, consider a fund manager that decides to change the position of a portfolio and enter (or exit) some market (in which case w_0 is the part of his wealth allocated to the new position). Another natural instance related to foreign exchange is when the player, for the purpose of emigrating to a foreign country, sells his local property in order to exchange the local currency received to the foreign one, etc.

3.3 Basic features of Bayesian results

The Bayesian approach derives search strategies that are dependent on a prior distribution of prices that is usually assumed to be known to the player. Also, it is typically assumed that this distribution is fixed throughout the search and that prices are independent observations drawn from this distribution. The theory developed is very rich and relies on mathematical tools from the theory of *optimal stopping* [12].

As may be expected, numerous variants of search problems have been studied. The reader is referred to the excellent surveys by Lippman and McCall that discuss Bayesian solutions for a wide array of search variants [35, 36]. One notable feature of Bayesian optimal search algorithms (applicable to many problem variants) is that they have the following structure: based on the problem parameters (in particular, the probability distribution assumed) there is a single fixed critical number called the *reservation price,* such that the optimal policy is to reject all prices below the reservation price and to accept any offer above it. Reservation prices change dynamically in the case of finite horizon and known duration (and they are nondecreasing in problems without recall; the case of recall is more complicated and not fully solved). Another issue of interest when pursuing Bayesian solutions is the expected number of price quotations required until stopping.

The least acceptable assumption of the Bayesian search models is that the probability distribution of prices is fully known to the player. Several models attempt to relax this assumption. For example, Rosenfield and Shapiro [41] studied cases where the distribution itself is a random variable and the player knows the probability distribution of some of its moments (say, the price distribution is unknown but known to be normal and the distribution of its mean is known). In various cases such relaxed models do not admit optimal solutions with a reservation price (see [35]).

3.4 Competitive search algorithms

In this section we describe various competitive solutions to variants of the search problem without recall in which the sampling cost is zero or is a fixed percentage of the return. With respect to the competitive ratio these two variants are readily equivalent. For the rest of this section assume that the sampling cost is zero. In all the variants we consider we only assume that prices are drawn from some finite interval $[m, M]$. Call the ratio $\varphi = M/m$ the *global fluctuation ratio*. In one of the variants the on-line players knows the values m and M and in another variant the player knows only φ. We seek to determine the optimal competitive performance in terms of these parameters.

Suppose that m and M are known to the player. The optimal deterministic solution is the following *reservation price policy* (denoted RPP): accept the first price greater than or equal $p^* = \sqrt{Mm}$. This strategy is $\sqrt{\varphi}$-competitive and its analysis is trivial. By "arbitrage argument" p^* should be chosen to equate the performance ratio, off-line to on-line, corresponding to the two events: (i) the maximum price encountered, $p_{\max}$, will be greater than p^*; and (ii) $p_{\max} \leq p^*$. It follows that p^* is the solution p of $M/p = p/m$. The above reservation policy is optimal for infinite and finite time horizon and when duration is known or unknown. It is not hard to see that if only φ is known to the player then no smaller competitive ratio than the trivial one, φ, is achievable by a deterministic algorithm.

A dramatic improvement is obtained by using randomization.[4] First we show how a simple randomized strategy due to Levin [34] attains a competitive ratio of $O(\log \varphi)$. For simplicity, assume first that $\varphi = 2^k$ (for some integer k). For $i = 0, 1, \ldots, k-1$ denote by $S(i)$ the deterministic reservation price policy with reservation price $m2^i$. Levin's randomized strategy, denoted S, is a uniform probability mixture over $\{S(i)\}_i$; that is, S chooses $S(i)$ with probability $1/k$. It is easy to show that S is $f(\varphi) \log \varphi$-competitive with $f(\varphi)$ greater than but approaching 1 as φ grows. Let $p_{\max}$ be the (posteriori) maximum price obtained and let j be an integer such that $m2^j \leq p_{\max} < m2^{j+1}$. The optimal off-line return is $e_{\max}$. For each $i \leq j$ the strategy $S(i)$ chooses a price not smaller than $m2^i$ and for all $i > j$, $S(i)$ obtains at least m. It follows that S will obtain at least $(2^{j+1} + (k - j - 1))m/k$ on average. The resulting competitive ratio is greater than but approaching $k = \log \varphi$.

Exactly the same bound holds even if the player does not know the values of m and M and knows only the global fluctuation φ. Here, however, the strategies $S(i)$ are set after the first price p_1 is revealed (in which case $S(i)$ has reservation price $p_1 2^i$). For an arbitrary φ (not a power of two) the same method gives $O(k)$-competitive performance where $b^k = \varphi$ (b real and k positive integer). Recall that for this variant (only φ is known) no deterministic strategy can achieve a competitive ratio smaller than φ.

[4] Note that in general randomization does not give the player any advantage if it is assumed that the price distribution is known and an average case performance is sought (i.e. the Bayesian approach).

This simple randomized strategy can be modified to work even when the player does not know φ. In this case a competitive ratio of $O(c(\varepsilon) \cdot \log^{1+\varepsilon}(\varphi))$ can be obtained for every positive ε. Here $c(x)$ is a monotone decreasing function and φ is the posteriori global fluctuation ratio.

For the variants where m and M are known or only φ is known to the player (known or unknown duration) the competitive ratio of $O(\log \varphi)$ is within a multiplicative factor of the best that can be obtained against an oblivious adversary. This is established in the paper [20] that presents strictly optimal algorithms for these variants. The presentation of the results of this paper is simplified when described as a one-way trading game. Consider a trader with initial wealth of 1 dollar. The trader is presented with a sequence of exchange rates $p_1, p_2, \ldots$, where p_i gives the exchange rate, yen per dollar for the ith period (say, day). It is assumed that all rates are drawn from $[m, M]$. On each day i the trader can exchange any amount s_i of his remaining dollars to yen. Upon completion the total return of the trader is the amount of yen he accumulated. It is assumed that arbitrary fractions of dollars can be traded.

Notice that in the search problem the on-line player must accept one price and in the one-way trading problem the trader can partition his initial wealth and trade the parts sequentially, each part at different exchange rate. Nevertheless, these problems are closely related, and in fact equivalent in the following sense. Any deterministic one-way trading algorithm can be interpreted as a randomized search algorithm and vice versa. This follows from the fact that any deterministic one-way trading algorithm that trades the initial wealth in parts is equivalent (in terms of returns) to a randomized trading algorithm and vice versa. Specifically, suppose that a deterministic algorithm trades a fraction s_i of its initial wealth at the ith period, $\sum_i s_i = 1$. The quantity s_i can be interpreted as the probability to trade the entire wealth at the ith period. Clearly, the average return of this randomized algorithm equals the return of the deterministic algorithm. Thus, the competitive ratio of the deterministic algorithm is exactly the average competitive ratio of the randomized algorithm against an oblivious adversary. On the other hand, using Kuhn's theorem of game theory [32], any randomized one-way trading algorithm is equivalent to a mixed randomized algorithm (i.e. a probability distribution over deterministic algorithms) and by linearity of expectation (and using the fact that the sum of all traded amounts is 1) this mixed algorithm is equivalent to a deterministic algorithm that trades the initial wealth in parts. It follows that an optimal deterministic one-way trading algorithm is an optimal randomized search algorithm. This implies that randomization cannot improve the competitive performance in one-way trading. In contrast randomization is advantageous for search. When fixed sampling costs are introduced (i.e. the player pays some constant for each price quotation he obtains) there is no longer an equivalence between deterministic one-way trading algorithms and randomized search (and one-way trading) algorithms. the reason is that the randomized algorithm may spend less on sampling on the average.

Threat-based algorithms The optimal performance is obtained by algorithms that obey the following *threat-based* policy. Let c be any competitive ratio that can be attained by a deterministic trading algorithm. Assume that c is known to the trader. For each such c the corresponding threat-based policy consists of the following two rules:

Rule 1 Consider trading dollars to yen only when the current rate is the highest seen so far;

Rule 2 Whenever you convert dollars, convert *just enough* to ensure that a competitive ratio c would be obtained if an adversary dropped the exchange rate to the minimum possible rate[5] and kept it there throughout the game.

Thus, algorithms prescribed by this policy convert dollars to yen based on the threat that the exchange rate will drop permanently to the minimum possible rate. For each attainable competitive ratio c the corresponding threat-based algorithm can be shown to be c-competitive. Intuitively, this statement can be justified as follows. Consider the first trade (exchange rate is p_1). Since the current exchange rate is the highest seen so far the algorithm considers a trade. Since the competitive ratio c is attainable (by some deterministic trading algorithm), there exists some $s \geq 0$ such that the ratio c will still be attainable if s dollars are traded to yen. Further, the chosen amount of dollars s is such that the ratio c is so far guaranteed even if there will be a permanent drop of the exchange rate and no further trades will be conducted (except for one last trade converting the remaining dollars with the minimum possible exchange rate). In particular, there is no need to consider any exchange rate which is smaller than p_1. Similar arguments can be used to justify the choice of the amounts chosen for the rest of the trades and may convince the reader that this policy induces a c-competitive algorithm. A sketch of a more formal analysis follows.

Assume the problem variant with known duration, n, and known m and M. We now show how the optimal threat-based algorithm (denoted THREAT) for this variant can be derived. Any exchange rate which is not a global maxima at the time it is revealed to the trader is ignored (Rule 1). Hence we can assume w.l.o.g. that the exchange rate sequence consists of an initial segment of successive maxima of length $k \leq n$. In order to realize a threat the adversary may choose $k < n$ and choose $p_{k+1} = p_{k+2} = \cdots = p_n = m$.

Initially, assume that the optimal competitive ratio attainable by THREAT, c^*, is known. For each $i = 0, 1, \ldots, n$ set D_i and Y_i to be the number of remaining dollars and the number of accumulated yen, respectively, just after the i th period. By assumption the trader starts with $D_0 = 1$ dollars and $Y_0 = 0$ yen. Let $s_i = D_{i-1} - D_i$ be the number of dollars traded at the ith period, $i = 1, 2, \ldots, n$. Thus, $Y_i = \sum_{j=1}^{i} s_j p_j$. The ratio c^* is attained by algorithm THREAT and therefore, by Rule 2, it must be that the amounts s_i are chosen such that

[5] The "minimum possible rate" is defined with respect to the information known to the trader. That is, it is m, if m is known, and it is p/φ, if only φ is known and p is highest price seen so far.

for any $1 \leq i \leq n$,

$$\frac{p_i}{Y_i + m \cdot D_i} = \frac{p_i}{(Y_{i-1} + s_i p_i) + m \cdot (D_{i-1} - s_i)} \leq c^* . \quad (1)$$

Here the denominator of the l.h.s. represents the return of THREAT if an adversary dropped the exchange rate to m, and the numerator of the l.h.s. is the return of OPT for such an exchange rate sequence. By Rule 2, THREAT must spend the minimal s_i that satisfies inequality (1). The l.h.s. is decreasing with s_i. Since s_i is chosen to achieve the target competitive ratio c^*, and since the trader must spend the minimum possible amount (in order to leave as much as possible for higher rates), we replace the inequality in (1) with equality. Solving the resulting equality for s_i we obtain

$$s_i = \frac{p_i - c^* \cdot (Y_{i-1} + mD_{i-1})}{c^* \cdot (p_i - m)} . \quad (2)$$

From (1) (with equality) we also obtain the following relation

$$Y_i + m \cdot D_i = p_i / c^* . \quad (3)$$

Closed form expressions for the s_i's can now be obtained. From (2) at $i = 1$ we get $s_i = \frac{1}{c^*} \frac{p_1 - mc^*}{e_1 - m}$. From (2) and (3), with $i - 1$ replacing i, we obtain $Y_{i-1} + mD_{i-1} = p_{i-1}/c^*$. Hence for $i > 1$ we have $s_i = \frac{1}{c^*} \frac{p_i - p_{i-1}}{p_i - m}$.

It remains of course to determine c^*, the optimal competitive ratio attainable by algorithm THREAT. For any sequence of k exchange rate maxima it must be that the s_i's satisfy $\sum_{1 \leq i \leq k} s_i \leq D_0 = 1$. If the value of k is known to the trader then the optimal choice of s_i's is such that there will be no dollars remaining after the last transaction. That is, the optimal competitive ratio has the property that $\sum_{1 \leq i \leq k} s_i = 1$. Substituting into this equation the expressions determined for the s_i's one can obtain

$$c^* = c^*(k, m, p_1, p_2, \ldots, p_k) = 1 + \frac{p_1 - m}{p_1} \cdot \sum_{i=2}^{k} \frac{p_i - p_{i-1}}{p_i - m} .$$

Denote by $c_n^*(m, M)$ the optimal competitive ratio for the n-day game. Thus,

$$c_n^*(m, M) = \max_{\substack{k \leq n \\ m \leq p_1 < p_2 < \cdots < p_k \leq M}} c^*(k, m, p_1, p_2, \ldots, p_k) .$$

We skip this maximization routine and now discuss the end result. An explicit expression for $c_n^*(m, M)$ cannot be obtained but it can be shown that $c_n^*(m, M)$ is the unique root, c^*, of the equation

$$c = n \cdot \left(1 - \left(\frac{m(c-1)}{M - m} \right)^{\frac{1}{n}} \right) . \quad (4)$$

Thus, $c_n^*(m, M)$ is the minimum competitive ratio attainable by algorithm THREAT for an n-day game. On the other hand, it can be shown that $c_n^*(m, M)$

is a lower bound on the competitive ratio of any randomized algorithm against an oblivious adversary for an n-day trading game [20]. Hence $c_n^*(m, M)$ is the exact competitive ratio for the trading problem of known duration and known bounds (m and M). The fact that randomization cannot help in this problem is perhaps somewhat surprising but clear given the fact that any deterministic trading algorithm that partitions the initial wealth and performs multiple trades is equivalent to a randomized one-way trading (and search) algorithm with the same performance.

The above solution implies a solution for one-way trading with unknown duration. On the one hand, the threat-based algorithm corresponding to the ratio $c_\infty^*(m, M) = \lim_{n\to\infty} c_n^*(m, M)$ can handle any finite number of days while attaining a competitive ratio of $c_\infty^*(m, M)$. On the other hand, when duration is not known (it is made known on-line on the last day) the adversary can choose an arbitrary large n thus forcing a competitive ratio approaching $c_\infty^*(m, M)$ (note that $c_n^*(m, M)$ is strictly increasing with n). $c_\infty^*(m, M)$ can be shown to be the unique root, c^*, of

$$c = \ln \frac{M - m}{m(c-1)}.$$

Clearly, $c_\infty^*(m, M) = \Theta(\ln \varphi)$.

The paper [20] studies two other variants of the one-way trading problem in which the trader knows only the global fluctuation $\varphi = M/m$ but not the actual values of m and M (known and unknown duration). The value of $c_n^*(\varphi)$, the optimal competitive ratio of this trading game (known duration, n-day game), is determined and is shown to be

$$c_n^*(\varphi) = \varphi\left(1 - (\varphi - 1)\left(\frac{\varphi - 1}{\varphi^{n/(n-1)} - 1}\right)^n\right) . \tag{5}$$

Here again, this optimal performance is attainable by algorithm THREAT. Not surprisingly $c_n^*(\varphi)$ is monotone increasing with n and φ and the optimal competitive ratio of the unknown duration game, $c_\infty^*(\varphi)$, can be shown to be

$$c_\infty^*(\varphi) = \varphi - \frac{\varphi - 1}{\varphi^{1/(\varphi-1)}} = \Theta(\ln \varphi).$$

In order to get some feel for the actual competitive ratios obtained it is interesting to observe some numerical examples. Consider Table 1 which summarizes the competitive ratio of the algorithms discussed in this section for some values of φ. We see that the optimal threat based algorithm is always significantly superior to all other algorithms. Note that the deterministic algorithm RPP is superior to Levin's algorithm for small values of φ. Nevertheless, recall that the growth rate of the competitive ratio of Levin's algorithms is almost the logarithm of the growth rate of the competitive ratio of the deterministic algorithm.

It is also interesting to consider the rate of increase of the optimal competitive ratio as a function of the number of trading days n. It is possible to show that this function grows very quickly to its asymptote. Nevertheless, there is still a slight advantage to play short games. For instance, already at the 20th period

Algorithm	Value of φ					
	1.5	2	4	8	16	32
RPP (m, M known)	1.22	1.41	2	2.82	4	5.65
LEVIN'S (only φ known)	1.50	2	2.66	3.42	4.26	5.16
THREAT (only φ known)	1.27	1.50	2.11	2.80	3.53	4.28
THREAT (m, M known)	1.15	1.28	1.60	1.97	2.38	2.83

Table 1. Numerical examples of competitive ratios for some search and one-way trading algorithms (unknown duration)

$c_n^*(1,2)$ almost approaches its asymptote, $c_\infty^*(1,2) \approx 1.278$ (which is equivalent to guaranteeing 78.2% of the optimal off-line return) at $n = 10$ the ratio achieved is 1.26 (79.3% of OPT) and at $n = 5$ the ratio is 1.24 (80.6% of OPT).

Other results The paper [20] studies two other variants of the trading game: a continuous time model and a model in which the adversary chooses a probability distribution that determines the exchange rates (this probability distribution is made known to the trader). We note that the continuous time model is significantly simpler to analyze but the precise formulation of the model is more involved.

4 Replacement problems

4.1 A general replacement problem

At each time the on-line player must be engaged in a single *activity*. Associated with each activity a is its *cost* $c(a)$ and its *flow rate* $f(a)$ (which may also be a function of time $f(a) = f(a,t)$). Throughout the time period where the player is engaged in the activity a he pays money at the rate $f(a)$. From time to time new activities are offered as possible replacement to the current activity. If at some time t the player chooses to replace the current activity a with a new one a', he pays a *replacement* (or *changeover*) *cost* of $r_t(a, a')$ where r_t is a real valued function parameterized by time (for $t = 0$ the replacement cost is $r_t(a')$ where a' is the initial activity chosen). For any sequence of activities $a_0, a_1, \ldots, a_k$ chosen at times $t_0, t_1, \ldots, t_k$, the total cost incurred by the player up to time T is

$$r_0(a_0) + \sum_{i=1}^{k} \left((t_{i+1} - t_i)f(a_i) + r_{t_i}(a_{i-1}, a_i)\right),$$

where $t_0 = 0$ and $t_{k+1} = T$.

Some problem variants This section discusses two special instances of the above general replacement problem that are described using the following notation. At the start of each period i the on-line player is presented with a finite set R_i, $i = 0, 1, \ldots$, containing some number of activities offered as a possible replacement to the current one. R_i is called the ith *replacement set*. We first consider a continuous time, finite horizon model such that at each time there may be only a single replacement alternative to the current one; that is, for all i, $|R_i| = 1$. Also in this simple variant all replacement costs equal a constant C; that is, for all activities a, b, $r_t(a, b) = C$. This problem variant is referred to as the *elementary replacement problem with finite continuous time horizon.*

In the second variant we assume discrete time and that for all activities a, b, $r_t(a, b) = c(b)$. Also we require that that for all i, $R_i \subseteq R_{i+1}$. That is, an activity that is once offered is always available thereafter (i.e. replacement options are permanent). This variant is referred to as the *discrete time replacement problem with multiple, permanent replacement options.* In contrast to the first variant, here we allow for multiple replacement alternatives and for varying replacement costs. We note that the assumption of permanent replacement options together with the assumption of discrete time simplify the problem. Indeed, despite the added complication that replacement costs may vary, this variant allows for somewhat easier analysis and much better competitive ratios. Finally, notice that the above replacement model captures also the leasing problem as a special case. Specifically, at each time there are two optional activities. One, with a small replacement cost that corresponds to renting and another, with a higher replacement cost that corresponds buying. It is also given that the game ends as soon as the player chooses the activity that corresponds to buying.

4.2 Applications of the replacement problem

The replacement problem has various interesting applications. In all of these applications the basic question is when to switch from one activity, investment, or facility, to another more rewarding one, when there is a cost associated with making the switch. Some striking examples of particular applications are the following.

Equipment and machine replacement Here the player needs to use some piece of equipment throughout the time horizon. For its regular use, the equipment incurs some operating, production, and/or maintenance cost. From time to time, due to *a priori* unknown economic events, technological improvements and/or equipment deterioration, the player can and may wish to switch to a different or newer equipment that incurs a lower operating cost (or higher payoff). Some examples of equipment for which this application is relevant are cars, computers, industrial machinery, etc. The same formulation applies of course to more abstract types of “equipment” such as jobs, etc. In many of these examples the operating cost can be approximated by a fixed rate payment flow (e.g. gasoline consumption rate, salary, etc). Other applications may require a more elaborate modeling. For instance, in order to model deterioration, the maintenance may be required to be some monotone nondecreasing function (rather than a constant).

Supplier replacement A firm is purchasing goods at a constant rate from one supplier. The cost of purchasing the same goods from other suppliers varies with time. The firm can switch to another supplier but at a certain cost. The cost of this switch-over can be approximated by some constant that accounts for the paperwork, the wasted time and possibly the costs involved in breaking the contract with the first supplier.

The menu cost problem Many firms are constantly faced with the problem of when to adjust prices of the goods or services they offer. Due to inflationary markets and/or other economic events the firm may wish to update its price menu to reflect their "real" values in order to increase its overall payoff. Each of these price adjustments which correspond to our (flow) changeovers, incurs some fixed cost to physically update the "menu", advertise, etc.

Mortgage refinancing In this application the flow rate corresponds to the mortgage payments rate which is based on a fixed interest rate (and the principal). Among the popular mortgages available in the north American markets are those where refinancing costs are fixed.[6] These fixed costs correspond to the paper work and time overhead required by the switch to another mortgage and thus are best modeled by a fixed changeover cost. In other kinds of (fixed rate) mortgages a changeover cost is some fixed percentage of the principal.

4.3 On some Bayesian solutions for the replacement problem

The literature related to on-line replacement problems is quite extensive. The typical assumption is that the flow rate function follows a particular (usually simple) stochastic process that may or may not be known to the on-line player. Let us describe two examples.

Derman [18] studies a simplified discrete time replacement problem where his analog of our flow rate function is a piecewise constant function in which the next value is determined via a simple one-stage Markov process.

Sheshinski and Weiss [43] study price adjustment policies solving the menu cost problem under the assumption that "real" prices are determined by the following two-state process. During each state the price level is changed at a fixed rate (in particular, they assume that in one state the price is fixed and in the second state, the price increases at a fixed rate). The duration of each state is an independent exponentially distributed random variable. Note that in both these examples (and in most other analyses of this kind) the optimal policy is heavily dependent on the stochastic assumptions.

[6] Such mortgages are sometimes called *zero point mortgages*.

4.4 Competitive replacement algorithms for the elementary continuous time, finite horizon problem

In this section we discuss competitive solutions for the elementary replacement problem with continuous time and finite horizon [22]. Recall that in this version of the replacement problem all changeover costs equal a constant C. Also in this variant there is only one possible replacement activity to the current one. Hence we denote the flow rate of the activity offered at time t by $f(t)$. The player starts with the initial activity paying at flow rate $f(0)$, and may choose any number k of *changeover times*, $0 < t_1 < t_2 < \cdots < t_k < T$. For each such changeover time t_i, the player pays a changeover cost C and throughout the interval $[t_i, t_{i+1})$, $i = 0, 1, \ldots, k$, his payment flow is at the rate $f(t_i)$. (By convention take $t_0 = 0$ and $t_{k+1} = T$.) For each particular choice of changeover times the *total cost* incurred by the player, comprised of payment flows and changeover costs, is

$$kC + \sum_{i=0}^{k}(t_{i+1} - t_i)f(t_i)$$

Any choice of the number k and of (k) changeover times is called a *replacement policy.* Of course we are interested in replacement policies that minimize the total cost. Given $f(t)$ and $C(t)$ it is straightforward to compute an optimal off-line replacement policy and $\text{OPT}(f)$, the optimal off-line cost, via (continuous) dynamic programming [5].

As in the search problem, we assume that the flow rate function is bounded such that for all t, $m \leq f(t) \leq M$ where $m, M \in \mathbb{R}$ and $0 \leq m < M$. Further, in this *on-line* variant of the replacement problem the player must determine his changeover times on-line without knowledge of future values of the flow rate function. Thus, we assume that at each time t the player knows f only over the interval $[0, t]$. Let S be any on-line replacement policy and denote its total cost with respect to the flow f by $S(f)$.

Problem reduction. By suitably scaling the time and cost axes, we may assume, that $C = T = 1$. Specifically, given an initial problem set-up with parameters m', M', T' and C' we map each flow rate $x' \in [m', M']$ to $x = x'T'/C'$ and each time $t' \in [0, T']$ to t'/T'. Thus, the new problem set-up is given by $M = M'T'/C'$, $m = m'T'/C'$ and $T = C = 1$. It is not hard to see that this scaling preserves the competitive ratio. After scaling, we further assume that $m + 1 < M$. For $M \leq m + 1$ the problem is trivial in the sense that the on-line player can always achieve a "perfect" competitive ratio of 1. For the rest of this section we consider the reduced replacement problem with $C = T = 1$ so that the only relevant parameters are m and M.

Some types of on-line replacement policies Perhaps the most naive on-line replacement policies that are still interesting are the following class of *time-independent* policies. A policy in this class is a sequence of constant changeover *thresholds* that are fixed over time independent of the flow rate function. Specifically, a time-independent policy is a decreasing sequence of real numbers,

$$M \geq M_1 > M_2 > \ldots > M_k \geq m \ .$$

The interpretation is that the on-line player changes over for the ith time when the flow rate decreases to the level of (or below) M_i.

A more sophisticated class of policies is the following class of *time-dependent* or *refusal* policies (we may use either name). A refusal policy is defined as a sequence $\{M_i(t)\}_{i=1}^k$ of functions such that $M_i(t) : [0,1] \rightarrow [m, M] \cup \{-1\}$. Each of these functions is non-increasing and for all i and t, $M_i(t) > M_{i+1}(t)$. Here again the interpretation is that the on-line player changes over for the ith time at the first instance t when $f(t) \leq M_i(t)$ but *refuses* to change over as long as $f(t) > M_i(t)$. A particular subclass of simple refusal strategies is the one where each M_i is a constant except for one step at some time b_i from which the function remains at -1; that is

$$M_i(t) = \begin{cases} M_i, \text{ if } t \leq b_i; \\ -1, \text{ otherwise.} \end{cases}$$

Call such a strategy a *constant threshold refusal policy.* Any such strategy is thus specified by the two sequences $\{M_i\}$ and $\{b_i\}$. Notice that a time-independent policy is a rudimentary form of a constant threshold refusal policy where the b_i's are all 1. Here we focus on time-independent and constant threshold refusal policies. Nevertheless, note that more sophisticated strategies would make use of the *history* of flow rates. As we shall see (and somewhat surprisingly), in all instances of the above replacement problem it is possible to obtain optimal or approximately optimal on-line performance using only time-independent and constant threshold refusal policies (and without resorting to history-dependent policies).

A lower bound A lower bound on the competitive ratio of any deterministic policy for this variant of the replacement problem is obtained in the paper [22]. We now sketch the essential ideas of this bound. Consider an adversary that can choose the flow rate functions from a restricted family consisting of functions that start at time zero at the rate M and then drop "instantaneously" (i.e. during an infinitesimally short time) and continuously to some rate μ chosen by the adversary and then "jump" back to the maximum possible rate M and remain there. We may assume that $\mu \leq M - 1$ since no sensible strategy will change over for any flow rate larger than $M - 1$ (the changeover cost is 1). It can be shown that against such functions the optimal on-line performance can be attained by a time-independent policy. Intuitively this is due to the fact that all replacement opportunities occur during an infinitesimally short time period.

For any choice of μ the optimal off-line cost is $\mu + 1$ since the optimal off-line algorithm changes over to the rate μ paying 1 for the changeover penalty. Let $S = \{M_i\}_{i=1}^k$ be a time-independent policy and assume that it is r-competitive. It can be shown that S must satisfy the following relations:

$$M_i + i = r(M_{i+1} + 1), \; i = 0, 1, \ldots, k \ . \tag{6}$$

(by convention we set $M_0 = M$ and $M_{k+1} = m$). Intuitively, the reason is that if an adversary wishes that μ will lie in some interval $[M_{i+1}, M_i]$, it will pick $\mu = M_{i+1} + \varepsilon$ (for some small ε) in which case the optimal off-line cost will be (arbitrarily close) to $M_{i+1} + 1$ and the on-line cost will be $M_i + i$ (the on-line strategy will change over i times). On the other hand, the strategy defined by the recurrence relation (6) is r-competitive provided that $M_{k+1} = m$. Based on properties of this recurrence relation it can be shown that if S is r-competitive then the number of thresholds in S is

$$k = \lceil (m+1)(r-1) \rceil \ . \tag{7}$$

A closed form expression for the recurrence relation (6) is

$$M_i = \left(a + \frac{r^2}{(r-1)^2}\right) r^{-i} + \frac{i}{r-1} - \frac{r^2}{(r-1)^2} \ . \tag{8}$$

Using (7) and (8) the *optimal* competitive ratio r^* can now be determined. It is the minimum r that solves the equation $M_{k+1} = m$. It can be shown that for a fixed m, $r^* = \Theta\left(\frac{\ln M}{\ln \ln M}\right)$. Since r^* is the optimal competitive ratio with respect to restricted flow rate functions it is a lower bound on the competitive ratio of any replacement policy against unrestricted flow rate functions.

A characterization of competitive refusal policies The paper [22] establishes a characterization theorem for constant threshold refusal policies that have a decreasing refusal time sequence $\{b_i\}$ (that is, if the policy refuses to change over for the first time, it will refuse to change over for the rest of the game). Given any refusal policy S (of the above type) with k changeover thresholds, and a real number $r > 1$ the theorem specifies the following two conditions, C1 and C2, such that they are both satisfied if and only if S is r-competitive.

C1 for all i and j with $0 \le i \le j \le k$,

$$M_i b_{j+1} + M_j(1 - b_{j+1}) + j \le r \cdot \min \begin{bmatrix} M_0, \\ M_{i+1} + 1, \\ M_0 b_{j+1} + m(1 - b_{j+1}) + 1, \\ M_{i+1} b_{j+1} + m(1 - b_{j+1}) + 2 \end{bmatrix} ;$$

C2 for all i and j with $0 \le i < j \le k$,

$$M_i b_j + M_j(1 - b_j) + j \le r \cdot \min \begin{bmatrix} M_0, \\ M_{i+1} + 1, \\ M_0 b_j + m(1 - b_j) + 1, \\ M_{i+1} b_j + m(1 - b_j) + 2 \end{bmatrix} .$$

The proof of this theorem is obtained by bounding from below the optimal off-line cost via a linear form of variables chosen by the adversary (these variables determine the flow function). It follows that the set of feasible choices for the adversary is a polytope. Using a convexity argument it is then sufficient to consider only corner points of this polytope. In each corner point most of the coordinates (i.e. variables) are zero. Then, with respect to the corner points we obtain simple expressions for the off-line costs (as they appear in the r.h.s of C1 and C2), in particular, each possible cost includes at most two changeovers.

Upper bounds Using the above characterization theorem it is possible to obtain various interesting upper bounds for this replacement problem. We now describe a policy that for every positive m achieves a competitive ratio that, for sufficiently large M, is within a constant factor of r^*.

Consider the following constant threshold time-independent policy, $\{M_i\}_{i=1}^k$, where the sequence of changeover thresholds, $\{M_i\}$, is defined by the following recurrence relation. For each $\rho > 1$, set $k = \lfloor \rho \rfloor$. Define

$$\begin{cases} M_0 &= M; \\ M_{i+1} &= \frac{M_i + k}{\rho} - 1, \quad \text{integer } i \geq 1 \end{cases} \tag{9}$$

It is possible to show that for every positive m, and for sufficiently large ρ the sequence defined by (9) decreases below m within k steps. Call a ρ for which $M_{k+1} \leq m$ and $M_k > m$ *good*. For each ρ, each $m > 0$ and each $M > m + 1$, let $S_\rho^*(m, M)$ denote the policy $\{M_i\}$ (as defined by (9)). Now, by considering $S_\rho^*(m, M)$ as a (degenerate) time-dependent policy, we can apply the above characterization conditions and prove that $S_\rho^*(m, M)$ is ρ-competitive for all good ρ and almost all values of M. Specifically, assume that $M > \lfloor \rho \rfloor / (\rho - 1) = \frac{k}{\rho - 1}$. This implies that $M_1 + 1 - M \leq 0$, which means that $\min\{M_i + 1, M\} = M_i + 1$. We will use this fact later. Let us now specialize the conditions C1 and C2 of the characterization result to the case where $b_{k+1} = 0$ and $b_i = 1$, $i = 1, 2, \ldots, k$. That is, when the (degenerate) refusal policy is a time-independent policy. For $r = \rho$, C1 and C2 reduce to the following condition: for all $0 \leq i \leq j \leq k$,

$$M_i + j \leq \rho \cdot \min\{M_0, M_{i+1} + 1\} .$$

But this condition readily holds by the definition of the M_i's and the fact that $M_1 + 1 \leq M$. For each (sufficiently large) M define $\rho(M)$ to be the minimum (infimum) good ρ. Then it is possible to show that for a fixed m, $\rho(M) = \Theta\left(\frac{\ln M}{\ln \ln M}\right)$ which has the same asymptotic growth as r^*, the lower bound for the problem (see above).

The paper [22] also considers also constant threshold refusal policy and proves its (strict) optimality whenever $\sqrt{M/(m+1)} \leq (m+2)/(m+1)$ or $m = 0$. The proof relies on the above characterization result but is much more involved.

4.5 Discrete time replacement with multiple, permanent replacement options

In this section we only state the results of a very recent paper by Azar, Bartal, Feuerstein, Fiat, Leonardi and Rosén [4] that study the discrete time replacement problem variant with multiple, permanent replacement options. Recall that in this problem variant we require that $|R_i| > 0$, and that for all i, $R_i \subseteq R_{i+1}$. (see the notation of Sec. 4.1). That is, an activity that is once offered is always available thereafter.

The convex problem variant An instance of the this replacement problem is called *convex* if for each i and $b_1, b_2 \in R_i$, if $f(b_1) < f(b_2)$ then $r(b_1) \geq r(b_2)$. For the convex variant Azar et al. obtain a simple 7-competitive algorithm.

The non-convex variant An instance of this problem which is not convex in the above sense is called *non-convex*. Non-convex instances turn out to be markedly harder. Azar et al. introduce an algorithm that attains the following competitive ratio:

$$O\left(\min\left\{\log(cr_{\max}), \log\log(cf_{\max}), \log(cn_{\max})\right\}\right) \ .$$

with $r_{\max}$ being the ratio between the maximum and minimum replacement costs, $f_{\max}$ is the ratio between the maximum and minimum flow rates, $n_{\max}$ is the total number of replacement alternatives presented to the player throughout the game, and c is some constant. Azar et al. also present a nemesis request sequence of replacement sets that forces the following competitive ratio on every on-line algorithm.

$$\Omega\left(\min\left\{\log(cr_{\max}), \frac{\log\log(cf_{\max})}{\log\log\log(cf_{\max})}, \frac{\log(cn_{\max})}{\log\log(cn_{\max})}\right\}\right) \ .$$

where c is some constant. Thus, as a function of $r_{\max}$ their algorithm attains a competitive ratio that is within a constant factor of the best possible.

5 Two-way trading and portfolio selection

5.1 The elementary portfolio selection problem

Consider a market of m securities. These can be stocks, bonds, foreign currencies or commodities. Let $\mathbf{p}_i = (p_{i1}, p_{i2}, \ldots, p_{im})$ denote a vector of *prices* where for each $j = 1, 2, \ldots, m$, p_{ij} denotes the number of units of the jth security that can be bought for one dollar at the start of the ith trading period, $i = 1, 2, \ldots$. The "local" currency, say *dollars*, may or may not be one of the m securities. This local currency is referred to as *cash*. The change in security prices during the ith period is represented as a vector $\mathbf{x}_j = (x_{i1}, x_{i2}, \ldots, x_{im})$ where for each i and j, $x_{ij} = p_{ij}/p_{i+1,j}$. The quantity x_{ij} is called the *price relative* of security j (of the ith period). Thus, an investment of d dollars in the jth security just before the start of the ith period yields dx_{ij} dollars by the end of the ith trading period.

An investment in the market, or *portfolio*, is specified as the proportion of dollar wealth currently invested in each of the m securities. Specifically, we represent a portfolio as a probability distribution $\mathbf{b} = (b_1, b_2, \ldots, b_m)$ where $b_i \geq 0$

and $\sum_i b_i = 1$. Consider a portfolio $\mathbf{b}_1$ invested just before the first period. By the start of the second period this portfolio yields

$$\mathbf{b}_1^t \cdot \mathbf{x}_1 = \sum_{j=1}^{m} b_{1j} x_{1j}$$

dollars per each initial dollar invested. At this stage the investment can be cashed and adjusted, say by reinvesting the entire current wealth in some other proportion $\mathbf{b}_2$, etc. Assuming an initial wealth of 1 dollar the *compounded return* of a sequence of portfolios, $B = \mathbf{b}_1, \mathbf{b}_2, \ldots, \mathbf{b}_n$ with respect to a sequence of market price relatives $X = \mathbf{x}_1, \ldots, \mathbf{x}_n$ is defined to be

$$R(B, X) = \prod_{i=1}^{n} \mathbf{b}_i^t \cdot \mathbf{x}_i = \prod_{i=1}^{n} \sum_{j=1}^{m} b_{ij} x_{ij} \ .$$

A *portfolio selection algorithm* is any sequence of portfolios specifying how to reinvest the current wealth from period to period. The compounded return of a portfolio selection algorithm ALG with respect to a market sequence X will be denoted by $\text{ALG}(X) = R(\text{ALG}, X)$. The *portfolio selection* problem revolves around the question of identifying and analyzing portfolio selection algorithms. Of course, here we shall mainly be concerned with on-line portfolio selection algorithms.

It is important to notice that the above basic portfolio selection problem is only a crude approximation of the corresponding real-life problem. For practical purposes perhaps the most important factor missing in this model is *transaction costs* (alternatively, *bid-ask spreads*). Also the above model makes the assumption that money and units of securities are arbitrarily divisible. Other important missing factors are taxes and interest rates that are typically important factors involved in investment planning. Finally, this model does not allow for many investment opportunities that exist in a modern market, starting with *short selling* and ending with a myriad of *derivative instruments* such as *futures, options*, etc. Nevertheless, this model is rich enough in itself and makes a reasonable foundation for studying some of the essential questions related to portfolio selection.

The two-way trading problem The special problem instance with $m = 2$ where one of the two securities is cash has received special attention and turns out to be sufficiently interesting in itself. We call this particular instance the *two-way trading problem.* Note that in the two-way trading problem we assume that the prices and price relatives of the cash are always 1. In the special case of the two-way trading problem we sometimes refer to portfolio selection algorithms simply as *(two-way) trading algorithms.*

5.2 Buy-and-hold vs. market timing

Financial agents study and use a large variety of portfolio selection strategies. On the one hand there are "slow" and almost static strategies typically used by mutual fund managers. These strategies select and buy some portfolio and hold it for quite long time. Such strategies rely on the natural tendency of securities to increase their value due to natural economic forces. For instance, stocks pay dividends and increase their prices as the underlying firms succeed in their businesses. Such "slow" strategies are generally called *buy-and-hold* (BAH). In contrast there are financial agents that use aggressive strategies that buy and sell securities very frequently, sometimes even many times during one day. Such strategies mainly attempt to take advantage of securities' price fluctuations and are called *market-timing* strategies. Of course, in the long run every buy-and-hold strategy is also a market timing strategy. After all, sooner or later owners of securities want to realize the monetary value of the assets they hold. Hence, whether a strategy is buy-and-hold or market timing is relative to the time horizon considered.

The fact that market timing strategies have the potential for enormous returns is not surprising. Consider the following illustration due to Shilling [44]. A one dollar portfolio invested in the Dow Jones Industrial Average in January 1946 was worth 116 dollars at the end of 1991 (including reinvestment of dividends but excluding tax deductions). This is equivalent to 11.2% compound annual gain. A market timing strategy that was lucky to be out of the market during the 50 weakest months during that period (spanning 552 months) but otherwise was fully invested using the same fixed portfolio would return 2541 dollars, or a 19% annual gain. Further, an off-line strategy which during the 50 weakest months was in a short position would have returned 44,967 dollars, or a 26.9% annual gain. The reader is referred to the comprehensive survey by Merton [37] for classical theories and ideas regarding buy-and-hold and market timing strategies.

The constant rebalanced algorithm An algorithm that is often used in practice and will be considered in this chapter is the following *constant rebalanced* algorithm that rebalances fixed portfolio $\mathbf{b} = (b_1, b_2, \ldots, b_m)$ at the start of each trading period. This algorithm, denoted $\text{CBAL}_{\mathbf{b}}$, is clearly a market timing strategy. The performance of the optimal off-line CBAL is always at least as good as that of the optimal off-line BAH but usually it is significantly better (See e.g. the examples in [13]). The reason is that the optimal off-line BAH only performs as the best security in the market but the optimal off-line CBAL also takes advantage of fluctuations in the market giving rise to exponential returns. Of course, CBAL makes as many as m transactions at the start of each trading period to adjust its portfolio whereas BAH performs a single transaction during the entire trading period. Although in the models considered here this large number of transactions is not a consideration, it definitely becomes significant when transaction costs are introduced.

5.3 On-line portfolio selection

Let ALG be an on-line portfolio selection. Recall that $\text{ALG}(X)$ denotes the compounded return of ALG with respect to the market sequence $X = \mathbf{x}_1, \ldots, \mathbf{x}_n$ of price relative vectors and starting with an initial wealth of 1 dollar. The competitive ratio of ALG is

$$\sup_X \frac{\text{OPT}(X)}{\text{ALG}(X)} ,$$

where OPT is an optimal off-line portfolio selection algorithm. By considering the simpler one-way trading problem (see Sect. 3.4), the paper [20] obtains a lower bound of $c_2^*(\varphi)^{n/2}$ for portfolio selection ($m = 2$) where the constant $c_2^*(\varphi) > 1$ is the optimal bound for a one-way 2-day trading game in which the player knows the global fluctuation ratio φ (see equation (5)). This simple lower bound is essentially obtained via a decomposition of the two-way trading problem to a sequence of one-way trading games and using the known lower bound for one-way trading games. Note that the same paper presents an upper bound of $c_\infty^*(\varphi)^n$ for the case $m = 2$, based again based on a straightforward decomposition of the two-way trading game into a sequence of one-way games.

Thus, for the on-line portfolio selection problem one cannot hope for competitive ratios that are sub-exponential in n, the number of trading periods. Also, to obtain bounded ratios one must assume lower and upper bounds on the securities' prices (or other equivalent constraints). Nevertheless, one should keep in mind that for typical market sequences the optimal off-line algorithm accrues astronomic returns (which are also typically exponential in n) so the above lower bound does not exclude possibilities of large returns to the on-line player. On the other hand, notice that competitive on-line strategies can lose money (i.e. end up the game with less dollars than its initial wealth) while the optimal off-line strategy will never lose.

5.4 Other performance measures

Worst case studies of on-line portfolio selection strategies consider performance measures other than the competitive ratio. These performance measures differ from the standard competitive ratio optimization in (some of) the following manners:

- Considering a restricted kind of (off-line) algorithm, for example, some kind of a "static" off-line algorithm (such as the optimal off-line CBAL) instead of the standard, unrestricted optimal off-line algorithm.
- Imposing more constraints on the adversary. For example, consider adversaries that must choose their worst case market sequence while conforming to some additional (statistical) parameters. Such adversaries are called *statistical adversaries* [40].[7]
- Using decision criteria other than the competitive ratio. For example, the *maximin* or the *minimax regret* (definitions follow).

[7] Notice that most of the adversaries that were considered so far in this chapter are "minimal" forms of statistical adversaries (e.g. some of the adversaries assumed in the one-way trading problem must conform to lower and upper bounds on prices).

- Instead of measuring the compounded return of the strategy (on-line or off-line), measuring returns via some utility functions. For example, one popular measure is the *exponential growth rate* defined (for an algorithm ALG and a market sequence X) to be $1/n \cdot \log(\text{ALG}(X))$.

Let us now define formally other measures that will be discussed later. As usual ALG denotes an on-line algorithm and OPT denotes an off-line algorithm. Let U denote any utility function. In the above example of the exponential growth rate the utility function is $U(x) = \log(x)/n$. With respect to ALG, OPT and U (and assuming profit maximization) we now distinguish between three adversaries. First, a *competitive ratio* adversary chooses a market sequence X^* such that

$$X^* = \arg\max_X U(\text{OPT}(X))/U(\text{ALG}(X)) \ .$$

The *maximin* adversary chooses a market sequence X^* such that

$$X^* = \arg\min_X U(\text{ALG}(X)) \ .$$

The *minimax-regret adversary* chooses X^* with

$$X^* = \arg\max_X \{U(\text{OPT}(X)) - U(\text{ALG}(X))\} \ .$$

Applying additional restrictions on the adversary simply means that we constrain the above maximizations (minimizations) with respect to these additional restrictions that specify what feasible market sequences are. Notice that the use of the exponential growth rate utility in conjunction with the minimax-regret adversary is equivalent to using the competitive ratio adversary with the identity utility function.

There are various motivations for the use of one (or a combination) of the above variations. For instance, one motivation is the desire to avoid the consideration of unrealistic adverse market sequences that are very unlikely to occur in practice. This motivation applies especially to the use of statistical adversaries. The use of different criteria other than the competitive ratio is clearly justified as it has never been successfully argued that the competitive ratio is a more suitable (worst-case) measure than others (see also [19]). Finally, note that most of the above variations apply in general to any on-line problem.

5.5 Results for the two-way trading problem

Here we focus on several results for the two-way trading problem. That is, the portfolio selection problem with $m = 2$ securities such that one of them is cash. In this special case, since the cash prices and price relatives are always 1 a market sequence is specified by a sequence of the security prices $p_1, p_2, \ldots$ (or price relatives).

Games against mean-variance statistical adversaries Raghavan [40] considers a market in which the security price sequence $p_1, p_2, \ldots$ maintains a mean of 1 and known standard deviation $\sigma \in [0,1]$. He also imposes the restrictions that (i) prices are drawn from the interval $[1-c, 1+c]$ where $c \leq 1$ is some constant and (ii) that $p_1 = p_n = 1$ where n is the number of trading periods. With respect to a (statistical) maximin adversary conforming to the above restrictions he obtains the following results.[8] First it is stated that any deterministic on-line algorithm cannot return more than $1+\sigma$. He then considers two strategies: CBAL and the following averaging strategy, denoted AVE. Assuming n trading periods AVE invests $1/n$ dollars in the security at the start of each trading period. The following bounds hold for AVE for each feasible market sequence $X = p_1, p_2, \ldots, p_n$,

$$1 + \sigma^2(1-c) \leq \text{AVE}(X) \leq 1 + \sigma^2 \ .$$

For $\text{CBAL}_\mathbf{b}$ that maintains a portfolio $\mathbf{b} = (1/2, 1/2)$ the following bounds are stated

$$1 + \sigma^3/3 \leq \text{CBAL}(X) \leq 1 + \sigma^3/2 \ .$$

It is stated that the $1 + \sigma^3/3$ lower bound holds for any constant-rebalanced strategy (under the same assumptions). The bounds for algorithms AVE and CBAL can be obtained as follows. Consider a price sequence $X = p_1, p_2, \ldots, p_n$. Any dollar investment of AVE at the ith trading period returns $1/p_i$ dollars (since $p_n = 1$). Since AVE invests $1/n$ dollars in each period the total return is $1/n \sum_i 1/p_i$. Hence, bounds on AVE's total return can be obtained by maximizing this expression over all price vectors satisfying the mean-variance (standard deviation) constraints. Similarly, obtaining the bounds for CBAL can be reduced to such constrained optimization. Note that the return of CBAL for the market sequence X is

$$\text{CBAL}(X) = 1/2^{n-1} \prod_{i=1}^{n-1} \left(\frac{p_i + p_{i+1}}{p_{i+1}} \right) \ .$$

5.6 Other statistical adversaries and "Money-Making" algorithms

Money-Making algorithms Define the *profit* of a trading strategy as its final compounded return minus its initial wealth (a negative profit is called *loss*). The realization that any competitive trading algorithm may end the trading game with a loss motivated the authors of [10] to study models with additional constraints that guarantee positive profit for the on-line player with respect to market sequences for which the optimal off-line algorithm accrues positive profits. An on-line trading algorithm that satisfies this property is called *money-making*. The models considered assume several kinds of statistical adversaries that must conform to prespecified constraints. Not surprisingly, in order to achieve the money-making property the assumptions on market sequences must be quite strong. Nevertheless, the consideration of such models leads to quite interesting

[8] The paper [40] does not contain proofs for the stated results.

results. For the rest of this section we consider finite market sequences prescribed by a sequence of security prices $p_1, p_2, \ldots, p_n$ or alternatively, by a sequence of the price relatives $x_1, x_2, \ldots, x_{n-1}$ where $x_i = p_i/p_{i+1}$, $i = 1, 2, \ldots, n-1$. When referring to a market sequence X we shall take the most convenient one of these representations.

The (n, ϕ)-adversary Fix some $n \geq 2$. Assume that each feasible price sequence is of length n and impose the restriction that the optimal off-line return associated with a feasible sequence is at least ϕ (clearly $\phi \geq 1$). The underlying assumption here is that true (n, ϕ) pairs exist in relevant real price sequences in the sense that such pairs can be statistically estimated from past markets with a reasonable degree of confidence. An adversary that is constrained to generate only such feasible sequences is called the (n, ϕ)-adversary. Let $X = x_1, x_2, \ldots, x_{n-1}$ be a feasible sequence of price relatives. It is not hard to see that the optimal off-line return ϕ is given by

$$\phi = \prod_{1 \leq i \leq n-1} \max\{1, x_i\} \ . \tag{10}$$

It follows that for any such sequence of prices (price relatives) an on-line player knowing ϕ and n at the start of the game, can determine at the start of the $(i+1)$st period, just after the $(i+1)$st price is revealed $(i = 1, 2, \ldots)$, what the optimal off-line return ϕ_{n-i} would be for a new $(n-i)$-period game starting in that period; that is, if the optimal off-line algorithm were to start in that period a "new game" (with initial wealth of 1) with respect to the suffix of the original market sequence, $p_{i+1}, p_{i+1}, \ldots, p_n$. Specifically, using (10) we have $\phi_{n-1} = \min\{\phi, \phi/x_1\}$, and

$$\phi_{n-j-1} = \min\{\phi_{n-j}, \phi_{n-j}/x_{j+1}\} \ .$$

Hence, against the (n, ϕ)-adversary the on-line player can track the "state" (i.e. current wealth accrued) of the optimal off-line algorithm with a delay of one day. This property allows for a dynamic programming derivation of the optimal on-line algorithm. Denote by $R_n(\phi)$ the return of the optimal on-line algorithm S^* (that knows the parameters n and ϕ). It is not hard to obtain the following recurrence relation for R_n.

$$\begin{aligned} R_n(\phi) &= \sup_{0 \leq b \leq 1} \inf_{x \leq \phi} \{(bx + 1 - b)\, R_{n-1}(\phi_{n-1})\} \ ; \qquad (11) \\ R_2(\phi) &= \phi \ . \end{aligned}$$

Notice that algorithm S^* attempts to choose its best investment b against the worst possible price relative x. The wealth obtained from the investment bx plus the remaining cash $1-b$ are then reinvested optimally with respect to an $(n-1)$-period game in which the optimal off-line return is ϕ_{n-1}, etc.

Although a closed form for R_n is probably beyond reach, it is not hard to prove by induction on n that S^* is money-making. On the other hand, it is not hard to obtain the following upper bound on $R_n(\phi)$ for any $\phi > 1$ and $n \geq 2$.

$$R_n(\phi) \leq \frac{1}{1-(1-1/\phi)^{n-1}} \quad . \tag{12}$$

This bound can be obtained by considering the following restricted version of the (n, ϕ)-adversary. In each period this adversary has two options: either to decrease the price by a factor of ϕ or to increase the price by a very large factor so that the dollar value of the previous investment becomes negligible. Once the adversary chooses the first option there would be no further downward fluctuations since the optimal off-line of ϕ has been realized. Hence, if this is the case, the game is over. The optimal on-line return $R'_n(\phi)$ for this restricted game is

$$R'_n(\phi) = \max_{0 \leq b \leq 1} \min\{b\phi + 1 - b,\, (1-b)R'_{n-1}(\phi)\} \;,$$

and its closed form is the r.h.s of (12) which clearly bounds above $R_n(\phi)$.

Using the approximation $(1-1/\phi)^{n-1} = (1-1/\phi)^{\phi\frac{n-1}{\phi}} \approx e^{-(n-1)/\phi}$, for large ϕ we have

$$R'_n(\phi) \approx 1/(1-e^{-(n-1)/\phi}) \,,$$

and then the following relations are obtained

- If $\phi = \omega(n)$ then $e^{-(n-1)/\phi} \approx 1-(n-1)/\phi$ and $R'_n(\phi) \approx \phi/(n-1)$;
- If $\phi = \Theta(n)$ then $R'_n(\phi) \approx 1/(1-e^{-c})$ where c is some positive constant;
- If $\phi = o(n)$ then $R'_n(\phi)$ approaches 1 as $n \to \infty$.

Hence, although S^* is "money-making," the optimal on-line return $R_n(\phi)$ against the (n, ϕ)-adversary can be a minuscule fraction of ϕ. In terms of competitiveness, the competitive ratio of S^* is not smaller than $\max\{n-1, \phi\}$ and no upper bound was determined.

The general "Money-Making" scheme against statistical adversaries The above derivation of an optimal on-line algorithm against the (n, ϕ)-adversary gives rise to a general scheme for obtaining money-making optimal on-line algorithms with respect to any statistical adversary that is at least as restricted as the (n, ϕ)-adversary. Specifically, for any collection of constraints C that subsume the (n, ϕ) constraint a similar relation as (11) obtains the corresponding optimal on-line algorithm when the constraints in C are appropriately included in the recurrence.

Weaker adversaries Against the (n, ϕ)-adversary the on-line player is forced to invest very small amounts in most trading periods since the adversary can depreciate the value of most investments by increasing the price arbitrarily. Theoretically, such threats can be made until the second last period. Such market sequences are of course unrealistic. By imposing additional constraints it is possible to reduce such threats. The following is a list of constraints that can substantially reduce these unrealistic threats: (i) upper bounds on price relatives; (ii) minimum and maximum bounds on prices; (ii) upper bounds on the length of runs of monotone increasing (decreasing) prices and (iii) other statistical parameters such as mean and higher moments of the empirical distribution observed. Note that the (n, ϕ) constraint is "minimal" in the sense that when one of these parameters is relinquished the money-making property cannot be achieved.

5.7 The fixed fluctuation model

In the *fixed fluctuation model* all price relatives x_i are in $\{\alpha, \alpha^{-1}\}$ where $\alpha > 1$. Assuming w.l.o.g. that the initial price is 1, it follows that all prices are in $\{\alpha^j : j \text{ integer}\}$. We now add the restriction that each feasible sequence of price relatives is of length n and the number of downward (i.e. profitable) α^{-1}-fluctuations is exactly k, $0 \leq k \leq n$. Hence, the number of upward, α-fluctuations is $n-k$. Thus, the "type" of each market sequence is specified by the number k. Call the adversary that produces such feasible sequences the (α, n, k)-adversary. Clearly this (α, n, k)-constraint subsumes the above (n, ϕ)-constraint since the implied optimal off-line return for each feasible market sequence is exactly $\phi = \alpha^k$. Hence, it is possible to obtain the optimal on-line trading algorithm (against the (α, n, k)-adversary) using the above "money-making" scheme. The paper [10] studies this optimal on-line algorithm and characterizes some of its properties.

On the fixed fluctuation model and time scaling Before we continue sketching the results of [10] let us take note of the practical relevancy of the fixed fluctuation model. Clearly "daily" price relatives are variable. Hence, to approximate the fixed fluctuation model one can use a *time scaling* approach where each trading period is of variable length such that the $(i+1)$st price "tick" occurs at the first time instance after the ith price tick when a price that approximates $p_i\alpha$ or p_i/α is encountered (with p_i denoting the price at the ith tick).[9] One advantage of this fixed fluctuation model is that the player may choose a suitable α that will filter out "noisy" fluctuations (i.e. very small fluctuations that should be avoided when transaction costs are taken into considerations).

The optimal strategy against the (α, n, k)-adversary Denote by $R(n, k) = R(\alpha, n, k)$ the return of the optimal on-line algorithm S^{**} against the (α, n, k)-adversary. Using the above scheme, a recurrence relation for $R(n, k)$ is easily obtained.

$$R(n,k) = \max_{0 \leq b \leq 1} \min \left\{ \begin{array}{l} (\alpha b + 1 - b)R(n-1, k-1), \\ (b\alpha^{-1} + 1 - b)R(n-1, k) \end{array} \right\} \tag{13}$$

[9] Of course, to allow for such time scaling the choice of α must be made in accordance with the special properties of the market in question.

$$R(n,0) = 1$$
$$R(n,n) = \alpha^n \ .$$

Here the b that minimizes (13) is the first investment made by algorithm S^{**}. Since the upper operand in the min operator in (13) is decreasing with b and the lower one is increasing with b, by an "arbitrage argument" the optimal b, denoted b^*, equates both operands. Therefore,

$$b^* = \frac{R(n-1,k) - R(n-1,k-1)}{(\alpha-1)R(n-1,k-1) - (\alpha^{-1}-1)R(n-1,k)} \ .$$

Substituting b^* for b in either operand of the min in (13) and rearranging we obtain the following expression for the reciprocal R^{-1}.

$$R^{-1}(n,k) = \frac{1}{\alpha+1}R^{-1}(n-1,k-1) + \frac{\alpha}{\alpha+1}R^{-1}(n-1,k) \ . \qquad (14)$$

It is possible to solve R^{-1} in a closed form (in terms of partial binomial sums) as follows. Artificially extend the domain of $R^{-1}(n,k)$ such that

$$R^{-1}(n,j) = \begin{cases} 1, & j \le 0; \\ \alpha^{n-2k}, & j \ge n. \end{cases} \qquad (15)$$

(It is possible to prove by induction on n that the extended recurrence is well defined.) Now consider the following graph (similar to the "Pascal triangle") that is called a *binomial tree*[10], corresponding to an expansion of the extended recurrence $R^{-1}(n,k)$. In this graph, each node is labeled by some pair (x,y) that corresponds to the value $R^{-1}(x,y)$. The root of this graph is (n,k) and it has two outgoing edges. One to its left child $(x-1,y-1)$ and the other, to its right child $(x-1,y)$. All the leaves are of the form $(1,k-(n-i))$, $i = 1,\ldots,n$ and the values for them are obtained by the extension (15). By (14), computing the value for each node (x,y) is computed from the values of its children, $(x-1,y-1)$ and $(x-1,y)$. Notice that in each node (x,y), x corresponds to the height of the node (with the leaves at level 1 and the root at level n) and y corresponds to the edge distance from the top-left to bottom-right diagonal (i.e. the sequence of edges connecting (n,k) to $(1,k)$). Notice that all the paths to the same leaf have the same number of left-right (and right-left) moves. In particular, the path to the leaf $(1,k-(n-i))$ has $n-i$ top-left edges and $(n-1)-(n-i) = i-1$ top-right edges (there is a total of $n-1$ edges in each path). By (14), when we calculate the value $R^{-1}(n,k)$ each left move contributes a factor $q = \frac{1}{1+\alpha}$ and each right move contributes a factor $p = \frac{\alpha}{1+\alpha}$ (notice that $q = 1-p$). Hence, the "weight" of the path to the leaf $(1,k-(n-i))$ is $p^{i-1}q^{n-i}$. Define

$$B_p\left\{ {n \atop k} \right\} = \sum_{0 \le i \le k} \binom{n}{i} p^i (1-p)^{n-i},$$

[10] The name *binomial tree* is typically used in finance [17]. Note that the resulting graph is not a tree in the graph-theoretic sense.

the partial binomial sum. Abbreviating $z_i = k - (n - i)$ we thus have

$$\begin{aligned}
R^{-1}(n,k) &= \sum_{\text{leaf } (1,z_i)} R^{-1}(1,z_i)\cdot[\#\text{ of paths to }(1,z_i)]\cdot[\text{weight of path to }(1,z_i)] \\
&= \sum_{1\le i\le n} R^{-1}(1,z_i)\binom{n-1}{i-1}p^{i-1}q^{n-i} \\
&= \sum_{z_i\le 0} R^{-1}(1,z_i)\binom{n-1}{i-1}p^{i-1}q^{n-i} \\
&\qquad + \sum_{z_i\ge 1} R^{-1}(1,z_i)\binom{n-1}{i-1}p^{i-1}q^{n-i} \\
&= \sum_{z_i\le 0} 1\binom{n-1}{i-1}p^{i-1}q^{n-i} + \sum_{z_i\ge 1}\alpha^{n-2k}\binom{n-1}{i-1}p^{i-1}q^{n-i} \\
&= \sum_{i=1}^{n-k}\binom{n-1}{i-1}p^{i-1}q^{n-i} + \alpha^{n-2k}\sum_{i=n-k+1}^{n}\binom{n-1}{i-1}p^{i-1}q^{n-i} \\
&= \sum_{i=0}^{n-k-1}\binom{n-1}{i}p^{i}q^{n-i-1} + \alpha^{n-2k}\cdot\sum_{i=0}^{n-k}\binom{n-1}{i}q^{i}p^{n-i-1} \\
&= B_p\left\{\begin{matrix} n-1 \\ n-k-1\end{matrix}\right\} + \alpha^{n-2k}\cdot B_q\left\{\begin{matrix} n-1 \\ n-k\end{matrix}\right\} \\
&= B_p\left\{\begin{matrix} n-1 \\ n-k-1\end{matrix}\right\} + \alpha^{n-2k}\cdot B_p\left\{\begin{matrix} n-1 \\ k-1\end{matrix}\right\}.
\end{aligned}$$

Using the above expression and tail estimates of the binomial distribution, Chou, Shrivastava and Sidney [11] obtained the following asymptotic behavior of $R(n,k)$. Let $c \in [0,1]$ and define

$$\gamma_x = \frac{x^x(1-x)^{(1-x)}(1+\alpha)}{\alpha^{(1-x)}} \quad . \tag{16}$$

Then, the following bounds[11] hold.

(i) if $0 \le c < p$, then $R(n,cn) \to 1$;
(ii) if $c = p$, then $R(n,cn) \to 2$;
(iii) if $p < c < q$, then $R(n,cn) = \Theta(\sqrt{n}\gamma_c^n)$;
(iv) if $c = q$, then $R(n,cn) \to 2\gamma_c^n = 2\alpha^{(2c-1)n}$;
(v) if $q < c \le 1$, then $R(n,cn) \to \alpha^{(2c-1)n}$ $(R(n,cn) \ge \alpha^{(2c-1)n})$.

This result entails an interesting corollary. Consider the (optimal off-line) BUY-AND-HOLD (BAH). If $c > 1/2$, BAH invests its entire wealth on the first period

[11] Cruder but similar bounds were obtained earlier in [10].

and cashes it at the end of the game. Otherwise, BAH keeps its wealth in cash. The return of BAH, $R(\text{BAH})$ is thus

$$R(\text{BAH}) = \begin{cases} 1, & \text{if } c \leq 1/2; \\ \alpha^{n(2c-1)}, & \text{if } c \geq 1/2. \end{cases} \tag{17}$$

It follows that algorithm S^{**} always outperforms BAH. Moreover, if $q < c < p$, S^{**} performs exponentially better than BAH. Whenever $c \in [0, 1/2]$ or $c \in [p, 1]$ this result readily follows from the bounds obtained for S^{**} and the return of BAH. Whenever $c \in (1/2, p)$, it is not hard to see that $\gamma_c > \alpha^{2c-1}$. (the function $f(c) = \gamma_c/\alpha^{2c-1} = (c/\alpha)^c(1-c)^{1-c}(1+\alpha)$ is strictly decreasing in $(1/2, p)$ and $f(p) = 1$).

Thus, whenever the market is "stable" in the sense that it does not exhibit a "major" trend, but it is nevertheless "active" (i.e. there are fluctuations) algorithm S^{**} performs remarkably well (in particular, relative to BAH). Moreover, even if the market exhibits a slight unfavorable trend (i.e. $c \in (q, 1/2)$), algorithm S^{**} still yields exponential returns. Notice that the size of this "profitable" interval (q, p) increases with α and thus can be controlled by the on-line player. Nevertheless, for larger values of α, because of the time scaling required to approximate the fixed fluctuation model, algorithm S^{**} may ignore many prices and therefore miss profitable transactions.

In [11], Chou et al. also compare these bounds on the performance of S^{**} to two more algorithms. First, they study the performance of the optimal constant rebalanced algorithm, denoted CBAL*, against the (α, n, k)-adversary. Second, they study the following distributional model corresponding to the constraint that feasible market sequences must be of type (n, k). Specifically, in this model a random price relative sequence is chosen uniformly among all the feasible sequences (of length n and with exactly k downward fluctuations). With respect to this model they determine the optimal trading algorithm (i.e. optimal with respect to average return). Consider first the constant rebalanced algorithm. Let b be the constant fraction of wealth invested by CBAL in the security. With respect to any feasible market sequence the return of CBAL is

$$(1 - b + b\alpha)^k(1 - b + b\alpha^{-1})^{n-k} \ .$$

The optimal value of b, denoted b^*, is easily determined to be $b^* = (\alpha c + c - 1)/(\alpha - 1)$ where $c = n/k$.

The distributional variant described above can be equivalently presented as follows. Define $l = n - k$. Given the (n, k) restriction, the probability of an α^{-1}-fluctuation in the first period is k/n and the probability of an α-fluctuation is l/n. After the first fluctuation was (randomly) obtained, the remaining sequence must be of type $(n-1, k-1)$ or $(n, k-1)$ depending on whether the first fluctuation was α^{-1} or α, respectively, etc. Denote by $R(n, l)$ the expected return of the optimal on-line algorithm when feasible markets are of type (n, k). Thus,

$$R(k, l) = \max_{0 \leq b \leq 1} \left\{ \frac{k}{n}(1 - b + b\alpha^{-1})R(k-1, l) + \frac{l}{n}(1 - b + b\alpha)R(k, l-1) \right\} .$$

Since the return is a maximum of a linear function of b, it follows that b is either 0 or 1. Note that b is only the investment for the first period. Nevertheless a non-trivial analysis presented in [11] shows that the optimal on-line algorithm is constant rebalanced. Therefore, given that the first investment is "all or nothing" the optimal on-line strategy is BUY-AND-HOLD which invests in the security iff $k \geq l$. We note that with respect to a sequence of price relatives generated randomly by a sequence of i.i.d. random variables it is known that the optimal algorithm is constant rebalanced [8].

The performances of the optimal constant rebalanced algorithm CBAL* and the optimal algorithm of the distributional problem variant (denoted DIST), plus the bounds for algorithm S^{**} are summarized in Table 2 (which also includes, for comparison, the return of the optimal off-line BAH).

Region of $c = \frac{k}{n}$	CBAL*	S^{**}	DIST	BAH
$(0, q)$	1	1	$\Theta(1)$	1
$c = q$	1	2	$\Theta(\sqrt{n})$	1
$(q, 1/2)$	γ_c	$\Theta(\sqrt{n} \cdot \gamma_c)$	$\Theta(\sqrt{n} \cdot \gamma_c)$	1
$[1/2, p)$	γ_c	$\Theta(\sqrt{n} \cdot \gamma_c)$	$\Theta(\sqrt{n} \cdot \gamma_c)$	$\alpha^{(2c-1)n}$
p	$\alpha^{(2c-1))n}$	$2\alpha^{(2c-1)n}$	$\Theta(\sqrt{n} \cdot \alpha^{(2c-1)n})$	$\alpha^{(2c-1)n}$
$(p, 1)$	$\alpha^{(2c-1)n}$	$\alpha^{(2c-1)n}$	$\Theta(\alpha^{(2c-1)n})$	$\alpha^{(2c-1)n}$

Table 2. Asymptotic performance of CBAL*, S^{**}, DIST and BAH for "trendy" and "non-trendy" markets

An immediate striking fact observed in this table is that the asymptotic returns of all three optimal on-line algorithms are similar. Denote the four regions of c, $(0, q), (q, 1/2), [1/2, p)$ and $(p, 1)$ by R1, R2, R3 and R4, respectively. For markets that exhibit significant trend (regions R1 and R4) the returns of all three algorithms are within a constant factor of BAH. In fact, in these regions the optimal constant rebalanced algorithm degenerates to BAH (and DIST acts in a similar way to BAH across all regions). For markets with no trend (regions R2 and R3) all three algorithms perform exponentially better than BAH.

These results also indicate that information about the "type" of the market can be almost as valuable as information about the distribution (note that in region R3, DIST outperforms S^{**} and CBAL by a constant factor).

These results, together with the fact that both CBAL* and S^{**} make transactions in almost every trading period, may indicate that the following is true whenever transaction costs are introduced. When the market is trendy, even if the trader has perfect prediction of the trends slope it may be beneficial to avoid market timing at all and follow the trend using BAH. On the other hand, the exponential advantage of market timing strategies in non-trendy markets may very well compensate for the losses incurred by transaction costs.

Finally, we note that Chou et al. [11] showed how to derive the return of S^{**} using a standard method called *binomial risk-neutral pricing* (see [26]) that is usually used to price options and other derivative instruments. Their results imply that the binomial risk-neutral pricing method is equivalent to worst case analysis.

On the empirical performance of S^{**} Chou reports on preliminary experimental results testing the performance of S^{**} [9]. He used two data sets, both of which include intra-day prices for both US dollars vs. Japanese yen, and US dollars vs. German Marks. Data set A spanned one month and included all price quotations[12] and data set B included prices of six minutes intervals during one year. In both cases α was set to $1 + 5/(\text{initial exchange rate})$ since almost all changes in this data were of 5 points[13]; since such changes are small relative to the exchange rates, the additive changes can be reasonably well approximated by multiplicative changes of α. The value of n varied between 100 and 1000, thus breaking the trading period into a sequence of short games with reinvestment. Finally, k was naively chosen to be $n/2$ for all games. The returns of S^{**} with respect to data set A were remarkably high (223% and 167% for USD/DM and USD/JY, respectively). The results with respect to data set B were marginal (104% and 111%). When transaction costs of 0.02% were introduced the returns against data set A remained very high (133% and 134%) but against data set B, S^{**} suffered severe losses (73% and 89%). Examinations of these data sets revealed that in data set A the assumption of $k = n/2$ was quite closely satisfied while in data set B it was violated many times. One not surprising drawback of S^{**} that was found in these experiments is that S^{**} reacts to and trades with every price tick.

In [11], Chou et al. report on extensive statistical tests measuring the trends exhibited in price sequences as a function of α and n. In particular, their goal was to measure the empirical distribution of trend types in terms of the above four regions (R1, R2, R3, and R4) with varying values of α and n. The data used included 486 of the 500 stocks comprising the Standards and Poors 500 index (S&P 500) spanning a time period of about 30 years. The tests reveal the following relations. Call regions R2 and R3 the "non-trendy" regions and regions R1 and R4 the "trendy" regions. As α or n grow the fraction of sequences (of length n) in the non-trendy regions grows (and the fraction of sequences in region R3 is always larger than in R2). For small values of α (e.g. $1.005 \le \alpha \le 1.05$) the majority of sequences are in the trendy regions. For larger values of α the majority of sequences are in the non-trendy regions. From these results it follows that with respect to the stock market, the practical advantage of market-timing strategies like S^{**} or constant rebalanced, using a fixed fluctuation model, can only be obtained when using large values of α and n (while using time scaling). On the other hand, if, in a particular market, such large values of α are not feasible, it is probably wiser to remain "static" and use BUY-AND-HOLD. Note

[12] In such data a new price "tick" occurs every 10–120 seconds.
[13] A *point* is the smallest unit used to measure prices.

that for large values of α (say, $\alpha \geq 1.01$) transaction points cannot occur too often (which may be considered as an advantage).

5.8 On-line portfolio selection, results for arbitrary m

The family of μ-weighted portfolio selection algorithms Cover and Ordentlich [15] define the following general class of *μ-weighted* on-line portfolio selection algorithms. Each algorithm in this class is specified by a probability measure μ over the set of all portfolios. The algorithm starts by hedging uniformly on all possible portfolios. Then it rebalances the weight it gives to various securities according to the past performance of constant rebalanced portfolios while putting more weight on the better performing ones. The probability measure μ provides an additional weighting mechanism that can favor particular portfolios.

The introduction of the μ-weighted algorithms requires some notation. Fix some positive integer m. Let $\mathcal{B}$ be the set of all possible portfolios over m securities (i.e. $\mathcal{B}$ is the $(m-1)$-dimensional simplex). With respect to a market sequence $X = \mathbf{x}_1, \ldots, \mathbf{x}_n$ define $X_i = \mathbf{x}_1, \ldots, \mathbf{x}_i$, $i = 1, 2, \ldots, n$, the prefix of X consisting of the first i market vectors. With respect to X_i define

$$R_i(\mathbf{b}) = R(\mathbf{b}, X_i) = \prod_{j=1}^{i} \mathbf{b}^t \cdot \mathbf{x}_j \ .$$

That is, $R_i(\mathbf{b})$ is the compounded return of a fixed portfolio $\mathbf{b}$ after i trading periods. By convention, set $R_0(\mathbf{b}) = 1$, the initial wealth. (Alternatively, $R_i(\mathbf{b}) = \text{CBAL}_{\mathbf{b}}(X_i)$.)

Fix some market sequence X and let μ be a probability measure over $\mathcal{B}$. An algorithm is a *μ-weighted portfolio selection algorithm* if its ith period portfolio $\mathbf{b}_i$ is specified by

$$\mathbf{b}_i = \frac{\int_{\mathcal{B}} \mathbf{b} R_{i-1}(\mathbf{b}) d\mu(\mathbf{b})}{\int_{\mathcal{B}} R_{i-1}(\mathbf{b}) d\mu(\mathbf{b})} \ . \tag{18}$$

Here the integral is Lebesgue integral. Note that when the probability measure μ has a density function the Lebesgue integral can be expressed in terms of ordinary (multiple) integrals. Clearly, any μ-weighted algorithm operates on-line.

Let X be any market sequence of length n and let $\mathbf{b}_1, \mathbf{b}_2, \ldots, \mathbf{b}_n$ be the portfolios obtained by a μ-weighted algorithm ALG_μ. The expression for the compounded return $\text{ALG}_\mu(X)$, of the μ-weighted algorithm ALG_μ, can be easily simplified as follows

$$\begin{aligned}
\text{ALG}_\mu(X) &= \prod_{1 \leq i \leq n} \mathbf{b}_i^t \cdot \mathbf{x}_i \\
&= \prod_{1 \leq i \leq n} \frac{\left(\int_{\mathcal{B}} \mathbf{b}^t R_{i-1}(\mathbf{b}) d\mu(\mathbf{b})\right) \cdot \mathbf{x}_i}{\int_{\mathcal{B}} R_{i-1}(\mathbf{b}) d\mu(\mathbf{b})}
\end{aligned}$$

$$= \prod_{1 \le i \le n} \frac{\left(\int_{\mathcal{B}} \mathbf{b}^t \cdot \mathbf{x}_i R_{i-1}(\mathbf{b}) d\mu(\mathbf{b})\right)}{\int_{\mathcal{B}} R_{i-1}(\mathbf{b}) d\mu(\mathbf{b})}$$

$$= \prod_{1 \le i \le n} \frac{\int_{\mathcal{B}} R_i(\mathbf{b}) d\mu(\mathbf{b})}{\int_{\mathcal{B}} R_{i-1}(\mathbf{b}) d\mu(\mathbf{b})} . \quad (19)$$

Hence, since the last product telescopes we have

$$\text{ALG}_\mu(X) = \int_{\mathcal{B}} R_n(\mathbf{b}) d\mu(\mathbf{b}) = \int_{\mathcal{B}} \text{CBAL}_{\mathbf{b}}(X)\mu(\mathbf{b}) . \quad (20)$$

Notice that although the μ-weighted algorithm adapts its portfolio on-line and gives more weight to the better performing (constant rebalanced) portfolios (see equation (18), its final return is simply a μ-average of the returns of all constant rebalanced portfolios in $\mathcal{B}$.

The uniform-weighted algorithm Denote by UNI the μ-weighted algorithm corresponding to the uniform distribution over $\mathcal{B}$. Denote by CBAL-OPT the optimal off-line constant rebalanced algorithm. That is, CBAL-OPT = $\text{CBAL}_{\mathbf{b}^*}$ where for any market sequence $X = \mathbf{x}_1, \ldots, \mathbf{x}_n$, the fixed portfolio $\mathbf{b}^*$ used by CBAL-OPT is

$$\mathbf{b}^* = \arg\max_{\mathbf{b}} \prod_{1 \le i \le n} \mathbf{b} \cdot \mathbf{x}_i .$$

Cover and Ordentlich show that with respect to the optimal off-line constant rebalanced algorithm, CBAL-OPT, and any market sequence X of length n the following holds.

$$\frac{\text{CBAL-OPT}(X)}{\text{UNI}(X)} \le \binom{n+m-1}{m-1} \le (n+1)^{m-1} . \quad (21)$$

Thus, in a restricted sense of competitiveness, algorithm UNI is $(n+1)^{m-1}$-competitive (with respect to CBAL-OPT). Stated equivalently (but perhaps more astoundingly) using the above exponential growth rate utility function and the minimax regret criterion, this result states that the minimax regret (of the exponential growth rates) of algorithm UNI is bounded above by $\frac{(m-1)\log(n+1)}{n}$. This means that the regret is diminishing when n grows.

The proof of the bound (21) is interesting and is based on deep ideas. Somewhat unexpectedly it is possible to transform the analysis into information theoretic grounds. Indeed, in proving this theorem we shall rely on some elementary results from information theory.

We now sketch the proof of this bound. Let $[m] = \{1, 2, \ldots, m\}$. Denote by $J_n = (j_1, j_2, \ldots, j_n)$ any sequence of indices from $[m]$ (i.e. $J_n \in [m]^n$). Let $X = \mathbf{x}_1, \ldots, \mathbf{x}_n$ be let $\mathbf{b}$ be any portfolio. The analysis relies on the following trick that gives a different representation for the compounded return of the constant rebalanced algorithm $\text{CBAL}_{\mathbf{b}}$.

$$\begin{aligned}
\text{CBAL}_{\mathbf{b}}(X) &= \prod_{1\leq i\leq n} \mathbf{b}^t \cdot \mathbf{x}_i \\
&= \prod_{1\leq i\leq n} \sum_{1\leq j\leq m} b_j x_j^i \\
&= \sum_{J_n\in[m]^n} \prod_{1\leq i\leq n} b_{j_i} x_{j_i}^i \ . \qquad (22)
\end{aligned}$$

Fix any market sequence X of length n. Let $\mathbf{b}^*$ be the portfolio used by CBAL-OPT. Using the representation (22) for CBAL-OPT and UNI we have

$$\text{CBAL-OPT}(X) = \sum_{J_n\in[m]^n} \prod_{i=1}^n b_{j_i}^* x_{j_i}^i \ ;$$

$$\text{UNI}(X) = \sum_{J_n\in[m]^n} \int_{\mathcal{B}} \prod_{i=1}^n b_{j_i} x_{j_i}^i d\mu(\mathbf{b}) \ .$$

where μ is a uniform probability measure over $\mathcal{B}$. The ratio of compound returns, $\text{CBAL-OPT}(X)/\text{UNI}(X)$, is now a ratio of two finite summations each involving $N = m^n$ non-negative terms. It is not hard to see that if $\alpha_1, \ldots, \alpha_N \geq 0$, $\beta_1, \ldots, \beta_N \geq 0$, then

$$\frac{\sum_{1\leq i\leq N} \alpha_i}{\sum_{1\leq i\leq N} \beta_i} \leq \max_j \frac{\alpha_j}{\beta_j} \ .$$

Using this upper bound on the ratio of sums we have

$$\begin{aligned}
\frac{\text{CBAL-OPT}(X)}{\text{UNI}(X)} &\leq \max_{J_n\in[m]^n : \Pi_i x_{j_i}>0} \frac{\prod_{i=1}^n b_{j_i}^* x_{j_i}^i}{\int_{\mathcal{B}} \prod_{i=1}^n b_{j_i} x_{j_i}^i d\mu(\mathbf{b})} \\
&= \max_{J_n\in[m]^n : \Pi_i x_{j_i}>0} \frac{\prod_{i=1}^n b_{j_i}^*}{\int_{\mathcal{B}} \prod_{i=1}^n b_{j_i} d\mu(\mathbf{b})} \\
&= \max_{J_n\in[m]^n} \frac{\prod_{i=1}^n b_{j_i}^*}{\int_{\mathcal{B}} \prod_{i=1}^n b_{j_i} d\mu(\mathbf{b})} \ . \qquad (23)
\end{aligned}$$

The r.h.s. of (23) can now be upper bounded as follows using the "method of types" of information theory (see Cover and Thomas [16] Chapt. 12). For each vector J_n and $r \in [m]$ denote by $\nu_r(J_n)$ the proportion of the number of occurrences of r in J_n (i.e. the number of occurrences divided by n). This way, the distribution $D(J_n) = (\nu_1(J_n), \nu_2(J_n), \ldots, \nu_m(J_n))$ specifies the "type" of the sequence J_n. Theorem 12.1.2 in [16] then states that for any distribution (portfolio) $\mathbf{b}$ and any such J_n

$$\prod_{1\leq i\leq n} b_{j_i} \leq 1/2^{nH(D)} \ ,$$

where
$H(D(J_n))$ is the Shannon entropy of D (i.e. $H(D) = \sum_i -\nu_i(J_n)\log\nu_i(J_n)$). Equality is obtained at $\mathbf{b} = D(J_n)$. This result upper bounds the numerator of (23). The integral in the denominator can be expressed in a closed form

$$\int_{\mathcal{B}} \prod_{1\leq i\leq n} b_{j_i} d\mu(\mathbf{b}) = 1/\left(\binom{n+m-1}{m-1} T(J_n)\right) ,$$

where $T(J_n)$ is the number of sequences of the type $D(J_n)$. Note that $\binom{n+m-1}{m-1}$ is the number of $D(J_n)$-types. Theorem 12.1.3 in [16] states that $T(J_n) \leq 2^{nH(D)}$. From these bounds it readily follows that

$$\frac{\text{BOPT}(X)}{\text{UNI}(X)} \leq \binom{n+m-1}{m-1} ,$$

and the upper bound $(n+1)^{m-1}$ on the r.h.s. follows from Theorem 12.1.1 in [16].

We note that Cover and Ordentlich obtain recursive procedures to compute the above algorithm but the space requirements of this procedure is $O(n^m)$ so it can only be applied to a small number of securities.

Dirichlet-weighted algorithms We now consider μ-weighted algorithms where μ is a Dirichlet distribution. A random vector $\mathbf{b} = (b_1, b_2, \ldots, b_m)$ has a *Dirichlet distribution,* [14] denoted Dirichlet($\boldsymbol{\alpha}$), with parametric vector $\boldsymbol{\alpha} = (\alpha_1, \alpha_2, \ldots, \alpha_m)$, $\alpha_i > 0$, if the probability density function of $\mathbf{b}$ with respect to $\boldsymbol{\alpha}$, $f_{\boldsymbol{\alpha}}(\mathbf{b})$, satisfies the following conditions. At any point $\mathbf{b} \in \mathbb{R}_m$ with $b_i \geq 0$, $\sum_i b_i = 1$,

$$f_{\boldsymbol{\alpha}}(\mathbf{b}) = \frac{\Gamma(\alpha_1 + \cdots + \alpha_m)}{\Gamma(\alpha_1)\cdots\Gamma(\alpha_m)} b_1^{\alpha_1 - 1} \cdots b_m^{\alpha_m - 1} ,$$

where Γ is the Gamma function.[15] At any other point $\mathbf{b}' \in \mathbb{R}_m$, $f_{\boldsymbol{\alpha}}(\mathbf{b}') = 0$. Thus $f_{\boldsymbol{\alpha}}(\mathbf{b})$ is positive only when the vector $\mathbf{b}$ is a distribution function, $\sum_i b_i = 1$. Hence, Despite its appearance, $f_{\boldsymbol{\alpha}}(\mathbf{b})$ is not an m-dimensional p.d.f. but rather gives the joint p.d.f. of any $(m-1)$-subset of the m random variables $b_1, b_2, \ldots, b_m$ (and one of them is set by the relation $\sum_i b_i = 1$). Notice that the Dirichlet$(1, 1, \ldots, 1)$ density is the uniform density over the $(m-1)$-dimensional simplex. Also, we note that the symmetric Dirichlet$(a, a, \ldots, a)$ density with $a < 1$ is strictly convex (with poles at the unit vectors). This means that if it is used by a μ-weighted algorithm it will give more weight to "extremal" portfolios, that invest in a small number of securities.

[14] The Dirichlet distribution is sometimes referred to as the *multinomial beta distribution.*

[15] The Gamma function is $\Gamma(x) = \int_0^\infty e^{-t} t^{x-1} dt$. It can be shown that $\Gamma(1) = 1$ and that $\Gamma(x+1) = x\Gamma(x)$. Thus if n is an integer, $n \geq 1$, $\Gamma(n+1) = n!$.

Denote by $\text{DIR}_{\boldsymbol{\alpha}}$ the μ-weighted portfolio selection algorithm with μ being Dirichlet($\boldsymbol{\alpha}$). Thus, $\text{DIR}_{(1,1,\ldots,1)}$ is exactly algorithm UNI. For the Dirichlet($\frac{1}{2},\ldots,\frac{1}{2}$)-weighted algorithm Cover and Ordentlich show that for any market sequence X,

$$\frac{\text{CBAL-OPT}(X)}{\text{DIR}_{(\frac{1}{2},\frac{1}{2},\ldots,\frac{1}{2})}(X)} \leq \frac{\Gamma(1/2)\Gamma(n+m/2)}{\Gamma(m/2)\Gamma(n+1/2)} \leq 2(n+1)^{(m-1)/2} \ .$$

Thus, $\text{DIR}_{(\frac{1}{2},\frac{1}{2},\ldots,\frac{1}{2})}$ is $\left(2(n+1)^{\frac{m-1}{2}}\right)$-competitive with respect to CBAL. Here again, stated in terms of the exponential growth rate utility function and the minimax regret criterion, these bound implies that that the minimax regret (of the exponential growth rates) of algorithm DIR is bounded above by $\frac{(m-1)\log(2n+2)}{2n}$.

This result for algorithm DIR can be proved following the same line as the proof of the bound for algorithm UNI but it is somewhat more technical (since the integrals involved are more complex).

Extremal mixture algorithms Consider Equation (22). This equation suggests the following interpretation. Fix some $J_n = (j_1, j_2, \ldots, j_n) \in [m]^n$ and consider the "extremal" algorithm, denoted EXT_{J_n}, that at time i invests the entire wealth in the j_ith security. Let $\mathbf{b} = (b_1, \ldots, b_m)$ and consider the constant rebalanced algorithm $\text{CBAL}_{\mathbf{b}}$. It is possible to represent $\text{CBAL}_{\mathbf{b}}$ in terms of extremal algorithms as follows. Partition $\text{CBAL}_{\mathbf{b}}$'s initial wealth into m^n portions, one for each of the m^n sequences J_n in $[m]^n$. The number of dollars in the part corresponding to J_n is $w(J_n) = \prod_{i=1}^n b_{j_i}$. Notice that

$$\sum_{J_n \in [m]^n} w(J_n) = 1 \ .$$

That is, $\{w(J_n)\}$ is a probability distribution (and a proper partition of the initial wealth). For each of the extremal algorithms EXT_{J_n} maintain a separate investment "account" starting with an initial wealth of $w(J_n)$. For a market sequence X, the wealth accrued by EXT_{J_n} is thus

$$\text{EXT}_{J_n}(X) = w(J_n) \prod_{i=1}^{n} x_{j_i}^i = \prod_{i=1}^{n} b_{j_i} x_{j_i}^i \ .$$

Hence, by (22) we see that the sum of wealths accrued by all the extremal algorithms is exactly $\text{CBAL}_{\mathbf{b}}(X)$.

Exactly as it is for the one-way (and search) algorithms (see Sect. 3) we can interpret (and apply) this algorithm as a (mixed) randomized algorithm. Simply choose the extremal algorithm EXT_{J_n} with probability $w(J_n)$. In this case, of course, $\text{CBAL}_{\mathbf{b}}(X)$ will be the expected return of this algorithm. From a computational point of view, operating the extremal mixture algorithm randomly saves most of the large memory required for operating the deterministic algorithm.

Using equation (20) we can interpret algorithms UNI and DIR (and any μ-weighted algorithm) analogously. All that is needed is to partition the initial wealth so that algorithm EXT_{J_n} is assigned

$$w(J_n) = \int_{\mathcal{B}} \prod_{i=1}^{n} b_{j_i} d\mu(\mathbf{b})$$

(with μ being the corresponding probability measure).

The above representation (interpretation) of the constant rebalanced (and μ-weighted algorithms) gives rise to the following general class of "extremal mixture" algorithms. For each n the *extremal mixture* strategy is specified by a probability distribution w over the set J_n and invests in the extremal algorithm EXT_{J_n} a fraction $w(J_n)$ of its initial wealth. From the above discussion it follows that for each n, an optimal on-line extremal mixture algorithm performs at least as well as the best μ-weighted algorithm.

For each n let OEXT_n be the extremal mixture algorithm specified by the following probability distribution over the extremal algorithms. For each $J_n \in [m]^n$, set $q(J_n) = \prod_{i=1}^{m} \nu_i(J_n)^{n_i(J_n)}$. The distribution w is then

$$w(J_n) = \frac{q(J_n)}{\sum_{J_n \in [m]^n} q(J_n)} .$$

In [39] Ordentlich and Cover prove that algorithm OEXT_n is an optimal extremal mixture algorithm and for each market sequence X of length n. Further, it is shown that for each such X,

$$\frac{\text{CBAL-OPT}(X)}{\text{OEXT}_n(X)} \leq \sum_{J_n \in [m]^n} \frac{n!}{\prod_{1 \leq i \leq m} n_i(J_n)!} \prod_{1 \leq i \leq m} \left(\frac{n_i(J_n)}{n}\right)^{n_i(J_n)} , \qquad (24)$$

and in the worst case this bound is tight.

Given the equivalence of deterministic and (mixed) randomized extremal mixture algorithms the ratio (24) is a lower bound on the competitive ratio of any randomized extremal mixture algorithm (and μ-weighted algorithm) against an oblivious adversary (with respect to CBAL-OPT).

For the case $m = 2$ this competitive ratio (24) of OEXT_n (with respect to CBAL) reduces to

$$\sum_{0 \leq k \leq n} \binom{n}{k} \cdot \left(\frac{k}{n}\right)^k \cdot \left(\frac{n-k}{n}\right)^{n-k} ,$$

and using Stirling approximation it can be shown to be approximately $\sqrt{\pi/2} \cdot \sqrt{n} \approx 1.253\sqrt{n}$. Note that the competitive ratio of the $\text{DIR}_{(\frac{1}{2},\ldots,\frac{1}{2})}$ algorithm for $m = 2$ is not larger than $2\sqrt{n+1}$ so the Dirichlet algorithm attains a competitive ratio that is within a factor of (approximately, for large n) 1.6 of algorithm OEXT_n.

Portfolio selection with side information Cover and Ordentlich [15] extended the above portfolio selection model to incorporate possible "side information" that the trader may have. Typically such "side information" may be based on various kinds of predictions of future values of market vectors. To model side information we consider an "oracle" that announces a number $\mathsf{info}_i \in \mathcal{I} = \{1, 2, \ldots, k\}$ at the start of the ith trading period. The number announced represents an abstract state of some prediction apparatus. For example, in the best case info_i identifies the best security in the ith period. Nevertheless, we assume that the on-line player does not know in advance the quality of the side information provided to him and moreover, that the quality of the side information may vary during the trading period. Thus, in order to benefit from such side information the trader must learn its quality during the trading period. Formally, with respect to an n-period trading game we define *side information* as any sequence $\mathsf{info}_1, \mathsf{info}_2, \ldots, \mathsf{info}_n$ with $\mathsf{info}_i \in \mathcal{I}$.

Let $X = \mathbf{x}_1, \ldots, \mathbf{x}_n$ be any market sequence and $I = \mathsf{info}_1, \mathsf{info}_2, \ldots, \mathsf{info}_n$, any side information. As usual, denote by I_i the prefix of I consisting of the first i elements. We denote by $\text{ALG}(X|I)$ the return of the algorithm ALG with respect to X given the side information I. Let $\mathbf{b}$ be any portfolio and $\ell \in \mathcal{I}$. With respect to the side information I_i and market sequence X_i define

$$R_i(\mathbf{b}|\ell) = \prod_{j \le i:\ \mathsf{info}_j = \ell}^{i} \mathbf{b}^t \mathbf{x}_j \ .$$

That is, $R_i(\mathbf{b}|\ell)$ is the compounded return of a constant rebalanced algorithm that is out of the market at all periods where the side information equals ℓ and is fully invested, using the portfolio $\mathbf{b}$, at all other periods.

Let μ be any probability measure. Using $R_i(\mathbf{b}|\cdot)$ we now define the *μ-weighted algorithm with side information* as the algorithm that at the ith period uses the portfolio

$$\mathbf{b}_i(\mathsf{info}_i) = \frac{\int_{\mathcal{B}} \mathbf{b} R_{i-1}(\mathbf{b}|\mathsf{info}_i) d\mu(\mathbf{b})}{\int_{\mathcal{B}} R_{i-1}(\mathbf{b}|\mathsf{info}_i) d\mu(\mathbf{b})} \ .$$

This is a generalization of the definition of the μ-weighted algorithm. Clearly, when $k = |\mathcal{I}| = 1$ (i.e. no side information) this definition reduces to (18).

In a way analogous to the simplification obtained by formula (19) we now simplify the expression for the compounded return $\text{ALG}_\mu(X|I)$

$$\begin{aligned} \text{ALG}_\mu(X|I) &= \prod_{i=1}^{n} \mathbf{b}_i^t(\mathsf{info}_i) \mathbf{x}_i \\ &= \prod_{j=1}^{k} \prod_{i \le n: \mathsf{info}_i = j} \mathbf{b}_i^t(j) \mathbf{x}_i \\ &= \prod_{j=1}^{k} \int_{\mathcal{B}} R_n(\mathbf{b}|j) d\mu(\mathbf{b}) \ . \end{aligned} \qquad (25)$$

Although it is probably impossible to determine the performance ratio of the μ-weighted algorithm with side information with respect to the optimal off-line constant rebalanced algorithm (it is heavily dependent on the side information), it is possible to determine the performance ratio with respect to a more powerful off-line algorithm that is also dependent on the side information. The advantage of this approach is that the dependence on the side information will "factor out" and we shall obtain a performance ratio that is only dependent on the cardinality of $\mathcal{I}$.

Fix $\mathcal{I}$. Let $B : \mathcal{I} \to \mathcal{B}$ be any mapping from side information to portfolios. Define the *state-constant rebalanced algorithm (with mapping B)*, denoted SCBAL_B, as the algorithm that at each period uses one of the portfolios $B(1), B(2), \ldots, B(k)$. Specifically, the algorithm invests according to the portfolio $B(\mathsf{info}_i)$ during the ith trading period (for which the side information is info_i). The return of algorithm SCBAL_B for a market sequence X and side information sequence I is thus given by

$$\text{SCBAL}_B(X|I) = \prod_{i=1}^{n} B^t(\mathsf{info}_i) \cdot \mathbf{x}_i \ .$$

The optimal off-line state-constant rebalanced algorithm, denoted SCBAL-OPT is a state-constant rebalanced algorithm that optimizes its choice of the mapping B based on advance knowledge of the market sequence X and the side information I. That is, this algorithm uses the portfolio B^* where

$$B^* = \arg\max_B \prod_{i=1}^{n} B^t(\mathsf{info}_i) \cdot \mathbf{x}_i \ .$$

Denote by $\text{UNI}(X|I)$ and $\text{SCBAL-OPT}(X|I)$ the compounded returns of the on-line uniform-weighted algorithm and the optimal off-line state-constant rebalanced algorithm, respectively, for a market sequence X given side information I. For each $j = 1, 2, \ldots, k$, set $n_j(I)$ to be the number j's in I. Now for each market sequence X and side information I,

$$\frac{\text{UNI}(X|I)}{\text{SCBAL-OPT}(X|I)} \le \prod_{j=1}^{k} (n_j(I)+1)^{(m-1)} \le (n+1)^{k(m-1)} \ .$$

To prove this bound, for each $j = 1, 2, \ldots, k$, denote by $X(j)$ the subsequence of X, $\mathbf{x}_{j_1}, \mathbf{x}_{j_2}, \ldots, \mathbf{x}_{j_l}$, where $\mathsf{info}_{j_r} = j$ for all $1 \le r \le l$. By the definition of SCBAL-OPT and equation (25) we have

$$\frac{\text{UNI}(X|I)}{\text{SCBAL-OPT}(X|I)} = \prod_{j=1}^{k} \frac{\text{UNI}(X(j))}{\text{SCBAL}_{B^*}(X(j))} \ .$$

Hence, by the known bound for algorithm UNI (21), applied with the sequences $X(j)$, we have

$$\prod_{j=1}^{k} \frac{\text{UNI}(X(j))}{\text{SCBAL}_{B^*}(X(j))} \le \prod_{j=1}^{k} (n_j(I)+1)^{(m-1)} \le \prod_{j=1}^{k} (n+1)^{(m-1)} \ .$$

With respect to a market sequence X and side information sequence I denote by $\text{DIR}_{(\frac{1}{2},\ldots,\frac{1}{2})}(X|I)$ the compounded return of the Dirichlet$(\frac{1}{2},\ldots,\frac{1}{2})$ algorithm with side information. A derivation similar to the above (for algorithm UNI) proves the following result. For each market sequence X and side information I,

$$\frac{\text{DIR}_{(\frac{1}{2},\ldots,\frac{1}{2})}(X|I)}{\text{SCBAL-OPT}(X|I)} \leq \prod_{j=1}^{k} (n_j(I)+1)^{\frac{m-1}{2}} \leq 2^k(n+1)^{\frac{k(m-1)}{2}} .$$

Other results There are quite a few other portfolio selection results that were not discussed in this chapter. Here we briefly mention several other known results. Cover and Gluss [14] consider a model where the set of possible market vectors of price relatives is finite, say with cardinality k. Using the game-theoretic approachability-excludability theorem of Blackwell [7] they obtain an on-line portfolio selection algorithm whose exponential growth rate approaches that of the optimal off-line constant rebalanced algorithm at (convergence) rate $(L\sqrt{k}+1)\sqrt{2L^2+1}\,/\sqrt{n}$ where L is a bound on the logarithm of the maximum price relative, and n is the length of the game.[16] Note that for the above fixed fluctuation model $k=2$ and $L=\alpha$. Thus, in the long run it is possible, in this model, to almost track the performance of the optimal off-line constant rebalanced algorithm and in particular to outperform the optimal off-line BAH strategy.

Cover and Ordentlich's results regarding the μ-weighted algorithms are a generalization of a result by Cover [13] where it is shown that the exponential growth rate of the uniform-weighted portfolio approaches, as n grows, the exponential growth rate of the optimal off-line constant rebalanced portfolio. This result assumes that all price relatives are bounded away from zero. In the same paper Cover presents a more refined analysis and bounds the competitive ratio of the uniform-weighted algorithm (with respect to CBAL) in terms of a sensitivity matrix A measuring the "empirical volatility" (in a sense defined in the paper) of prices exhibited in the market sequence presented to the algorithm. In particular the upper bound on the competitive ratio is $\sqrt{|A|}\,(n/(2\pi))^{(m-1)/2}\,/(m-1)!$. The value of the determinant $|A|$ can be bounded by a constant if the prices of all securities are sufficiently volatile so that CBAL gives positive weights to all securities. Cover also provides some interesting experimental results demonstrating the performance of this algorithm. In [28], Jamshidian analyzes the performance of the uniform-weighted algorithm in a continuous time model. The results obtained are similar to those obtained by Cover [13].

Awerbuch et al. [3] consider the following setting. The on-line player wishes to invest his entire wealth in one of the m securities, hoping that the chosen security will be a "winner" that yields high dividends. The decision is irreversible and after the player has chosen one security the game is essentially over. If D is the posteriori dividend return of the best security then the optimal expected

[16] Similar results can be obtained using the results of the recent paper by Hart and Mass-Collel [25].

return of the player is trivially $\Theta(D/m)$. Now consider a game such that in each period, each of the securities may issue a dividend of exactly 1 dollar. Under the assumption that the yield of the best security is $D \geq 3 \log m$, and that D is known to the on-line player Awerbuch et al. provide a selection strategy obtaining a return of at least $D/3 \log m$ with probability at least $1-(3 \log m)/D-2/m$. The strategy is the following. For each security j, $j = 1, 2, \ldots, m$ set $d(j, i)$ to be the cumulative number of dividends issued by the jth security within the first i periods. At the $(i+1)$st period choose the jth security with probability $m^{3d(j,i)/D-2}$. This basic result is extended for multiple choices of securities. Finally, in the case where D is not known in advance, the player can retain the same yield but the probability of success is decreased.

Auer, Cesa-Bianchi, Freund and Schapire [2] consider the following adversarial variant of the *multi-armed bandit* problem. The on-line player must repeatedly choose one among m slot machines. The player's goal is to maximize the total reward in a sequence of n trials (no sampling costs are assumed). Using the minimax regret adversary, that before each round selects an m-ary vector of the current rewards, and assuming that all individual rewards are in $[a, b]$, Auer et al. prove that expected regret (obtained by a randomized algorithm) is at most $O\left((b-a)n^{\frac{2}{3}}(m \log m)^{\frac{1}{3}}\right)$. They also prove a lower bound of $\Omega(\sqrt{nm})$ on the regret of any algorithm.

6 Concluding remarks and directions for future research

The competitive approach is shown to be productive for a variety of financial problems. Although the full extent of its theoretical value and its empirical relevance are yet to be determined, it is already quite clear, based on the examples in this chapter, that this approach gives rise to interesting and non-obvious algorithms and analyses.

One common argument against the use of competitive algorithms is that they are inherently risk-averse as they are optimized with respect to worst case event sequences. This argument is certainly valid in cases where decision makers possess reliable information about stochastic processes that determine the uncertainties. In this case the use of competitive algorithms that blatantly ignore this information may lead to inferior performance relative to Bayesian algorithms. Nevertheless, in many instances decision makers do not possess such information. In some cases, by investing sufficient resources they may acquire such information while in other situations, often due to the complexities of event sequences not much can be learned on the underlying stochastic process. Whatever the reason for the absence of such information, competitive algorithms offer reasonable initial solutions upon which more elaborate algorithms may be constructed after additional information is determined. Another important argument in favor of competitive algorithms (and in fact any worst-case approach) is that risk-averse economic decision makers may prefer somewhat inferior but guaranteed performance to a better average performance.

At the time of writing the work presented here is the state of the art in this area. Therefore, almost any improvement on the stated bounds and any extensions of the models considered will be desirable advancements. There are also several other attractive financial problems that can and should be studied. We conclude this chapter with formulations of three on-line financial problems that to the best of our knowledge have not yet been studied using the competitive approach (variants of all these problems have been extensively studied within the Bayesian approach, see references below).

- *Inventory management.* The theory of on-line inventory management is concerned with decision problems of when and how much to buy of certain commodities whose future demand, for production or trade, is uncertain. For concreteness, consider the following problem formulation. The on-line player runs a store that sells a certain commodity. The player maintains an inventory in which he stores some items of this commodity. During each period a *demand* for some number of items of this commodity is placed. For each item sold the player earns some price c per item. At the end of each period the player must decide on a number of items to order in order to replenish his inventory. For each item ordered he pays some wholesale price w. The items ordered arrive after a delay of d periods. For each item stored in the inventory the player pays, per period, some holding cost h. The player's objective is to issue orders so that the total profit is maximized.
  The literature on Bayesian inventory management is vast. The reader is referred to [42, 46]. We note that preliminary competitive results for this problem were obtained by Karp, Ostrovski and Rabani [31].
- *Insurance problems.* In insurance problems the on-line player must decide whether or not to purchase an insurance contract guaranteeing that the insurer will pay the player a certain amount of money provided that a stipulated event will occur. For instance, consider the following elementary one-stage, two-state insurance problem. There are two possible states of nature, one of which will prevail. In the first state the player is endowed with wealth w_1 whereas in the second, disaster state, the player is endowed with wealth w_2, $w_2 < w_1$. Before the state of nature is made known, the player can guarantee a compensation of c dollars, in the case where the second state of nature prevails, by paying the insurance company an amount of βc where $\beta < 1$. The amounts w_i, and β are fixed and known to the on-line player before the decision is made. The player's goal is to maximize his wealth. This elementary one-stage two-state formulation can be generalized to a multi-stage multiple-state insurance game.
  Insurance problems are most fundamental to the economics of uncertainty. For Bayesian solutions to this problem the reader is referred to [36] and references therein.
- *Consumption problems.* In consumption problems the on-line player must repeatedly decide how much of his current wealth to consume and how much to save for future consumption. His goal is to maximize his total, lifetime consumption. The uncertainties in this problem may arise in one or more of

the following ways. First, the player's future income may vary in some unknown manner. Second, future investment appreciation (depreciation) rates may fluctuate in some uncertain way. Finally, the player's life is unknown. Consider the following basic consumption problem. The player starts with some initial wealth w_0. At the start of the jth period, $j = 1, 2, \ldots$, the player receives an income s_i and must choose some amount c from his current wealth $s_i + w_{j-1}$ that he immediately consumes. The rest of the money is allocated for investment and grows at rate $1 + i_j$ throughout the jth period so that the player's wealth at the end of the jth period is $w_j = (1 + i_j)(s_i + w_{j-1})$. At the end of some period T, unknown in advance, the game ends. The player's goal is to maximize his total consumption throughout the game. In the "full uncertainty" variant of this formulation the sequence $\{(s_i, i_j)\}_j$ and the number T are revealed on-line.

For some Bayesian results concerning variants of this problem the reader is referred to [33, 38, 36].

Acknowledgments

My thanks to Thomas Cover and Erik Ordentlich for helping in obtaining the bibliographic material. I thank Allan Borodin, Andrew Chou, Vincent Feltkamp and Erik Ordentlich for many helpful discussions and comments that helped me understand the material and greatly improved the presentation. I also thank Robert Aumann, Dean Foster, Sergiu Hart, Sageev Oore, Motty Perry, Adi Rosén and Asher Wolinsky for their useful remarks. Finally, I thank Brenda Brown who proof read several versions of this article.

References

1. M. Ajtai, N. Megiddo, and O. Waarts. Improved algorithms and analysis for secretary problems and generalizations. In *Proceedings of the 36th Annual Symposium on Foundations of Computer Science*, pages 473–482, 1995.
2. P. Auer, N. Cesa-Bianchi, Y. Freund, and R.E. Schapire. Gambling in a rigged casino: The adversarial multi-armed bandit problem. In *Proceedings of the 36th Annual Symposium on Foundations of Computer Science*, pages 322–331, 1995.
3. B. Awerbuch, Y. Azar, A. Fiat, and T. Leighton. Making commitments in the face of uncertainty: How to pick a winner almost every time. In *Proceedings of the 28th Annual ACM Symposium on Theory of Computing*, 1996.
4. Y. Azar, Y. Bartal, E. Feuerstein, A. Fiat, S. Leonardi, and A. Rosén. On capital investment. In *Proceedings of ICALP 96*, 1996.
5. R. Bellman. Equipment replacement policy. *Journal of the Society for Industrial and Applied Mathematics*, 3:133–136, 1955.
6. S. Ben-David, A. Borodin, R. Karp, G. Tardos, and A. Widgerson. On the power of randomization in on-line algorithms. In *Proc. 22nd Symposium on Theory of Algorithms*, pages 379–386, 1990.
7. D. Blackwell. An analog of the minimax theorem for vector payoffs. *Pacific J. Math.*, 6:1–8, 1956.

8. L. Breiman. Optimal gambling systems for favorable games. In *Proceedings of the Fourth Berkeley Symposium on Mathematical Statistics and Probability*, volume 1, pages 65–78. Univ. of California Press, Berkeley, 1961.
9. A. Chou. Optimal trading strategies vs. a statistical adversary. Master's thesis, Massachusetts Institute of Technology, 1994.
10. A. Chou, J. Cooperstock, R. El-Yaniv, M. Klugerman, and T. Leighton. The statistical adversary allows optimal money-making trading strategies. In *Proceedings of the 6th Annual ACM-SIAM Symposium on Discrete Algorithms*, 1995.
11. A. Chou, A. Shrivastava, and R. Sidney. On the power of magnitude. to be published, May 1995.
12. Y.S. Chow, H. Robbins, and D. Siegmund. *The Theory of Optimal Stopping.* Dover Publications, Inc., 1971.
13. T.M. Cover. Universal portfolios. *Mathematical Finance*, 1(1):1–29, January 1991.
14. T.M. Cover and D.H. Gluss. Empirical Bayes stock market portfolios. *Advances in Applied Mathematics*, 7:170–181, 1986.
15. T.M. Cover and E. Ordentlich. Universal portfolios with side information. *IEEE Transactions on Information Theory*, March 1996.
16. T.M. Cover and J.A. Thomas. *Elements of Information Theory.* John Wiley & Sons, Inc., 1991.
17. J. Cox and M. Rubinstein. *Options Markets.* Prentice Hall, Inc., 1985.
18. C. Derman. On optimal replacement rules when changes of the state are Markovian. In R. Bellman, editor, *Mathematical Optimization Techniques.* 1963.
19. R. El-Yaniv. On the decision theoretic foundations of the competitive ratio, June 1996. unpublished manuscript.
20. R. El-Yaniv, A. Fiat, R. Karp, and G. Turpin. Competitive analysis of financial games. In *Proc. 33rd Symposium on Foundations of Computer Science*, pages 327–333, 1992.
21. R. El-Yaniv, R. Kaniel, and N. Linial. On the equipment rental problem. to be published, August 1993.
22. R. El-Yaniv and R.M. Karp. Nearly optimal competitive online replacement policies. Submitted to *Matematics of Operations Research.* A preliminary version of this article appeard in the proceedings of the *2nd Israel Symposium on Theory of Computing and Systems*, 1996.
23. P.R. Freeman. The secretary problem and its extensions. *International Statistical Review*, 51:189–206, 1983.
24. R.L. Graham. Bounds for certain multiprocessor anomalies. *Bell System Technical Journal*, 45:1563–1581, 1966.
25. S. Hart and A. Mas-Colell. Correlated learning. unpublished manuscript, March 1996.
26. J. Hull. *Options, Futures, and other Derivative Securities.* Prentice Hall, Inc., 1993.
27. S. Irani and D. Ramanathan. The problem of renting versus buying. unpublished manuscript, August 1994.
28. F. Jamshidian. Asymptotically optimal portfolios. *Mathematical Finance*, 2(2):131–150, 1992.
29. D.S. Johnson. *Near-Optimal Bin Packing Algorithms.* PhD thesis, Massachusetts Institute of Technology, 1973.
30. D.S. Johnson, A. Demers, J.D. Ullman, M.R. Garey, and R.L. Graham. Worst-case performance bounds for simple one-dimensional packing algorithms. *SIAM Journal on Computing*, 3:299–325, 1974.

31. R.M. Karp, R. Ostrovski, and Y. Rabani, 1993. Personal communication.
32. H.W. Kuhn. Extensive games and the problem of information. In H.W. Kuhn and A.W. Tucker, editors, *Contribution to the Theory of Games*, volume II, pages 193–216. Princeton University Press, 1953.
33. D. Levhari and L.J. Mirman. Saving and uncertainty with an uncertain horizon. *Journal of Political Economy*, 85:265–281, 1977.
34. L. Levin, July 1994. personal communication.
35. S.A. Lippman and J.J. McCall. The economics of job search: A survey. *Economic Inquiry*, XIV:155–189, June 1976.
36. S.A. Lippman and J.J. McCall. The economics of uncertainty: Selected topics and probabilistic methods. In K.J. Arrow and M.D. Intriligator, editors, *Handbook of Mathematical Economics*, volume 1, chapter 6, pages 211–284. North-Holland, 1981.
37. R.C. Merton. On the microeconomic theory of investment under uncertainty. In K.J. Arrow and M.D. Intriligator, editors, *Handbook of Mathematical Economics*, volume 2, chapter 13, pages 601–669. North-Holland, 1981.
38. L.J. Mirman. Uncertainty and optimal consumption decisions. *Econometrica*, 39:179–185, 1971.
39. E. Ordentlich and T.M. Cover. On-line portfolio selection. In *COLT 96*, 1996. A journal version is to be submitted to *Mathematics of Operations Research.*
40. P. Raghavan. A statistical adversary for on-line algorithms. DIMACS *Series in Discrete Mathematics and Theoretical Computer Science*, 7:79–83, 1992.
41. D.B. Rosenfield and R.D. Shapiro. Optimal price search with Bayesian extensions. *Journal of Economic Theory*, 1981.
42. H.E. Scarf. A survey of analytic techniques in inventory theory. In H.E. Scarf, D.M. Gilford, and M.W. Shelly, editors, *Multistage Inventory Models and Techniques*, chapter 7, pages 185–225. Stanford University Press, 1963.
43. E. Sheshinski and Y. Weiss. Optimum pricing policy under stochastic inflation. In E. Sheshinski and Y. Weiss, editors, *Optimal Pricing, Inflation and the Cost of Price Adjustment.* The MIT Press, 1993.
44. A.G. Shilling. Market timing: Better than a buy-and-hold strategy. *Financial Analysts Journal*, pages 46–50, March-April 1992.
45. D. Sleator and R. E. Tarjan. Amortized efficiency of list update and paging rules. *Communications of the ACM*, 28:202–208, 1985.
46. N.L. Stokey and R.E. Lucas, Jr. *Recursive Methods in Economic Dynamics*, chapter 13, pages 389–413. Harvard University Press, 1989.
47. A.C. Yao. New algorithms for bin packing. *Journal of the ACM*, 27:207–227, 1980.

16

On the Performance of Competitive Algorithms in Practice

ANNA R. KARLIN

1 Introduction

In this chapter we briefly describe a number of empirical studies of competitive algorithms: applications of ski-rental including cache coherence, virtual circuit holding times and mobile computing, paging, routing and admission control.

An algorithm is said to be *on-line* if it receives its input in a sequence of requests, and must service each request as it arrives, without any knowledge of the future. Many fundamental problems in computer science are inherently on-line, including memory management, processor scheduling, load balancing and routing. A commonly used method of analyzing on-line algorithms is *competitive analysis* [23], where the performance of the on-line algorithm is compared to the performance of the optimal off-line algorithm.

An on-line algorithm is said to be *c-competitive* if on each input its performance is within a factor of c of the performance of the optimal off-line algorithm on that input. Competitive analysis of algorithms has advantages over other more traditional methods of evaluating algorithms: Worst-case analysis can be overly pessimistic, while average case analysis requires a statistical model of the input. It is difficult to devise realistic statistical models, since input patterns tend to change dynamically with applications and over time.

Note that competitive analysis is still a worst-case measure, since we ask that the algorithm have a cost at most c times optimal on *every* input. Another way of saying this, is that the performance of the algorithm must be within a factor of c on inputs generated by a *worst-case adversary.*

The cornerstone of competitive analysis, the paging problem, is of long-standing interest to the operating systems and architecture communities. The setting for this problem is a two-level store consisting of a fast memory (the *cache*) that can hold k pages, and a slow memory that can store n pages. The n pages in slow memory represent the virtual memory pages. A *paging algorithm* is presented with a sequence of requests to virtual memory pages. If the page requested is in fast memory (a *hit*), no cost is incurred; but if not (a *fault*), the algorithm must bring it into the fast memory at unit cost. The algorithm must decide which of the k pages currently in fast memory to evict in order to make room for the newly requested page.

Many other problems that have been studied by the competitive algorithms community have direct bearing on computer systems. In this article we briefly describe several empirical studies of competitive algorithms. We begin with a number of applications of the "ski-rental problem" including cache coherence, virtual circuit holding times and mobile computing. We then touch on experimental work on paging, and on routing and admission control. These performance studies lead to the following conclusions: First, algorithms optimized to work against the worst-case adversary tend not to work well in practice. Second, the standard competitive algorithms give insight leading to the "correct" algorithm. Finally, nearly all successful empirical studies use algorithms that adapt in some way to the input distributions.

2 Applications of ski rental

2.1 The ski rental problem

We begin by reviewing the ski rental problem.[1] Suppose you are about to go skiing for the first time in your life. Naturally, you ask yourself whether to rent skis or to buy them. Renting skis costs, say, $30, whereas buying skis costs, say $300. If you knew how many times you would go skiing in the future (ignoring complicating factors such as inflation, and changing models of skis), then your choice would be clear. If you knew you would go at least 10 times, you would be financially better off by buying skis right from the beginning, whereas if you knew you would go less than 10 times, you would be better off renting skis every time.

Alas, the future is unclear, and you must make a decision nonetheless – you are faced with an *on-line problem*, and must deal with events (in this case ski trips) as they arrive without knowledge (or with only limited knowledge) of future events.

For the ski-rental problem, we readily observe that if an adversary determines exactly at what point you will never go skiing again, it can ensure that you pay at least twice what you could have paid, regardless of your strategy. Indeed, suppose your algorithm is to continue renting until you have skied j times and

[1] This analogy was originally suggested by Larry Rudolph in the context of the work on competitive snoopy caching [16].

then to buy skis. The adversary will then simply wait until you buy skis, and then make sure you never ski again. Since you pay $30j + 300$, and you could have paid $\min(30j, 300)$, you pay at least twice the optimal off-line cost.

There is a well-known on-line algorithm that is guaranteed to achieve this bound: rent until you've skied 10 times, then buy. In so doing, if you ski j times, where $j \leq 10$, you pay $30j$, exactly the optimal off-line cost. If $j > 10$, you pay $2 \cdot 300$, twice the optimal off-line cost.

The principle underlying this algorithm is often referred to as *the ski principle*. In essence, the ski principle says that once the algorithm has incurred enough total cost by executing a number of cheap actions, the algorithm can afford to take a more expensive action, which can be amortized against the collection of cheap actions.

2.2 Examples of ski rental in computer systems

A number of on-line decision-making problems arise in real systems that are analogous to the ski rental problem just described. We briefly survey some of these problems.

Waiting for a lock: Spin or block?

In a shared memory multiprocessor, the consistency of shared data is protected by a lock. There are two common ways of handling the situation where a thread attempts to acquire a lock that is held by another thread. The first is *blocking*: the thread relinquishes the processor, after enqueuing itself to be awakened when the lock becomes free. The second is *spinning*: the thread continues to execute on its processor, repeatedly testing to see if the lock is available. A third alternative is to spin for a while and then block.

If the thread blocks, processor time is consumed by software overhead, which usually includes saving the thread's state, enqueuing the blocked thread, scheduling a ready thread and restoring the thread's state when it is awakened. This time is the cost of a *context switch*. On the other hand, if the thread spins, processor time is wasted probing the lock, when in principle other threads are available to run that could actually do useful work. If the goal is to maximize processor utilization (minimize time spent on context switch overhead or useless spinning), the optimal off-line algorithm is to spin if the lock will be released within the time it would take to perform a context switch, and to block if the lock will be held longer.

An on-line algorithm which is 2-competitive can be obtained by applying the balancing principle: the strongly competitive algorithm spins up to C time units and then blocks, where C is the time to perform a context switch.

Virtual circuit holding times: Hold or release?

Connection-oriented networks such as those using ATM must be able to communicate with existing networks such as the Internet which uses the connectionless

Internet Protocol (IP). Therefore, mechanisms must be devised to carry IP traffic over ATM networks. The fundamental issue is how to carry datagrams over virtual circuits. It is clear that the arrival of an IP datagram should cause a virtual circuit to be opened, if one is not open already. However, it isn't clear how to handle the open circuit thereafter. It may be desirable to keep it open for some time, to amortize the cost of opening the circuit over many packets. Unfortunately, the IP datagrams do not carry information about the length and rate of higher layer conversations. and so it is unknown whether the two endpoints of a virtual circuit will communicate again soon. After each transmission, the ATM adaptation layer must choose between holding the virtual circuit and releasing it. The optimal policy depends on the pricing model. For the purposes of this discussion, assume that the cost of setting up a virtual circuit is $C, and that the cost of holding the circuit open is $1 per time unit. The optimal off-line policy is to hold the circuit open if the next packet will arrive less than C time units later and to release it immediately otherwise.

An on-line algorithm which is 2-competitive is to hold the circuit open for C time units. If a new packet arrives in that time, start over. Otherwise after the C time units, close the circuit.

Cache coherency: Update or invalidate?

In a shared memory multiprocessor such as a snoopy cache multiprocessor system, each processor has a cache in which it stores blocks of data. Each cache is connected to a bus used to communicate with other caches and with main memory. A property of such systems is that shared data must be maintained in a consistent state. When a processor updates data in its local cache which is shared with other processor(s), it can maintain coherency either by updating data in the other caches with the new value (by sending an update message out over the bus), or by invalidating the data in the other caches (by sending an invalidate message over the bus). After an invalidation the data is private to the processor, and so future updates by that processor to that particular data item do not generate any further bus traffic. Since the bus can service only one request at a time, inefficient use of the bus may cause a processor to idle while its cache is waiting for the bus. Therefore, it is desirable to design coherency protocols that minimize bus traffic.

The costs for the various operations are as follows: If a processor issues a read for a variable not present in its cache, the block in which that variable resides must be brought into the cache at a cost of $B + 1$ bus cycles, where B is the number of variables per cache block. A write to a variable in some block costs one bus cycle if the block is shared (to perform the update or invalidate) and zero bus cycles if the block is private to the writing processor.

Consider two processors executing a sequence of reads and writes to a cache block. The optimal off-line algorithm will immediately invalidate the block in one processor, whenever the other processor executes a write run (a sequence of writes to the block with no intermingled reads or writes by the first processor) of length at least $B+1$. Whenever write runs have length at most B, updates will be propagated.

A 2-competitive algorithm can be obtained by applying the ski principle: each processors updates will be propagated up to B times, and the next time that processor issues an update, it will invalidate the block in the other cache. Note that the count is reset whenever the other processor accesses the block.

Mobile computing: To spin or not to spin

Power is a precious resource for mobile (portable) computers. Therefore, a natural question is whether or not to leave the disk spinning in between accesses to data. Imagine that leaving the disk spinning costs one dollar per time unit (a rough measure of the cost of the power consumed). Clearly, this is wasteful if no accesses to disk are being made. When the disk is off, there is no cost. However, if a memory reference is made that forces the disk to be turned on, there is a restart cost of r dollars (measuring the power consumed to restart it and the inconvenience of the delay the user encounters waiting for the disk to start spinning again).

The optimal off-line algorithm identifies those intervals of time in which no disk access occurs for at least r time units, and turns off the disk at the beginning of those intervals. It is easy to see that an adversary can force any on-line algorithm to spend twice the optimal number of dollars. An on-line algorithm that turns off the disk whenever r units of time pass with no disk accesses is obviously 2-competitive [8].

2.3 Results of empirical studies

For of the ski-rental analogues just discussed, an empirical study was performed in which the 2-competitive and other on-line algorithms were evaluated [8, 15, 20, 1, 9]. The two strategies that had been primarily used in real systems prior to these studies corresponded either to the "Always-rent" strategy or the "Buy-immediately" strategy, neither of which is competitive.

A very interesting phenomenon was common to all of the empirical studies: in every case, it was found that the competitive algorithm had uniformly good performance, but sometimes not better than the best of the Always-rent or Buy-immediately strategies.

For example, Eggers and Katz [9] studied the performance of three cache coherence protocols: an update protocol (corresponding to the Always-rent policy), an invalidate protocol (corresponding to the Buy-immediately policy), and a competitive protocol. For applications with active sharing, the update protocol performed best, followed by the competitive protocol and then the invalidate protocol. For applications with little sharing, the order was reversed. For any given application, the competitive algorithm performed nearly as well as the best protocol, but never performed best. The bimodal distribution of the data suggested that better performance would be obtained by an adaptive protocol, that dynamically switches between an update and an invalidate protocol depending on the degree of sharing.

Another point of commonality between all the studies was that in each case, studies of real input traces showed that inputs are likely to come from a "nice" but *unpredictable* distribution. In particular, in all of the studies it was the case that in any given run of the ski-rental experiment, the number of ski-trips was either quite small relative to the ratio between the purchase cost and the rental cost, or else was much larger than this ratio. For example, measurements of programs running on shared-memory multiprocessors showed that different locks had vastly different waiting-time distributions (the waiting-time distribution of a lock corresponds to the number of ski trips) [15]. For the vast majority of locks, the lock-waiting time was either significantly more than a context switch or less than a third of a context switch. Moreover, the lock-waiting behavior of each individual lock was consistent from experiment to experiment. (Each experiment corresponds to an instance of the ski-rental problem, and the experiment ends when skis are purchased.)

Interestingly, this corresponds to my observations from real life skiing as well – a person is quite likely to go skiing only once, but once a person has gone skiing three or four times, they are likely to ski many times.

Similar observations were made in the empirical evaluation of holding time policies in IP-over-ATM networks [20]. In addition, this study showed that different Internet "conversations" had substantially different interpacket arrival characteristics, but that each individual conversation had very consistent inter-arrival characteristics.

The observations made in these studies suggest that it makes sense to assume that, in ski-rental parlance, there is some nice probability distribution such that p_i is the probability that there will be i ski trips. Given this assumption, a natural algorithm is to dynamically determine the probability distribution p_i and converge to the optimal on-line behavior against that distribution.

Ski-rental against a distribution

Suppose that p_i is the probability that i ski trips will be taken. Given knowledge of p_i, it is quite easy to choose the right time to buy skis, which we call τ, so that the expected total cost is minimized – one must choose τ so that

$$\sum_{i<\tau} p_i i C_r + \sum_{i\geq\tau} p_i(\tau C_r + C_b)$$

is minimized, where C_r is the cost to rent skis and C_b is the cost of buying skis. It was shown that the expected competitive ratio of the algorithm that rents up the τ-th ski trip is at most $e/(e-1) \approx 1.58$ times that of the optimal algorithm for any distribution p_i [17].

Of course, in practice, one cannot expect the distribution p_i to be *a priori* known. However, there are a number of reasonable ways to learn and approximate it. Such algorithms were evaluated experimentally for several of the problems we have discussed. Given our previous discussion, the results were unsurprising: for all of the different problems, the on-line algorithms that learned and adapted to the input distribution p_i outperformed all other algorithms, and had performance very close to that of the optimal off-line algorithm.

Measurements

We focus on one example to give the flavor of the results. For the problem of spinning vs blocking on a lock in a shared memory multiprocessor, the performance of the following algorithms was measured [15]:

1. Always block immediately.
2. Always spin.
3. Optimal off-line.
4. The 2-competitive algorithm, where a process spins up to a context switch and then blocks. We call this algorithm A_C.
5. The 3-competitive algorithm, where a process spins up to half of a context switch and then blocks. We call this algorithm $A_{C/2}$.
6. Optimal on-line given knowledge of the distribution of lock-waiting times for each lock.
7. Use the last three waiting times to compute the optimal threshold: If the last three waiting times were i, j and k time units, the assumption is made that $P_L(i) = P_L(j) = P_L(k) = 1/3$, and the optimal on-line algorithm for inputs from that distribution is used.
8. An adaptive random walk algorithm, which operates as follows. Initially the spin threshold is set to the context switch time C. If the last waiting time was greater than C, the spin threshold is decremented by some amount (with a minimum of 0). Otherwise, the threshold is incremented by some amount, with a maximum of C.

The following table gives the approximate competitive ratios of the various algorithms on the data collected. Only average synchronization costs are reported here, so a value of 1.3 for an algorithm means that it spent 30% more time spinning and blocking than the optimal off-line algorithm.

Algorithm	Min	Max	Comment
Block	1.6	7.2	high variance
Spin	1.3	3.5	typically between 2.3 and 3.4
$A_{C/2}$	1.3	1.5	
A_C	1.2	1.7	average approx 1.6
Optimal On-line	1.1	1.2	
3-samples	1.2	1.4	
Random Walk	1.1	1.6	average approx 1.3

The table shows that the competitive algorithms performed as well or better than the better of the pure block or pure spin strategies, but somewhat worse than the adaptive algorithms.

3 Paging

One of the shortcomings of competitive analysis is the fact that it is still a worst-case measure. There is an implicit assumption that the algorithm operates against a worst-case adversary, one that chooses the request sequence in order to maximize the ratio between the on-line cost and the off-line cost. In "real life", this does not happen. In every ski-rental analogue we discussed, this was an overly pessimistic assumption.

This worst-case measure has two important negative side-effects. First, the theoretical results themselves are overly pessimistic. Algorithms which are known to have excellent performance in practice have high competitive ratios. The canonical example of this is the paging problem. Sleator and Tarjan [23] showed that no deterministic paging algorithm can achieve a competitiveness better than k, where k is the number of pages of fast memory. They also showed that some practical algorithms, such as first-in-first-out (FIFO) and least-recently-used (LRU) are k-competitive, and hence best possible in their model.

These results conflict with practical experience on paging in at least two ways. First FIFO and LRU have the same competitiveness, even though in practice LRU usually outperforms FIFO. Secondly LRU usually incurs much less than k times the optimal number of faults, even though its competitiveness is k.

The reason for the practical success of LRU has long been known: most programs exhibit *locality of reference* [7]. This means that when a page is referenced by a program, the next request is very likely to come from some small set of pages. Indeed, a two-level store is only useful if request sequences are *not* arbitrary.

Motivated by these observations, Borodin, Irani, Raghavan and Schieber [6] proposed a technique for incorporating locality of reference into the traditional Sleator-Tarjan framework. Their notion of an *access graph* limits the set of request sequences the adversary is allowed to make. An access graph is a graph that has a vertex for every page that the program can reference. Locality of reference is imposed by the edge relation – the pages that can be referenced after a page p are just the neighbors of p in the graph or the page itself. An algorithm is c-competitive within the access graph model if on each request sequence that corresponds to a walk on the access graph its performance is within a factor of c of the performance of the optimal off-line algorithm on that input.

An undirected access graph might be a suitable model when the page reference patterns are governed by the data structures used by the program, whereas a directed access graph might be a more suitable model for program flow.

Several authors have designed competitive algorithms for the access graph model [6, 12, 10]. For example, for the undirected case, it was shown that an algorithm called FAR, which, roughly speaking, attempts to evict the page furthest from the page fault in the access graph (subject to certain "marking" constraints), has the best possible competitive ratio to within a constant factor.

In a recent empirical study, Fiat and Rosen [11] evaluated a set of paging heuristics motivated by the access graph model and the FAR algorithm. In their work, it is not assumed that the access graph is known, but rather that it is learned and built dynamically. Their algorithms are truly on-line and deviate slightly from the strict competitive behavior of FAR. Interestingly, their experiments suggest that the heuristics they've designed outperform LRU and FAR fairly consistently, over a wide range of cache sizes and programs. These results reconfirm our observation that algorithms which adapt to input distributions and don't worry excessively about worst-case inputs outperform their competitors.

4 IP paging

Similar phenomena have been observed for another type of paging problem. Recall the problem of setting up virtual circuits in order to carry IP traffic over ATM networks. Another interesting pricing model that has been considered is one in which a host pays a fixed charge for each virtual circuit set up, but there is a limit k on the maximum number of virtual circuits that a host may have open simultaneously. This leads to the question of which virtual circuit to replace when a packet arrives that needs to be sent between two hosts that have no virtual circuit currently open. The goal is to make these choices so as to minimize the total number of virtual circuits a host ever sets up.

This is a standard paging problem, where a communication between a pair of hosts corresponds to a page, and a reference to that page occurs when a packet arrives for that particular pair of hosts.

As previously mentioned, in this context as well Keshav *et al* [20] found that the traffic exhibited locality of reference and that the individual conversations had characteristics that were fairly constant over time.

In paging parlance, a reasonable probabilistic model for this situation is that for each page there is a arbitrary probability distribution that characterizes the inter-reference time for that page, and that these probability distributions are independent for different pages. Lund, Phillips and Reingold [21] studied this problem theoretically and proposed the Median Algorithm: on a fault, evict the page for which the median of the distribution on the remaining time until the next request is largest.

They were able to show that the Median algorithm has performance within a factor of 5 of that of the optimal on-line algorithm for this type of distributional paging.

The empirical study of Keshav *et al* went on to empirically compare the performance of LRU, the Median algorithm and several other commonly used strategies for this variant of the IP-over-ATM virtual circuit setup problem, using real network traces. They found that uniformly, the Median algorithm had the best performance of all the algorithms. Again we find that algorithms which adapt to input distributions outperform their competitors.

Lund, Phillips and Reingold went on to remove the restriction in their theoretical result that the probability distributions for different pages are independent. They were able to design an algorithm whose performance is within a constant factor of that of the optimal on-line algorithm for page requests coming from arbitrary probability distributions.

5 Virtual circuit routing and admission control

Our final example concerns empirical studies of virtual circuit routing and admission control for Future Broadband Integrated Services Digital Networks (B-ISDN). These networks will support services such as video-on-demand that require quality-of-service guarantees, and therefore reservation of resources. Creating a virtual circuit requires reservation of bandwidth on some path between the endpoints of the connection. Admission control and routing strategies have been proposed for addressing the following two questions:

- What path should be used to route a given circuit?
- Which circuits should be routed and which ones should be rejected?

A series of papers [22, 2, 3, 4] proposed competitive strategies for these problems. Roughly, the idea behind the competitive strategies is to assign each link a "length" that is an exponential function of its current congestion. The new circuit is routed along the shortest path with respect to this length. The circuit is rejected if the length of the shortest path exceeds some predefined threshold associated with this circuit.

An empirical study [14] of these strategies showed that in practice its performance was not sufficiently good. In fact, the resulting performance was observed to be always worse than the performance of a simple greedy min-hop strategy. Again, the reason is that the worst-case aspect of competitive analysis leads to an overly conservative admission control strategy, where requests can be rejected even if they involve links that are essentially unused.

This led Kamath, Palmon and Plotkin [19] to reconsider the theoretical question and determine the best on-line algorithm if one makes limited distributional assumptions about the input requests. By assuming that requests for virtual circuits between any two points arrive according to a Poisson process of unknown rates (and chosen by an adversary) and where the circuit holding times are exponentially distributed, they were able to develop a theoretical algorithm with significantly stronger performance guarantees. Empirical studies of this modified strategy [14] showed that it outperforms strategies that are commonly used in practice such as min-hop and reservation-based algorithms.

6 Conclusions

There are several obvious conclusions that can be drawn from the empirical studies we have mentioned.

Competitive analysis is overly pessimistic. Competitive algorithms with high competitive ratios such as LRU tend to perform better in practice than the theory would suggest. However, better performance yet can be obtained by designing algorithms that are optimal against the particular input distributions that are faced in practice. Indeed, the explanation of the strong performance of some competitive strategies, such as the Move-To-Front strategy for reorganizing linked lists, and splay tree algorithms, is that they have this adaptive characteristic naturally built into them.

Studies of real traces for problems ranging from paging to routing and admission control to mobile computing indicate that there *is* structure underlying the input sequences, however that structure can not be predicted. Therefore, no detailed assumptions can be made about the input distribution as in typical average case analysis. There is a need for further research into competitive algorithms that assume that inputs come from *some* distribution, but don't know the parameters of the distribution, and adapt to it.

Nearly all successful empirical studies used algorithms that did adapt in some way to input distributions. However, it was the traditional competitive algorithms that provided the insight that led to these algorithms. Indeed, in some cases there has been a nice feedback loop between the empirical studies and the theory, where the former suggested the type of probability distributions that should be considered, and these were then attacked theoretically [19, 21].

In my opinion, the way of the future for competitive analysis is the design of universal algorithms that are competitive against large classes of input distributions and therefore have small competitive ratios and excellent performance in practice.

References

1. C. Anderson and A. Karlin. Two Adaptive Hybrid Cache Coherency Protocols. In *Proceedings of the 1996 Second International Symposium on High-Performance Computer Architecture*, San Jose, February 1996.
2. J. Aspnes, Y. Azar, A. Fiat, S. Plotkin, O. Waarts. On-line load balancing with applications to machine scheduling and virtual circuit routing. In *Proc. 25th Annual ACM Symposium on Theory of Computing*, 1993.
3. B. Awerbuch, Y. Azar, S. Plotkin. Throughput-competitive on-line routing. In *Proceedings of 34th Annual Symposium on Foundations of Computer Science*, 1993.
4. B. Awerbuch, Y. Azar, S. Plotkin, O. Waarts. Competitive Routing of Virtual Circuits with Unknown Durations. In *Proceedings of the Fifth Annual ACM-SIAM Symposium on Discrete Algorithms*, 1994.
5. L.A. Belady. A study of replacement algorithms for virtual storage computers. *IBM Systems Journal*, 5:78–101, 1966.
6. A. Borodin, S. Irani, P. Raghavan, and B. Schieber. Competitive Paging with Locality of Reference. In *Proc. 23rd Annual ACM Symposium on Theory of Computing*, pages 249–259, 1991.
7. P.J. Denning. Working sets past and present. *IEEE Trans. Software Engg.*, SE-6:64–84, 1980.

8. F. Douglis, P. Krishnan, B. Bershad. Adaptive Disk Spin-down Policies for Mobile Computers. In *Proceedings of Second USENIX Symposium on Mobile and Location-Independent Computing*, 1995.
9. S.J. Eggers and R.H. Katz. Evaluating the performance of four snooping cache coherency protocols. In *Proceedings of 16th Annual International Symposium on Computer Architecture*, 1989.
10. A. Fiat and A. Karlin Randomized and Multipointer Paging with Locality of Reference. In *Proceedings of 27th ACM Symposium on Theory of Computing*, May, 1995.
11. A. Fiat and Z. Rosen Experimental Studies of Access Graph Based Heuristics: Beating the LRU Standard? In *Proceedings of the Eighth Annual ACM-SIAM Symposium on Discrete Algorithms*, 1997.
12. S. Irani, A.R. Karlin and S. Phillips. Strongly competitive algorithms for paging with locality of reference. In *Proceedings of the Third Annual ACM-SIAM Symposium on Discrete Algorithms*, 1992.
13. P.A. Franaszek and T.J. Wagner. Some distribution-free aspects of paging performance. *Journal of the ACM*, 21:31–39, 1974.
14. R. Gawlick, A. Kamath, S. Plotkin, K. Ramakrishnan. Routing and Admission Control in General Topology Networks. Stanford Technical Report STAN-CS-TR-95-1548.
15. A.R. Karlin, K. Li, M. Manasse, and S. Owicki. Empirical studies of competitive spinning for shared memory multiprocessors. In *Proceedings of the 13th ACM Symposium on Operating System Principles*, 1991.
16. A. R. Karlin, M. S. Manasse, L. Rudolph, and D.D. Sleator. Competitive snoopy caching. *Algorithmica*, 3(1):70–119, 1988.
17. A. R. Karlin, M. S. Manasse, L. McGeogh, S. Owicki. Randomized Competitive Algorithms for Nonuniform Problems. *Algorithmica,* Vol. 11, No. 1, January, 1994.
18. E. Koutsoupias and C. Papadimitriou. Beyond competitive analysis. In *Proceedings of 35th Annual Symposium on Foundations of Computer Science*, 1994.
19. A. Kamath, O. Palmon, S. Plotkin. Routing and Admission Control in General Topology Networks with Poisson Arrivals. In *Proceedings of the 7th ACM-SIAM Symposium On Discrete Algorithms*, 1996.
20. S. Keshav, C. Lund, S. Phillips, N. Reingold, H. Saran. An Empirical Evaluation of Virtual Circuit Holding Time Policies in IP-over-ATM Networks. AT&T Bell Laboratories Technical Memorandum 39199-11.
21. C. Lund, S. Phillips and N. Reingold. IP-paging and distributional paging. In *Proceedings of 35th Annual Symposium on Foundations of Computer Science*, 1994.
22. S. Plotkin. Competitive Routing of Virtual Circuits in ATM Networks. Survey, invited paper to IEEE Journal on Selected Areas in Communications.
23. D.D. Sleator and R.E. Tarjan. Amortized efficiency of list update and paging rules. *Communications of the ACM*, 28:202–208, February 1985.
24. J.R. Spirn. *Program Behavior: Models and Measurements.* Elsevier Computer Science Library. Elsevier, Amsterdam, 1977.

17

Competitive Odds and Ends

AMOS FIAT
GERHARD J. WOEGINGER

1 What it really means

The verdict is still uncertain on the real significance of competitive analysis with respect to real algorithms. Chapter 16 by Anna Karlin deals with experimental results and seems to suggest that the competitive approach, while not necessarily resulting in good algorithms, does provide insight into the underlying problem.

The nature of competitive analysis is that good algorithms must be highly adaptive and must restructure their behavior to meet with changing circumstances. This seems to be a highly desirable property of algorithms in any case. Because any algorithm with a good competitive ratio is forcibly adaptive, one can gain insight as to how one can adapt. It seems that the problem with competitive analysis is, once again, that it is a worst case measure. However, there are interesting possibilities here. It seems that if one could restrict the adversary to a limited class that more closely model potential real behavior, then good competitive algorithms with respect to this limited class of adversaries may prove to be directly relevant to the real world environment. Some discussion of these limited adversary models appears in several earlier chapters and in a subsequent section of this chapter.

There are general methodologies to perform the deterministic or randomized (weighted) merger of several different competitive strategies [18, 5]. Thus, this suggests that practically significant algorithms could be devised that have suboptimal yet bounded worst case guarantees in the more unlikely scenarios yet behave significantly better in more typical scenarios. Potentially, good algorithms will be jointly optimized for distributional assumptions (for the "typical" case) and for distribution-independent worst case behavior.

2 Refining and modifying the rules

The primary motivation for competitive analysis is to refine the discrimination between the quality of different algorithms that behave equally well (or bad) in a worst case setting. *E.g.*, both Least-Recently-Used and Last-In-First-Out paging algorithms are equally good from the perspective of standard worst case analysis, clearly a nonsensical observation.

Unfortunately, in many cases we observe that "standard" competitive analysis does not discriminate between different algorithms either. Typically, this will happen when we observe a lower bound on the competitive ratio of any algorithm for the problem that matches the upper bound that can be obtained by a completely trivial algorithm. We call this phenomena the "triviality barrier". Examples of this abound, most noteworthy is that both Least-Recently-Used and First-In-First-Out paging algorithms have a competitive ratio of k, the number of page slots in memory. However, even if the triviality barrier has been hit, the question of "which is the better algorithm?" has not disappeared.

The underlying problem with the triviality barrier is exactly the same problem that we had with standard worst case analysis, the adversary is so powerful and the worst case it can force is so bad that meaningful information about the real quality of the various algorithms is lost. However, the real motivation of the entire study is to compare two real algorithms, not to compare the algorithm to some imaginary all powerful adversary. The adversary is simply a technical aid so that we can make this comparison.

2.1 The access graph model

One possible refinement to the standard competitive measure is to limit the adversary in some way. The access graph model for paging has been successful in discriminating between LRU and FIFO (LRU is better, this matches practical experience too). The access graph model due to Borodin, Irani, Raghavan and Schieber [11] is described in detail in Chapter 3.

2.2 Competing with a distributed algorithm

In a distributed setting, the problem to be solved should be solved by a distributed algorithm running on some network. However, in standard competitive analysis, this algorithm must compete against all algorithms, the class of all algorithms includes algorithms that "know" the future and "know" exactly what is going on in the entire network.

One possible extension to the competitive measure is to ensure that the algorithm is only competing against other *distributed* algorithms [2]. Any distributed algorithm must pay for the distributed control messages that allow it to "know" what is going on in the network. Work regarding this type of analysis is described in detail in Chapter 6.

2.3 The value of information in a distributed setting

Specific studies on the value of information in distributed settings were studied in Deng and Papadimitriou and later by Irani and Rabani [16, 30, 20, 21]. The model assumes a set of agents who are given the input to some optimization problem, but every agent gets only part of the input. A strategy for the agents, S, is a directed graph G and a set of algorithms, one for every agent. The arcs of the digraph represent knowledge flow in that the head of the arc knows the input associated with itself and the input associated with the tail of the arc. A strategy S is c-competitive in this model if the value of the optimization problem computed (distributively) by the agents under strategy S is no worse than c times the optimal value of the optimization problem.

Various results have been obtained in this model for problems such as load balancing, routing, and linear programming. Results have been obtained under the assumption that the graph is fixed in advance or that additional arcs can be added (at some cost). For a fixed knowledge graph G and for the load balancing problem, [21] show that the the competitive ratio is between $\sqrt{\alpha(G)}$ and $2\sqrt{\phi(G)}$, where $\alpha(G)$ is the size of a maximum independent set of G and $\phi(G)$ is the size of the minimum clique cover in G.

2.4 Loosely competitive algorithms

The concept of loosely competitive algorithms is due to Young [33, 34, 35]. In [34, 35] loosely competitive is defined in the context of a specific problem (paging and file caching). Here, we try to give the spirit of the definition that should be usable for many other cost problems.

To motivate the definition of loosely competitive we make two observations. One observation is that if the total cost to our algorithm is sufficiently small then the ratio with the optimal cost could be irrelevant. Generally for cost problems, the cost functions used (number of page faults, time to completion, etc.) ignores some hidden setup cost or inherent cost to process the input.

Another observation is that we do not expect the problem parameters chosen to be worst possible. If the problem instance is generated by an adversary that is oblivious to our specific specific parameters then there is no reason to believe that this is so. Thus, we might be interested in a typical value of the problem parameters rather than a specific value.

Chapter 3 defines loosely competitive algorithms in the context of paging and shows that one can get constant competitive ratios in this model, analogous results hold for file caching [35].

2.5 The diffuse adversary

While competitive analysis assumes we know nothing about the input distribution, and classical distributional assumptions assume that we know everything about the distribution, the diffuse adversary model of Koustupias and Papadimitriou [26] assumes that we know something about the distribution.

One assumes that the distribution D is in a known class Δ of possible distributions. The *performance ratio* for a given class of distributions Δ is defined as

$$R(\Delta) = \min_{A} \max_{D \in \Delta} \frac{E(Cost_A(x))}{E(Cost_{OPT}(x))}.$$

Again, Chapter 3 discusses the diffuse adversary in the context of paging, for a specific class of distributions.

2.6 Competing on unequal terms

In the original Sleator-Tarjan paper on paging, they consider the competitive ratio when the paging algorithm has k pages and the competing optimal algorithm has some $h \leq k$ pages.

In several cases, when hitting the triviality barrier, considering what happens if the on-line algorithm has some advantage over the competition may lead to interesting results. Kalyanasundaram and Pruhs [23] compete against slower machines, as do Phillips *et. al.* [31]. Less memory is available to the adversary when competing on distributed paging in general networks [4].

2.7 The statistical adversary

Borodin et. al. [12] consider a new model for dynamic packet routing, where packets are continuously injected into a network. Unlike most previous work that deals with probabilistic assumptions on the packet injection behavior, this paper introduces an adversary model, and limits the adversary behavior in the spirit of restricting the adversary to "interesting behavior".

An adversary injects packets into the network. In this model the packet routing issue is not addressed, but it is assumed that the path to be taken by the packet is predetermined by the adversary. One rule that restricts the adversary is that the paths requested at a given time step cannot share any edges. Such a set of non-intersecting paths is called a request. For $\epsilon > 0$, an adversary is said to inject at rate $1 - \epsilon$ if during r consecutive steps, the adversary can make at most $r(1 - \epsilon)$ requests.

Fix the network and a *queuing policy*, a queuing policy specifies, for every edge and every time step, which of the waiting packets is to proceed along the edge. A queuing policy is said to be stable if, against any deterministic adversary, there is a constant C (which may depend on the network size) such that the number of packets in the network is at all times bounded by C.

Andrews *et. al.* [3] show that some greedy strategies are stable on every network when the adversary rate is less than 1. They also extend the set of possible adversaries to include some burstiness.

Rabani and Tardos [32] and Ostrovsky and Rabani [29] give good solutions to the static routing problem (close to minimal maximal delay) which can also be generalized to the dynamic case in the statistical adversary model.

2.8 Obtaining benefit with high probability

Given a randomized competitive algorithm for a cost problem, the Markov inequality always allows us to give some bound on the probability that our cost is much larger than the expectation.

For benefit problems, there is no analog to the Markov inequality. Generally, a competitive ratio of c does not rule out the possibility that with probability $1 - 1/c$ the benefit attained by the algorithm is close to zero. In fact, the classify and randomly select paradigm (discussed in Chapter 11) that is used by many competitive algorithms for benefit problems seems to impose this property on the algorithm.

Leonardi et. al. [27] deal specifically with the problem of call control on trees and meshes (accepting the largest number of edge disjoint paths). They give variants of previously known algorithms that attain optimal competitive ratios and modify them to ensure a constant fraction of the expected benefit with at least constant probability. Alternately, at the expense of some increase in the competitive ratio, they get a constant fraction of the expected benefit with probability asymptotically approaching one as appropriate problem parameters approach infinity.

One could hope that it may be possible possible to use this approach in a more general setting.

3 More discussion on metrical task systems and the k-server problem

3.1 Universal algorithms for task systems

If we consider a metrical task system to consist of a metric on the n configurations and allow an arbitrary set of tasks then no algorithm can be better than $2n - 1$ competitive. The work function algorithm (described in Chapter 4) attains this ratio.

As noted by Burley and Irani [14] the metrical task system formulation can be used to model many other problems *if* part of the task system description is the set of valid task requests. Thus, the natural question is "given a metrical task system (including the valid set of tasks), what is the competitive ratio attainable?". Ideally, one could come up with a universal algorithm, that given a description of the task system, attains the very best competitive ratio.

In [14] it is shown that the decision problem (is the competitive ratio less than c?) can be computationally very difficult (PSPACE-complete). However, they also give a simple greedy algorithm that attains a competitive ratio within a factor of $O(\log n)$ for any uniform metrical task system, where the word uniform refers to the fact that the distance between any two configurations is exactly one. Clearly, it would be exceedingly interesting to extend this result.

3.2 Some comments on randomized algorithms for task systems

A widely believed conjecture (at least by the authors) is that for for every metrical task system on n points (allowing any set of tasks), the randomized competitive ratio is $O(\log n)$. Another conjecture is that for any metric space there exists a set of tasks such the the competitive ratio is $\Omega(\log n)$. Unfortunately, a previously widely believed conjecture was that the randomized competitive ratio is exactly equal to H_k, this conjecture is now known to be false [24, 22].

For a uniform metric space, both (current) conjectures are known to be true [13] For a set of equally spaced points on the line, Karloff and Foster [10] give an upper bound of $O(2^{\sqrt{\log n \log\log n}})$. From Karloff, Rabani, Ravid, [25] and from Blum *et. al.* [9] we know that the competitive ratio is $\Omega(\sqrt{\log n \log\log n})$ for any metric space.

We remark that the k server problem can be phrased as a task system with a specific set of tasks, and the lower bound also gives a lower bound on the randomized k-server competitive ratio on $n = k + 1$ equally spaced points. This statement does not follow immediately from the statement on the task system (the other direction follows immediately).

Irani and Seiden [22] show that for every metric space the competitive ratio is no more than $e/(e-1)n$ (improving on the $2n-1$ deterministic lower bound), while Bartal *et. al.* [7] reduce this dramatically to $O(\log^6 n)$. This latter result makes use of the probabilistic approximation of metric spaces in [6].

Lund and Reingold [28] solve a linear programming problem to obtain an upper bound on the best possible competitive ratio. This upper bound can also be achieved constructively. They give a linear program which provides a lower bound and give sufficient conditions for these bounds to agree. If the bounds agree then the algorithm is clearly optimal.

3.3 Variants on metrical task systems and k-server problems

Many variants on the seminal metrical task system and k server formulations have been studied. This section contains a short (and definitely non-exhaustive) list of some of these problems and references the relevant publications.

Relaxed metrical task systems In [8] a generalization of metrical task systems called relaxed task systems is considered. Relaxed metrical task systems are also considered in A metrical task system has a distance function between configurations $dist(C, C')$, and a cost of service when in a configuration $cost(C, e)$. An event e that arrives when the task system is in configuration C can be serviced by first moving to configuration C' (at a cost of $dist(C, C')$), and then paying an additional $cost(C, e)$. The relaxed task system assumes that actually changing configurations is more expensive ($D \cdot dist(C, C')$) than simply servicing the request while staying in configuration C (at a cost of $dist(C, C') + cost(C', e)$).

The main theorem states that given a c-competitive task system, one can translate the algorithm into a c^2 deterministic algorithm for the relaxed version of the task system and into an $O(c)$ competitive randomized algorithm for the relaxed task system against oblivious adversaries.

Relaxed task systems are useful for the analysis of the k-migration problem, a variant on the k server problem where service can be provided without moving the server. This is discussed in Chapter 5.

The weighted cache problem The weighted cache problem is like the k-server problem but the distance required to move a server to point p depends only on point p and not on the position that the server is currently located. This problem is covered in Chapter 3. Almost nothing is known about the randomized competitive ratio.

The weighted server problem The weighted server problem is a generalization of the k-server problem. Every server is assigned a positive weight. The cost to move a server from point p to point q is the distance between p and q times the weight of the server. This problem was introduced in [19] where both lower bounds and upper bounds are given. This problem can be used to model situations where the write time to memory depends on the type of memory used.

The k-client problem This problem is from [1]. In some sense it is the dual of the k server problem. There is one server but multiple clients for server services. This can be used to model multi-threaded access to a disk drive. k different threads require disk access and the algorithm must decide which of the threads is to be serviced next.

To minimize the makespan, [1] give a $2k-1$ competitive algorithm. The lower bound is only $\log k/2$ so there is a large gap here.

The related problem of multi-threaded paging was considered by [17].

Yet another version of this problem is where k completely different points can be requested every time step. (For the k client problem all the points are the same as at the last step except for the point serviced). The (single) server has to go to one of these points. This variant is also known as a metrical service system and is equivalent to the general framework of layered graph traversal.

The dynamic servers problem The dynamic servers problem [15] is defined as follows. At any point in time, one or more servers are located at points in a metric space. As in the k-server problem, requests are points in the metric space and a request is serviced by moving a server to the request location as a cost equal to the distance moved. Requests arrive on-line and are intermixed with "rent" events. On a rent event the algorithm is charged rental cost equal to the number of currently active servers. A server can be moved from point to point, a server can be cloned so that two servers occupy the same location, and two servers at the same location can be merged.

[15] give a deterministic $O(\min\{\log n, \log \rho\})$-competitive algorithm where n is the number of requests and ρ is the normalized diameter of the metric space.

They also give a lower bound of $\Omega(\log\log\rho/\log\log\log\rho)$ on the competitive ratio of any deterministic algorithm for the problem.

The dynamic servers problem has been used to model data structures and algorithms for the maintenance of kinematic structures.

4 More open problems

Open problems are littered throughout the chapters of this book. In this section we will give a very partial list of well defined open problems that we consider to be rather interesting (ignoring the truly infamous Splay tree and k-server conjectures). Such a list is most definitely highly subjective and controversial and the (cowardly) authors agree in advance that all (future) comments critical of this list are obviously true. This is yet another example of competitive behavior. Certainly, should the authors ever solve any problem whatsoever trivial as it might be, then the reader should consider this problem as though it were on the list even if it is not. In the following, we assume an oblivious adversary whenever considering a randomized competitive algorithm.

1. Metrical Task Systems: Given a metric space of configurations, and some fixed set of tasks, can one come up with a universal algorithm that is within $O(1)$ of the best possible competitive ratio? What about randomized algorithms? Prove or disprove that the randomized competitive ratio of any task system is $O(\log n)$, where n is the number of states in the system.
2. Randomized k-server: Prove or disprove that the randomized competitive ratio of the k-server problem is $\Theta(\log k)$.
3. Paging and Caching: Can one come up with a universal strongly competitive algorithm for the directed access graph model? Randomized? What are randomized competitive ratios for the file caching and weighted cache problems? Prove or disprove that the deterministic and randomized competitive ratios of the distributed paging problem are $\emptyset(m)$ and $O(\log m)$ respectively, where m is the total number of page slots in the system.
4. Navigation and Exploration: What are the deterministic and randomized competitive ratios for point-to-point navigation without a map where arbitrary convex obstacles are present?
5. Virtual Circuit Routing: What competitive ratios can be achieved with only local decision making? This question can also be addressed in the context of specific architectures. What is the randomized competitive ratio for the throughput version of virtual circuit routing when all calls require bandwidth one and all edges have constant capacity $c \geq 2$.

References

1. H. Aborzi, E. Torng, P. Uthaisombut, and S. Wagner. The k-client problem. In *Proc. of the 8th Annual ACM-SIAM Symposium on Discrete Algorithms*, pages 73–82, 1997.

2. M. Ajtai, J. Aspnes, C. Dwork, and O. Waarts. A theory of competitive analysis for distributed algorithms. In *Proc. 35th Symp. of Foundations of Computer Science*, pages 401–411, 1994.
3. M. Andrews, B. Awerbuch, A. Fernández, J. Kleinberg, T. Leighton, and Z. Liu. Universal stability results for greedy contention-resolution protocols. In *Proc. 37th IEEE Symposium on Foundations of Computer Science*, pages 380–389, 1996.
4. B. Awerbuch, Y. Bartal, and A. Fiat. Distributed paging for general networks. In *Proceedings of the 7th Annual ACM-SIAM Symp. on Discrete Algorithms*, pages 574–583, 1996.
5. Y. Azar, A. Broder, and M. Manasse. On-line choice of on-line algorithms. In *Proc. 4th ACM-SIAM Symposium on Discrete Algorithms*, pages 432–440, 1993.
6. Y. Bartal. Probabilistic approximation of metric spaces and its algorithmic applications. In *Proc. 37th IEEE Symposium on Foundations of Computer Science*, pages 184–193, 1996.
7. Y. Bartal, A. Blum, C. Burch, and A. Tomkins. A polylog(n)-competitive algorithm for metrical task systems. In *Proceedings of the 29th annual ACM symposium on theory of computing*, pages 711–719, 1997.
8. Y. Bartal, M. Charikar, , and P. Indyk. On page migration and other relaxed task systems. In *Proc. of the 8th Annual ACM-SIAM Symposium on Discrete Algorithms*, pages 43–52, 1997.
9. A. Blum, H. Karloff, Y. Rabani, and M. Saks. A decomposition theorem and lower bounds for randomized server problems. In *Proceedings of the 33rd Annual IEEE Symposium on Foundations of Computer Science*, pages 197–207, 1992.
10. A. Blum, P. Raghavan, and B. Schieber. Navigating in unfamiliar geometric terrain. In Lyle A. McGeoch and Daniel D. Sleator, editors, *On-line Algorithms*, volume 7 of *DIMACS Series in Discrete Mathematics and Theoretical Computer Science*, pages 151–156. AMS/ACM, February 1991.
11. A. Borodin, S. Irani, P. Raghavan, and B. Schieber. Competitive paging with locality of reference. In *Proc. 23rd ACM Symposium on Theory of Computing*, pages 249–259, 1991.
12. A. Borodin, J. Kleinberg, P. Raghavan, M. Sudan, and D.P. Williamson. Adversarial queueing theory. In *Proceedings of the 28th Annual ACM Symposium on Theory of Computing*, 1997.
13. A. Borodin, N. Linial, and M. Saks. An optimal online algorithm for metrical task systems. In *Proc. 19th Annual ACM Symposium on Theory of Computing*, pages 373–382, 1987.
14. W.R. Burley and S. Irani. On algorithm design for metrical task systems. In *Proceedings of the 6th Annual ACM-SIAM Symposium on Discrete Algorithms*, pages 420–429, 1995.
15. M. Charikar, D. Halperin, , and R. Motwani. The dynamic servers problem. In *Proc. of the 9th Annual ACM-SIAM Symposium on Discrete Algorithms*, pages 410–419, 1998.
16. X. Deng and C.H. Papadimitriou. Competitive distributed decision making. In *International Federation for Infomation Processing, A-12*, pages 250–256, 1992.
17. E. Feuerstein. *On-line pagine of Structured Data and Multi-threaded Paging.* PhD thesis, University of Rome, 1995.
18. A. Fiat, D. Foster, H. Karloff, Y. Rabani, Y. Ravid, and S. Vishwanathan. Competitive algorithms for layered graph traversal. In *Proc. 32nd IEEE Symposium on Foundations of Computer Science*, pages 288–297, 1991.

19. A. Fiat and M. Ricklin. Competitive algorithms for the weighted server problem. In *Theoretical Computer Science*, volume 130, pages 85–99, 1994.
20. S. Irani and Y. Rabani. On the value of information in coordination games. In *Proc. 34th IEEE Symposium on Foundations of Computer Science*, 1993.
21. S. Irani and Y. Rabani. On the value of coordination in distributed decision making. *SIAM J. Comput.*, 25(3):498–519, 1996.
22. S. Irani and S.S. Seiden. Randomized algorithms for metrical task systems. *Theoretical Computer Science*, 194(1-2):163–182, March 1998.
23. B. Kalyanasundaram and K. Pruhs. Speed is as powerful as clairvoyance. In *Proc. 36th IEEE Symposium on Foundations of Computer Science*, pages 214–221, 1995.
24. A.R. Karlin, M.S. Manasse, L.A. McGeoch, and S. Owicki. Competitive randomized algorithms for non-uniform problems. In *Proceedings of the 1st Annual ACM-SIAM Symposium on Discrete Algorithms*, 1990.
25. H. Karloff, Y. Rabani, and Y. Ravid. Lower bounds for randomized k-server and motion planning algorithms. In *Proceedings of the 23rd Annual ACM Symposium on Theory of Computing*, pages 278–288, 1991.
26. E. Koutsoupias and C.H. Papadimitriou. Beyond competitive analysis. In *Proc. 35th IEEE Symposium on Foundations of Computer Science*, pages 394–400, 1994.
27. S. Leonardi, A. Marchetti-Spaccamela, A. Presciutti, and A. Rosen. On-line randomized call control revisited. In *Proc. of the 9th Annual ACM-SIAM Symposium on Discrete Algorithms*, pages 323–332, 1998.
28. C. Lund and N. Reingold. Linear programs for randomized on-line algorithms. In *Proceedings of the 5th Annual ACM-SIAM Symposium on Discrete Algorithms*, pages 382–391, 1994.
29. R. Ostrovsky and Y. Rabani. Universal $o(congenstion + dilation + \log^{1+\epsilon} n)$ local control packet switching algorithms. In *Proceedings of the 29th Annual ACM Symposium on Theory of Computing*, pages 644–653, 1997.
30. C.H. Papadimitriou and M. Yannakakis. Linear programming without the matrix. In *Proc. 25th ACM Symposium on Theory of Computing, San Diego*, pages 121–129, 1993.
31. C.A. Phillips, C. Stein, E. Torng, and J. Wein. Optimal time-critical scheduling via resource augmentation. In *Proceedings of the 29th Annual ACM Symposium on Theory of Computing*, pages 140–149, 1997.
32. Y. Rabani and E. Tardos. Distributed packet switching in arbitrary networks. In *Proceedings of the 28th Annual ACM Symposium on Theory of Computing*, pages 366– 367, 1996.
33. N.E. Young. On-line caching as cache size varies. In *Proc. of the 2nd Annual ACM-SIAM Symposium on Discrete Algorithms*, pages 241–250, 1991.
34. N.E. Young. The k-server dual and loose competitiveness for paging. *Algorithmica*, 11(6):525–541, 1994.
35. N.E. Young. On-line file caching. In *Proc. of the 9th Annual ACM-SIAM Symposium on Discrete Algorithms*, pages 82–86, 1998.

Bibliography on Competitive Algorithms

MAREK CHROBAK
JOHN NOGA

K. Abrahamson. On achieving consensus using a shared memory. In *Proc. 7th Symp. on Principles of Distributed Computing*, pages 291–302, 1988.

N. Abramson. *Information Theory and Coding.* McGraw-Hill, New York, 1983.

D. Achlioptas, M. Chrobak, and J. Noga. Competitive analysis of randomized paging algorithms. In *Proc. 4th European Symp. on Algorithms*, volume 1136 of *Lecture Notes in Computer Science*, pages 419–430. Springer, 1996.

A. Aggarwal, B. Alpern, A. K. Chandra, and M. Snir. A model for hierarchical memory. In *Proc. 19th Symp. Theory of Computing*, pages 305–313, 1987.

A. Aggarwal, A. Bar-Noy, D. Coppersmith, R. Ramaswami, B. Schieber, and M. Sudan. Efficient routing and scheduling algorithms for optical networks. In *Proc. 6th Symp. on Discrete Algorithms*, pages 412–423, 1995.

A. Aggarwal and A. K. Chandra. Virtual memory algorithms. In *Proc. 20th Symp. Theory of Computing*, pages 173–185, 1988.

A. Aggarwal, A. K. Chandra, and M. Snir. Hierarchical memory with block transfer. In *Proc. 28th Symp. Foundations of Computer Science*, pages 204–216, 1987.

M. Ajtai, J. Aspnes, C. Dwork, and O. Waarts. A theory of competitive analysis for distributed algorithms. In *Proc. 35th Symp. Foundations of Computer Science*, pages 401–411, 1994.

M. Ajtai, J. Komlós, and G. Tusnády. On optimal matching. *Combinatorica*, 4:259–264, 1984.

M. Ajtai, N. Megiddo, and O. Waarts. Improved algorithms and analysis for secretary problems and generalizations. In *Proc. 36th Symp. Foundations of Computer Science*, pages 473–482, 1995.

S. Albers. The influence of lookahead in competitive paging algorithms. In *Proc. 1st European Symp. on Algorithms*, volume 726 of *Lecture Notes in Computer Science*, pages 1–12. Springer, 1993.

S. Albers. A competitive analysis of the list update problem with lookahead. In *Proc. 19th Symp. on Mathematical Foundations of Computer Science*, Lecture Notes in Computer Science, pages 201–210. Springer, 1994.

S. Albers. Improved randomized on-line algorithms for the list update problem. In *Proc. 6th Symp. on Discrete Algorithms*, pages 412–419, 1995.

S. Albers. Better bounds for online scheduling. In *Proc. 29th Symp. Theory of Computing*, pages 130–139, 1997.

S. Albers and M. Henzinger. Exploring unknown environments. In *Proc. 29th Symp. Theory of Computing*, pages 416–425, 1997.

S. Albers and H. Koga. New on-line algorithms for the page replication problem. Technical report, MPI Informatik Saarbrücken, 1993.

S. Albers and M. Mitzenmacher. Average case analyses of list update algorithms, with applications to data compression. In *Proc. 23rd International Colloquium on Automata, Languages, and Programming*, volume 1099 of *Lecture Notes in Computer Science*, pages 514–525. Springer, 1996.

S. Albers, B. von Stengel, and R. Werchner. A combined BIT and TIMESTAMP algorithm for the list update problem. *Information Processing Letters*, 56:135–139, 1995.

H. Alborzi, E. Torng, P. Uthaisombut, and S. Wagner. The k-client problem. In *Proc. 8th Symp. on Discrete Algorithms*, pages 73–82, 1997.

P. Algoet. Universal schemes for prediction, gambling and portfolio selection. *The Annals of Probability*, 20:901–941, 1992.

B. Allen and I. Munro. Self-organizing binary search trees. *Journal of the ACM*, 25:526–535, 1978.

N. Alon and Y. Azar. On-line steiner trees in the Euclidean plane. *Discrete Computational Geometry*, 10:113–121, 1993.

N. Alon, Y. Azar, J. Csirik, L. Epstein, S. V. Sevastianov, A. P. A. Vestjens, and G. J. Woeginger. On-line and off-line approximation algorithms for vector covering problems. *Algorithmica*, 21:104–118, 1998.

N. Alon, G. Kalai, M. Ricklin, and L. Stockmeyer. Lower bounds on the competitive ratio for mobile user tracking and distributed job scheduling. *Theoretical Computer Science*, 130:175–201, 1994.

N. Alon, R. M. Karp, D. Peleg, and D. West. A graph-theoretic game and its applications to the k-server problem. In L. A. McGeoch and D. D. Sleator, editors, *On-line Algorithms*, volume 7 of *DIMACS Series in Discrete Mathematics and Theoretical Computer Science*, pages 1–10. AMS/ACM, 1991.

N. Alon, R. M. Karp, D. Peleg, and D. West. A graph-theoretic game and its application to the k-server problem. *SIAM Journal on Computing*, 24:78–100, 1995.

L. Alonso, E. M. Reingold, and R. Schott. Determining the majority. *Information Processing Letters*, 47:253–255, 1993.

C. Anderson and A. Karlin. Two adaptive hybrid cache coherency protocols. In *Proc. 2nd International Symp. on High-Performance Computer Architecture*, 1996.

E. J. Anderson, P. Nash, and R. R. Weber. A counterexample to a conjecture on optimal list ordering. *Journal on Applied Probability*, 19:730–732, 1982.

R. Anderson and H. Woll. Wait-free parallel algorithms for the union-find problem. In *Proc. 23rd Symp. Theory of Computing*, pages 370–380, 1991.

M. Andrews. Constant factor bounds for on-line load balancing on related machines. Manuscript, 1996.

M. Andrews, B. Awerbuch, A. Fernández, J. Kleinberg, T. Leighton, and Z. Liu. Universal stability results for greedy contention-resolution protocols. In *Proc. 37th Symp. Foundations of Computer Science*, pages 380–389, 1996.

M. Andrews, M. Goemans, and L. Zhang. Improved bounds for on-line load balancing. In *Proc. 2nd Conference on Computing and Combinatorics*, 1996.

D. Angluin. Queries and concept learning. *Machine Learning*, 2:319–342, 1988.

D. Angluin, J. Westbrook, and W. Zhu. Robot navigation with range queries. In *Proc. 28th Symp. Theory of Computing*, pages 469–478, 1996.

A. Apostolico and M. Crochemore. Optimal canonization of all substrings of a string. *Information and Computation*, 95:76–95, 1991.

C. R. Aragon and R. G. Seidel. Randomized search trees. In *Proc. 30th Symp. Foundations of Computer Science*, pages 540–545, 1989.

R. Armstrong, D. Freitag, T. Joachims, and T. Mitchell. Webwatcher: A learning apprentice for the world wide web. In *1995 AAAI Spring Symp. on Information Gathering from Heterogeneous Distributed Environments*, 1995.

S. Arora, T. Leighton, and B. Maggs. On-line algorithms for path selection in a nonblocking network. In *Proc. 22nd Symp. Theory of Computing*, pages 149–58, 1990.

J. A. Aslam and A. Dhagat. On-line algorithms for 2-coloring hypergraphs via chip games. *Theoretical Computer Science*, 112:355–369, 1993.

A. H. Aslidis. Minimization of overstowage in containership operations. *Operations Research*, 90:457–471, 1990.

J. Aspnes, Y. Azar, A. Fiat, S. Plotkin, and O. Waarts. On-line load balancing with applications to machine scheduling and virtual circuit routing. *Journal of the ACM*, 44:486–504, 1997.

J. Aspnes and W. Hurwood. Spreading rumors rapidly despite an adversary. In *Proc. 15th Symp. on Principles of Distributed Computing*, pages 143–151, 1996.

J. Aspnes and O. Waarts. Modular competitiveness for distributed algorithms. In *Proc. 28th Symp. Theory of Computing*, pages 237–246, 1996.

S. F. Assmann. *Problems in Discrete Applied Mathematics.* PhD thesis, MIT, Cambridge, MA, 1983.

S. F. Assmann, D. S. Johnson, D. J. Kleitman, and J. Y.-T. Leung. On a dual version of the one-dimensional bin packing problem. *Journal of Algorithms*, 5:502–525, 1984.

H. Attiya and O. Rachman. Atomic snapshots in $o(n \log n)$ operations. In *Proc. 12th Symp. on Principles of Distributed Computing*, pages 29–40, 1993.

P. Auer and N. Cesa-Bianchi. On-line learning with malicious noise and the closure algorithm. In *Proc. 4th International Workshop on Analogical and Inductive Inference: Algorithmic Learning Theory*, pages 229–247, 1994.

P. Auer, N. Cesa-Bianchi, Y. Freund, and R. E. Schapire. Gambling in a rigged casino: The adversarial multi-armed bandit problem. In *Proc. 36th Symp. Foundations of Computer Science*, pages 322–331, 1995.

P. Auer, P. M. Long, W. Maass, and G. J. Woeginger. On the complexity of function learning. *Machine Learning*, 18:187–230, 1995.

P. Auer and M. K. Warmuth. Tracking the best disjunction. In *Proc. 36th Symp. Foundations of Computer Science*, pages 312–321, 1995.

Y. Aumann. Efficient asynchronous consensus with the weak adversary scheduler. In *Proc. 16th Symp. on Principles of Distributed Computing*, 1997.

Y. Aumann and M. A. Bender. Efficient asynchronous consensus with the value-oblivious adversary scheduler. In *Proc. 23rd International Colloquium on Automata, Languages, and Programming*, volume 1099 of *Lecture Notes in Computer Science*. Springer, 1996.

G. Ausiello, E. Feuerstein, S. Leonardi, L. Stougie, and M. Talamo. Serving requests with on-line routing. In *Proc. 4th Scandinavian Workshop on Algorithm Theory*, Lecture Notes in Computer Science, pages 37–48. Springer, 1994.

G. Ausiello, E. Feuerstein, S. Leonardi, L. Stougie, and M. Talamo. Competitive algorithms for the on-line traveling salesman problem. In *Proc. 4th Workshop on Algorithms and Data Structures*, volume 955 of *Lecture Notes in Computer Science*. Springer, 1995.

G. Ausiello, G. F. Italiano, A. M. Spaccamela, and U. Nanni. On-line computation of minimal and maximal length paths. *Theoretical Computer Science*, 95:245–261, 1992.

A. Avidor, Y. Azar, and J. Sgall. Ancient and new algorithms for load balancing in the l_p norm. In *Proc. 9th Symp. on Discrete Algorithms*, pages 426–435, 1998.

M. Avriel and M. Penn. Stowage planning for container ships to reduce the number of shifts. Technical report, Faculty of Industrial Engineering and Management, Technion - Israel Institute of Technology, 1993.

M. Avriel, M. Penn, and N. Shpirer. Container ship stowage problem: Complexity and applications to coloring of circle graphs. Technical report, Faculty of Industrial Engineering and Management, Technion - Israel Institute of Technology, 1994.

B. Awerbuch and Y. Azar. Local optimization of global objectives: competitive distributed deadlock resolution and resource allocation. In *Proc. 35th Symp. Foundations of Computer Science*, pages 240–249, 1994.

B. Awerbuch and Y. Azar. Competitive multicast routing. *Wireless Networks*, 1:107–114, 1995.

B. Awerbuch, Y. Azar, and Y. Bartal. On-line generalized steiner problem. In *Proc. 7th Symp. on Discrete Algorithms*, pages 68–74, 1996.

B. Awerbuch, Y. Azar, A. Blum, and S. Vempala. Improved approximation guarantees for minimum weight k-trees and prize-collecting salesmen. In *Proc. 27th Symp. Theory of Computing*, 1995.

B. Awerbuch, Y. Azar, and A. Fiat. Packet routing via min-cost circuit routing. In *Proc. 4th Israel Symp. on Theory of Computing and Systems*, pages 37–42, 1996.

B. Awerbuch, Y. Azar, A. Fiat, and T. Leighton. Making commitments in the face of uncertainty: How to pick a winner almost every time. In *Proc. 28th Symp. Theory of Computing*, pages 519–530, 1996.

B. Awerbuch, Y. Azar, A. Fiat, S. Leonardi, and A. Rosén. On-line competitive algorithms for call admission in optical networks. In *Proc. 4th European Symp. on Algorithms*, volume 1136 of *Lecture Notes in Computer Science*, pages 431–444. Springer, 1996.

B. Awerbuch, Y. Azar, E. F. Grove, M.-Y. Kao, P. Krishnan, and J. S. Vitter. Load balancing in the l_p norm. In *Proc. 36th Symp. Foundations of Computer Science*, pages 383–391, 1995.

B. Awerbuch, Y. Azar, and S. Plotkin. Throughput-competitive online routing. In *Proc. 34th Symp. Foundations of Computer Science*, pages 32–40, 1993.

B. Awerbuch, Y. Azar, S. Plotkin, and O. Waarts. Competitive routing of virtual circuits with unknown duration. In *Proc. 5th Symp. on Discrete Algorithms*, pages 321–327, 1994.

B. Awerbuch, Y. Bartal, and A. Fiat. Competitive distributed file allocation. In *Proc. 25th Symp. Theory of Computing*, pages 164–173, 1993.

B. Awerbuch, Y. Bartal, and A. Fiat. Heat & dump: Competitive on-line routing. In *Proc. 34th Symp. Foundations of Computer Science*, pages 22–31, 1993.

B. Awerbuch, Y. Bartal, and A. Fiat. Distributed paging for general networks. In *Proc. 7th Symp. on Discrete Algorithms*, pages 574–583, 1996.

B. Awerbuch, Y. Bartal, A. Fiat, and A. Rosén. Competitive non-preemptive call control. In *Proc. 5th Symp. on Discrete Algorithms*, pages 312–320, 1994.

B. Awerbuch, M. Betke, R. Rivest, and M. Singh. Piecemeal graph learning by a mobile robot. In *Proc. 10th Conf. on Computational Learning Theory*, pages 321–328, 1997.

B. Awerbuch, R. Gawlick, T. Leighton, and Y. Rabani. On-line admission control and circuit routing for high performance computing and communication. In *Proc. 35th Symp. Foundations of Computer Science*, pages 412–423, 1994.

B. Awerbuch, S. Kutten, and D. Peleg. Competitive distributed job scheduling. In *Proc. 24th Symp. Theory of Computing*, pages 571–580, 1992.

B. Awerbuch and T. Leighton. Improved approximation algorithms for the multicommodity flow problem and local competitive routing in dynamic networks. In *Proc. 26th Symp. Theory of Computing*, pages 487–496, 1994.

B. Awerbuch and M. Saks. A dining philosophers algorithm with polynomial response time. In *Proc. 31st Symp. Foundations of Computer Science*, pages 65–74, 1990.

B. Awerbuch and T. Singh. Online algorithms for selective multicast and maximal dense trees. In *Proc. 29th Symp. Theory of Computing*, pages 354–362, 1997.

Y. Azar, Y. Bartal, E. Feuerstein, A. Fiat, S. Leonardi, and A. Rosén. On capital investment. In *Proc. 23rd International Colloquium on Automata, Languages, and Programming*, volume 1099 of *Lecture Notes in Computer Science*. Springer, 1996.

Y. Azar, A. Broder, and M. Manasse. On-line choice of on-line algorithms. In *Proc. 4th Symp. on Discrete Algorithms*, pages 432–440, 1993.

Y. Azar, A. Z. Broder, and A. R. Karlin. On-line load balancing. *Theoretical Computer Science*, 130:73–84, 1994.

Y. Azar and L. Epstein. On two dimensional packing. In *Proc. 5th Scandinavian Workshop on Algorithm Theory*, volume 1097 of *Lecture Notes in Computer Science*. Springer, 1996.

Y. Azar and L. Epstein. On-line load balancing of temporary tasks on identical machines. In *Proc. 5th Israel Symp. on Theory of Computing and Systems*, pages 119–125, 1997.

Y. Azar, B. Kalyanasundaram, S. Plotkin, K. Pruhs, and O. Waarts. On-line load balancing of temporary tasks. In *Proc. 3rd Workshop on Algorithms and Data Structures*, Lecture Notes in Computer Science, pages 119–130. Springer, 1993.

Y. Azar, J. Naor, and R. Rom. The competitiveness of on-line assignments. *Journal of Algorithms*, 18:221–237, 1995.

Y. Azar and O. Regev. On-line bin stretching. Manuscript, 1997.

R. Bachrach and R. El-Yaniv. Online list accessing algorithms and their applications: Recent empirical evidence. In *Proc. 8th Symp. on Discrete Algorithms*, pages 53–62, 1997.

R. Baeza-Yates, J. Culberson, and G. Rawlins. Searching in the plane. *Information and Computation*, 106:234–252, 1993.

R. Baeza-Yates and R. Schott. Parallel searching in the plane. *Computational Geometry: Theory and Applications*, 5:143–154, 1995.

R. A. Baeza-Yates, J. C. Culberson, and G. J. E. Rawlins. Searching with uncertainty. In *Proc. 1st Scandinavian Workshop on Algorithm Theory*, Lecture Notes in Computer Science, pages 176–189. Springer, 1988.

J.-C. Bajard, J. Duprat, S. Kla, and J.-M. Muller. Some operators for on-line radix-2 computations. *Journal on Parallel and Distributed Computing*, 22:336–345, 1994.

B. S. Baker. A new proof for the First-Fit decreasing bin-packing algorithm. *Journal of Algorithms*, 6:49–70, 1985.

B. S. Baker, D. J. Brown, and H. P. Katseff. A 5/4 algorithm for two-dimensional packing. *Journal of Algorithms*, 2:348–368, 1981.

B. S. Baker, D. J. Brown, and H. P. Katseff. Lower bounds for two-dimensional packing algorithms. *Acta Informatica*, 8:207–225, 1982.

B. S. Baker and E. G. Coffman. A tight asymptotic bound for Next-Fit-Decreasing bin-packing. *SIAM Journal on Algebraic and Discrete Methods*, 2:147–152, 1981.

B. S. Baker, E. G. Coffman, and R. L. Rivest. Orthogonal packings in two dimensions. *SIAM Journal on Computing*, 9:846–855, 1980.

B. S. Baker and J. S. Schwartz. Shelf algorithms for two-dimensional packing problems. *SIAM Journal on Computing*, 12:508–525, 1983.

G. R. Baliga and A. M. Shende. On space bounded server algorithms. In *Proc. 5th International Conference on Computing and Information*, pages 77–81, 1993.

E. Bar-Eli, P. Berman, A. Fiat, and P. Yan. Online navigation in a room. In *Proc. 3rd Symp. on Discrete Algorithms*, pages 237–249. ACM/SIAM, 1992.

E. Bar-Eli, P. Berman, A. Fiat, and P. Yan. Online navigation in a room. *Journal of Algorithms*, 17:319–341, 1994.

A. Bar-Noy, R. Canetti, S. Kutten, Y. Mansour, and B. Schieber. Bandwidth allocation with preemption. In *Proc. 27th Symp. Theory of Computing*, pages 616–625, 1995.

A. Bar-Noy, F. K. Hwang, I. Kessler, and S. Kutten. A new competitive algorithm for group testing. *Discrete Applied Mathematics*, 52:29–38, 1994.

A. Bar-Noy, R. Motwani, and J. Naor. The greedy algorithm is optimal for online edge coloring. *Information Processing Letters*, 44:251–253, 1992.

A. Bar-Noy and B. Schieber. The Canadian traveller problem. In *Proc. 2nd Symp. on Discrete Algorithms*, pages 261–270, 1991.

Y. Bartal. Probabilistic approximation of metric spaces and its algorithmic applications. In *Proc. 37th Symp. Foundations of Computer Science*, pages 184–193, 1996.

Y. Bartal, A. Blum, C. Burch, and A. Tomkins. A polylog(n)-competitive algorithm for metrical task systems. In *Proc. 29th Symp. Theory of Computing*, pages 711–719, 1997.

Y. Bartal, M. Charikar, and P. Indyk. On page migration and other related task systems. In *Proc. 8th Symp. on Discrete Algorithms*, pages 43–52, 1997.

Y. Bartal, M. Chrobak, and L. L. Larmore. A randomized algorithm for two servers on the line. Unpublished Manuscript, 1998.

Y. Bartal, A. Fiat, H. Karloff, and R. Vohra. New algorithms for an ancient scheduling problem. *Journal Computer Systems Science*, 51:359–366, 1995.

Y. Bartal, A. Fiat, and S. Leonardi. Lower bounds for on-line graph problems with application to on-line circuit and optical routing. In *Proc. 28th Symp. Theory of Computing*, pages 531–540, 1996.

Y. Bartal, A. Fiat, and Y. Rabani. Competitive algorithms for distributed data management. In *Proc. 24th Symp. Theory of Computing*, pages 39–50, 1992.

Y. Bartal and E. Grove. The harmonic k-server algorithm is competitive. To appear in Journal of the ACM.

Y. Bartal, H. Karloff, and Y. Rabani. A better lower bound for on-line scheduling. *Information Processing Letters*, 50:113–116, 1994.

Y. Bartal and S. Leonardi. On-line routing in all-optical networks. In *Proc. 24th International Colloquium on Automata, Languages, and Programming*, volume 1256 of *Lecture Notes in Computer Science*, pages 516–526. Springer, 1997.

Y. Bartal, S. Leonardi, A. Marchetti-Spaccamela, J. Sgall, and L. Stougie. Multiprocessor scheduling with rejection. In *Proc. 7th Symp. on Discrete Algorithms*, pages 95–103, 1996.

Y. Bartal and A. Rosén. The distributed k-server problem — a competitive distributed translator for k-server algorithms. In *Proc. 33rd Symp. Foundations of Computer Science*, pages 344–353, 1992.

J. J. Bartholdi, J. H. vande Vate, and J. Zhang. Expected performance of the shelf heuristic for two-dimensional packing. *Operations Research Letters*, 8:11–16, 1989.

S. Baruah, G. Koren, D. Mao, B. Mishra, A. Raghunathan, L. Rosier, D. Shasha, and F. Wang. On the competitiveness of on-line real-time task scheduling. *Real-Time Systems*, 4:125–144, 1992.

S. Baruah, G. Koren, B. Mishra, A. Raghunatan, L. Roiser, and D. Shasha. On-line scheduling in the presence of overload. In *Proc. 32nd Symp. Foundations of Computer Science*, pages 100–110, 1991.

S. K. Baruah, J. Haritsa, and N. Sharma. On-line scheduling to maximize task completions. In *Proc. Real-Time Systems Symp.*, pages 228–236, 1994.

S. K. Baruah and M. E. Hickey. Competitive on-line scheduling of imprecise computations. In *Proc. 29th Hawaii International Conference on System Sciences*, pages 460–468, 1996.

R. D. Barve, E. F. Grove, and J. S. Vitter. Application-controlled paging for a shared cache. In *Proc. 36th Symp. Foundations of Computer Science*, pages 204–213, 1995.

G. D. Battista and R. Tamassia. On-line graph algorithms with SPQR-trees. In *Proc. 17th International Colloquium on Automata, Languages, and Programming*, Lecture Notes in Computer Science, pages 598–611. Springer, 1990.

P. Bay and G. Bilardi. Deterministic on-line routing on area-universal networks. In *Proc. 31st Symp. Foundations of Computer Science*, pages 297–306, 1990.

D. Bean. Effective coloration. *Journal of Symbolic Logic*, 41:469–480, 1976.

R. Beigel and W. I. Gasarch. The mapmaker's dilemma. *Discrete Applied Mathematics*, 34:37–48, 1991.

J. Bekesi, G. Galambos, U. Pferschy, and G. J. Woeginger. The fractional greedy algorithm for data compression. *Computing*, 56:29–46, 1996.

L. A. Belady. A study of replacement algorithms for virtual storage computers. *IBM Systems Journal*, 5:78–101, 1966.

J. Bell and G. Gupta. Evaluation of self-adjusting binary search tree techniques. *Software — Practice and Experience*, 23:369–382, 1993.

T. Bell and D. Kulp. Longest-match string searching for Ziv-Lempel compression. *Software – Practice and Experience*, 23:757–771, 1993.

R. Bellman. Equipment replacement policy. *Journal of the SIAM*, 3:133–136, 1955.

S. Ben-David and A. Borodin. A new measure for the study of on-line algorithms. *Algorithmica*, 11:73–91, 1994.

S. Ben-David, A. Borodin, R. M. Karp, G. Tardos, and A. Wigderson. On the power of randomization in on-line algorithms. *Algorithmica*, 11:2–14, 1994.

M. A. Bender and D. K. Slonim. The power of team exploration: two robots can learn unlabeled directed graphs. In *Proc. 35th Symp. Foundations of Computer Science*, pages 75–85, 1994.

J. Bentley and J. Saxe. An analysis of two heuristics for the Euclidean traveling salesman problem. In *Proc. Allerton Conference on Communication, Control, and Computing*, pages 41–49, 1980.

J. L. Bentley, K. L. Clarkson, and D. B. Levine. Fast linear expected-time algorithms for computing maxima and convex hulls. In *Proc. 1st Symp. on Discrete Algorithms*, pages 179–187, 1990.

J. L. Bentley, D. S. Johnson, F. T. Leighton, and C. C. McGeoch. An experimental study of bin packing. In *Proc. 21st Allerton Conf. on Communication, Control, and Computing*, pages 51–60, 1983.

J. L. Bentley, D. S. Johnson, F. T. Leighton, C. C. McGeoch, and L. A. McGeoch. Some unexpected expected behavior results for bin packing. In *Proc. 16th Symp. Theory of Computing*, pages 279–288, 1984.

J. L. Bentley and C. C. McGeoch. Amortized analyses of self-organizing sequential search heuristics. *Communications of the ACM*, 28:404–411, 1985.

J. L. Bentley, D. Sleator, R. E. Tarjan, and V. K. Wei. A locally adaptive data compression scheme. *Communications of the ACM*, 29:320–330, 1986.

P. Berman, A. Blum, A. Fiat, H. Karloff, A. Rosén, and M. Saks. Randomized robot navigation algorithms. In *Proc. 7th Symp. on Discrete Algorithms*, pages 75–84, 1996.

P. Berman, M. Charikar, and M. Karpinski. On-line load balancing for related machines. In *Proc. 5th Workshop on Algorithms and Data Structures*, volume 1272 of *Lecture Notes in Computer Science*, pages 116–125. Springer, 1997.

P. Berman and C. Coulston. On-line algorithms for steiner tree problems. In *Proc. 29th Symp. Theory of Computing*, pages 344–353, 1997.

P. Berman, H. Karloff, and G. Tardos. A competitive algorithm for three servers. In *Proc. 1st Symp. on Discrete Algorithms*, pages 280–290, 1990.

P. Berman and M. Karpinski. Randomized navigation to a wall through convex obstacles. Technical report, Bonn University, 1994. Tech. Rep. 85118-CS.

M. Bern, D. Greene, and A. Raghunathan. On-line algorithms for cache sharing. In *Proc. 25th Symp. Theory of Computing*, pages 422–430, 1993.

M. Bern, D. H. Greene, A. Raghunathan, and M. Sudan. On-line algorithms for locating checkpoints. *Algorithmica*, 11:33–52, 1994.

N. Bhatnagar and J. Mostow. On-line learning from search failures. *Machine Learning*, 15:69–117, 1994.

J. R. Bitner. Heuristics that dynamically organize data structures. *SIAM Journal on Computing*, 8:82–110, 1979.

D. Black, A. Gupta, and W.-D. Weber. Competitive management of distributed shared memory. In *Proc. Spring Compcon, IEEE Computer Society*, pages 184–190, 1989.

D. L. Black and D. D. Sleator. Competitive algorithms for replication and migration problems. Technical Report CMU-CS-89-201, Department of Computer Science, Carnegie-Mellon University, 1989.

D. Blackwell. An analog of the minimax theorem for vector payoffs. *Pacific Journal of Mathematics*, 6:1–8, 1956.

D. Blitz, A. van Vliet, and G. J. Woeginger. Lower bounds on the asymptotic worst-case ratio of online bin packing algorithms. Unpublished manuscript, 1996.

A. Blum. Learning boolean functions in an infinite attribute space. *Machine Learning*, 9:373–386, 1992.

A. Blum. Separating distribution-free and mistake-bound learning models over the boolean domain. *SIAM Journal on Computing*, 23:990–1000, 1994.

A. Blum. Empirical support for Winnow and Weighted-Majority based algorithms: results on a calendar scheduling domain. In *Proc. 12th International Conference on Machine Learning*, pages 64–72, 1995.

A. Blum and C. Burch. On-line learning and the metrical task system problem. In *Proc. 10th Conf. on Computational Learning Theory*, pages 45–53, 1997.

A. Blum and P. Chalasani. An on-line algorithm for improving performance in navigation. In *Proc. 34th Symp. Foundations of Computer Science*, pages 2–11, 1993.

A. Blum, P. Chalasani, D. Coppersmith, B. Pulleyblank, P. Raghavan, and M. Sudan. The minimum latency problem. In *Proc. 26th Symp. Theory of Computing*, pages 163–171, 1994.

A. Blum, L. Hellerstein, and N. Littlestone. Learning in the presence of finitely or infinitely many irrelevant attributes. *Journal Computer Systems Science*, 50:32–40, 1995.

A. Blum and A. Kalai. Universal portfolios with and without transaction costs. In *Proc. 10th Conf. on Computational Learning Theory*, pages 309–313, 1997.

A. Blum, H. Karloff, Y. Rabani, and M. Saks. A decomposition theorem and lower bounds for randomized server problems. In *Proc. 33rd Symp. Foundations of Computer Science*, pages 197–207, 1992.

A. Blum, P. Raghavan, and B. Schieber. Navigating in unfamiliar geometric terrain. *SIAM Journal on Computing*, 26:110–137, 1997.

R. Board. The online graph bandwidth problem. *Information and Computation*, 100:178–201, 1992.

J.-D. Boissonnat, O. Devillers, R. Schott, M. Teillaud, and M. Yvinec. Applications of random sampling to online algorithms in computational geometry. *Discrete Computational Geometry*, 8:51–71, 1992.

R. Boppana and A. Floratos. Load balancing in the Euclidean norm. Manuscript, 1997.

A. Borodin and R. El-Yaniv. Call admission and circuit-routing. In *Online Computation and Competitive Analysis.* Cambridge University Press, 1997.

A. Borodin, S. Irani, P. Raghavan, and B. Schieber. Competitive paging with locality of reference. *Journal Computer Systems Science*, 50:244–258, 1995.

A. Borodin, J. Kleinberg, P. Raghavan, M. Sudan, and D. P. Williamson. Adversarial queueing theory. In *Proc. 28th Symp. Theory of Computing*, pages 376–385, 1996.

A. Borodin, N. Linial, and M. Saks. An optimal online algorithm for metrical task system. *Journal of the ACM*, 39:745–763, 1992.

L. Breiman. Optimal gambling systems for favorable games. In *Proc. 4th Berkeley Symp. on Mathematical Statistics and Probability*, volume 1, pages 65–78. Univ. of California Press, Berkeley, 1961.

D. Breslauer. An on-line string superprimitivity test. *Information Processing Letters*, 44:345–347, 1992.

D. Breslauer. On competitive on-line paging with lookahead. In *Proc. 13th Symp. on Theoretical Aspects of Computer Science*, pages 593–603, 1996.

A. R. Brown. *Optimum packing and Depletion.* American Elsevier, New York, 1971.

D. J. Brown. A lower bound for on-line one-dimensional bin packing algorithms. Technical Report R-864, Coordinated Sci. Lab., Urbana, Illinois, 1979.

D. J. Brown, B. S. Baker, and H. P. Katseff. Lower bounds for two-dimensional packing algorithms. *Acta Informatica*, 8:207–225, 1982.

J. L. Bruno and P. J. Downey. Probabilistic bounds for dual bin packing. *Acta Informatica*, 22:333–345, 1985.

J. Brzustowski. Can you win at TETRIS? Master's thesis, The University of British Columbia, 1992.

N. H. Bshouty, Z. Chen, and S. Homer. On learning discretized geometric concepts. In *Proc. 35th Symp. Foundations of Computer Science*, pages 54–63, 1994.

H. Burgiel. How to lose at TETRIS. Technical report, The Geometry Center, Minneapolis, MN, 1996.

W. Burley. Traversing layered graphs using the work function algorithm. Technical Report CS93-319, Department of Computer Science and Engineering, University of California at San Diego, 1993.

W. Burley. Exploring-backtracing games and an application to layered graph traversal. unpublished manuscript, 1994.

W. Burley. Toward an optimal online algorithm for layered graph traversal. Technical Report CS94-382, Department of Computer Science and Engineering, University of California at San Diego, 1994.

W. R. Burley. Traversing layered graphs using the work function algorithm. *Journal of Algorithms*, 20:479–511, 1996.

W. R. Burley and S. Irani. On algorithm design for metrical task systems. In *Proc. 6th Symp. on Discrete Algorithms*, pages 420–429, 1995.

M. Burrows and D. J. Wheeler. A block-sorting lossless data compression algorithm. Technical Report 124, DEC SRC, 1994.

S. K. Buruah and L. E. Rosier. Limitations concerning on-line scheduling algorithms for overloaded real-time systems. In *Proc. Workshop Real Time Programming*, pages 123–125, 1992.

P. J. Burville and J. F. C. Kingman. On a model for storage and search. *Journal on Applied Probability*, 10:697–701, 1973.

A. R. Calderbank, E. G. Coffman, and L. Flatto. Sequencing problems in two-server systems. *Mathematics of Operations Research*, 10:585–598, 1985.

R. Canetti and S. Irani. Bounding the power of preemption in randomized scheduling. In *Proc. 27th Symp. Theory of Computing*, pages 616–615, 1995.

P. Cao, E. W. Felten, and K. Li. Application-controlled file caching policies. In *Proc. for the Summer USENIX Conference*, 1994.

P. Cao and S. Irani. Cost-aware WWW proxy caching algorithms. Technical report, 1343, Dept. of Computer Sciences, University of Wisconsin-Madison, 1997.

N. Cesa-Bianchi, Y. Freund, D. P. Helmbold, D. Haussler, R. E. Schapire, and M. K. Warmuth. How to use expert advice. In *Proc. 25th Symp. Theory of Computing*, pages 382–391, 1993.

N. Cesa-Bianchi, Y. Freund, D. P. Helmbold, and M. Warmuth. On-line prediction and conversion strategies. In *Computational Learning Theory: Eurocolt '93*, volume New Series Number 53 of *The Institute of Mathematics and its Applications Conference Series*, pages 205–216, Oxford, 1994. Oxford University Press.

S. Chakrabarti, C. A. Phillips, A. S. Schulz, D. B. Shmoys, C. Stein, and J. Wein. Improved scheduling algorithms for minsum criteria. In *Proc. 23rd International Colloquium on Automata, Languages, and Programming*, volume 1099 of *Lecture Notes in Computer Science*, pages 646–657. Springer, 1996.

K. F. Chan and T. W. Lam. An on-line algorithm for navigating in an unknown environment. *International Journal of Computational Geometry and Applications*, 3:227–244, 1993.

B. Chandra. Does randomization help in online bin packing? *Information Processing Letters*, 43:15–19, 1992.

B. Chandra and S. Vishwanathan. Constructing reliable communication networks of small weight. *Journal of Algorithms*, 18:159–175, 1995.

T. Chandra. Polylog randomized wait-free consensus. In *Proc. 15th Symp. on Principles of Distributed Computing*, 1996.

E.-C. Chang, W. Wang, and M. S. Kankanhalli. Multidimensional online bin packing: An algorithm and its average-case analysis. *Information Processing Letters*, 48:121–125, 1993.

M. Charikar, D. Halperin, and R. Motwani. The dynamic servers problem. In *Proc. 9th Symp. on Discrete Algorithms*, pages 410–419, 1998.

R. Chaudhuri and H. Hoft. Splaying a search tree in preorder takes linear time. *SIGACT News*, 24:88–93, 1993.

F. Chauny, R. Loulou, S. Sadones, and F. Soumis. A two-phase heuristic for strip packing: algorithm and probabilistic analysis. *Operations Research Letters*, 6:25–33, 1987.

R. P. Cheetham, J. B. Oommen, and D. Ng. Adaptive structuring of binary search trees using conditional rotations. *IEEE Transactions on Knowledge and Data Engineering*, 5:695–704, 1993.

C. Chekuri, R. Motwani, B. Natarajan, and C. Stein. Approximation techniques for average completion time scheduling. In *Proc. 8th Symp. on Discrete Algorithms*, pages 609–618, 1997.

B. Chen, A. van Vliet, and G. J. Woeginger. Lower bounds for randomized online scheduling. *Information Processing Letters*, 51:219–222, 1994.

B. Chen, A. van Vliet, and G. J. Woeginger. New lower and upper bounds for on-line scheduling. *Operations Research Letters*, 16:221–230, 1994.

B. Chen, A. van Vliet, and G. J. Woeginger. An optimal algorithm for preemptive on-line scheduling. *Operations Research Letters*, 18:127–131, 1995.

B. Chen and A. P. A. Vestjens. Scheduling on identical machines: How good is lpt in an on-line setting? *Operations Research Letters*, 21:165–169, 1998.

B. Chen, A. P. A. Vestjens, and G. J. Woeginger. On-line scheduling of two-machine open shops where jobs arrive over time. *Journal of Combinatorial Optimization*, 1:355–365, 1997.

B. Chen and G. J. Woeginger. A study of on-line shop scheduling problems on two machines. In D.-Z. Du and P. M. Pardalos, editors, *Minmax Problems and Applications*, pages 97–107. Kluewer, 1995.

B.-C. Chien, R.-J. Chen, and W.-P. Yang. Competitive analysis of the on-line algorithms for multiple stacks systems. In *Proc. 3rd International Symp. on Algorithms and Computation*, Lecture Notes in Computer Science, pages 78–87. Springer, 1992.

Y. Cho and S. Sahni. Bounds for list schedules on uniform processors. *SIAM Journal on Computing*, 9:91–103, 1980.

B. Chor, A. Israeli, and M. Li. Wait–free consensus using asynchronous hardware. *SIAM Journal on Computing*, 23:701–712, 1994.

A. Chou. Optimal trading strategies vs. a statistical adversary. Master's thesis, Massachusetts Institute of Technology, 1994.

A. Chou, J. Cooperstock, R. El-Yaniv, M. Klugerman, and T. Leighton. The statistical adversary allows optimal money-making trading strategies. In *Proc. 6th Symp. on Discrete Algorithms*, 1995.

A. Chou, A. Shrivastava, and R. Sidney. On the power of magnitude. To appear.

Y. S. Chow, H. Robbins, and D. Siegmund. *The Theory of Optimal Stopping.* Dover Publications, Inc., 1971.

M. Chrobak, H. Karloff, T. H. Payne, and S. Vishwanathan. New results on server problems. *SIAM Journal on Discrete Mathematics*, 4:172–181, 1991.

M. Chrobak and L. L. Larmore. A new approach to the server problem. *SIAM Journal on Discrete Mathematics*, 4:323–328, 1991.

M. Chrobak and L. L. Larmore. A note on the server problem and a benevolent adversary. *Information Processing Letters*, 38:173–175, 1991.

M. Chrobak and L. L. Larmore. On fast algorithms for two servers. *Journal of Algorithms*, 12:607–614, 1991.

M. Chrobak and L. L. Larmore. An optimal online algorithm for k servers on trees. *SIAM Journal on Computing*, 20:144–148, 1991.

M. Chrobak and L. L. Larmore. HARMONIC is three-competitive for two servers. *Theoretical Computer Science*, 98:339–346, 1992.

M. Chrobak and L. L. Larmore. Metrical service systems: Deterministic strategies. Technical Report UCR-CS-93-1, Department of Computer Science, University of California at Riverside, 1992. Submitted.

M. Chrobak and L. L. Larmore. Metrical service systems: Randomized strategies. manuscript, 1992.

M. Chrobak and L. L. Larmore. The server problem and on-line games. In *DIMACS Series in Discrete Mathematics and Theoretical Computer Science*, volume 7, pages 11–64, 1992.

M. Chrobak and L. L. Larmore. Generosity helps or an 11-competitive algorithm for three servers. *Journal of Algorithms*, 16:234–263, 1994.

M. Chrobak, L. L. Larmore, C. Lund, and N. Reingold. A better lower bound on the competitive ratio of the randomized 2-server problem. to appear inInformation Processing Letters.

M. Chrobak, L. L. Larmore, N. Reingold, and J. Westbrook. Page migration algorithms using work functions. In *Proc. 4th International Symp. on Algorithms and Computation*, Lecture Notes in Computer Science, pages 406–415. Springer, 1993.

M. Chrobak and J. Noga. Competitive algorithms for multilevel caching and relaxed list update. In *Proc. 9th Symp. on Discrete Algorithms*, pages 87–96, 1998.

M. Chrobak and J. Noga. LRU is better than FIFO. In *Proc. 9th Symp. on Discrete Algorithms*, pages 78–81, 1998.

M. Chrobak and M. Ślusarek. On some packing problems relating to Dynamical Storage Allocations. *RAIRO Journal on Information Theory and Applications*, 22:487–499, 1988.

F. R. K. Chung, M. R. Garey, and D. S. Johnson. On packing two-dimensional bins. *SIAM Journal on Algebraic and Discrete Methods*, 3:66–76, 1982.

F. R. K. Chung, R. L. Graham, and M. E. Saks. A dynamic location problem for graphs. *Combinatorica*, 9:111–131, 1989.

F. R. K. Chung, D. J. Hajela, and P. D. Seymour. Self-organizing sequential search and Hilbert's inequality. In *Proc. 17th Symp. Theory of Computing*, pages 217–223, 1985.

E. G. Coffman. *Computer and Job-Shop Scheduling Theory.* John Wiley, 1976.

E. G. Coffman. An introduction to proof techniques for packing and sequencing algorithms. In M. A. H. D. et al, editor, *Deterministis and Stochastic Scheduling*, pages 245–270. Reidel Publishing Co., Amsterdam, 1982.

E. G. Coffman. An introduction to combinatorial models of dynamic storage allocation. *SIAM Reviews*, 25:311–325, 1983.

E. G. Coffman, C. A. Courcoubetis, M. R. Garey, D. S. Johnson, L. A. McGeogh, P. W. Shor, R. R. Weber, and M. Yannakakis. Fundamental discrepancies between average-case analyses under discrete and continuous distributions - a bin packing case study. In *Proc. 23rd Symp. Theory of Computing*, pages 230–240, 1991.

E. G. Coffman, M. R. Garey, and D. S. Johnson. Dynamic bin packing. *SIAM Journal on Computing*, 12:227–258, 1983.

E. G. Coffman, M. R. Garey, and D. S. Johnson. Approximation algorithms for bin packing: a survey. In D. Hochbaum, editor, *Approximation algorithms.* PWS Publishing Company, 1996.

E. G. Coffman, M. R. Garey, D. S. Johnson, and R. E. Tarjan. Performance bounds for level oriented two-dimensional packing algorithms. *SIAM Journal on Computing*, 9:808–826, 1980.

E. G. Coffman and E. N. Gilbert. Dynamic, First Fit packings in two- or more dimensions. *Information and Control*, 61:1–14, 1984.

E. G. Coffman, M. Hofri, K. So, and A. C.-C. Yao. A stochastic model of bin packing. *Information and Control*, 44:105–115, 1980.

E. G. Coffman, D. S. Johnson, P. W. Shor, and R. R. Weber. Markov chains, computer proofs, and Best Fit packing. In *Proc. 25th Symp. Theory of Computing*, pages 412–421, 1993.

E. G. Coffman, D. S. Johnson, P. W. Shor, and R. R. Weber. Bin packing with discrete item sizes, part ii: average case behaviour of First Fit. Unpublished manuscript, 1996.

E. G. Coffman and G. S. Lueker. *Probabilistic analysis of Packing and Partitioning Algorithms.* John Wiley, New York, 1991.

E. G. Coffman, G. S. Lueker, and A. H. G. R. Kan. Asymptotic methods in the probabilistic analysis of sequencing and packing heuristics. *Management Science*, 34:266–290, 1988.

E. G. Coffman and P. W. Shor. A simple proof of the $O(sqrt(nlog^3/4))$ up-right matching bound. *SIAM Journal on Discrete Mathematics*, 4:48–57, 1991.

E. G. Coffman and P. W. Shor. Packing in two dimensions: asymptotic average-case analysis of algorithms. *Algorithmica*, 9:253–277, 1993.

E. G. Coffman, K. So, M. Hofri, and A. C.-C. Yao. A stochastic model of bin packing. *Information and Control*, 44:105–115, 1980.

D. Cohen and M. L. Fredman. Weighted binary trees for concurrent searching. *Journal of Algorithms*, 20:87–112, 1996.

R. Cole. On the dynamic finger conjecture for splay trees. Part 2: Finger searching. Technical Report 472, Courant Institute, NYU, 1989.

R. Cole. On the dynamic finger conjecture for splay trees. In *Proc. 22nd Symp. Theory of Computing*, pages 8–17, 1990.

R. Cole and R. Hariharan. Tighter bounds on the exact complexity of string matching. In *Proc. 33rd Symp. Foundations of Computer Science*, pages 600–609, 1992.

R. Cole, R. Hariharan, M. Paterson, and U. Zwick. Tighter lower bounds on the exact complexity of string matching. *SIAM Journal on Computing*, 24:30–45, 1995.

R. Cole, R. Hariharan, M. S. Paterson, and U. Zwick. Which patterns are hard to find? (string matching). In *Proc. 2nd Israel Symp. on Theory of Computing and Systems*, pages 59–68, 1993.

R. Cole, B. Mishra, J. Schmidt, and A. Siegel. On the dynamic finger conjecture for splay trees. part 1: Splay sorting log n-block sequences. Technical Report 471, Courant Institute, NYU, 1989.

R. Cole and A. Raghunathan. Online algorithms for finger searching. In *Proc. 31st Symp. Foundations of Computer Science*, pages 480–489, 1990.

L. Colussi, Z. Galil, and R. Giancarlo. On the exact complexity of string matching. In *Proc. 31st Symp. Foundations of Computer Science*, pages 135–144, 1990.

D. Coppersmith, P. G. Doyle, P. Raghavan, and M. Snir. Random walks on weighted graphs and applications to on-line algorithms. *Journal of the ACM*, 40:421–453, 1993.

D. Coppersmith and P. Raghavan. Multidimensional online bin packing: Algorithms and worst case analysis. *Operations Research Letters*, 8:17–20, 1989.

T. Cormen, C. Leiserson, and R. Rivest. *Introduction to Algorithms.* McGraw-Hill, New York, NY, 1990.

C. Courcoubetis and R. R. Weber. Necessary and sufficient conditions for stability of a bin packing system. *Journal on Applied Probability*, 23:989–999, 1986.

C. Courcoubetis and R. R. Weber. Stability of online packing with random arrivals and long-run average constraints. *Prob. Eng. Inf. Sci.*, 4:447–460, 1990.

T. M. Cover. Universal portfolios. *Mathematical Finance*, 1:1–29, 1991.

T. M. Cover and D. H. Gluss. Empirical Bayes stock market portfolios. *Advances in Applied Mathematics*, 7:170–181, 1986.

T. M. Cover and E. Ordentlich. Universal portfolios with side information. *IEEE Transactions on Information Theory*, 42, 1996.

T. M. Cover and J. A. Thomas. *Elements of Information Theory.* John Wiley & Sons, Inc., 1991.

J. Cox and M. Rubinstein. *Options Markets.* Prentice Hall, Inc., 1985.

C. A. Crane. Linear lists and priority queues as balanced binary trees. Technical Report STAN-CS-72-259, Dept. of Computer Science, Stanford University, 1972.

J. Csirik. Bin-packing as a random walk: a note on Knoedels paper. *Operations Research Letters*, 5:161–163, 1986.

J. Csirik. An online algorithm for variable-sized bin packing. *Acta Informatica*, 26:697–709, 1989.

J. Csirik. The parametric behaviour of the First Fit decreasing bin-packing algorithm. *Journal of Algorithms*, 15:1–28, 1993.

J. Csirik and J. B. G. Frenk. A dual version of bin packing. *Algorithms Review*, 1:87–95, 1990.

J. Csirik, J. B. G. Frenk, A. M. Frieze, G. Galambos, and A. H. G. R. Kan. A probabilistic analysis of the Next Fit Decreasing bin-packing heuristic. *Operations Research Letters*, 5:233–236, 1986.

J. Csirik, J. B. G. Frenk, G. Galambos, and A. H. G. R. Kan. Probabilistic analysis of algorithms for dual bin packing problems. *Journal of Algorithms*, 12:189–203, 1991.

J. Csirik, J. B. G. Frenk, and M. Labbe. Two dimensional rectangle packing: on line methods and results. *Discrete Applied Mathematics*, 45:197–204, 1993.

J. Csirik and G. Galambos. An O(n) bin packing algorithm for uniformly distributed data. *Computing*, 36:313–319, 1986.

J. Csirik and G. Galambos. On the expected behaviour of the NF algorithm for a dual bin packing problem. *Acta Cybernetica*, 8:5–9, 1987.

J. Csirik, G. Galambos, and G. Turán. A lower bound on online algorithms for decreasing lists. In *Proc. of EURO VI*, 1984.

J. Csirik and B. Imreh. On the worst-case performance of the NkF bin packing heuristic. *Acta Cybernetica*, 9:89–105, 1989.

J. Csirik and D. S. Johnson. Bounded space on-line bin packing: best is better than first. In *Proc. 2nd Symp. on Discrete Algorithms*, pages 309–319, 1991.

J. Csirik and E. Máté. The probabilistic behaviour of the NFD bin-packing heuristic. *Acta Cybernetica*, 7:241–246, 1986.

J. Csirik and V. Totik. Online algorithms for a dual version of bin packing. *Discrete Applied Mathematics*, 21:163–167, 1988.

J. Csirik and A. van Vliet. An online algorithm for multidimensional bin packing. *Operations Research Letters*, 13:149–158, 1993.

J. Csirik and G. J. Woeginger. Shelf algorithms for on-line strip packing. *Information Processing Letters*, 63:171–175, 1997.

Y. Dai, H. Imai, K. Iwano, and N. Katoh. How to treat delete requests in semi-online problems. In *Proc. 4th International Symp. on Algorithms and Computation*, Lecture Notes in Computer Science, pages 48–57. Springer, 1993.

F. d'Amore, A. Marchetti-Spaccamela, and U. Nanni. Competitive algorithms for the weighted list update problem. In *Proc. 2nd Workshop on Algorithms and Data Structures*, Lecture Notes in Computer Science, pages 240–248. Springer, 1991.

P. Dasgupta, P. P. Chakrabarti, and S. C. DeSarkar. A near optimal algorithm for the extended cow-path problem in the presence of relative errors. In *Proc. 15th Conference Foundations of Software Technology and Theoretical Computer Science*, pages 22–36, 1995.

A. Datta, C. Hipke, and S. Schuierer. Competitive searching in polygons — beyond generalized streets. In *Proc. 6th International Symp. on Algorithms and Computation*, Lecture Notes in Computer Science, pages 32–41. Springer, 1995.

A. Datta and C. Icking. Competitive searching in a generalized street. In *Proc. 10th Symp. on Computational Geometry*, pages 175–182, 1994.

E. Davis and J. M. Jaffe. Algorithms for scheduling tasks on unrelated processors. *Journal of the ACM*, 28:721–736, 1981.

A. S. de Loma. New results on fair multi-threaded paging. In *Proc. 1st Argentine Workshop on Theoretical Informatics*, pages 111–122, 1997.

X. Deng and P. Dymond. On multiprocessor system scheduling. In *Proc. 8th Symp. on Parallel Algorithms and Architectures*, pages 82–88, 1996.

X. Deng, N. Gu, T. Brecht, and K. C. Lu. Preemptive scheduling of parallel jobs on multiprocessors. In *Proc. 7th Symp. on Discrete Algorithms*, pages 159–167, 1996.

X. Deng, T. Kameda, and C. Papadimitriou. How to learn an unknown environment. In *Proc. 32nd Symp. Foundations of Computer Science*, pages 298–303, 1991.

X. Deng and E. Koutsoupias. Competitive implementations of parallel programs. In *Proc. 4th Symp. on Discrete Algorithms*, pages 455–461, 1993.

X. Deng, E. Koutsoupias, and P. MacKenzie. Competitive implementation of parallel programs. To appear in Algorithmica, 1998.

X. Deng and S. Mahajan. Infinite games, randomization, computability, and applications to online problems. In *Proc. 23rd Symp. Theory of Computing*, pages 289–298, 1991.

X. Deng and S. Mahajan. Server problems and resistive spaces. *Information Processing Letters*, 37:193–196, 1991.

X. Deng and A. Mirzaian. Robot mapping: foot-prints vs tokens. In *Proc. 4th International Symp. on Algorithms and Computation*, Lecture Notes in Computer Science, pages 353–362. Springer, 1993.

X. Deng and C. Papadimitriou. Exploring an unknown graph. In *Proc. 31st Symp. Foundations of Computer Science*, pages 355–361, 1990.

X. Deng and C. H. Papadimitriou. Competitive distributed decision making. *Algorithmica*, 16:133–150, 1996.

P. J. Denning. Working sets past and present. *IEEE Transactions on Software Engineering*, 6:64–84, 1980.

C. Derman. On optimal replacement rules when changes of the state are Markovian. In R. Bellman, editor, *Mathematical Optimization Techniques.* University of California Press,, 1963.

M. Dertouzos and A. Mok. Multiprocessor on-line scheduling with release dates. *IEEE Transactions on Software Engineering*, 15:1497–1506, 1997.

A. DeSantis, G. Markowsky, and M. Wegman. Learning probabilistic prediction functions. In *Proc. 29th Symp. Foundations of Computer Science*, pages 110–119, 1988.

P. F. Dietz, J. I. Seiferas, and J. Zhang. A tight lower found for on-line monotonic list labeling. In *Proc. 4th Scandinavian Workshop on Algorithm Theory*, Lecture Notes in Computer Science, pages 131–142. Springer, 1994.

E. Dinitz. Maintaining the 4-edge-connected components of a graph on-line. In *Proc. 2nd Israel Symp. on Theory of Computing and Systems*, pages 88–97, 1993.

H. N. Djidjev, G. E. Pantziou, and C. D. Zaroliagis. On-line and dynamic algorithms for shortest path problems. Technical report, MPI Informatik Saarbrücken, 1994.

H. N. Djidjev, G. E. Pantziou, and C. D. Zaroliagis. On-line and dynamic shortest paths through graph decompositions. Technical report, MPI Informatik Saarbrücken, 1994.

F. Douglis, P. Krishnan, and B. Bershad. Adaptive disk spin-down policies for mobile computers. In *Proc. 2nd USENIX Symp. on Mobile and Location-Independent Computing*, 1995.

L. W. Dowdy and D. V. Foster. Comparative models of the file assignment problem. *ACM Computing Surveys*, 14:287–313, 1982.

D.-Z. Du and F. K. Hwang. Competitive group testing. In L. A. McGeoch and D. D. Sleator, editors, *On-line Algorithms*, volume 7 of *DIMACS Series in Discrete Mathematics and Theoretical Computer Science*, pages 125–134. AMS/ACM, 1991.

D.-Z. Du and H. Park. On competitive group testing. *SIAM Journal on Computing*, 23:1019–1025, 1994.

G. Dudek, K. Romanik, and S. Whitesides. Localizing a robot with minimum travel. In *Proc. 6th Symp. on Discrete Algorithms*, pages 437–446, 1995.

J. Dugundji and A. Granas. *Fixed point theory.* Polish Scientific Publishers, 1982.

J. Edmonds, D. Chinn, T. Brecht, and X. Deng. Non-clairvoyant multiprocessor scheduling of jobs with changing execution characteristics. In *Proc. 29th Symp. Theory of Computing*, pages 120–129, 1997.

S. J. Eggers and R. H. Katz. Evaluating the performance of four snooping cache coherency protocols. In *Proc. 16th International Symp. on Computer Architecture*, 1989.

R. El-Yaniv. On the decision theoretic foundations of the competitive ratio. unpublished manuscript, 1996.

R. El-Yaniv. There are infinitely many competitive-optimal online list accessing algorithms. Manuscript, 1996.

R. El-Yaniv, A. Fiat, R. M. Karp, and G. Turpin. Competitive analysis of financial games. In *Proc. 33rd Symp. Foundations of Computer Science*, pages 327–333, 1992.

R. El-Yaniv, R. Kaniel, and N. Linial. On the equipment rental problem. to be published, 1993.

R. El-Yaniv and R. M. Karp. The mortgage problem. In *Proc. 2nd Israel Symp. on Theory of Computing and Systems*, pages 304–312, 1993. Submitted to Mathematics of Operations Research.

R. El-Yaniv and R. M. Karp. Nearly optimal competitive online replacement policies. *Mathematics of Operations Research*, 22:814–839, 1997.

R. El-Yaniv and J. Kleinberg. Geometric two-server algorithms. *Information Processing Letters*, 53:355–358, 1995.

P. Elias. Universal codeword sets and the representation of the integers. *IEEE Transactions on Information Theory*, 21:194–203, 1975.

E. A. Emerson and C. S. Jutla. Tree automata, mu-calculus and determinacy. In *Proc. 32nd Symp. Foundations of Computer Science*, pages 368–377, 1991.

L. Epstein, J. Noga, S. S. Seiden, J. Sgall, and G. J. Woeginger. Randomized on-line scheduling for two related machines. Work in progress, 1997.

M. D. Ercegovac and T. Lang. On-line scheme for computing rotation factors. *Journal on Parallel and Distributed Computing*, 5:209–227, 1988.

V. Estivill-Castro. The design of competitive algorithms via genetic algorithms. In *Proc. 5th Int. Conference on Computing and Information*, pages 305–309, 1993.

V. Estivill-Castro and M. Sherk. Competitiveness and response time in on-line algorithms. In *Proc. 2nd International Symp. on Algorithms and Computation*, Lecture Notes in Computer Science, pages 284–293. Springer, 1991.

S. Even and B. Monien. On the number of rounds needed to disseminate information. In *Proc. 1st Symp. on Parallel Algorithms and Architectures*, 1989.

U. Faigle, R. Garbe, and W. Kern. Randomized online algorithms for maximizing busy time interval scheduling. *Computing*, 56:95–104, 1996.

U. Faigle, W. Kern, and G. Turán. On the performane of online algorithms for partition problems. *Acta Cybernetica*, 9:107–119, 1989.

U. Faigle and W. M. Nawijn. Note on scheduling intervals on-line. *Discrete Applied Mathematics*, 58:13–17, 1995.

M. Feder, N. Merhav, and M. Gutman. Universal prediction of individual sequences. *IEEE Transactions on Information Theory*, 38:1258–1270, 1992.

A. Feldmann, M.-Y. Kao, J. Sgall, and S.-H. Teng. Optimal online scheduling of parallel jobs with dependencies. In *Proc. 25th Symp. Theory of Computing*, pages 642–651. ACM, 1993.

A. Feldmann, B. Maggs, J. Sgall, D. Sleator, and A. Tomkins. Competitive analysis of call admission algorithms that allow delay. Technical Report CMU-CS-95-102, Carnegie-Mellon University, 1995.

A. Feldmann, J. Sgall, and S.-H. Teng. Dynamic scheduling on parallel machines. In *Proc. 32nd Symp. Foundations of Computer Science*, pages 111–120. IEEE, 1991.

A. Feldmann, J. Sgall, and S.-H. Teng. Dynamic scheduling on parallel machines. *Theoretical Computer Science*, 130:49–72, 1994.

S. Felsner. On-line chain partitions of orders. To appear in Theoretical Computer Science.

E. Feuerstein. *On-line paging of Structured Data and Multi-threaded Paging.* PhD thesis, University of Rome, 1995.

E. Feuerstein and A. S. de Loma. On multi-threaded paging. In *Proc. 7th International Symp. on Algorithms and Computation*, Lecture Notes in Computer Science, pages 417–426. Springer, 1996.

A. Fiat, D. Foster, H. Karloff, Y. Rabani, Y. Ravid, and S. Vishwanathan. Competitive algorithms for layered graph traversal. In *Proc. 32nd Symp. Foundations of Computer Science*, pages 288–297, 1991.

A. Fiat and A. Karlin. Randomized and multipointer paging with locality of reference. In *Proc. 27th Symp. Theory of Computing*, pages 626–634, 1995.

A. Fiat, R. Karp, M. Luby, L. A. McGeoch, D. Sleator, and N. E. Young. Competitive paging algorithms. *Journal of Algorithms*, 12:685–699, 1991.

A. Fiat and M. Mendel. Truly online paging with locality of reference. In *Proc. 38th Symp. Foundations of Computer Science*, pages 326–335, 1997.

A. Fiat, Y. Rabani, and Y. Ravid. Competitive k-server algorithms. *Journal Computer Systems Science*, 48:410–428, 1994.

A. Fiat, Y. Rabani, Y. Ravid, and B. Schieber. A deterministic $O(k^3)$-competitive k-server algorithm for the circle. *Algorithmica*, 11:572–578, 1994.

A. Fiat and M. Ricklin. Competitive algorithms for the weighted server problem. *Theoretical Computer Science*, 130:85–99, 1994.

A. Fiat and Z. Rosen. Experimental studies of access graph based heuristics: Beating the lru standard? In *Proc. 8th Symp. on Discrete Algorithms*, pages 63–72, 1997.

A. Fiat and G. J. Woeginger. On-line scheduling on a single machine: Minimizing the total completion time. Technical Report Woe-04, TU Graz, Austria, 1997.

M. J. Fischer and L. J. Stockmeyer. Fast on-line integer multiplication. In *Proc. 5th Symp. Theory of Computing*, pages 67–72, 1973.

D. C. Fisher. Next-Fit packs a list and its reverse into the same number of bins. *Operations Research Letters*, 7:291–293, 1988.

P. Flajolet, D. Gardy, and L. Thimonier. Birthday paradox, coupon collectors, caching algorithms and self-organizing search. *Discrete Applied Mathematics*, 39:207–229, 1992.

D. P. Foster and R. V. Vohra. Probabilistic analysis of a heuristic for the dual bin packing problem. *Information Processing Letters*, 31:287–290, 1989.

D. P. Foster and R. V. Vohra. A randomization rule for selecting forecasts. *Operations Research*, 41:704–709, 1993.

P. A. Franaszek and T. J. Wagner. Some distribution-free aspects of paging performance. *Journal of the ACM*, 21:31–39, 1974.

A. Frank. Packing paths, cuts, and circuits — a survey. In B. Korte, L. Lovász, H. Proemel, and A. Schrijver, editors, *Paths, Flows, and VLSI-Layout*, pages 49–100. Springer-Verlag, 1990.

G. N. Frederickson. Probabilistic analysis for simple one- and two-dimensional bin packing algorithms. *Information Processing Letters*, 11:156–161, 1980.

G. N. Frederickson. Self-organizing heuristics for implicit data structures. *SIAM Journal on Computing*, 13:277–291, 1984.

G. N. Frederickson. Data structures for on-line updating of minimum spanning trees with applications. *SIAM Journal on Computing*, 14:781–798, 1985.

G. N. Frederickson and M. A. Srinivas. On-line updating of solutions to a class of matroid intersection problems. *Information and Computation*, 74:113–139, 1987.

M. L. Fredman, D. S. Johnson, L. A. McGeoch, and G. Ostheimer. Data structures for traveling salesman. In *Proc. 5th Symp. on Discrete Algorithms*, pages 145–154. ACM, 1994.

P. R. Freeman. The secretary problem and its extensions. *International Statistical Review*, 51:189–206, 1983.

J. B. G. Frenk. On banach algebras, renewal measures and regenerative processes. Technical Report Tract No. 38, CWI, 1987.

Y. Freund. Predicting a binary sequence almost as well as the optimal biased coin. In *Proc. 9th Conf. on Computational Learning Theory*, pages 89–98, 1996.

Y. Freund and R. Schapire. Game theory, on-line prediction and boosting. In *Proc. 9th Conf. on Computational Learning Theory*, pages 325–332, 1996.

Y. Freund, R. Schapire, Y. Singer, and M. Warmuth. Using and combining predictors that specialize. In *Proc. 29th Symp. Theory of Computing*, pages 334–343, 1997.

J. Friedman and N. Linial. On convex body chasing. *Discrete Computational Geometry*, 9:293–321, 1993.

D. K. Friesen and M. A. Langston. A storage-size selection problem. *Information Processing Letters*, 18:295–296, 1984.

D. K. Friesen and M. A. Langston. Variable sized bin packing. *SIAM Journal on Computing*, 15:222–230, 1986.

H. Fukushima. Expected costs in some classes of binary search trees. *Journal of Information and Optimization Sciences*, 5:183–197, 1984.

H. Fukushima and M. Kimura. Self-organizing binary search tree with counting field. In *Proc. International Conference on Cybernetics and Society*, pages 1498–1502, New York, NY, USA, 1978. IEEE.

H. Fukushima and M. Kimura. Expected costs in some types of self-organizing binary search tree. *Record of Electrical and Communication Engineering Conversazione Tohoku University*, 52:18–26, 1983.

G. Gal. *Search Games.* Academic Press, 1980.

G. Galambos. Parametric lower bounds for online bin packing. *SIAM Journal on Algebraic and Discrete Methods*, 7:362–367, 1986.

G. Galambos. Notes on Lee's harmonic fit algorithm. *Annales Univ. Sci. Budapest., Sect. Comp.*, 9:121–126, 1988.

G. Galambos. A 1.6 lower bound for the two-dimensional online rectangle bin packing. *Acta Cybernetica*, 10:21–24, 1991.

G. Galambos and J. B. G. Frenk. A simple proof of Liang's lower bound for online bin packing and the extension to the parametric case. *Discrete Applied Mathematics*, 41:173–178, 1993.

G. Galambos, H. Kellerer, and G. J. Woeginger. A lower bound for online vector packing algorithms. *Acta Cybernetica*, 10:23–34, 1994.

G. Galambos and A. van Vliet. Lower bounds for 1-, 2-, and 3-dimensional online bin packing algorithms. *Computing*, 52:281–297, 1994.

G. Galambos and G. J. Woeginger. An on-line scheduling heuristic with better worst case ratio than graham's list scheduling. *SIAM Journal on Computing*, 22:349–355, 1993.

G. Galambos and G. J. Woeginger. Repacking helps in bounded space online bin packing. *Computing*, 49:329–338, 1993.

G. Galambos and G. J. Woeginger. Online bin packing - a restricted survey. *ZOR – Mathematical Methods of Operations Research*, 42:25–45, 1995.

Z. Galil and G. F. Italiano. Maintaining the 3-edge connected components of a graph online. *SIAM Journal on Computing*, 22:11–28, 1993.

R. G. Gallager. *Information Theory and Reliable Communication.* John Wiley and Sons, New York, 1968.

G. Gambosi, A. Postiglione, and M. Talamo. New algorithms for on-line bin packing. In *Proc. 1st Italian Conference on Algorithms and Complexity*, 1990.

J. A. Garay and I. S. Gopal. Call preemption in communications networks. In *Proc. INFOCOM '92*, pages 1043–1050, 1992.

J. A. Garay, I. S. Gopal, S. Kutten, Y. Mansour, and M. Yung. Efficient on-line call control algorithms. In *Proc. 2nd Israel Symp. on Theory of Computing and Systems*, pages 285–293, 1993.

M. R. Garey and R. L. Graham. Bounds on multiprocessor scheduling with resource constraints. *SIAM Journal on Computing*, 4:187–200, 1975.

M. R. Garey, R. L. Graham, D. S. Johnson, and A. C. C. Yao. Resource constrained scheduling as generalized bin packing. *Journal of Combinatorial Theory (Series A)*, 21:257–298, 1976.

M. R. Garey and D. S. Johnson. *Computers and Intractability: A Guide to the theory of of NP-Completeness.* Freeman and Company, San Francisco, 1979.

M. R. Garey and D. S. Johnson. Approximation algorithms for bin packing problems - a survey. In G. Ausiello and M. Lucertini, editors, *Analysis and Design of Algorithms in Combinatorial Optimization*, pages 147–172. Springer, New York, 1981.

R. Gawlick, A. Kamath, S. Plotkin, and K. Ramakrishnan. Routing and admission control in general topology networks. Technical Report STAN-CS-TR-95-1548, Stanford Technical Report, 1995.

J. Gergov. Approximation algorithms for dynamic storage allocation. In *Proc. 4th European Symp. on Algorithms*, volume 1136 of *Lecture Notes in Computer Science*, pages 52–61. Springer, 1996.

O. Gerstel and S. Kutten. Dynamic wavelength allocation in WDM ring networks. In *Proc. ICC '97*, 1997.

O. Gerstel, G. H. Sasaki, and R. Ramaswami. Dynamic channel assignment for WDM optical networks with little or no wavelength conversion. In *Proc. 34th Allerton Conference on Communication, Control, and Computing*, 1996.

A. Goel, M. R. Henzinger, and S. Plotkin. Online throughput-competitive algorithms for multicast routing and admission control. In *Proc. 9th Symp. on Discrete Algorithms*, pages 97–106, 1998.

M. X. Goemans. Improved approximation algorithms for scheduling with release dates. In *Proc. 8th Symp. on Discrete Algorithms*, pages 591–598, 1997.

B. L. Golden, L. Levy, and R. Vohra. The orienteering problem. *Naval Research Logistics*, 34:307–318, 1987.

M. J. Golin. Phd thesis. Technical Report CS-TR-266-90, Department of Computer Science, Princeton University, 1990.

G. H. Gonnet, J. I. Munro, and H. Suwanda. Towards self-organizing linear search. In *Proc. 20th Symp. Foundations of Computer Science*, pages 169–174, 1979.

T. F. Gonzales. The on-line d-dimensional dictionary problem. In *Proc. 3rd Symp. on Discrete Algorithms*, pages 376–385, 1992.

T. F. Gonzales and D. B. Johnson. A new algorithm for preemptive scheduling of trees. *Journal of the ACM*, 27:287–312, 1980.

R. Graham and P. Hell. On the history of the minimum spanning tree problem. *Annals of the History of Computing*, 7:43–57, 1985.

R. L. Graham. Bounds for certain multiprocessing anomalies. *Bell System Technical Journal*, 45:1563–1581, 1966.

R. L. Graham. Bounds on multiprocessing timing anomalies. *SIAM Journal on Applied Mathematics*, 17:263–269, 1969.

P. E. Green. *Fiber-Optic Communication Networks.* Prentice Hall, 1992.

D. Grinberg, S. Rajagopalan, R. Venkatesan, and V. K. Wei. Splay trees for data compression. In *Proc. 6th Symp. on Discrete Algorithms*, pages 522–530, 1995.

H. Groemer. Covering and packing properties of bounded sequences of convex sets. *Mathematika*, 29:18–31, 1982.

E. Grove. The harmonic k-server algorithm is competitive. In *Proc. 23rd Symp. Theory of Computing*, pages 260–266, 1991.

E. Grove, M.-Y. Kao, P. Krishnan, and J. Vitter. Online perfect matching and mobile computing. In *Proc. 4th Workshop on Algorithms and Data Structures*, volume 955 of *Lecture Notes in Computer Science*. Springer, 1995.

E. F. Grove. Online binpacking with lookahead. In *Proc. 6th Symp. on Discrete Algorithms*, pages 430–436, 1995.

L. J. Guibas, R. Motwani, and P. Raghavan. The robot localization problem in two dimensions. In *Proc. 3rd Symp. on Discrete Algorithms*, pages 259–268, 1992.

R. Gwangsoo. An approximation algorithm for the reversible shortest common superstring problem. *Journal of the Korea Information Science Society*, 21:1261–1268, 1994.

A. Gyárfás. *On Ramsey covering numbers*, pages 217–227. North-Holland/ American Elsevier, 1975.

A. Gyárfás. Problems from the world surrounding perfect graphs. *Zastowania Matematyki Applicationes Mathemacticae*, 19:413–441, 1985.

A. Gyárfás and J. Lehel. On-line and first-fit colorings of graphs. *Journal of Graph Theory*, 12:217–227, 1988.

A. Gyárfás and J. Lehel. Effective on-line coloring of P_5-free graphs. *Combinatorica*, 11:181–184, 1991.

A. Gyárfás, E. Szemerédi, and Z. Tuza. Induced subtrees in graphs of large chromatic number. *Discrete Mathematics*, 30:235–244, 1980.

S. Halfin. Next Fit bin packing with random piece sizes. *Journal on Applied Probability*, 26:503–511, 1989.

L. Hall, D. Shmoys, and J. Wein. Scheduling to minimize average completion time: Off-line and on-line algorithms. In *Proc. 7th Symp. on Discrete Algorithms*, pages 142–151, 1996.

L. A. Hall, A. S. Schulz, D. B. Shmoys, and J. Wein. Scheduling to minimize average completion time: Off-line and on-line approximation algorithms. *Mathematics of Operations Research*, 22:513–544, 1997.

M. Halldórsson. *Frugal Algorithms for the Independent Set and Graph Coloring Problems.* PhD thesis, Rutgers University, 1991.

M. M. Halldórsson. Parallel and on-line graph coloring algorithms. In *Proc. 3rd International Symp. on Algorithms and Computation*, Lecture Notes in Computer Science, pages 61–70. Springer, 1992.

M. M. Halldorsson and M. Szegedy. Lower bounds for on-line graph coloring. *Theoretical Computer Science*, 130:163–174, 1994.

S. Hart and A. Mas-Colell. Correlated learning. unpublished manuscript, 1996.

J. J. J. Hebrard. An algorithm for distinguishing efficiently bit-strings by their subsequences. *Theoretical Computer Science*, 82:35–49, 1991.

D. Helmbold, R. Sloan, and M. K. Warmuth. Learning nested differences of intersection closed concept classes. *Machine Learning*, 5:165–196, 1990.

W. J. Hendricks. An extension of a theorem concerning an intersting Markov chain. *Journal on Applied Probability*, 10:886–890, 1973.

W. J. Hendricks. An account of self-organizing systems. *SIAM Journal on Computing*, 5:715–723, 1976.

M. C. Herbordt, J. C. Corbett, C. C. Weems, and J. Spalding. Practical algorithms for online routing on fixed and reconfigurable meshes. *Journal on Parallel and Distributed Computing*, 20:341–356, 1994.

M. Herlihy. Wait-free synchronization. *ACM Trans. Prog. Lang. Syst.*, 13:124–149, 1991.

J. H. Hester and D. S. Hirschberg. Self-organizing linear search. *ACM Computing Surveys*, 17:295–312, 1985.

D. Hochbaum, editor. *Approximation Algorithms for NP-hard Problems.* PWS Publishing Company, 1997.

U. Hoffman. A class of simple stochastic online bin packing algorithms. *Computing*, 29:227–239, 1982.

F. Hoffmann, C. Icking, R. Klein, and K. Kriegel. A competitive strategy for learning a polygon. In *Proc. 8th Symp. on Discrete Algorithms*, pages 166–174, 1997.

M. Hofri. Two-dimensional packing: Expected performance of simple level algorithms. *Information and Control*, 45:1–17, 1980.

M. Hofri. A probabilistic analysis of the Next-Fit bin packing algorithm. *Journal of Algorithms*, 5:547–556, 1984.

M. Hofri. *Probabilistic analysis of algorithms.* Springer, New York, 1987.

M. Hofri and S. Kahmi. A stochastic analysis of the NFD bin packing algorithm. *Journal of Algorithms*, 7:489–509, 1986.

K. S. Hong and J. Y.-T. Leung. On-line scheduling of real-time tasks. *IEEE Transactions on Computing*, 41:1326–1331, 1992.

J. A. Hoogeveen and A. P. A. Vestjens. Optimal on-line algorithms for single-machine scheduling. In *Proc. 5th Conf. Integer Programming and Combinatorial Optimization*, pages 404–414, 1996.

R. R. Howell and M. K. Venkatrao. On non-preemptive scheduling of recurring tasks using inserted idle times. *Information and Computation*, 117:50–62, 1995.

S. O. Hoyland. Bin-packing in 1.5 dimensions. In *Proc. 1st Scandinavian Workshop on Algorithm Theory*, Lecture Notes in Computer Science. Springer, 1988.

X. D. Hu, P. D. Chen, and F. K. Hwang. A new competitive algorithm for the counterfeit coin problem. *Information Processing Letters*, 51:213–218, 1994.

L. C. K. Hui and C. Martel. Unsuccessful search in self-adjusting data structures. *Journal of Algorithms*, 15:477–481, 1993.

L. C. K. Hui and C. U. Martel. On efficient unsuccessful search. In *Proc. 3rd Symp. on Discrete Algorithms*, pages 217–227, 1992.

L. C. K. Hui and C. U. Martel. Randomized competitive algorithms for successful and unsuccessful search on self-adjusting linear lists. In *Proc. 4th International Symp. on Algorithms and Computation*, Lecture Notes in Computer Science, pages 426–435. Springer, 1993.

L. C. K. Hui and C. U. Martel. Analyzing deletions in competitive self-adjusting linear list algorithms. In *Proc. 5th International Symp. on Algorithms and Computation*, Lecture Notes in Computer Science, pages 433–441. Springer, 1994.

J. Hull. *Options, Futures, and other Derivative Securities.* Prentice Hall, Inc., 1993.

C. A. J. Hurkens, L. Lovasz, A. Schrijver, and E. Tardos. How to tidy up your set-sytem? *Coll. Math. Societatis Janos Bolyai*, 52:309–313, 1987.

O. H. Ibarra, T. Jiang, and H. Wang. Some results concerning 2-d online tessellation acceptors and 2-d alternating finite automata. In *Proc. 16th Symp. on Mathematical Foundations of Computer Science*, Lecture Notes in Computer Science, pages 221–230. Springer, 1991.

C. Icking and R. Klein. Searching for the kernel of a polygon- a competitive strategy. In *Proc. 11th Symp. on Computational Geometry*, pages 258–266, 1995.

C. Icking, R. Klein, and L. Ma. How to look around a corner. In *Proc. 5th Canadian Conference on Computational Geometry*, pages 443–448, 1993.

R. Idury and A. Schaffer. A better lower bound for on-line bottleneck matching. manuscript, 1992.

K. K. II and D. Wood. A note on some tree similarity measures. *Information Processing Letters*, 15:39–42, 1982.

M. Imase and B. M. Waxman. Dynamic steiner tree problem. *SIAM Journal on Discrete Mathematics*, 4:369–384, 1991.

K. Inoue, I. Takanami, and R. Vollmar. Alternating on-line turing machines with only universal states and small space bounds. *Theoretical Computer Science*, 41:331–339, 1985.

S. Irani. Two results on the list update problem. *Information Processing Letters*, 38:301–306, 1991.

S. Irani. Coloring inductive graphs on-line. *Algorithmica*, 11:53–72, 1994.

S. Irani. Corrected version of the split algorithm. Manscript, 1996.

S. Irani. Page replacement with multi-size pages and applications to web caching. In *Proc. 29th Symp. Theory of Computing*, pages 701–710, 1997.

S. Irani, A. R. Karlin, and S. Phillips. Strongly competitive algorithms for paging with locality of reference. *SIAM Journal on Computing*, 25:477–497, 1996.

S. Irani and V. Leung. Scheduling with conflicts, and applications to traffic signal control. In *Proc. 7th Symp. on Discrete Algorithms*, pages 85–94, 1996.

S. Irani and V. Leung. Probabilistic analysis for scheduling with conflicts. In *Proc. 8th Symp. on Discrete Algorithms*, pages 286–295, 1997.

S. Irani and Y. Rabani. On the value of coordination in distributed decision making. *SIAM Journal on Computing*, 25:498–519, 1996.

S. Irani and D. Ramanathan. The problem of renting versus buying. unpublished manuscript, 1994.

S. Irani, N. Reingold, D. Sleator, and J. Westbrook. Randomized competitive algorithms for the list update problem. In *Proc. 2nd Symp. on Discrete Algorithms*, pages 251–260, 1991.

S. Irani and R. Rubinfeld. A competitive 2-server algorithm. *Information Processing Letters*, 39:85–91, 1991.

S. Irani and S. Seiden. An almost optimal algorithm for uniform task systems. Unpublished manuscript, 1995.

S. Irani and S. S. Seiden. Randomized algorithms for metrical task systems. *Theoretical Computer Science*, 194:163–182, 1998.

A. Ito, K. Inoue, and I. Takanami. Deterministic two-dimensional on-line tessellation acceptors are equivalent to two-way two-dimensional alternating finite automata through 180 degrees-rotation. *Theoretical Computer Science*, 66:273–287, 1989.

Z. Ivkovic and E. Lloyd. Fully dynamic algorithms for bin packing: Being myopic helps. In *Proc. 1st European Symp. on Algorithms*, volume 726 of *Lecture Notes in Computer Science*, pages 224–235. Springer, 1993.

Z. Ivkovic and E. Lloyd. Partially dynamic bin packing can be solved within 1+eps in (amortized) polylogaritmic time. Technical report, Dept. Comp. and Inf. Sc., University of Delaware, Newark, 1994.

Z. Ivkovic and E. Lloyd. A fundamental restriction on fully dynamic maintenance of bin packing. *Information Processing Letters*, 59:229–232, 1996.

F. Jamshidian. Asymptotically optimal portfolios. *Mathematical Finance*, 2:131–150, 1992.

R. Janardan. On maintaining the width and diameter of a planar point-set online. In *Proc. 2nd International Symp. on Algorithms and Computation*, Lecture Notes in Computer Science, pages 137–149. Springer, 1991.

J. Januszewski and M. Lassak. Online covering by boxes and by covex bodies. *Bull. Pol. Academy Mathematics*, 42:69–76, 1994.

J. Januszewski and M. Lassak. Online covering the unit cube by cubes. *Discrete Computational Geometry*, 12:433–438, 1994.

J. Januszewski and M. Lassak. Efficient on-line covering of large cubes by convex bodies of at most unit diameters. *Bull. Pol. Acad. Math.*, 43:305–315, 1995.

J. Januszewski and M. Lassak. Online covering the unit square by squares and the three-dimensional unit cube by cubes. *Demonstratio Math.*, 28:143–149, 1995.

J. Januszewski and M. Lassak. Online packing the unit cube by cubes. manuscript, University of Bydgoszcz, Poland, 1996.

J. Januszewski, M. Lassak, G. Rote, and G. J. Woeginger. Online q-adic covering by the method of the n-th segment and its application to online covering by cubes. *Contributions to Algebra and Geometry*, 37:51–65, 1996.

S. R. Jernigan, S. Ramaswamy, and K. S. Barber. On-line scheduling using a distributed simulation technique for intelligent manufacturing systems. In *Proc. International Conference on Systems, Man and Cybernetics. Intelligent Systems for the 21st Century*, pages 2159–2164, 1995.

J. Jeuring. The derivation of on-line algorithms, with an application to finding palindromes. *Algorithmica*, 11:146–184, 1994.

T. Jiang, O. H. Ibarra, and H. Wang. Some results concerning 2-d on-line tessellation acceptors and 2-d alternating finite automata. *Theoretical Computer Science*, 125:243–257, 1994.

P. Johansen. On-line string matching with feedback. *Theoretical Computer Science*, 141:53–67, 1995.

D. S. Johnson. *Near-optimal bin packing algorithms.* PhD thesis, MIT, Cambridge, MA, 1973.

D. S. Johnson. Fast algorithms for bin packing. *Journal Computer Systems Science*, 8:272–314, 1974.

D. S. Johnson, A. Demers, J. D. Ullman, M. R. Garey, and R. L. Graham. Worst-case performance bounds for simple one-dimensional packing algorithms. *SIAM Journal on Computing*, 3:256–278, 1974.

D. W. Jones. Application of splay trees to data compression. *Communications of the ACM*, 31:996–1007, 1988.

G. S. Jr. and F. W. Dauer. A probabilistic model of a self-organizing file system. *SIAM Journal on Applied Mathematics*, 15:874–888, 1967.

B. Kalyanasundaram and K. Pruhs. Visual searching and mapping. In L. A. McGeoch and D. D. Sleator, editors, *On-line Algorithms*, volume 7 of *DIMACS Series in Discrete Mathematics and Theoretical Computer Science*, pages 157–162. AMS/ACM, 1991.

B. Kalyanasundaram and K. Pruhs. A competitive analysis of nearest neighbor based algorithms for searching unknown scenes. In *Proc. 24th Symp. Theory of Computing*, pages 147–157, 1992.

B. Kalyanasundaram and K. Pruhs. A competitive analysis of algorithms for searching unknown scenes. *Computational Geometry: Theory and Applications*, 3:139–155, 1993.

B. Kalyanasundaram and K. Pruhs. Online weighted matching. *Journal of Algorithms*, 14:478–488, 1993.

B. Kalyanasundaram and K. Pruhs. Constructing competitive tours from local information. *Theoretical Computer Science*, 130:125–138, 1994.

B. Kalyanasundaram and K. Pruhs. Fault-tolerant scheduling. In *Proc. 26th Symp. Theory of Computing*, pages 115–124, 1994.

B. Kalyanasundaram and K. Pruhs. The online transportation problem. In *Proc. 3rd European Symp. on Algorithms*, volume 979 of *Lecture Notes in Computer Science*, pages 484–493. Springer, 1995.

B. Kalyanasundaram and K. Pruhs. Speed is as powerful as clairvoyance. In *Proc. 36th Symp. Foundations of Computer Science*, pages 214–221, 1995.

B. Kalyanasundaram and K. Pruhs. An optimal deterministic algorithm for online b-matching. manuscript, 1996.

B. Kalyanasundaram and K. R. Pruhs. Fault-tolerant real-time scheduling. In *Proc. 5th European Symp. on Algorithms*, volume 1284 of *Lecture Notes in Computer Science*, pages 296–307. Springer, 1997.

A. Kamath, O. Palmon, and S. Plotkin. Routing and admission control in general topology networks with poisson arrivals. In *Proc. 7th Symp. on Discrete Algorithms*, pages 269–278, 1996.

Y. C. Kan and S. M. Ross. Optimal list orders under partial memory constraints. *Journal on Applied Probability*, 17:1004–1015, 1980.

A. Kanevsky, R. Tamassia, G. D. Battista, and J. Chen. On-line maintenance of the four-connected components of a graph. In *Proc. 32nd Symp. Foundations of Computer Science*, pages 793–801, 1991.

M.-Y. Kao, Y. Ma, M. Sipser, and Y. Yin. Optimal constructions of hybrid algorithms. In *Proc. 5th Symp. on Discrete Algorithms*, pages 372–381, 1994.

M.-Y. Kao, J. H. Reif, and S. R. Tate. Searching in an unknown environment: an optimal randomized algorithm for the cow-path problem. *Information and Computation*, 131:63–79, 1996.

M.-Y. Kao and S. R. Tate. Online matching with blocked input. *Information Processing Letters*, 38:113–116, 1991.

M.-Y. Kao and S. R. Tate. On-line difference maximization. In *Proc. 8th Symp. on Discrete Algorithms*, pages 175–182, 1997.

D. Karger, C. Stein, and J. Wein. Scheduling algorithms. In M. J. Atallah, editor, *Handbook of Algorithms and Theory of Computation*. CRC Press, 1997.

D. R. Karger, S. J. Phillips, and E. Torng. A better algorithm for an ancient scheduling problem. *Journal of Algorithms*, 20:400–430, 1996.

A. Karlin, K. Li, M. Manasse, and S. Owicki. Empirical studies of competitive spinning for shared memory multiprocessors. In *Proc. 13th ACM Symp. on Operating Systems Principles*, 1991.

A. Karlin, M. Manasse, L. McGeoch, and S. Owicki. Randomized competitive algorithms for nonuniform problems. *Algorithmica*, 11, 1994.

A. Karlin, M. Manasse, L. Rudolph, and D. Sleator. Competitive snoopy caching. *Algorithmica*, 3:79–119, 1988.

A. Karlin, S. Phillips, and P. Raghavan. Markov paging. In *Proc. 33rd Symp. Foundations of Computer Science*, pages 208–217, 1992.

H. Karloff, Y. Rabani, and Y. Ravid. Lower bounds for randomized k-server and motion-planning algorithms. *SIAM Journal on Computing*, 23:293–312, 1994.

N. Karmarkar. Probabilistic analysis of some bin packing algorithms. In *Proc. 23rd Symp. Foundations of Computer Science*, pages 107–118, 1982.

R. M. Karp, M. Luby, and A. Marchetti-Spaccamela. Probabilistic analysis of multi-dimensional binpacking problems. In *Proc. 16th Symp. Theory of Computing*, pages 289–298, 1984.

R. M. Karp, R. Ostrovski, and Y. Rabani, 1993. Personal communication.

R. M. Karp, U. V. Vazirani, and V. V. Vazirani. An optimal algorithm for on-line bipartite matching. In *Proc. 22nd Symp. Theory of Computing*, pages 352–358, Baltimore, Maryland, 1990.

M. Kearns. Efficient noise-tolerant learning from statistical queries. In *Proc. 25th Symp. Theory of Computing*, pages 392–401, 1993.

M. Kearns, M. Li, L. Pitt, and L. Valiant. On the learnability of boolean formulae. In *Proc. 19th Symp. Theory of Computing*, pages 285–295, New York, New York, 1987. To appear in Journal of the ACM.

M. Kearns, R. Schapire, and L. Sellie. Toward efficient agnostic learning. *Machine Learning*, 17:115–142, 1994.

H. Kellerer, V. Kotov, M. G. Speranza, and Z. Tuza. Semi on-line algorithms for the partition problem. To appear.

C. Kenyon. Best-fit bin-packing with random order. In *Proc. 7th Symp. on Discrete Algorithms*, pages 359–364, 1996.

C. Kenyon, Y. Rabani, and A. Sinclair. Biased random walks, lyapunov functions, and stochastic analysis of best fit bin packing. In *Proc. 7th Symp. on Discrete Algorithms*, pages 351–358, 1996.

S. Keshav, C. Lund, S. Phillips, N. Reingold, and H. Saran. An empirical evaluation of virtual circuit holding time policies in ip-over-atm networks. Technical Report Technical Memorandum 39199-11, AT&T Bell Laboratories, 1994.

S. Khuller, S. G. Mitchell, and V. V. Vazirani. On-line algorithms for weighted bipartite matching and stable marriages. *Theoretical Computer Science*, 127:255–267, 1994.

H. A. Kierstead. On-line coloring k-colorable graphs. To appear in *Israel J. of Math.*

H. A. Kierstead. An effective version of Dilworth's theorem. *Transactions of the AMS*, 268:63–77, 1981.

H. A. Kierstead. Recursive colorings of highly recursive graphs. *Canadian Journal of Mathematics*, 33:1279–1290, 1981.

H. A. Kierstead. An effective version of Hall's theorem. *Proceedings of the AMS*, 88:124–128, 1983.

H. A. Kierstead. The linearity of first-fit coloring of interval graphs. *SIAM Journal on Discrete Mathematics*, 1:526–530, 1988.

H. A. Kierstead. A polynomial time approximation algorithm for dynamic storage allocation. *Discrete Applied Mathematics*, 88:231–237, 1991.

H. A. Kierstead and K. Kolossa. On-line coloring of perfect graphs. To appear in Combinatorica.

H. A. Kierstead and S. G. Penrice. Recent results on a conjecture of Gyárfás. *Congressus Numerantium*, 79:182–186, 1990.

H. A. Kierstead and S. G. Penrice. Radius two trees specify χ-bounded classes. *Journal of Graph Theory*, 18:119–129, 1994.

H. A. Kierstead, S. G. Penrice, and W. T. Trotter. On-line graph coloring and recursive graph theory. *SIAM Journal on Discrete Mathematics*, 7:72–89, 1994.

H. A. Kierstead, S. G. Penrice, and W. T. Trotter. First-fit and on-line coloring of graphs which do not induce P_5. *SIAM Journal on Discrete Mathematics*, 8:485–498, 1995.

H. A. Kierstead and J. Qin. Coloring interval graphs with First-Fit. *SIAM Journal on Discrete Mathematics*, 8:47–57, 1995.

H. A. Kierstead and W. T. Trotter. An extremal problem in recursive combinatorics. *Congressus Numerantium*, 33:143–153, 1981.

H. A. Kierstead and W. T. Trotter. On-line graph coloring. In L. A. McGeoch and D. D. Sleator, editors, *On-line Algorithms*, volume 7 of *DIMACS Series in Discrete Mathematics and Theoretical Computer Science*, pages 85–92. AMS/ACM, 1991.

H. A. Kierstead and W. T. Trotter. Planar graph coloring with an uncooperative partner. In W. T. Trotter, editor, *Planar Graphs*, pages 85–93. Amer. Math. Soc. series in Discrete Mathematics and Theoretical Computer Science, vol. 9, 1993.

T. Kilburn, D. Edwards, M. Lanigan, and F. Sumner. One-level storage system. *IRE Transactions Elect. Computers*, 37:223–235, 1962.

D. Kimber and P. M. Long. On-line learning of smooth functions of a single variable. *Theoretical Computer Science*, 148:141–156, 1995.

N. G. Kinnersley and M. A. Langston. Online variable-sized bin packing. *Discrete Applied Mathematics*, 22:143–148, 1988.

R. Klein. Walking an unknown street with bounded detour. *Computational Geometry: Theory and Applications*, 1:325–351, 1992.

J. Kleinberg and R. Rubenfield. Short paths in expander graphs. In *Proc. 37th Symp. Foundations of Computer Science*, pages 86–95, 1996.

J. Kleinberg and E. Tardos. Disjoint paths in densely embedded graphs. In *Proc. 36th Symp. Foundations of Computer Science*, pages 52–61, 1995.

J. M. Kleinberg. The localization problem for mobile robots. In *Proc. 35th Symp. Foundations of Computer Science*, pages 521–531, 1994.

J. M. Kleinberg. A lower bound for two-server balancing algorithms. *Information Processing Letters*, 52:39–43, 1994.

J. M. Kleinberg. On-line search in a simple polygon. In *Proc. 5th Symp. on Discrete Algorithms*, pages 8–15, 1994.

W. F. Klostermeyer. Optimizing searching with self-adjusting trees. *Journal of Information and Optimization Sciences*, 13:85–95, 1992.

W. F. Klostermeyer. On-line scheduling with optimal worst-case response time. *Journal of Information and Optimization Sciences*, 16:449–60, 1995.

W. Knoedel. A bin packing algorithm with complexity o(nlogn) and performance 1 in the stochastic limit. In *Proc. 10th Symp. on Mathematical Foundations of Computer Science*, Lecture Notes in Computer Science, pages 369–378. Springer, 1981.

W. Knoedel. Ueber das mittlere Verhalten von online Packungsalgorithmen. *EIK*, 19:427–433, 1983.

D. E. Knuth. Optimum binary search trees. *Acta Informatica*, pages 14–25, 1971.

H. Koga. Randomized on-line algorithms for the page replication problem. In *Proc. 4th International Symp. on Algorithms and Computation*, Lecture Notes in Computer Science, pages 436–445. Springer, 1993.

G. Koren and D. Shasha. Moca: a multiprocessor on-line competitive algorithm for real-time system scheduling. *Theoretical Computer Science*, 128:75–97, 1994.

G. Koren and D. Shasha. d^{over}: an optimal on-line scheduling algorithm for overloaded uniprocessor real-time systems. *SIAM Journal on Computing*, 24:318–39, 1995.

G. Koren, D. Shasha, and S.-C. Huang. Moca: A multiprocessor on-line competitive algorithm for real-time system scheduling. In *Proc. Real-Time Systems Symp.*, pages 172–181, 1993.

P. Kornerup and D. W. Matula. An on-line arithmetic unit for bit-pipelined rational arithmetic. *Journal on Parallel and Distributed Computing*, 5:310–330, 1988.

L. T. Kou and G. Markowsky. Multidimensional bin packing algorithms. *IBM Journal on Research and Development*, 21:443–448, 1977.

E. Koutsoupias and C. Papadimitriou. Beyond competitive analysis. In *Proc. 35th Symp. Foundations of Computer Science*, pages 394–400, 1994.

E. Koutsoupias and C. Papadimitriou. On the k-server conjecture. *Journal of the ACM*, 42:971–983, 1995.

E. Koutsoupias and C. Papadimitriou. The 2-evader problem. *Information Processing Letters*, 57:249–252, 1996.

K. Krause, L. L. Larmore, and D. J. Volper. Packing items from a triangular distribution. *Information Processing Letters*, 25:351–361, 1987.

P. Krishnan and J. S. Vitter. Optimal prediction for prefetching in the worst case. In *Proc. 5th Symp. on Discrete Algorithms*, pages 392–401, 1994.

H. Kuhn. Extensive games and the problem of information. In H. Kuhn and A. Tucker, editors, *Contributions to the Theory of Games*, pages 193–216. Princeton University Press, 1953.

W. Kuperberg. Online covering a cube by a sequence of cubes. *Discrete Computational Geometry*, 12:83–90, 1994.

J. Labetoulle, E. L. Lawler, J. K. Lenstra, and A. H. G. R. Kan. Preemptive scheduling of uniform machines subject to release dates. In W. R. Pulleyblank, editor, *Progress in Combinatorial Optimization*, pages 245–261. Academic Press, 1984.

K. Lam, M. K. Sui, and C. T. Yu. A generalized counter scheme. *Theoretical Computer Science*, 16:271–278, 1981.

L. L. Larmore and B. Schieber. On-line dynamic programming with applications to the prediction of RNA secondary structure. *Journal of Algorithms*, 12:490–515, 1991.

M. Lassak. Online covering a box by cubes. *Contributions to Algebra and Geometry*, 36:1–7, 1995.

M. Lassak and J. Zhang. An on-line potato-sack theorem. *Discrete Computational Geometry*, 6:1–7, 1991.

C. C. Lee and D. T. Lee. A simple online bin packing algorithm. *Journal of the ACM*, 32:562–572, 1985.

C. C. Lee and D. T. Lee. Robust online bin packing algorithms. Technical Report 83-03-FC-02, Department of Electrical Engineering and Computer Science, Northwestern University, Evanston, IL, 1987.

D. Lee and M. Yannakakis. Online minimization of transition systems. In *Proc. 24th Symp. Theory of Computing*, pages 264–274, 1992.

S. K. Lee. On-line multiprocessor scheduling algorithms for real-time tasks. In *Proc. Region 10's 9th International Conference. Theme: Frontiers of Computer Technology*, pages 607–611, 1994.

F. T. Leighton, B. M. Maggs, A. G. Ranade, and S. B. Rao. Randomized routing and sorting on fixed-connection networks. *Journal of Algorithms*, 17:157–205, 1994.

F. T. Leighton and P. Shor. Tight bounds for minimax grid matching with applications to the average case analysis of algorithms. *Combinatorica*, 9:161–187, 1989.

T. Leighton, F. Makedon, S. Plotkin, C. Stein, E. Tardos, and S. Tragoudas. Fast approximation algorithms for multicommodity flow problems. In *Proc. 23rd Symp. Theory of Computing*, pages 101–111, 1991.

T. Leighton and E. J. Schwabe. Efficient algorithms for dynamic allocation of distributed memory. In *Proc. 32nd Symp. Foundations of Computer Science*, pages 470–479, 1991.

S. Leonardi and A. Marchetti-Spaccamela. On-line resource management with applications to routing and scheduling. In *Proc. 22nd International Colloquium on Automata, Languages, and Programming*, volume 944 of *Lecture Notes in Computer Science*, pages 303–314. Springer, 1995.

S. Leonardi, A. Marchetti-Spaccamela, A. Presciutti, and A. Rosèn. On-line randomized call-control revisited. In *Proc. 9th Symp. on Discrete Algorithms*, pages 323–332, 1998.

S. Leonardi and D. Raz. Approximating total flow time on parallel machines. In *Proc. 29th Symp. Theory of Computing*, pages 110–119, 1997.

D. Levhari and L. J. Mirman. Saving and uncertainty with an uncertain horizon. *Journal of Political Economy*, 85:265–281, 1977.

K. Li and K. H. Cheng. Generalized First-Fit algorithms in two and three dimensions. *International Journal on Foundations of Computer Science*, 2:131–150, 1990.

K. Li and K. H. Cheng. Heuristic algorithms for online packing in three dimensions. *Journal of Algorithms*, 13:589–605, 1992.

R. Li and L. Shi. An on-line algorithm for some uniform processor scheduling. In *Proc. 1st International Conference on Computing and Combinatorics*, pages 627–632, 1995.

T. Li, W. Zhao, and N. Carter. A VLSI architecture for on-line scheduler in real-time systems. In *Proc. International Conference Computing and Information*, pages 201–206, 1989.

F. M. Liang. A lower bound for online bin packing. *Information Processing Letters*, 10:76–79, 1980.

A. M. Liao. Three priority queue applications revisited. *Algorithmica*, 7:415–427, 1992.

S. A. Lippman and J. J. McCall. The economics of job search: A survey. *Economic Inquiry*, XIV:155–189, 1976.

S. A. Lippman and J. J. McCall. The economics of uncertainty: Selected topics and probabilistic methods. In K. J. Arrow and M. D. Intriligator, editors, *Handbook of Mathematical Economics*, volume 1, chapter 6, pages 211–284. North-Holland, 1981.

R. J. Lipton and A. Tomkins. Online interval scheduling. In *Proc. 5th Symp. on Discrete Algorithms*, pages 302–311, 1994.

N. Littlestone. Learning quickly when irrelevant attributes abound: A new linear-threshold algorithm. *Machine Learning*, 2:285–318, 1988.

N. Littlestone. A mistake-bound version of Rivest's decision-list algorithm. Personal communication, 1989.

N. Littlestone. Redundant noisy attributes, attribute errors, and linear-threshold learning using winnow. In *Proc. 4th Conf. on Computational Learning Theory*, pages 147–156, Santa Cruz, California, 1991. Morgan Kaufmann.

N. Littlestone, P. M. Long, and M. K. Warmuth. On-line learning of linear functions. In *Proc. 23rd Symp. Theory of Computing*, pages 465–475. ACM Press, New York, NY, 1991.

N. Littlestone and M. K. Warmuth. The weighted majority algorithm. *Information and Computation*, 108:212–261, 1994.

D. D. E. Long and M. N. Thakur. Scheduling real-time disk transfers for continuous media applications. In *Proc. 12th Symp. on Mass Storage Systems. Putting all that Data to Work*, pages 227–232, 1993.

A. Lopez-Ortiz and S. Schuierer. Going home through an unknown street. In *Proc. 4th Workshop on Algorithms and Data Structures*, volume 955 of *Lecture Notes in Computer Science*, pages 135–146. Springer, 1995.

M. C. Loui and D. R. Luginbuhl. Optimal online simulations of tree machines by random access machines. *SIAM Journal on Computing*, 21:959–971, 1992.

L. Lovász, M. Saks, and W. T. Trotter. An on-line graph coloring algorithm with sublinear performance ratio. *Discrete Mathematics*, 75:319–325, 1989.

T. Lu, P.-F. Jia, and B. Zhang. On-line opportunistic scheduling in multirobot assembly system. In *Proc. 2nd Asian Conference on Robotics and Its Applications*, pages 608–611, 1994.

M. Luby, J. Naor, and A. Orda. Tight bounds for dynamic storage allocation. In *Proc. 5th Symp. on Discrete Algorithms*, pages 724–732, 1994.

J. M. Lucas. The rotation graph of binary trees is hamiltonian. *Journal of Algorithms*, 8:503–535, 1987.

J. M. Lucas. Arbitrary splitting in splay trees. Technical Report DCS-TR-234, Rutgers University, 1988.

J. M. Lucas. Canonical forms for competitive binary search tree algorithms. Technical Report DCS-TR-250, Rutgers University, 1988.

J. M. Lucas. On the competitiveness of splay trees: Relations to the union-find problem. In L. A. McGeoch and D. D. Sleator, editors, *On-line Algorithms*, volume 7 of *DIMACS Series in Discrete Mathematics and Theoretical Computer Science*, pages 95–124. AMS/ACM, 1991.

F. Luccio and L. Pagli. On the upper bound on the rotation distance of binary trees. *Information Processing Letters*, 31:57–60, 1989.

G. S. Lueker. An average-case analysis of bin packing with uniformly distributed item sizes. Technical Report 181, Dept. Information and Computer Science, University of California at Irvine, 1982.

G. S. Lueker. Bin-packing with items uniformly distributed over intervals [a,b]. In *Proc. 24th Symp. Foundations of Computer Science*, pages 289–297, 1983.

V. J. Lumelsky and A. A. Stepanov. Dynamic path planning for a mobile automaton with limited information on the environment. *IEEE Transactions on Automation and Control*, AC-31:1058–1063, 1986.

V. J. Lumelsky and A. A. Stepanov. Path-planning strategies for a point mobile automaton moving amidst unknown obstacles of arbitrary shape. *Algorithmica*, 2:403–430, 1987.

C. Lund, S. Phillips, and N. Reingold. Ip over connection-oriented networks and distributional paging. In *Proc. 35th Symp. Foundations of Computer Science*, pages 424–435, 1994.

C. Lund and N. Reingold. Linear programs for randomized on-line algorithms. In *Proc. 5th Symp. on Discrete Algorithms*, pages 382–391, 1994.

C. Lund, N. Reingold, J. Westbrook, and D. Yan. On-line distributed data management. In *Proc. 2nd European Symp. on Algorithms*, volume 855 of *Lecture Notes in Computer Science*, pages 202–214. Springer, 1994.

Y. Ma and S. Plotkin. Improved lower bounds for load balancing of tasks with unknown duration. *Information Processing Letters*, 62:31–34, 1997.

E. Makinen. On top-down splaying. *BIT*, 27:330–339, 1987.

E. Makinen. On the rotation distance of binary trees. *Information Processing Letters*, 26:271–272, 1988.

M. Manasse, L. A. McGeoch, and D. Sleator. Competitive algorithms for server problems. *Journal of Algorithms*, 11:208–230, 1990.

A. B. Manaster and J. G. Rosenstein. Effective matchmaking (recursion theoretic aspects of a theorem of Philip Hall). *Proceedings of the London Mathematics Society*, 25:615–654, 1972.

A. B. Manaster and J. G. Rosenstein. Effective matchmaking and k-chromatic graphs. *Proceedings of the AMS*, 39:371–378, 1973.

U. Manber and G. Myers. Suffix arrays: a new method for on-line string searches. *SIAM Journal on Computing*, 22:935–948, 1993.

Y. Mansour and B. Schieber. The intractability of bounded protocols for online sequence transmission over nonfifo channels. *Journal of the ACM*, 39:783–99, 1992.

W. Mao. Best-k-Fit bin packing. *Computing*, 50:265–270, 1993.

W. Mao. Tight worst-case performance bounds for Next-k-Fit bin packing. *SIAM Journal on Computing*, 22:46–56, 1993.

W. Mao, R. K. Kincaid, and A. Rifkin. On-line algorithms for a single machine scheduling problem. In *Impact of Emerging Technologies on Computer Science and Operations Research*, pages 157–173, 1995.

M. V. Marathe, H. B. H. III, and S. S. Ravi. Efficient approximation algorithms for domatic partition and on-line coloring of circular arc graphs. In *Proc. 5th International Conference on Computing and Information*, pages 26–30, 1993.

A. Marchetti-Spaccamela and C. Vercellis. Efficient on-line algorithms for the knapsack problem. In *Proc. 14th International Colloquium on Automata, Languages, and Programming*, volume 267 of *Lecture Notes in Computer Science*, pages 445–456. Springer, 1987.

A. Marchetti-Spaccamela and C. Vercellis. Stochastic on-line knapsack problems. *Mathematical Programming*, 68:73–104, 1995.

C. Martel. Self-adjusting multi-way search trees. *Information Processing Letters*, 38:135–141, 1991.

J. Matousek. On-line computation of convolutions. *Information Processing Letters*, 32:15–16, 1989.

T. Matsumoto. Competitive analysis of the Round Robin algorithm. In *Proc. 3rd International Symp. on Algorithms and Computation*, Lecture Notes in Computer Science, pages 71–77. Springer, 1992.

E. W. Mayr and R. Werchner. Optimal routing of parentheses on the hypercube. *Journal on Parallel and Distributed Computing*, 26:181–192, 1995.

J. McCabe. On serial files with relocatable records. *Operations Research*, 12:609–618, 1965.

L. McGeoch and D. Sleator. A strongly competitive randomized paging algorithm. *Journal of Algorithms*, 6:816–825, 1991.

L. A. McGeoch and D. D. Sleator, editors. *On-line Algorithms*, volume 7 of *DIMACS Series in Discrete Mathematics and Theoretical Computer Science*. AMS/ACM, February 1991.

N. Megiddo, S. L. Hakimi, M. R. Garey, D. S. Johnson, and C. H. Papadimitriou. The complexity of searching a graph. *Journal of the ACM*, 35:18–44, 1988.

K. Mehlhorn. Nearly optimal binary search trees. *Acta Informatica*, 5:287–295, 1975.

K. Mehlhorn. Dynamic binary search. *SIAM Journal on Computing*, 8:175–198, 1979.

A. Mei and Y. Igarashi. Efficient strategies for robot navigation in unknown environment. *Information Processing Letters*, 52:51–56, 1994.

A. Meir and L. Moser. On packing of squares and cubes. *Journal of Combinatorial Theory*, 5:126–134, 1968.

A. A. Melkman. On-line construction of the convex hull of a simple polyline. *Information Processing Letters*, 25:11–12, 1987.

N. Merhav and M. Feder. Universal sequential learning and decisions from individual data sequences. In *Proc. 5th Conf. on Computational Learning Theory*, pages 413–427. ACM Press, New York, NY, 1992.

R. C. Merton. On the microeconomic theory of investment under uncertainty. In K. J. Arrow and M. D. Intriligator, editors, *Handbook of Mathematical Economics*, volume 2, chapter 13, pages 601–669. North-Holland, 1981.

M. Mihail, C. Kaklamanis, and S. Rao. Efficient access to optical bandwidth. In *Proc. 36th Symp. Foundations of Computer Science*, pages 548–557, 1995.

L. J. Mirman. Uncertainty and optimal consumption decisions. *Econometrica*, 39:179–185, 1971.

A. Moffat, G. Eddy, and O. Petersson. Splaysort: fast, versatile, practical. *Australian Computer Science Communications*, 15:595–604, 1993.

A. Moffat and G. Port. A fast algorithm for melding splay trees. In *Proc. 1st Workshop on Algorithms and Data Structures*, Lecture Notes in Computer Science, pages 450–459. Springer, 1989.

R. Motwani, S. Phillips, and E. Torng. Non-clairvoyant scheduling. *Theoretical Computer Science*, 130:17–47, 1994.

R. Motwani and P. Raghavan. *Randomized Algorithms*. Cambridge University Press, 1997.

R. Motwani, V. Saraswat, and E. Torng. Online scheduling with lookahead: multipass assembly lines. manuscript, 1994.

F. D. Murgolo. An efficient approximation scheme for variable-sized bin packing. *SIAM Journal on Computing*, 16:149–161, 1987.

J. V. Neumann and O. Morgenstern. *Theory of games and economic behavior*. Princeton University Press, 1st edition, 1944.

H. L. Ong, M. J. Magazine, and T. S. Wee. Probabilistic analysis of bin packing heuristics. *Operations Research*, 5:983–998, 1984.

B. J. Oommen, S. S. Iyengar, S. V. Rao, and R. L. Kashyap. Robot navigation in unknown terrains using learned visibility graphs. Part I: The disjoint convex obstacle case. Technical Report SCS-TR-86, School of Computer Science, Carleton University, 1986.

E. Ordentlich and T. M. Cover. On-line portfolio selection. In *COLT 96*, pages 310–313, 1996.

G. Oriolo, M. Vendittelli, and G. Ulivi. On-line map building and navigation for autonomous mobile robots. In *Proc. 1995 International Conference on Robotics and Automation*, pages 2900–2906, 1995.

R. Ostrovsky and Y. Rabani. Universal $o(congenstion+dilation+\log^{1+\epsilon} n)$ local contro l packet switching algorithms. In *Proc. 29th Symp. Theory of Computing*, pages 644–653, 1997.

C. H. Papadimitriou and M. Yannakakis. Shortest paths without a map. *Theoretical Computer Science*, 84:127–150, 1991.

C. H. Papadimitriou and M. Yannakakis. Linear programming without the matrix. In *Proc. 25th Symp. Theory of Computing*, pages 121–129, 1993.

C. V. Papadopoulos. A dynamic algorithm for online scheduling of parallel processes. In *Proc. 6th European Conference on Parallel Architectures and Languages*, pages 601–610, 1994.

B. Patt-Shamir and S. Rajsbaum. A theory of clock synchronization. In *Proc. 26th Symp. Theory of Computing*, pages 810–819, 1994.

A. Pedrotti. Analysis of a list-update strategy. *Information Processing Letters*, 52:115–21, 1994.

S. G. Penrice. On-line algroithms for ordered sets and comparability graphs. *Discrete Applied Mathematics*, 60:319–329, 1995.

G. C. Pflug. On-line optimization of simulated markovian processes. *Mathematics of Operations Research*, 15:381–395, 1990.

C. Phillips, C. Stein, E. Torng, and J. Wein. Optimal time-critical scheduling via resource augmentation. In *Proc. 29th Symp. Theory of Computing*, pages 140–149, 1997.

S. Phillips and J. Westbrook. On-line load balancing and network flow. In *Proc. 25th Symp. Theory of Computing*, pages 402–411, 1993.

S. J. Phillips. *Theory and Applications of Online Algorithms.* PhD thesis, Department of Computer Science, Stanford University, 1993.

S. Plotkin. Competitive routing in atm networks. *IEEE Transactions on Communication*, pages 1128–1136, 1995.

S. Plotkin and Y. Ma. An improved lower bound for load balancing of tasks with unknown duration. Manuscript, 1998.

H. Prodinger. Einige Bemerkungen zu einer Arbeit von W. Knoedel ueber das mittlere Verhalten von online Packungsalgorithmen. *EIK*, 21:3–7, 1985.

K. R. Pruhs. Average case scalable on-line algorithms for fault replacement. *Information Processing Letters*, 52:131–136, 1994.

S. Przybylski. *Cache and Memory Hierarchy Design: A Performance-Directed Approach.* Morgan Kaufmann, 1990.

S. Przybylski, S. Horowitz, and M. Henessy. Performance tradeoffs in cache design. In *Proc. 15th International Symp. on Computer Architecture*, pages 290–298, 1988.

Y. Rabani and E. Tardos. Distributed packet switching in arbitrary networks. In *Proc. 28th Symp. Theory of Computing*, pages 366–367, 1996.

P. Raghavan. A statistical adversary for on-line algorithms. DIMACS *Series in Discrete Mathematics and Theoretical Computer Science*, 7:79–83, 1992.

P. Raghavan and M. Snir. Memory versus randomization in on-line algorithms. *IBM Journal on Research and Development*, 38:683–707, 1994.

P. Raghavan and E. Upfal. Efficient routing in all-optical networks. In *Proc. 26th Symp. Theory of Computing*, pages 133–143, 1994.

G. Ramalingam and T. Reps. On competitive on-line algorithms for the dynamic priority-ordering problem. *Information Processing Letters*, 51:155–161, 1994.

P. Ramanan. Average case analysis of the Smart Next Fit algorithm. *Information Processing Letters*, 31:221–225, 1989.

P. Ramanan, D. J. Brown, C. C. Lee, and D. T. Lee. Online bin packing in linear time. *Journal of Algorithms*, 10:305–326, 1989.

P. Ramanan and K. Tsuga. Average case analysis of the modified harmonic algorithm. *Algorithmica*, 4:519–533, 1989.

H. Ramesh. On traversing layered graphs on-line. *Journal of Algorithms*, 18:480–512, 1995.

S. Ramos-Thuel and J. P. Lehoczky. On-line scheduling of hard deadline aperiodic tasks in fixed-priority systems. In *Proc. Real-Time Systems Symp.*, pages 160–171, 1993.

S. Rao, P. Sadayappan, F. Hwang, and P. Shor. The rectilinear steiner arborescence problem. *Algorithmica*, pages 277–288, 1992.

N. Reingold and J. Westbrook. Optimum off-line algorithms for the list update problem. Technical Report YALEU/DCS/TR-805, Yale University, 1990.

N. Reingold, J. Westbrook, and D. D. Sleator. Randomized competitive algorithms for the list update problem. *Algorithmica*, 11:15–32, 1994.

C. Rey and R. Ward. On determining the on-line minimax linear fit to a discrete point set in the plane. *Information Processing Letters*, 24:97–101, 1987.

D. J. Reyniers. Coordinated search for an object hidden on the line. *European Journal of Operational Research*, 95:663–670, 1996.

W. T. Rhee and M. Talagrand. The complete convergence of best fit decreasing. *SIAM Journal on Computing*, 18:909–918, 1989.

W. T. Rhee and M. Talagrand. The complete convergence of first fit decreasing. *SIAM Journal on Computing*, 18:919–938, 1989.

W. T. Rhee and M. Talagrand. Dual bin packing of items of random size. *Mathematical Programming*, 58:229–242, 1993.

W. T. Rhee and M. Talagrand. On line bin packing with items of random size. *Mathematics of Operations Research*, 18:438–445, 1993.

W. T. Rhee and M. Talagrand. On line bin packing with items of random sizes — ii. *SIAM Journal on Computing*, 22:1251–1256, 1993.

M. B. Richey. Improved bounds for harmonic-based bin packing algorithms. *Discrete Applied Mathematics*, 34:203–227, 1991.

R. Rivest. On self-organizing sequential search heuristics. *Communications of the ACM*, 19:63–67, 1976.

R. L. Rivest. Learning decision lists. *Machine Learning*, 2:229–246, 1987.

H. Robbins. Asymptotically subminimax solutions of compound statistical decision problems. In *Proc. 2nd Berkeley Symp. Math. Statist. Prob.*, pages 131–148, 1951.

J. M. Robson. An estimate of the store size necessary for dynamic storage allocation. *Journal of the ACM*, 18:416–423, 1971.

J. M. Robson. Bounds for some functions concerning dynamic storage allocation. *Journal of the ACM*, 21:491–499, 1974.

J. M. Robson. Worst case fragmentation of first-fit and best-fit storage allocation strategies. *The Computer Journal*, 20:242–244, 1977.

D. B. Rosenfield and R. D. Shapiro. Optimal price search with Bayesian extensions. *Journal of Economic Theory*, 1981.

D. Rosenkrantz, R. Stearns, and P. Lewis. An analysis of several heuristics for the traveling salesman problem. *SIAM Journal on Computing*, 6:563–581, 1977.

M. Rothschild. Searching for the lowest price when the distribution of prices is unknown. *Journal of Political Economy*, 82:689–711, 1974.

S. Sahni and Y. Cho. Nearly on line scheduling of a uniform processor system with release times. *SIAM Journal on Computing*, 8:275–285, 1979.

M. Saks, N. Shavit, and H. Woll. Optimal time randomized consensus — making resilient algorithms fast in practice. In *Proc. 2nd Symp. on Discrete Algorithms*, pages 351–362, 1991.

M. E. Saks and M. Werman. On computing majority by comparisons. *Combinatorica*, 11:383–387, 1991.

L. A. Savchenko. General self-organizing sequential search procedure. *Programming and Computer Software*, 7:282–285, 1981.

H. E. Scarf. A survey of analytic techniques in inventory theory. In H. E. Scarf, D. M. Gilford, and M. W. Shelly, editors, *Multistage Inventory Models and Techniques*, chapter 7, pages 185–225. Stanford University Press, 1963.

C. Scheurich and M. Dubois. Dynamic page migration in multiprocessors with distributed global memory. *IEEE Transactions on Computing*, 38:1154–1163, 1989.

J. H. Schmerl. Recursive colorings of graphs. *Canadian Journal of Mathematics*, 32:821–830, 1980.

B. Schoemakers. Systematic analysis of splaying. *Information Processing Letters*, 45:41–50, 1993.

A. S. Schulz and M. Skutella. Scheduling-LPs bear probabilities. In *Proc. 5th European Symp. on Algorithms*, volume 1284 of *Lecture Notes in Computer Science*, pages 416–429. Springer, 1997.

C. Schwarz and M. Smid. An O(n logn loglogn) algorithm for the on-line closest pair problem. In *Proc. 3rd Symp. on Discrete Algorithms*, pages 280–285. ACM-SIAM, 1992.

C. Schwarz, M. Smid, and J. Snoeyink. An optimal algorithm for the on-line closest-pair problem. *Algorithmica*, 12:18–29, 1994.

S. Seiden. Unfair problems and randomized algorithms for metrical task systems. Unpublished manuscript, 1996.

S. Seiden. Randomized algorithms for that ancient scheduling problem. In *Proc. 5th Workshop on Algorithms and Data Structures*, volume 1272 of *Lecture Notes in Computer Science*, pages 210–223. Springer, 1997.

S. S. Seiden. Randomized online scheduling with delivery times. Submitted.

S. S. Seiden. More multiprocessor scheduling with rejection. Technical Report Woe-16, TU Graz, Austria, 1997.

S. S. Seiden. Randomized preemptive online interval scheduling. Technical Report Woe-08, TU Graz, Austria, 1997.

J. Sgall. On-line scheduling of parallel jobs. In *Proc. 19th Symp. on Mathematical Foundations of Computer Science*, Lecture Notes in Computer Science, pages 159–176. Springer, 1994.

J. Sgall. On-line scheduling on parallel machines. Technical Report Technical Report CMU-CS-94-144, Carnegie-Mellon University, Pittsburgh, PA, USA, 1994.

J. Sgall. Randomized on-line scheduling of parallel jobs. *Journal of Algorithms*, 21:149–175, 1996.

J. Sgall. A lower bound for randomized on-line multiprocessor scheduling. *Information Processing Letters*, 63:51–55, 1997.

S. D. Shapiro. Performance of heuristic bin packing algorithms with segments of random length. *Information and Control*, 35:146–148, 1977.

G. Shedler and C. Tung. Locality in page reference strings. *SIAM Journal on Computing*, 1:218–241, 1972.

C. N. Shen and G. Nagy. Autonomous navigation to provide long distance surface traverses for mars rover sample return mission problems. In *Proc. IEEE Symp. on Intelligent Control*, pages 362–367, 1989.

M. Sherk. Self-adjusting k-ary search trees. *Journal of Algorithms*, 19:25–44, 1995.

E. Sheshinski and Y. Weiss. Optimum pricing policy under stochastic inflation. In E. Sheshinski and Y. Weiss, editors, *Optimal Pricing, Inflation and the Cost of Price Adjustment*. The MIT Press, 1993.

W.-K. Shih and J. W. S. Liu. On-line scheduling of imprecise computations to minimize error. In *Real-Time Systems Symp.*, pages 280–289, 1992.

A. G. Shilling. Market timing: Better than a buy-and-hold strategy. *Financial Analysts Journal*, pages 46–50, 1992.

D. B. Shmoys, J. Wein, and D. P. Williamson. Scheduling parallel machines on-line. In L. A. McGeoch and D. D. Sleator, editors, *On-line Algorithms*, volume 7 of *DIMACS Series in Discrete Mathematics and Theoretical Computer Science*, pages 163–166. AMS/ACM, 1991.

D. B. Shmoys, J. Wein, and D. P. Williamson. Scheduling parallel machines on-line. *SIAM Journal on Computing*, 24:1313–1331, 1995.

P. W. Shor. The average-case analysis of some online algorithms for bin packing. *Combinatorica*, 6:179–200, 1986.

P. W. Shor. How to pack better than Best-Fit: Tight bounds for average-case on-line bin packing. In *Proc. 32nd Symp. Foundations of Computer Science*, pages 752–759, 1991.

J. Shtarkov. Universal sequential coding of single measures. *Problems of Information Transmission*, pages 175–185, 1987.

K. Sikkel. Parallel on-line parsing in constant time per word. *Theoretical Computer Science*, 120:303–310, 1993.

H. J. Sips and H. X. Lin. A new model for on-line arithmetic with an application to the reciprocal calculation. *Journal on Parallel and Distributed Computing*, 8:218–230, 1990.

D. Sleator and R. E. Tarjan. Amortized efficiency of list update and paging rules. *Communications of the ACM*, 28:202–208, 1985.

D. D. Sleator and R. E. Tarjan. Self-adjusting binary search trees. *Journal of the ACM*, 32:652–686, 1985.

D. D. Sleator, R. E. Tarjan, and W. P.Thurston. Rotation distance, triangulations, and hyperbolic geometry. In *Proc. 18th Symp. Theory of Computing*, pages 122–135, 1986.

D. D. K. D. B. Sleator. A 2.5 times optimal algorithm for packing in two dimensions. *Information Processing Letters*, 10:37–40, 1976.

M. Ślusarek. A coloring algorithm for interval graphs. In *Proc. 14th Symp. on Mathematical Foundations of Computer Science*, volume 379 of *Lecture Notes in Computer Science*, pages 471–480. Springer, 1989.

M. Ślusarek. Optimal on-line coloring of circular arc graphs. *RAIRO Journal on Information Theory and Applications*, 29:423–429, 1995.

M. Smid. Dynamic rectangular point location, with an application to the closest pair problem. *Information and Computation*, 116:1–9, 1995.

A. J. Smith. Cache memories. *ACM Computing Surveys*, 14:6–27, 1982.

A. J. Smith. Cache evaluation and the impact of workload choice. In *Proc. 12th International Symp. on Computer Architecture*, pages 64–73, 1985.

A. J. Smith. Bibliography and readings on CPU cache memories and related topics. *Computer Architecture News*, 14:22–42, 1986.

J. R. Spirn. *Program Behavior: Models and Measurements.* Elsevier Computer Science Library, 1977.

A. Steinberg. A strip-packing algorithm with absolute performance bound 2. *SIAM Journal on Computing*, 26:401–409, 1997.

N. L. Stokey and J. R. E. Lucas. *Recursive Methods in Economic Dynamics*, chapter 13, pages 389–413. Harvard University Press, 1989.

L. Stougie and A. P. A. Vestjens. Randomized on-line scheduling: How low can't you go? Manuscript, 1997.

D. P. Sumner. *Subtrees of a graph and chromatic number*, pages 557–576. John Wiley & Sons, 1981.

R. Sundar. Twists, turns, cascades, deque conjecture, and scanning theorem. In *Proc. 30th Symp. Foundations of Computer Science*, pages 555–559, 1989.

R. Sundar. Twists, turns, cascades, deque conjecture, and scanning theorem. Technical Report 427, Courant Institute, New York University, 1989.

T. Takaoka. An on-line pattern matching algorithm. *Information Processing Letters*, 22:329–330, 1986.

R. E. Tarjan. *Data Structures and Network Algorithms.* SIAM, Philadelphia, 1983.

R. E. Tarjan. Amortized computational complexity. *SIAM Journal on Algebraic and Discrete Methods*, 6:306–318, 1985.

R. E. Tarjan. Sequential access in splay trees takes linear time. *Combinatorica*, 5:367–378, 1985.

B. Teia. A lower bound for randomized list update algorithms. *Information Processing Letters*, 47:5–9, 1993.

A. Tenenbaum. Simulations of dynamic sequential search algorithms. *Communications of the ACM*, 21:790–79, 1978.

A. M. Tenenbaum and R. M. Nemes. Two spectra of self-organizing sequential search. *SIAM Journal on Computing*, 11:557–566, 1982.

P. Tetali. Design of on-line algorithms using hitting times. In *Proc. 5th Symp. on Discrete Algorithms*, pages 402–410, 1994.

E. Torng. A unified analysis of paging and caching. In *Proc. 36th Symp. Foundations of Computer Science*, pages 194–203, 1995.

Y. Tsai, C. Tang, and Y. Chen. Average performance of a greedy algorithm for the on-line minimum matching problem on Euclidean space. *Information Processing Letters*, 51:275–282, 1994.

Y. T. Tsai and C. Y. Tang. The competitiveness of randomized algorithms for on-line steiner tree and on-line spanning tree problems. *Information Processing Letters*, 48:177–182, 1993.

G. Turan. On the greedy algorithm for an edge partitioning problem. *Coll. Math. Societatis Janos Bolyai*, 44:405–423, 1984.

J. S. Turner. Approximation algorithms for the shortest common superstring problem. *Information and Computation*, 83:1–20, 1989.

E. Ukkonen. On-line construction of suffix trees. *Algorithmica*, 14:249–260, 1995.

L. G. Valiant. A theory of the learnable. *Communications of the ACM*, 27:1134–1142, 1984.

J. van Leeuwen, editor. *Handbook of theoretical computer science, Vol A, Algorithms and Complexity.* The MIT Press, 1990.

A. van Vliet. An improved lower bound for online bin packing algorithms. *Information Processing Letters*, 43:277–284, 1992.

A. van Vliet. *Lower and upper bounds for online bin packing and scheduling heuristics.* PhD thesis, Erasmus University, Rotterdam, The Netherlands, 1995.

S. S. Venkatesh. Directed drift: a new linear threshold algorithm for learning binary weights online. *Journal Computer Systems Science*, 46:198–217, 1993.

A. P. A. Vestjens. Scheduling uniform machines on-line requires nondecreasing speed ratios. Technical Report Memorandum COSOR 94-35, Eindhoven University of Technology, 1994.

A. P. A. Vestjens. *On-line Machine Scheduling.* PhD thesis, Eindhoven University of Technology, The Netherlands, 1997.

S. Vishwanathan. Randomized on-line graph coloring. *Journal of Algorithms*, 13:657–669, 1992.

J. S. Vitter. Two papers on dynamic huffman codes. Technical Report CS-85-13, Brown University Computer Science, Providence. R.I., 1986.

A. V. Vliet. On the asymptotic worst case behaviour of harmonic fit. *Journal of Algorithms*, 20:113–136, 1996.

V. Vovk. Aggregating strategies. In *Proc. 3rd Conf. on Computational Learning Theory*, pages 371–383. Morgan Kaufmann, 1990.

V. G. Vovk. A game of prediction with expert advice. In *Proc. 8th Conf. on Computational Learning Theory*, pages 51–60, 1995.

Q. Wang and K. H. Cheng. A heuristic of scheduling parallel tasks and its analysis. *SIAM Journal on Computing*, 21:281–294, 1992.

J. Westbrook. Randomized algorithms for multiprocessor page migration. *SIAM Journal on Computing*, 23:951–965, 1994.

J. Westbrook. Load balancing for response time. In *Proc. 3rd European Symp. on Algorithms*, volume 979 of *Lecture Notes in Computer Science*, pages 355–368. Springer, 1995.

J. Westbrook and R. E. Tarjan. Maintaining bridge-connected and biconnected components on-line. *Algorithmica*, 7:433–464, 1992.

J. Westbrook and D. Yan. Linear bounds for on-line steiner problems. *Information Processing Letters*, 55:59–63, 1995.

J. Westbrook and D. C. K. Yan. Greedy algorithms for the on-line steiner tree and generalized steiner problems. In *Proc. 3rd Workshop on Algorithms and Data Structures*, Lecture Notes in Computer Science, pages 621–633. Springer, 1993.

J. Westbrook and D. C. K. Yan. Lazy and greedy: On-line algorithms for steiner problems. In *Proc. 3rd Workshop on Algorithms and Data Structures*, Lecture Notes in Computer Science. Springer, 1993.

R. Wilber. Lower bounds for accessing binary search trees with rotations. *SIAM Journal on Computing*, 18:56–67, 1989.

G. Wilfong. On-line algorithms for compressing planar curves. In *Proc. 8th Symp. on Discrete Algorithms*, pages 158–165, 1997.

H. S. Witsenhausen. On Woodall's interval problem. *Journal of Combinatorial Theory (Series A)*, 21:222–229, 1976.

I. H. Witten and T. Bell. The Calgary/Canterbury text compression corpus. Anonymous ftp from ftp.cpsc.ucalgary.ca /pub/text.compression/corpus/ text.compression.corpus.tar.Z.

G. J. Woeginger. Improved space for bounded-space online bin packing. *SIAM Journal on Discrete Mathematics*, 6:575–581, 1993.

G. J. Woeginger. On-line scheduling of jobs with fixed start and end times. *Theoretical Computer Science*, 130:5–16, 1994.

O. Wolfson. A distributed algorithm for adaptive replication data. Technical Report CUCS-057-90, Department of Computer Science, Columbia University, 1990.

D. R. Woodall. Problem no. 4, combinatorics. In *Proc. British Combinatorial Conference*. Cambridge University Press, 1973.

D. R. Woodall. The bay restaurant - a linear storage problem. *American Mathematical Monthly*, 81:240–246, 1974.

J. Xu. Online multiversion database concurrency control. *Acta Informatica*, 29:121–160, 1992.

K. Yamanishi. A loss bound model for on-line stochastic prediction algorithms. *Information and Computation*, 119:39–54, 1995.

A. C. C. Yao. New algorithms for bin packing. *Journal of the ACM*, 27:207–227, 1980.

A. C. C. Yao. Probabilistic computations: Towards a unified measure of complexity. In *Proc. 11th Symp. Theory of Computing*, 1980.

S.-M. Yoo and H. Y. Youn. An on-line scheduling and allocation scheme for real-time tasks in 2d meshes. In *Proc. 7th Symp. on Parallel and Distributed Processing*, pages 630–637, 1995.

N. E. Young. *Competitive paging and dual-guided algorithms for weighted caching and matching.* PhD thesis, Princeton University, 1991. Tech. Report CS-TR-348-91.

N. E. Young. On-line caching as cache size varies. In *Proc. 2nd Symp. on Discrete Algorithms*, pages 241–250, 1991.

N. E. Young. The k-server dual and loose competitiveness for paging. *Algorithmica*, 11:525–541, 1994.

N. E. Young. Bounding the diffuse adversary. In *Proc. 9th Symp. on Discrete Algorithms*, pages 420–425, 1998.

N. E. Young. Online file caching. In *Proc. 9th Symp. on Discrete Algorithms*, pages 82–86, 1998.

T. Yuba and M. Miyakawa. Performance studies of optimum sequence trees and self-organizing sequence trees. *Transactions of the Institute of Electronics and Communication Engineers of Japan*, E63(8):629–630, 1980.

G. Zhang. A tight worst-case performance bound for AFB-k. Technical Report 015, Institute of Applied Mathematics, Academia Sinica, Beijing, China, 1994.

G. Zhang. Worst-case analysis of the FFH algorithm for online variable-sized bin packing. *Computing*, 56:165–172, 1996.

Lecture Notes in Computer Science

For information about Vols. 1–1386

please contact your bookseller or Springer-Verlag

Vol. 1387: C. Lee Giles, M. Gori (Eds.), Adaptive Processing of Sequences and Data Structures. Proceedings, 1997. XII, 434 pages. 1998. (Subseries LNAI).

Vol. 1388: J. Rolim (Ed.), Parallel and Distributed Processing. Proceedings, 1998. XVII, 1168 pages. 1998.

Vol. 1389: K. Tombre, A.K. Chhabra (Eds.), Graphics Recognition. Proceedings, 1997. XII, 421 pages. 1998.

Vol. 1390: C. Scheideler, Universal Routing Strategies for Interconnection Networks. XVII, 234 pages. 1998.

Vol. 1391: W. Banzhaf, R. Poli, M. Schoenauer, T.C. Fogarty (Eds.), Genetic Programming. Proceedings, 1998. X, 232 pages. 1998.

Vol. 1392: A. Barth, M. Breu, A. Endres, A. de Kemp (Eds.), Digital Libraries in Computer Science: The MeDoc Approach. VIII, 239 pages. 1998.

Vol. 1393: D. Bert (Ed.), B'98: Recent Advances in the Development and Use of the B Method. Proceedings, 1998. VIII, 313 pages. 1998.

Vol. 1394: X. Wu. R. Kotagiri, K.B. Korb (Eds.), Research and Development in Knowledge Discovery and Data Mining. Proceedings, 1998. XVI, 424 pages. 1998. (Subseries LNAI).

Vol. 1395: H. Kitano (Ed.), RoboCup-97: Robot Soccer World Cup I. XIV, 520 pages. 1998. (Subseries LNAI).

Vol. 1396: E. Okamoto, G. Davida, M. Mambo (Eds.), Information Security. Proceedings, 1997. XII, 357 pages. 1998.

Vol. 1397: H. de Swart (Ed.), Automated Reasoning with Analytic Tableaux and Related Methods. Proceedings, 1998. X, 325 pages. 1998. (Subseries LNAI).

Vol. 1398: C. Nédellec, C. Rouveirol (Eds.), Machine Learning: ECML-98. Proceedings, 1998. XII, 420 pages. 1998. (Subseries LNAI).

Vol. 1399: O. Etzion, S. Jajodia, S. Sripada (Eds.), Temporal Databases: Research and Practice. X, 429 pages. 1998.

Vol. 1400: M. Lenz, B. Bartsch-Spörl, H.-D. Burkhard, S. Wess (Eds.), Case-Based Reasoning Technology. XVIII, 405 pages. 1998. (Subseries LNAI).

Vol. 1401: P. Sloot, M. Bubak, B. Hertzberger (Eds.), High-Performance Computing and Networking. Proceedings, 1998. XX, 1309 pages. 1998.

Vol. 1402: W. Lamersdorf, M. Merz (Eds.), Trends in Distributed Systems for Electronic Commerce. Proceedings, 1998. XII, 255 pages. 1998.

Vol. 1403: K. Nyberg (Ed.), Advances in Cryptology – EUROCRYPT '98. Proceedings, 1998. X, 607 pages. 1998.

Vol. 1404: C. Freksa, C. Habel. K.F. Wender (Eds.), Spatial Cognition. VIII, 491 pages. 1998. (Subseries LNAI).

Vol. 1405: S.M. Embury, N.J. Fiddian, W.A. Gray, A.C. Jones (Eds.), Advances in Databases. Proceedings, 1998. XII, 183 pages. 1998.

Vol. 1406: H. Burkhardt, B. Neumann (Eds.), Computer Vision – ECCV'98. Vol. I. Proceedings, 1998. XVI, 927 pages. 1998.

Vol. 1407: H. Burkhardt, B. Neumann (Eds.), Computer Vision – ECCV'98. Vol. II. Proceedings, 1998. XVI, 881 pages. 1998.

Vol. 1409: T. Schaub, The Automation of Reasoning with Incomplete Information. XI, 159 pages. 1998. (Subseries LNAI).

Vol. 1411: L. Asplund (Ed.), Reliable Software Technologies – Ada-Europe. Proceedings, 1998. XI, 297 pages. 1998.

Vol. 1412: R.E. Bixby, E.A. Boyd, R.Z. Ríos-Mercado (Eds.), Integer Programming and Combinatorial Optimization. Proceedings, 1998. IX, 437 pages. 1998.

Vol. 1413: B. Pernici, C. Thanos (Eds.), Advanced Information Systems Engineering. Proceedings, 1998. X, 423 pages. 1998.

Vol. 1414: M. Nielsen, W. Thomas (Eds.), Computer Science Logic. Selected Papers, 1997. VIII, 511 pages. 1998.

Vol. 1415: J. Mira, A.P. del Pobil, M.Ali (Eds.), Methodology and Tools in Knowledge-Based Systems. Vol. I. Proceedings, 1998. XXIV, 887 pages. 1998. (Subseries LNAI).

Vol. 1416: A.P. del Pobil, J. Mira, M.Ali (Eds.), Tasks and Methods in Applied Artificial Intelligence. Vol.II. Proceedings, 1998. XXIII, 943 pages. 1998. (Subseries LNAI).

Vol. 1417: S. Yalamanchili, J. Duato (Eds.), Parallel Computer Routing and Communication. Proceedings, 1997. XII, 309 pages. 1998.

Vol. 1418: R. Mercer, E. Neufeld (Eds.), Advances in Artificial Intelligence. Proceedings, 1998. XII, 467 pages. 1998. (Subseries LNAI).

Vol. 1419: G. Vigna (Ed.), Mobile Agents and Security. XII, 257 pages. 1998.

Vol. 1420: J. Desel, M. Silva (Eds.), Application and Theory of Petri Nets 1998. Proceedings, 1998. VIII, 385 pages. 1998.

Vol. 1421: C. Kirchner, H. Kirchner (Eds.), Automated Deduction – CADE-15. Proceedings, 1998. XIV, 443 pages. 1998. (Subseries LNAI).

Vol. 1422: J. Jeuring (Ed.), Mathematics of Program Construction. Proceedings, 1998. X, 383 pages. 1998.

Vol. 1423: J.P. Buhler (Ed.), Algorithmic Number Theory. Proceedings, 1998. X, 640 pages. 1998.

Vol. 1424: L. Polkowski, A. Skowron (Eds.), Rough Sets and Current Trends in Computing. Proceedings, 1998. XIII, 626 pages. 1998. (Subseries LNAI).

Vol. 1425: D. Hutchison, R. Schäfer (Eds.), Multimedia Applications, Services and Techniques – ECMAST'98. Proceedings, 1998. XVI, 532 pages. 1998.

Vol. 1427: A.J. Hu, M.Y. Vardi (Eds.), Computer Aided Verification. Proceedings, 1998. IX, 552 pages. 1998.

Vol. 1429: F. van der Linden (Ed.), Development and Evolution of Software Architectures for Product Families. Proceedings, 1998. IX, 258 pages. 1998.

Vol. 1430: S. Trigila, A. Mullery, M. Campolargo, H. Vanderstraeten, M. Mampaey (Eds.), Intelligence in Services and Networks: Technology for Ubiquitous Telecom Services. Proceedings, 1998. XII, 550 pages. 1998.

Vol. 1431: H. Imai, Y. Zheng (Eds.), Public Key Cryptography. Proceedings, 1998. XI, 263 pages. 1998.

Vol. 1432: S. Arnborg, L. Ivansson (Eds.), Algorithm Theory – SWAT '98. Proceedings, 1998. IX, 347 pages. 1998.

Vol. 1433: V. Honavar, G. Slutzki (Eds.), Grammatical Inference. Proceedings, 1998. X, 271 pages. 1998. (Subseries LNAI).

Vol. 1434: J.-C. Heudin (Ed.), Virtual Worlds. Proceedings, 1998. XII, 412 pages. 1998. (Subseries LNAI).

Vol. 1435: M. Klusch, G. Weiß (Eds.), Cooperative Information Agents II. Proceedings, 1998. IX, 307 pages. 1998. (Subseries LNAI).

Vol. 1436: D. Wood, S. Yu (Eds.), Automata Implementation. Proceedings, 1997. VIII, 253 pages. 1998.

Vol. 1437: S. Albayrak, F.J. Garijo (Eds.), Intelligent Agents for Telecommunication Applications. Proceedings, 1998. XII, 251 pages. 1998. (Subseries LNAI).

Vol. 1438: C. Boyd, E. Dawson (Eds.), Information Security and Privacy. Proceedings, 1998. XI, 423 pages. 1998.

Vol. 1439: B. Magnusson (Ed.), System Configuration Management. Proceedings, 1998. X, 207 pages. 1998.

Vol. 1441: W. Wobcke, M. Pagnucco, C. Zhang (Eds.), Agents and Multi-Agent Systems. Proceedings, 1997. XII, 241 pages. 1998. (Subseries LNAI).

Vol. 1442: A. Fiat. G.J. Woeginger (Eds.), Online Algorithms. XVIII, 436 pages. 1998.

Vol. 1443: K.G. Larsen, S. Skyum, G. Winskel (Eds.), Automata, Languages and Programming. Proceedings, 1998. XVI, 932 pages. 1998.

Vol. 1444: K. Jansen, J. Rolim (Eds.), Approximation Algorithms for Combinatorial Optimization. Proceedings, 1998. VIII, 201 pages. 1998.

Vol. 1445: E. Jul (Ed.), ECOOP'98 – Object-Oriented Programming. Proceedings, 1998. XII, 635 pages. 1998.

Vol. 1446: D. Page (Ed.), Inductive Logic Programming. Proceedings, 1998. VIII, 301 pages. 1998. (Subseries LNAI).

Vol. 1447: V.W. Porto, N. Saravanan, D. Waagen, A.E. Eiben (Eds.), Evolutionary Programming VII. Proceedings, 1998. XVI, 840 pages. 1998.

Vol. 1448: M. Farach-Colton (Ed.), Combinatorial Pattern Matching. Proceedings, 1998. VIII, 251 pages. 1998.

Vol. 1449: W.-L. Hsu, M.-Y. Kao (Eds.), Computing and Combinatorics. Proceedings, 1998. XII, 372 pages. 1998.

Vol. 1450: L. Brim, F. Gruska, J. Zlatuška (Eds.), Mathematical Foundations of Computer Science 1998. Proceedings, 1998. XVII, 846 pages. 1998.

Vol. 1451: A. Amin, D. Dori, P. Pudil, H. Freeman (Eds.), Advances in Pattern Recognition. Proceedings, 1998. XXI, 1048 pages. 1998.

Vol. 1452: B.P. Goettl, H.M. Halff, C.L. Redfield, V.J. Shute (Eds.), Intelligent Tutoring Systems. Proceedings, 1998. XIX, 629 pages. 1998.

Vol. 1453: M.-L. Mugnier, M. Chein (Eds.), Conceptual Structures: Theory, Tools and Applications. Proceedings, 1998. XIII, 439 pages. (Subseries LNAI).

Vol. 1454: I. Smith (Ed.), Artificial Intelligence in Structural Engineering. XI, 497 pages. 1998. (Subseries LNAI).

Vol. 1456: A. Drogoul, M. Tambe, T. Fukuda (Eds.), Collective Robotics. Proceedings, 1998. VII, 161 pages. 1998. (Subseries LNAI).

Vol. 1457: A. Ferreira, J. Rolim, H. Simon, S.-H. Teng (Eds.), Solving Irregularly Structured Problems in Prallel. Proceedings, 1998. X, 408 pages. 1998.

Vol. 1458: V.O. Mittal, H.A. Yanco, J. Aronis, R-. Simpson (Eds.), Assistive Technology in Artificial Intelligence. X, 273 pages. 1998. (Subseries LNAI).

Vol. 1459: D.G. Feitelson, L. Rudolph (Eds.), Job Scheduling Strategies for Parallel Processing. Proceedings, 1998. VII, 257 pages. 1998.

Vol. 1460: G. Quirchmayr, E. Schweighofer, T.J.M. Bench-Capon (Eds.), Database and Expert Systems Applications. Proceedings, 1998. XVI, 905 pages. 1998.

Vol. 1461: G. Bilardi, G.F. Italiano, A. Pietracaprina, G. Pucci (Eds.), Algorithms – ESA'98. Proceedings, 1998. XII, 516 pages. 1998.

Vol. 1462: H. Krawczyk (Ed.), Advances in Cryptology - CRYPTO '98. Proceedings, 1998. XII, 519 pages. 1998.

Vol. 1464: H.H.S. Ip, A.W.M. Smeulders (Eds.), Multimedia Information Analysis and Retrieval. Proceedings, 1998. VIII, 264 pages. 1998.

Vol. 1465: R. Hirschfeld (Ed.), Financial Cryptography. Proceedings, 1998. VIII, 311 pages. 1998.

Vol. 1466: D. Sangiorgi, R. de Simone (Eds.), CONCUR'98: Concurrency Theory. Proceedings, 1998. XI, 657 pages. 1998.

Vol. 1467: C. Clack, K. Hammond, T. Davie (Eds.), Implementation of Functional Languages. Proceedings, 1997. X, 375 pages. 1998.

Vol. 1469: R. Puigjaner, N.N. Savino, B. Serra (Eds.), Computer Performance Evaluation. Proceedings, 1998. XIII, 376 pages. 1998.

Vol. 1473: X. Leroy, A. Ohori (Eds.), Types in Compilation. Proceedings, 1998. VIII, 299 pages. 1998.

Vol. 1475: W. Litwin, T. Morzy, G. Vossen (Eds.), Advances in Databases and Information Systems. Proceedings, 1998. XIV, 369 pages. 1998.

Vol. 1482: R.W. Hartenstein, A. Keevallik (Eds.), Field-Programmable Logic and Applications. Proceedings, 1998. XI, 533 pages. 1998.